Handbook of Research on Safe Disposal Methods of Municipal Solid Wastes for a Sustainable Environment

Prateek Srivastava
Chitkara University Institute of Engineering and Technology, Chitkara University, India

Durgaprasad Ramteke
Fibre and Particle Engineering, Faculty of Technology, University of Oulu, Finland

Ankush Kumar Bedyal
Department of Physics, School of Sciences, Cluster University of Jammu, India

Monika Gupta
Chitkara Business School, Chitkara University, India

Jasminder Kaur Sandhu
Chitkara University Institute of Engineering and Technology, Chitkara University, India

IGI Global
PUBLISHER of TIMELY KNOWLEDGE

A volume in the Practice, Progress, and Proficiency in Sustainability (PPPS) Book Series

Published in the United States of America by
IGI Global
Engineering Science Reference (an imprint of IGI Global)
701 E. Chocolate Avenue
Hershey PA, USA 17033
Tel: 717-533-8845
Fax: 717-533-8661
E-mail: cust@igi-global.com
Web site: http://www.igi-global.com

Library of Congress Cataloging-in-Publication Data

Names: Srivastava, Prateek, 1990- editor.
Title: Handbook of research on safe disposal methods of municipal solid wastes for a sustainable environment / edited by Prateek Srivastava, Durgaprasad Ramteke, Ankush Kumar Bedyal, Monika Gupta, Jasminder Kaur Sandhu.
Description: Hershey, PA : Engineering Science Reference, [2023] | Includes bibliographical references and index. | Summary: "There are several research papers in the scientific literature about specific waste streams, to quantify their effect on the environment in a wide range of fields. However, till now, no systematic efforts have been made to come out with a book which exclusively covers waste management principles and strategies in different fields and corresponding applications. We will try to club different types of waste and waste management strategies in this book so that a reader can look at the range of materials in a single book. The book chapters including results and case studies can be of reference for academicians, researchers and stakeholders for quantifying the comprehensive impacts and for planning integrated solid waste collection and treatment systems, for improving sustainability at a global level. The main objectives are: 1. To improve and manage solid waste systems through the application of modern management techniques. 2. To provide information about the application of waste management strategies in different interdisciplinary areas such as municipal solid waste, plastic waste, bio-medical wastes, hazardous wastes, e-wastes, soil remediation technologies, environmental impact assessment and many more. 3. To focus on the waste management strategies for a sustainable environment that have usually emerged in a decade or two. 4. To cover the content from basic fundamental to complex application logically for a better understanding of the reader. 5. To present the latest information on the increase in solid and hazardous wastes due to economic growth, urbanization, and industrialization. To discuss the main challenge for governments of different nations and international organizations to develop environment-friendly strategies for efficient waste management to save human beings and the environment from its adverse effects"-- Provided by publisher.
Identifiers: LCCN 2023001148 (print) | LCCN 2023001149 (ebook) | ISBN 9781668481172 (h/c) | ISBN 9781668481196 (eISBN)
Subjects: LCSH: Refuse and refuse disposal. | Sustainable engineering.
Classification: LCC TD791 .S25 2023 (print) | LCC TD791 (ebook) | DDC 628.4/40286--dc23/eng/20230131
LC record available at https://lccn.loc.gov/2023001148
LC ebook record available at https://lccn.loc.gov/2023001149

This book is published in the IGI Global book series Practice, Progress, and Proficiency in Sustainability (PPPS) (ISSN: 2330-3271; eISSN: 2330-328X)

British Cataloguing in Publication Data
A Cataloguing in Publication record for this book is available from the British Library.

Practice, Progress, and Proficiency in Sustainability (PPPS) Book Series

Ayman Batisha
International Sustainability Institute, Egypt

ISSN:2330-3271
EISSN:2330-328X

Mission

In a world where traditional business practices are reconsidered and economic activity is performed in a global context, new areas of economic developments are recognized as the key enablers of wealth and income production. This knowledge of information technologies provides infrastructures, systems, and services towards sustainable development.

The **Practices, Progress, and Proficiency in Sustainability (PPPS) Book Series** focuses on the local and global challenges, business opportunities, and societal needs surrounding international collaboration and sustainable development of technology. This series brings together academics, researchers, entrepreneurs, policy makers and government officers aiming to contribute to the progress and proficiency in sustainability.

Coverage

- Knowledge clusters
- Innovation Networks
- Global Content and Knowledge Repositories
- Intellectual Capital
- Eco-Innovation
- Socio-Economic
- Strategic Management of IT
- E-Development
- ICT and knowledge for development
- Technological learning

IGI Global is currently accepting manuscripts for publication within this series. To submit a proposal for a volume in this series, please contact our Acquisition Editors at Acquisitions@igi-global.com or visit: http://www.igi-global.com/publish/.

The Practice, Progress, and Proficiency in Sustainability (PPPS) Book Series (ISSN 2330-3271) is published by IGI Global, 701 E. Chocolate Avenue, Hershey, PA 17033-1240, USA, www.igi-global.com. This series is composed of titles available for purchase individually; each title is edited to be contextually exclusive from any other title within the series. For pricing and ordering information please visit http://www.igi-global.com/book-series/practice-progress-proficiency-sustainability/73810. Postmaster: Send all address changes to above address.

Titles in this Series

For a list of additional titles in this series, please visit: www.igi-global.com/book-series

Artificial Intelligence Tools and Technologies for Smart Farming and Agriculture Practices
Rajeev Kumar Gupta (Pandit Deendayal Energy University, India) Arti Jain (Jaypee Institute of Information Technology, India) John Wang (Montclair State University, USA) Santosh Kumar Bharti (Pandit Deendayal Energy University, India) and Samir Patel (Pandit Deendayal Energy University, India)
Engineering Science Reference • copyright 2023 • 303pp • H/C (ISBN: 9781668485163) • US $225.00 (our price)

Financing Regions Toward Sustainability in the Midst of Climate Change Risks and Uncertainty
Beata Zofia Filipiak (University of Szczecin, Poland) Dominika Kordela (University of Szczecin, Poland) and Izabela Nawrolska (University of Szczecin, Poland)
Information Science Reference • copyright 2023 • 330pp • H/C (ISBN: 9781668476208) • US $225.00 (our price)

Pragmatic Legal and Policy Implications of Environmental Lawmaking
Hussein Movahedian (Islamic Azad University, UAE & Department of Private Law, Islamic Studies and Law Faculty, Imam Sadiq University, Tehran, Iran) and Nima Norouzi (Islamic Azad University, UAE & Law and Political Science Department, University of Tehran, Tehran, Iran)
Information Science Reference • copyright 2023 • 269pp • H/C (ISBN: 9781668441589) • US $215.00 (our price)

Leadership and Governance for Sustainability
Neeta Baporikar (Namibia University of Science and Technology, Namibia & SP Pune University, India)
Information Science Reference • copyright 2023 • 350pp • H/C (ISBN: 9781668497111) • US $250.00 (our price)

Implications of the COVID-19 Pandemic and the Russia-Ukraine Crisis on the Agricultural Sector
Vítor João Pereira Domingues Martinho (ESAV, CERNAS-IPV Research Centre, Polytechnic Institute of Viseu, Portugal)
Engineering Science Reference • copyright 2023 • 515pp • H/C (ISBN: 9781668489239) • US $215.00 (our price)

Innovations in Environmental Legislation and Justice Environmental and Water-Energy-Food Nexus Laws
Nima Norouzi (Islamic Azad University, UAE & Law and Political Science Department, University of Tehran, Tehran, Iran) and Hussein Movahedian (Islamic Azad University, UAE & Department of Private Law, Islamic Studies and Law Faculty, Imam Sadiq University, Tehran, Iran)
Information Science Reference • copyright 2023 • 268pp • H/C (ISBN: 9781668471883) • US $215.00 (our price)

Sustainable Approaches and Strategies for E-Waste Management and Utilization
A. M. Rawani (National Institute of Technology, Raipur, India) Mithilesh Kumar Sahu (O.P. Jindal University, India) Siddharth S. Chakarabarti (O.P. Jindal University, India) and Ajit Kumar Singh (GLA University, India)
Engineering Science Reference • copyright 2023 • 300pp • H/C (ISBN: 9781668475737) • US $240.00 (our price)

IGI Global
PUBLISHER of TIMELY KNOWLEDGE

701 East Chocolate Avenue, Hershey, PA 17033, USA
Tel: 717-533-8845 x100 • Fax: 717-533-8661
E-Mail: cust@igi-global.com • www.igi-global.com

List of Contributors

Table of Contents

Detailed Table of Contents

multifaceted technology, as evidenced by a thorough examination of the relevant literature. It will alter the competitive landscape by shifting the advantage away from large corporations and toward small and medium-sized enterprises (SMEs) in emerging marketsand developed markets. As human and technological skills advance quickly, businesses may be able to profit.

Autonomous systems that cannot adapt to real-world environments have proliferated due to rapid technological improvement. These programmes will free people from repetitive, inefficient chores. The monotonous, dusty, and dangerous conditions of the unmanned aerial vehicle (UAV) pose a severe threat to human health and safety. Autonomous systems improve supply chain, tracking, and hazardous climate control. This chapter proposes merging compression and track architecture to improve UAV performance. The UAV and pipe increase trackwheel friction. The UAV doesn't move in the tube. Cleaning commences when sensors detect the mouth. Dirt sensors prevent the washing process from working. This chapter focuses on drain cleaning automation. Device automation addresses mobility and space. This study supports this method for garbage disposal and filtering. The technology removes manual cleaning and relies on human control of system movement.

Electronic waste is one of the most challenging problems faced by the globe today. In order to solve it in an efficient and effective manner, this chapter highlights the role of the circular economy and its elements in effectively recycling and managing electronic waste. With the help of extant literature, the research digs deep into the concept of electronic waste and circular economy and how the elements of this economy can be combined to effectively manage electronic waste. This is followed by the development of a conceptual framework that encompasses all important stakeholders including suppliers, manufacturers, and customers for effective management of electronic waste. The research also emphasizes the adoption of a holistic and all-encompassing policy framework for effectively managing electronic waste and also details significant implications for manufacturers and policymakers.

In the present scenario, the e-waste from the various electronic sectors has been increasing due to increased utilization of electric components. In this chapter, the bioleaching(biomining) process of a computer printed circuit board (CPCB) is illustrated to extract the metal components. Basic concepts for e-waste management, their impacts, and various e-waste treatment methods have been explained. The various existing conventional metal extraction methods for the wasted CPCB have also been explored. Definitions, types, cryogenic bioleaching (biomining), influencing factors, and procedures of the bioleaching process have been illustrated. The microbiological methods for the processing of e-waste, the selection of process parameters, and the optimization or maximization of metal extraction processes were demonstrated to promote the e-waste management processes.

United Nations SDGs (sustainable development goals) have prompted health-care organisations in and around this world to implement health care initiatives that aim to improve their long-term sustainability. Analysis of 637 documents from the Scopus database was retrieved using bibliometric methods for the existing study. Research on health care sustainability has been documented over almost the past 12 years (2010-November, 2022) in this review. Results have indicated that the intellectual structure of this knowledge body comprises of five clusters. First, the authors saw that this review database had 22 writers listed as authors. This indicates that a high number of academicians are interested in this topic. But at the same time, it is also evident that majority of contributors produce only few publications on health care sustainability.

Technology is being used to reduce the need for in-person medical appointments and save on the cost of treatment, which is a boon for human civilization in the context of the pandemic. The Central Pollution Control Board's existing rules provide a framework for managing healthcare waste with respect for the safety of the public. Bio-waste management strategies using IoT and IoMT techniques are illustrated in this chapter.

Swapan Banerjee, Department of Nutrition, Seacom Skills University, India
Soumen Ghosh, Agricultural and Food Engineering Department, Indian Institute of Technology, Kharagpur, India
Gourav Dhar Bhowmick, Agricultural and Food Engineering Department, Indian Institute of Technology, Kharagpur, India
Digvijay Pandey, Department of Technical Education, IET, Dr. A.P.J. Abdul Kalam Technical University, Lucknow, India
Binay Kumar Pandey, Department of Information Technology, College of Technology, Govind Ballabh Pant University of Agriculture and Technology, Pantnagar, India
Pankaj Dadheech, Swami Keshvanand Institute of Technology, Management, and Gramothan, India

Biofuels, such as coal and oil, have been the most suitable and sufficient substitutes for other conventional fuels in the energy sector for the last few years. For decades, entire populations have suffered from deforestation, global warming, and carbon monoxide issues. Biofuel production is an eco-friendly process that still has health hazards for manufacturing unit workers. That is due to flammable and combustible raw materials and their chemical reactions. Bioethanol is a flammable biofuel prepared by cellulose items and the fermentation of grains. Biodiesel is another combustible liquid-based biofuel produced by alcohol and glyceride in vegetable oil duly catalyzed by strong alkaline like caustic soda. Biofuel plants or trees are the biggest isoprene producers, which meet the ozone layer in the earth's atmosphere. While combining with ozone always causes many health hazards for manufacturing units' employees despite several safety measures. One of the significant consequences is that some factory workers often have asthma, allergy, and lung disorders due to ozone attacks.

Preface

Municipal solid waste (MSW) management is a significant environmental challenge facing many countries worldwide. Safe disposal methods for MSW are essential to mitigate the impact of waste on the environment and human health. Some of the sustainable methods of MSW disposal are: Landfilling, Incineration, Composting, Recycling, Bioreactor Landfills, and Waste-to-Energy. Overall, the key to sustainable MSW disposal is to use a combination of methods that maximize resource recovery, minimize environmental impact, and protect human health.

In the present scenario, the amount of e-waste generated by the various electronic sectors is increasing due to the increased utilization of electronic components. This chapter focuses on the bioleaching (biomining) process for computer printed circuit board (CPCB) to extract metal components. Basic concepts for e-waste management, their impacts, and various e-waste treatment methods have been explained. The conventional metal extraction methods for wasted CPCB have also been explored. Definitions, types, cryogenic bioleaching (biomining), influencing factors, and procedures of the bioleaching process have been illustrated. The microbiological methods for the processing of E-waste, the selection of process parameters, and the optimization or maximization of metal extraction processes were demonstrated to promote the E-waste management processes. Overall, the bioleaching process has shown promising results in extracting metals from e-waste in an efficient and sustainable manner. It offers a potential solution for the extraction of valuable metals from e-waste while also minimizing environmental impacts. Further research is needed to optimize the bioleaching process for e-waste management and improve its economic viability.

Waste management represents a challenge due to the rapid increase in waste production and the emergence of new waste types. Overcoming the issue involves using innovative technologies such as nanotechnology. Nanotechnology uses nanomaterials, which are materials that have at least one dimension less than 100 nm. Due to their small size, these materials increase reactivity in processes such as adsorption and oxidation/reduction. The application of nanotechnologies is significant in the production of new materials to replace current raw materials, and in providing novel solutions for waste recycling and disposal. Furthermore, nanofiltration is effective in the treatment of metals, toxic waste, and nonbiodegradable materials of leachate. Nanomaterials, however, represent a safety risk for the environment, and a serious threat to human health due to their small size and long suspension time. Chapter 1 deals with the use of nanotechnology in waste management, including reduction, recycling, treatment, and disposal phases.

Population growth, urbanisation, industry, modernization, and digitalization increase residential, industrial, commercial, mining, radioactive, agricultural, hospital, and electronic wastes in the 21st century. Waste management is becoming the biggest global challenge. Waste management includes collecting, transporting, sorting, destroying, processing, recycling, controlling, monitoring, and regulating garbage, sewage, and other waste. Waste management preserves the environment, prevents pollution, and protects health. Global waste management is modern. Biological reprocessing, recycling, composting, waste-to-energy, bioremediation, incineration, pyrolysis, plasma gasification, ocean/sea disposal, etc.Waste management enhances life. This ensures future peace and wellness. Global health depends on waste management. This optimises waste management. Chapter 2 discusses worldwide garbage management. It also offers the best waste management approach by critically reviewing previous researchers' findings.

Chapter 3 highlights the issue of household solid waste mismanagement in Rajshahi city of Bangladesh and its impacts on the environment and public health. The study collected data through personal in-depth interviews with 120 households in the city to investigate the situation and effects of household waste management. The findings of the study reveal that the city mainly produces food waste, which is collected by scavengers employed through city corporations. However, the waste is poorly managed due to a lack of categorization and recycling technologies. As a result, waste odor and disease infections are the most common impacts of household waste mismanagement in Rajshahi city, as reported by the respondents. The study concludes that the lack of proper knowledge and technology for waste management and recycling is one of the significant reasons for the bad impacts of waste mismanagement in the city. The city authority practices unplanned open dumping and burning systems, which further worsen the situation. The study recommends the reorganization of the waste management system in the city by installing proper dustbins and implementing efficient waste management and recycling technologies. By doing so, the negative impacts of household waste mismanagement on the environment and public health can be minimized.

Chapter 4 highlights the challenges and barriers to the adoption of circular economy principles for electronic waste management, including lack of awareness, inadequate infrastructure, and inadequate policies and regulations. The research concludes that the adoption of circular economy principles can significantly reduce the negative impact of electronic waste on the environment, and a coordinated effort from all stakeholders is essential to ensure the success of electronic waste management programs. Moreover, it is recommended that policymakers should adopt a circular economy-based approach to electronic waste management that considers the entire life cycle of electronic products, from design to disposal, and encourages the use of sustainable materials, and encourages the adoption of sustainable business models that promote the efficient use of resources.

Autonomous systems that cannot adapt to real-world environments have proliferated due to rapid technological improvement. These programs will free people from repetitive, inefficient chores. The monotonous, dusty, and dangerous conditions of the Unmanned Aerial Vehicle (UAV) pose a severe threat to human health and safety. Autonomous systems improve supply chain, tracking, and hazardous climate control. This paper proposes merging compression and track architecture to improve UAV performance. The UAV and pipe increase trackwheel friction. The UAV doesn't move in the tube. Cleaning commences when sensors detect the mouth. Dirt sensors prevent the washing process from working. This article focuses on drain cleaning automation. Device automation addresses mobility and space. Chapter 5 supports this method for garbage disposal and filtering. The technology removes manual cleaning and relies on human control of system movement.

Electronic waste is one of the most challenging problems faced by the globe today. In order to solve it in an efficient and effective manner, chapter 6 highlights the role of the circular economy and its elements in effectively recycling and managing electronic waste. With the help of extant literature, the research digs deep into the concept of electronic waste and circular economy and how the elements of this economy can be combined to effectively manage electronic waste. This is followed by the development of a conceptual framework that encompasses all important stakeholders including suppliers, manufacturers, and customers for effective management of electronic waste. The research also emphasizes the adoption of a holistic and all-encompassing policy framework for effectively managing electronic waste and also details significant implications for manufacturers and policymakers.

Nanotechnology has the potential to revolutionize waste management by providing innovative solutions for reducing, recycling, treating, and disposing of waste. Nanomaterials can increase the reactivity of processes such as adsorption and oxidation/reduction, which can enhance the efficiency of waste treatment technologies. For instance, nanofiltration is an effective technology for treating metals, toxic waste, and non-biodegradable materials in leachate. Moreover, nanotechnology can be used to produce new materials to replace current raw materials and improve the efficiency of waste recycling. The use of nanomaterials in waste treatment also has the potential to reduce energy consumption, waste volume, and environmental impact. For instance, the use of nanocatalysts in the treatment of wastewater can reduce energy consumption, increase treatment efficiency, and lower the cost of treatment. However, the use of nanomaterials in waste management raises concerns about the potential environmental and health risks associated with these materials. Nanomaterials can remain suspended in the environment for extended periods and can potentially interact with living organisms, leading to unintended consequences. Therefore, it is essential to develop safe and sustainable nanotechnologies for waste management and address ethical issues related to their use. In conclusion, nanotechnology can be a promising solution for the challenges faced in waste management, including the treatment of different waste types, reduction of waste volume, and development of new materials. However, it is crucial to ensure the safety and sustainability of nanotechnologies and address ethical concerns in their use.

Chapter 7 discusses the current status of solid waste management in India, highlighting the challenges faced by the government and local authorities in managing the rapidly increasing amount of waste generated by urban centers. The chapter explores various aspects of solid waste management, including waste generation, collection, transportation, treatment, and disposal. The chapter provides an in-depth analysis of the various technologies and processes available for waste treatment and management, such as composting, incineration, pyrolysis, and gasification. It also highlights the importance of source segregation and waste minimization in achieving sustainable waste management practices. In addition, the chapter discusses the potential benefits of waste-to-energy and waste-to-resource conversion technologies, which can help reduce the burden on landfills and generate renewable energy sources. The chapter emphasizes the need for effective policies and regulations to promote the adoption of sustainable waste management practices and create a favorable environment for private sector investment in the waste management sector. Overall, the chapter provides a valuable insight into the challenges and opportunities associated with solid waste management in India and offers a roadmap for achieving sustainable waste management practices in the country.

Waste has become a major concern in the current era. Most nations across the globe have also become more concerned about a safe and healthy environment, and they are working on the issue continuously. Therefore, waste management has emerged as a major challenge and smart waste management is proving very helpful in dealing with waste, especially in developing countries. However, the implementation of smart waste management systems is not without its challenges. Chapter 8 aims to identify and analyze the key issues and challenges in smart waste management.

The reuse and recirculation of products and materials are the basis of the concept of the circular economy, which is highlighted in chapter 9. A circular economy entails markets that give incentives to reusing products, rather than scrapping them and then extracting new resources. The circular economy model has become highly relevant in recent years, with the electronics industry being one of the sectors that have considered its application. Electronic waste (e-waste), that is, waste arising from end-of-life electronic products similar to computers and mobile phones, is one of the swift-growing waste aqueducts in the world moment. The developing world has come to the primary destination for the habituated electrical and electronic outfit (EEE) exported by the advanced world, making waste operations critical. Indeed, though numerous studies were conducted on waste operation, veritably many are conducted in developing countries that are significant donors of used EEE. India is among the top five-waste-producing countries in the world with an estimated periodic product of 2 million tons.

In the present scenario, the e-waste from the various electronic sectors has been increasing due to increased utilization of electric components. In chapter 10, the bioleaching(biomining) process of a computer printed circuit board (CPCB) is illustrated to extract the metal components. Basic concepts for e-waste management, their impacts, and various e-waste treatment methods have been explained. The various existing conventional metal extraction methods for the wasted CPCB have also been explored. Definitions, types, cryogenic bioleaching (biomining), influencing factors, and procedures of the bioleaching process have been illustrated. The microbiological methods for the processing of E-waste, the selection of process parameters, and the optimization or maximization of metal extraction processes were demonstrated to promote the E-waste management processes.

United Nations SDGs (Sustainable Development Goals) have prompted health-care organizations in and around this world to implement health care initiatives that aim to improve their long-term sustainability. Analysis of 637 documents from the Scopus database was retrieved using bibliometric methods for chapter 11. Research on health care sustainability has been documented over the past 12 years (November 2010-2022) in this review. Results have indicated that the intellectual structure of this knowledge body comprises of 5 clusters. First, we saw that our review database had 22 writers listed as authors. This indicates that a high number of academicians are interested in this topic. But at the same time, it is also evident that the majority of contributors produce only few publications on health care sustainability.

Municipal waste management is a critical issue for sustainable development, and traditional methods are becoming unsustainable. Biogas plants offer a promising solution using organic waste to produce biogas for electricity or fuel. This review examines the benefits and challenges of biogas plants for municipal waste management, including the complex biochemical process, factors affecting efficiency, and challenges like high costs, etc. Biogas production diverts waste from landfills, reduces greenhouse gas emissions, and creates economic opportunities. Successful case studies from Germany, India, South Africa, etc. are also highlighted in chapter 12. Future research directions are also discussed to optimize the process, assess environmental impacts, and evaluate economic viability. Overall, biogas plants represent a promising approach to sustainable municipal waste management that can contribute to the transition toward a more circular and sustainable economy.

In chapter 13, natural gas is a fossil fuel that produces fewer greenhouse gas emissions than oil and coal, making it a cleaner energy source. It is also a flexible fuel that can be used in many different ways, including heating homes and buildings, generating electricity, and fueling vehicles. In addition, natural gas is relatively cheap and abundant in many regions, making it an attractive option for energy production. This chapter compares the utilization of different fuels. While natural gas has the potential to play an important role in meeting our energy needs, further research is needed to fully understand its benefits and limitations. Collaboration between academics, industry, and government stakeholders can help ensure that the use of natural gas is guided by informed decision-making and sustainable practices.

In chapter 14, eco-friendly environment concern contributes to the survival of life on earth. Monitoring chemicals produced by growth requires a multidisciplinary scientific approach. Climate change is a result of pollution and overuse, particularly in the air. Although government and industry have taken initiatives to limit air pollution, it is still necessary to check the quality of the air at ground level. The goal of this endeavor is to build intercommunication-based individual vehicle air pollution monitoring and identification by Internet of Things (IoT). Idling and accelerating conditions were recorded using tailpipe emission sensing. Authorities can act and assure upkeep if they are able to identify a car at a busy intersection. Mobile connectivity enabled route and vehicle identification. In India, fewer people will die from air pollution because to data gathered in the cloud.

Chapter 15 discusses the various methods adopted by the Indian Government to manage plastic waste, such as setting up plastic waste management facilities, promoting the use of biodegradable and compostable bags, and encouraging the use of plastic waste in road construction. The government has also launched campaigns to create awareness about the harmful effects of plastic waste and the importance of recycling. Overall, the Indian Corporate Sector and the Indian Government are taking significant initiatives towards managing plastic waste and promoting sustainability practices. However, there is a need for more awareness and adoption of sustainable practices by consumers to effectively manage plastic waste and protect the environment.

Chapter 16 highlights the need for effective policies and regulations to be implemented to manage mobile phone waste. These policies should focus on reducing the use of hazardous materials in mobile phone production, promoting the reuse and recycling of mobile phones, and ensuring safe disposal of mobile phone waste. The study also suggests that the development of innovative technologies such as bioreactors, membrane technologies, and supercritical fluid extraction can be used to recover valuable materials from mobile phone waste, thereby reducing the need for virgin materials and reducing the environmental impact of mobile phone production. The study concludes by emphasizing the importance of public awareness campaigns to promote responsible disposal of mobile phone waste and encourage consumers to adopt sustainable practices in mobile phone usage and disposal.

Artificial intelligence (AI) and machine learning techniques have been used in critical water and wastewater treatment applications, leading to optimization, modeling, and automation of processes, as focused on in chapter 17. These techniques have been applied in different water environments, such as surface water, groundwater, drinking water, sewage, and seawater, to evaluate water quality. Prediction of wastewater characteristics in wastewater treatment plants (WWTPs) using machine learning algorithms can significantly reduce the cost, energy consumption, and number of samples required.

The importance of reuse and recycling of the waste at all age group and gender plays a prominent role towards conserving the environment. In India 19.2% of population belongs to the age group of 15-24 years that is considered as youth population. The research attempts to understand the practice of reuse and recycle of waste among the male and female youth population. Chapter 18 investigates and document upon the five waste materials that can be recycled that is textile, paper, plastic, glass and e waste. The study is conducted in Bhopal city through direct and focus group discussion. The primary survey is conducted in institutes and schools of Bhopal to capture the ground realities. The outcome of the research depicts the variation is reuse and recycle among the male and female group of the youth population.

The global accumulation of electronic waste is a growing concern in terms of the health and safety of our planet as highlighted in chapter 19. Electronic waste has a greater potential to create environmental contamination and human health problems due to the way it is often disposed of, making it a worldwide problem. 'Sustainable consumption and production' may help nations solve the problem of electronic waste management by forcing them to make fundamental changes in the way their communities manufacture and use commodities. Governments, NGOs, businesses, and citizens all work together to change unsustainable consumption and production habits. The effects of e-waste on human health and the natural environment will be discussed, as will various situations involving this kind of garbage. Developed and developing nations' (namely India's) approaches to managing discarded electrical and electronic equipment will be examined, as will appropriate pathways concerning recyclable components using extended producer responsibility that led to the creation of a circular economy.

In chapter 20, solid waste disposal is a significant issue that affects the environment and water resources. When solid waste is disposed of improperly, it can leach down into deeper layers of water bodies, contaminating the water with dissolved organic and inorganic constituents from the waste. This results in the formation of polluted liquid, commonly known as leachate, which increases in concentration as it seeps deeper into the soil. Leachate is brownish to black in color and has a rotten smell. It contains high levels of organic pollutants and ammoniacal nitrogen, which can be toxic to human health and aquatic animals. The toxicity levels of leachate depend on the type of waste present in the solid waste, such as plastic or metals. Unselective and mismanaged waste disposal also introduces various toxic compounds, including heavy metals, into the environment. This not only degrades the environment but also contributes to climatic changes by emitting greenhouse gases. Therefore, proper solid waste management practices are essential to minimize the impact on water resources and the environment. The management practices include proper collection, segregation, transportation, and disposal of solid waste. It is essential to develop and implement effective waste management policies to ensure that waste is disposed of properly and to prevent environmental pollution.

Chapter 21 provides valuable insights into the importance of efficient E-waste management and the need for a comprehensive approach to address this issue. The suggested measures and recommendations can serve as a guide for policymakers and stakeholders to ensure sustainable waste management practices, and to promote a cleaner and healthier environment for future generations.

The handling of genetic residues, biohazardous waste, and other non-shrinkable materials such as e-waste is essential for the preservation of Mother Nature and human wellbeing. Chapter 22 will bring forth different methods of managing biomedical wastes, inspecting the roots of the spread of the hazardous virus, and protecting the front-line warriors. To properly react to this disaster, each administration must create a backup strategy based on regional circumstances and the seriousness of the coronavirus's spread. Technology is being used to reduce the need for in-person medical appointments and save on the cost of treatment, which is a boon for human civilization in the context of the pandemic. The Central Pollution Control Board's existing rules provide a framework for managing healthcare waste with respect for the safety of the public. Bio-waste management strategies using IoT and IoMT techniques are illustrated in this chapter.

Biofuels, such as coal and oil, have been the most suitable and sufficient substitutes for other conventional fuels in the energy sector for the last few years, as highlighted in chapter 23. For decades, entire populations have suffered from deforestation, global warming, and carbon monoxide issues. Biofuel production is an eco-friendly process that still has health hazards for manufacturing unit workers. That is due to flammable and combustible raw materials and their chemical reactions. Bioethanol is a flammable biofuel prepared by cellulose items and the fermentation of grains. Biodiesel is another combustible liquid-based biofuel produced by alcohol and glyceride in vegetable oil duly catalyzed by strong alkaline like caustic soda. Biofuel plants or trees are the biggest isoprene producers, which meet the ozone layer in the earth's atmosphere. While combining with ozone always causes many health hazards for manufacturing-units employees despite several safety measures. One of the significant consequences is that some factory workers often have asthma, allergy, and lung disorders due to ozone attacks.

This book provides a comprehensive overview of the safe disposal methods of municipal solid waste (MSW) for a sustainable environment. It covers a range of topics related to MSW management, including landfilling, incineration, composting, recycling, bioreactor landfills, and waste-to-energy. The book also addresses issues such as the impact of MSW on the environment and human health, the economics of MSW management, and the policies and regulations governing MSW management. In addition to traditional MSW, the book also covers the growing problem of electronic waste (e-waste), which has become a global concern due to its hazardous nature and potential impact on the environment and human health. The book explores strategies for managing e-waste in a sustainable manner, including recycling and safe disposal methods. Overall, this book is an excellent resource for anyone interested in waste management principles and strategies, including environmentalists, government officials, researchers, scholars, academicians, practitioners, instructors, and students. It provides valuable insights into the challenges and solutions related to MSW and e-waste management and highlights the importance of sustainable waste management practices for a cleaner and healthier environment.

Prateek Srivastava
Chitkara University Institute of Engineering and Technology, Chitkara University, India

Durgaprasad Ramteke
Fibre and Particle Engineering, Faculty of Technology, University of Oulu, Finland

Ankush Kumar Bedyal
DDepartment of Physics, School of Sciences, Cluster University of Jammu, India

Monika Gupta
Chitkara Business School, Chitkara University, India

Jasminder Kaur Sandhu
Chitkara University Institute of Engineering and Technology, Chitkara University, India

Acknowledgment

We recognize the chapter authors for their excellent work, which has made the book possible. Collaborative efforts are crucial in achieving success, and we express our gratitude towards the chapter authors. Without the contributions of the chapter authors, the book may not have been possible. Their hard work, dedication, and expertise have been essential in shaping the book's content and ensuring its quality. Their contributions are a testament to the importance of collaboration and teamwork in achieving a shared goal. Without their contributions, this book would not have been possible.

The authors have shown appreciation towards the anonymous reviewers who provided constructive and valuable comments on the chapters. Peer review is a critical process in academic publishing, and the efforts of the anonymous reviewers are essential in ensuring the quality of the work.

We express our gratitude towards the editors and production group at IGI Global, who played a significant role in bringing the book to fruition. Their efforts in editing and production are essential in ensuring the final book is of high quality and meets the necessary standards. Furthermore, we thank all our friends and colleagues who provided support and encouragement during the editing process. It is essential to have a supportive network of people who can provide motivation and reminders that there is more to life than work. Taking breaks and engaging in activities that bring joy and fulfilment can help improve overall well-being and enhance productivity in the long run.

Thanks to many anonymous reviewers for carefully reviewing all chapters, and providing constructive and valuable comments. At last, we are grateful to the editors and production group at IGI Global. Special thanks are given to a number of friends and colleagues for their support and encouragement during the editing of the book. We will not forget their gentle reminding messages that there are more beautiful things in life beyond working.

Chapter 1
Waste Management Using Nanotechnology

Mahmoud M. Elwaheidi
https://orcid.org/0000-0003-4863-3184
Department of Geology and Geophysics, College of Science, King Saud University, Saudi Arabia

ABSTRACT

Waste management represents a challenge due to the rapid increase in waste production and the emergence of new waste types. Overcoming the issue involves using innovative technologies such as nanotechnology. Nanotechnology uses nanomaterials, which are materials that have at least one dimension less than 100 nm. Due to their small size, these materials increase reactivity in processes such as adsorption and oxidation/reduction. The application of nanotechnologies is significant in the production of new materials to replace current raw materials, and in providing novel solutions for waste recycling and disposal. Furthermore, nanofiltration is effective in the treatment of metals, toxic waste, and nonbiodegradable materials of leachate. Nanomaterials, however, represent a safety risk for the environment, and a serious threat to human health due to their small size and long suspension time. This chapter deals with the use of nanotechnology in waste management, including reduction, recycling, treatment, and disposal phases.

INTRODUCTION

Waste is an unavoidable by-product of numerous human activities. The rapid urbanization, prosperity, and the increase of population have caused an enormous increase in the production of several types of waste materials. In fact, the global production of waste is expected to reach 2.59 and 3.40 billion tonnes per year by 2030 and 2050, respectively (Kaza et al., 2018). In the USA, the municipal waste materials generated in 2018 was 265.22 million tonnes (EPA, 2020). Waste is generated from several sources such as residential, commercial, construction and demolition, and industrial. For example, solid waste is produced at every stage of industrial processes. Most industrial waste sources are petrochemicals, chemical and pharmaceutical industries, fertilizer and agricultural chemical production, plastic manufacturing and many others.

DOI: 10.4018/978-1-6684-8117-2.ch001

Generally, waste can be classified into two main categories: hazardous waste and non- hazardous waste (Arni & Elwaheidi, 2020; Reddy, 2011). Hazardous waste consists of more than a single chemical or substance that has specific characteristics such as ignitability, corrosively, reactivity and toxicity. On the other hand, non-hazardous waste includes mineral processing materials, coal ash, oil, and other waste materials that lack the characteristics of hazardous waste.

Various components of waste cause health and environmental problems. Health problems include infections and chronic diseases, which are the result of the exposure to toxic chemicals and biological contaminants. The environmental impacts of solid waste include pollution, acidification, and water toxicity. Therefore, procedures and practices were established for efficient management of urban solid waste and to use it to extract reusable components and produce energy (Reddy, 2011). Waste management, therefore, implies the use of adequate technologies whose application depends on local, regional, and national capabilities that include regulations, policies, economic priorities, and practical local limits (infrastructure) (Arni & Elwaheidi, 2020).

The standard management practices include source reduction, recycling, and processing/disposal through different technologies such as composting, combustion/incineration, gasification, anaerobic digestion, and landfill (Arni & Elwaheidi, 2020). Nanotechnology is an emerging technology that is being employed everywhere in many sectors. The sectors include electronics (laser diodes and fiber optics transistors), medicine (drug delivery, test systems, and diagnostic systems), energy (lighting, fuel cells and solar cells), cosmetics (sunscreen, lipsticks and creams), textile (surface coating and coloring), food-drink (as packaging additives), military (neutralization of chemical weapons), automotive (catalysts, painting and sensors), and construction (insulation and flame retardants) (Bayda et al., 2019; Imani & Safaei, 2019; Safaei et al., 2019; Solano et al., 2021). The technology can be likely utilized in environmental protection and particularly in waste management. It is an interdisciplinary science that includes nanophysics, nanochemistry, nanomaterials and several other sciences. The technology uses nanomaterials, which are materials that have at least one dimension less than 100 nm.

Due to their small size, these materials increase the reactivity in processes such as adsorption, oxidation/reduction, and as catalysis of organic and inorganic contaminant materials (Dermatas et al., 2018). Typical applications of nanotechnology include nanocomposite packaging systems, which are used in meal bags to decrease the amount of waste (Pushparaj et al., 2022). Numerous additional applications of nanotechnology in waste management are reported in the scientific literature (Part et al., 2018; Walser et al., 2012).

This chapter investigates the potential role of nanotechnology in the management of waste. In particular, employment of nanotechnology in enhancing waste reduction and recycling will be discussed, with a focus on use in treatment and disposal phases of waste management.

WASTE MANAGEMENT

Globally recognized sustainable waste management strategies aim primarily to prevent waste generation and to reduce its harmfulness. However, where this is not possible, waste materials should be subjected to the "4Rs" processes: reduce, reuse, recycle, and resource recovery, which collectively ensure waste reduction. Landfill disposal is considered if none of the 4Rs is applicable and thus ranked as lowest in priority (Arni & Elwaheidi, 2020; Kaufman & Themelis, 2009; Rosenzweig et al., 2018). Figure 1a illustrates the steps of waste pre-treatment processes, starting from post-consumer waste until waste pre-

Figure 1. (a) Illustration of the steps of waste pre-treatment, and (b) treatment major processes applied to solid waste

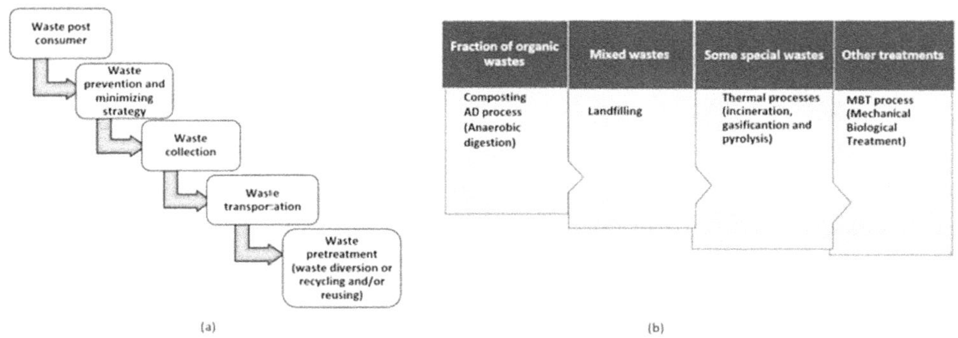

treatment, and Figure 1b shows the processes applied to organic waste. While the first steps of treatment enforce the use of low to intermediate technological level, the steps of thermal processes, Mechanical Biological Treatment (MBT) process and Anaerobic Digestion (AD) process, oblige the use of sophisticated technology and incur higher costs.

Waste disposal strategies include landfilling, composting and incineration. Figure 2 shows the sequential processes of solid waste treatment for the scope of recycling and recovery of materials and energy. Waste is collected in several types of containers such as bins and plastic drums. Waste collection and transport constitutes the major solid waste management cost. Recycling involves converting waste into a usable raw material, which extends the life of materials, reduces resources consumption and avoids disposal costs. The processes adapted for the recovery of materials and energy are divided into biologi-

Figure 2. A diagram representation illustrating the sequential processes of waste treatment using recycling and recovery of material and energy

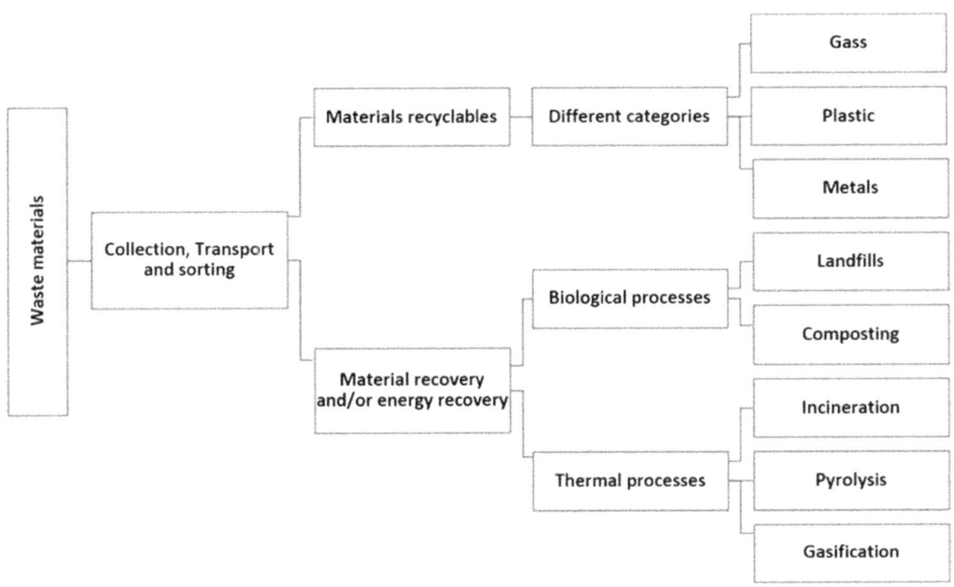

cal, such as composting and landfilling, and thermal, such as incineration, pyrolysis, and gasification (Rosenzweig et al., 2018).

NANOTECHNOLOGY

Nanotechnology is the science of manipulating and controlling matter at atomic or molecular scale. Nanomaterials (NMs) represent the core of nanotechnology and are considered the key in many applications of nanotechnology. The unique physicochemical properties of NMs make them highly effective in waste management since they have much larger surface areas than bulk particles. Furthermore, NMs work with chemical groups to increase the affinity toward certain chemical compounds. NMs can also work as recyclable legends for toxic metal ions, organic and inorganic solutes/anions in aqueous solutions (Cloete, 2010). In addition, NMs provide the possibility to develop efficient water-purification catalysts.

NMs have unique characteristics and properties such as extreme small size, large surface, high reactivity, superparamagnetic behavior, and conductivity (Kumar, 2018). NMs have dimensions of less than 100 nm. The nanoscale has a size between 1 nm and 100 nm (~1-100 nm). NMs are used for waste management in reverse osmosis (<1nm), nanofiltration (1nm), ultrafiltration (10nm), and microfiltration (100nm) methods (Kumar, 2018; Part et al., 2018). In the following sections, classifications of NMs based on several criteria, NMs sources of production, and typical example of application in waste treatment are presented.

NMs Classifications

NMs can be classified based on their: dimensions, morphology, composition, uniformity, and agglomeration. Based on dimensionality, NMs could be of 0D (such as nanopores), 1D (such as nanowires and nanotubes), 2D (such as nanofilms, nanoplates, and coatings), and 3D (such as nanocrystal and liposome) (Figure 3). As per their morphology, NMs can have different morphology based on the aspect ratio of particles; high ratio (such as nanowires and nanotubes) and low ratio (such as spherical, cubic oval, prism, and pillar shapes) (Kumar, 2018).

The chemical composition of NMs is made of one or several materials. Natural NMs are formed of diverse types of materials, while pure NMs have a single composition that can be synthesized in vitro. According to their properties, NMs can be formed in different forms such as a suspension/colloid, solid, aerosol, or agglomerated form. Agglomerated NMs synthesized due to the van der Waals forces and Brownian motion (Borbón, 2011; Tiruvalam et al., 2011).

Based on their use, NM can be further classified into nine basic groups (Dhawan et al., 2018; Sitharaman, 2016); these are: (i) metallic nanomaterials such as gold and iron oxide NMs, (ii) ceramic nanomaterials such as alumina and zirconia, (iii) semiconductor-based nanomaterials such as cadmium sulfide and cadmium selenide, (iv) organic/carbon-based nanomaterials such as carbon nanotubes (CTs) and fullerenes, (v) organic-inorganic hybrid nanomaterials, which have different types based on the source of the core and the shell, (vi) silica-based nanomaterials such as silica nanotubes and solid nanospheres, (vii) polymeric nanomaterials and nanoconjugates such as such as alumoxane and carbon nanotubes, (viii) biological nanomaterials, and (ix) biologically directed/self-assembled nanobiomaterials.

Figure 3. Classification of nanomaterials based on dimension

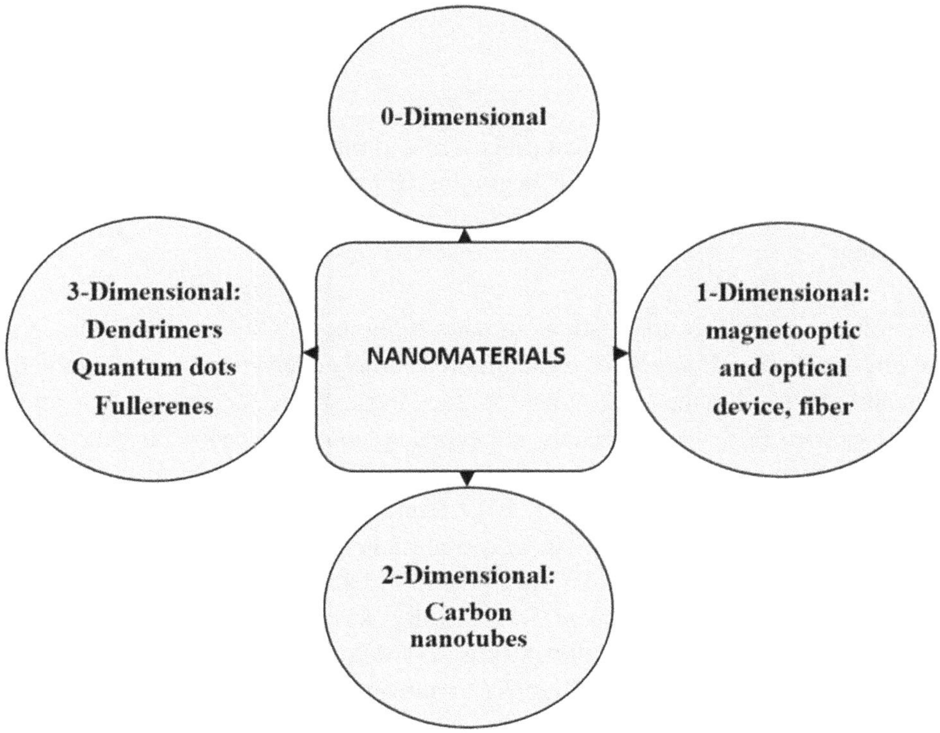

Sources of NMs

NMs are produced from three diverse sources, these are:

Engineered NMS

Man-made NMs that are produced for commercial purposes using physical, chemical and biological methods (Topmiller & Dunn, 2013). Typical examples of physical methods are laser pyrolysis, vapor decomposition, and electrospraying, while thermal decomposition and precipitation and hydrothermal synthesis are examples of wet chemical methods. Examples of biological methods include plant extract, microorganism and bio assisted biogenesis.

At least 95% of engineered NMs occur in mixed urban solid waste, packaging waste, waste of electrical and electronic equipment (WEEE), textile waste, construction & demolition waste (CDW) and battery waste. These engineered NMs include Nano-Ag, Nano-TiO2, Nano-ZnO and CNT (carbon nanotube) (Rajkovic et al., 2020).

Incidental NMs

NMs that are produced non-intentionally by anthropogenic activities, either directly or indirectly. These are also produced as by-products by various industrial processes, and combustion processes such as fuel combustion. Certain types (such as fullerenes) are also released from burning gas, biomass, and as a by-product of corrosion products.

Natural NMs

NMs produced naturally via biogeochemical processes without any anthropogenic intervention. These have been found in the earth system since its beginning (Barceló & Farré, 2012; Kumar et al., 2021).

Nanotoxicology

Despite their numerous fields of applications in nanotechnology, NMs small size, large surface area, and exclusive physicochemical properties make them extremely oversensitive leading them to represent a threat to human health and to the environment. In fact, these materials could pose a potential toxicity risk that varies based on their characteristics and exposure context. Specifically, the engineered NMs are more dangerous since any change in their properties, such as the regional surface, affects their solubility and chemical toxicity (Erdogan & Kara, 2019; Tonelli et al., 2015). The actual risk in using the engineered NMs lies in the fact that these are new materials that human beings and living things may not have effective immune system to resist their potential risks. However, several other factors might contribute to the increase of the environmental and health risks of NMs. These include the lack of public knowledge on environmental toxicity and exposure to NMs, lack of knowledge on the NMs entering mechanism into the environment and their transfer to humans, and the lack of familiarity with the risks and the valid ways to minimize them. In fact, NMs could easily enter the body through the skin, mucous, and respiratory tract, causing unknown toxins that might threaten human health.

In addition, nanoparticles accumulate in environmental and biological fluids due to their high ionic strength. Also, metal-based nanoparticles that are used in the drug delivery system have a negative environmental impact. Furthermore, polyvinyl pyrrolidone coated silver nanoparticles are transferred from E. coli to Caenorhabditis elegans via the food chain and gets accumulated inside the digestive system. Small coated silver nanoparticles (25 nm) are easily accumulated in the food chain imposing toxicity to higher trophic levels (Tomar et al., 2020).

So far, however, the toxicological risks of NMs are not fully understood since NMs behavior in the environment is fully different from regular materials and they may show more reactive and toxic effects compared to other waste materials in the environment (Njuguna et al., 2021). Consequently, using NMs implies that the associated potential hazards should be taken seriously. In this context, the proper management of NMs, their complete cycle should be tracked to ensure the monitoring and control of their impacts from manufacturing stage to dumping one. Specific attention should be given to the surface coating materials than core materials because the surface area of NMs is the factor controlling the increases of reactivity, which leads to the increase of NMs toxic impact (Kumar et al., 2021). Therefore, proper regulations and guidelines for the assessment of NMs life cycle based on their long-term benefits compared to their risk impacts on the environment must be adapted.

APPLICATIONS

The application of nanotechnologies is significant in waste management and disposal, which comes at two levels. The first is the production of new materials to replace current raw materials, which result in the generation of new waste such as nanocomposites that can be used to improve biodegradables such as plastic wrap for food. The second level is the application of novel solutions and facilities for recycling and disposal, such as devices, nanoprocessing and or the improvement of existing processes in treatment, or to reduce the amount of generated waste (Baldé et al., 2015; Bayda et al., 2019; Forti et al., 2020; Hornak, 2021; Parthasarathy, 2021; Safaei et al., 2019).

Nanotechnology Processes

Nanotechnology uses nanofilters, nanoparticles, nanophotocatalyst, nanoclays, and carbon nanotubes and nanosensors in waste treatment processes, and adsorption techniques like nano-activated carbon (Kumar et al., 2021; Qu et al., 2013; Robinson, 2007; Shekdar, 2009; Spellman, 2018). In addition, nanofiltration has been considered effective in treatment of toxic waste and nonbiodegradable materials in waste and their leachate treatments, such as arsenic, xenobiotic organic compounds, ammonia, inorganic macro components, heavy metals, anions and cations, arsenic, uranium, chromium, and pathogens from wastewater (Anand & Singh, 2014; Li et al., 2016; Ngo et al., 2009). Nanofilters such as nanomembrane filtration can be used to remove a wide range of pollutants such as biological fouling, chemical oxygen demand (COD) and ammonium. The percentage of COD removal from leachate by nanofilters has a range 60 to 70%, while the ammonium reached 50% (Taran et al., 2021).

Nanoparticles such as zinc and silver coated on a filter are used in the fabrication of catalysts. The nanophotocatalyst (as TiO_2) and nonporous catalysts are used to improve the absorption of dyes and heavy metals from wastewater (Kyung et al., 2005; Liao et al., 2012; Narasaiah & Mandal, 2019; et al., 2006). In the case of plastic waste, which forms a substantial part of urban solid wastes, extensive research is in progress to find alternative plastic products at nano levels. These will improve the biodegradability of plastic, such as polymer nanocomposites. Furthermore, the application of nanomaterials-based adsorption, membrane filtration and separation and photocatalysis can be used in ex-situ waste management technologies (Dermatas et al., 2018).

Using nanotechnology fabrication such as nano filter and Ag, Cu, ZnO, TiO_2 nanoparticles (nano-catalysts), carbon nanotubes etc., will increase waste treatment efficiency and provide an advanced way for waste disposal and treatment. However, the risks emerging from nanomaterials waste need to be taken into consideration. This implies collection of data and information on the hazardousness of nanomaterials, which can be emitted from waste and have serious environmental effects. For example, after the combustion of urban solid waste, the incineration plant can still have crystalline nano-CeO_2 particles (Adrados et al., 2012; Mahapatra et al., 2018; Walser et al., 2012).

Synthesis of nanoparticles and nanostructures is a fundamental aspect of nanotechnology. In fact, new physical properties and applications of nanomaterials are made possible when nanostructured materials are generated with desired size, shape, morphology, crystal structure and chemical composition. Synthesized nanoparticles could be manufactured using two main approaches, these are the top-down and the bottom-up. The top-down approach involves breaking down a bulk material into large pieces to obtain the nanosized particles and structures. This is a simple approach that, however, has the problem of surface structure imperfection. For example, synthesized nanowires using lithography do not have

Table 1. A comparative summary of the basic approaches utilized for producing synthesis nanoparticles

Top-down	Bottom-up
Smaller objects are generated from larger objects.	Small components are assembled into complex nanostructures.
Based on molecular recognition principle.	Uses layer-by-layer self-assembly.
Uses microfabrication method which convert thin substrate into a complex structure that has the required shape and order.	It has high precision accuracy and therefore can manufacture small structures.
The process key is to control nucleation and growth.	The process key is to control energy input and contamination.
Conducted in gas or liquid phases.	Conducted in solids phase.
Examples: Thermal evaporation; laser ablation; sputtering.	Examples: Hydrothermal; Sol-gel.

smooth surfaces and might contain impurities and defects. In the alternative bottom-up approach, single atoms and molecules are assembled into larger nanoparticles. It has the advantage of being an economical approach because it creates less waste (Abid et al., 2022; Kumar et al., 2021). Table 1 illustrates the basic characteristics of the two approaches and provides a few examples.

Examples of Waste Treatment Applications

In the following sections, typical example applications of nanotechnology mechanisms in the treatment, removal, and recycling of waste materials are provided.

Heavy Metals Remediation

Heavy metals are naturally occurring metallic elements and metalloids that have higher density than water (Skoulidas et al., 2006). The increasing use of heavy metals in industrial, agricultural, and urbanization activities has contributed to the production of waste materials that are rich with metals such as arsenic, cadmium, lead, and mercury. Heavy metals are challenging environmental pollutants and a source of risk to health because they are nondegradable substances and remain for a long time in the contaminated environment (Azeh Engwa et al., 2019; Duffus, 2002; He et al., 2005; Hillie & Hlophe, 2007; Sheet et al., 2014; Tchounwou et al., 2012).

Usually, chemical, physical and biological methods such as heat, precipitation, bioleaching, ion exchange, and chemical extraction processes are used for heavy metals remediation. However, these processes lack efficiency, have unstable by-products, and not cost convenient (Gunatilake, 2015; Kumar et al., 2021). Therefore, nanotechnology provides an effective alternative technology for the remediation of heavy metal waste. In the following, typical nanotechnology approaches that are often applied for the remediation of heavy metals, and their characteristics are presented:

Carbon-based NMs. Characterized by high resistance to corrosion, remarkably high thermal and electrical conductivity, and stability. Carbon nanotubes, in particular, are highly effective in heavy metal removal since they have large specific surface area, high adsorption capacity, and fast adsorption properties (Ali et al., 2019; Mallikarjunaiah et al., 2020).

Silica-based NMs. Non-toxic, have a light weight and a controllable morphology. These properties make them valid adsorbents for heavy metal removal. Furthermore, these NMs can be used as protective shields at low pH solutions (Bhunia et al., 2018; Wang et al., 2018).

Nanocomposites. Hybrid materials in which one or more nano dimension phases are embedded in a metal, ceramic, and polymer matrix (Ajayan et al., 2006; Sen, 2020). Inorganic (non-polymer) nanocomposites consist of carbon and natural components of zeolite, bentonite, and montmorillonite. These composites have excellent performance for remediation of heavy metals as compared to pure carbon-based materials such carbon nanotube, graphene, and graphene oxides. Bentonite and zeolite are widely used in heavy metal treatment due to their high adsorption, high ion and cation exchange properties, and chemical stability. Polymer nanocomposites are also used for heavy metal remediation due to their excellent mechanical strength, degradable properties and being eco-friendly (Kosuge et al., 2016; Zhao et al., 2018).

Another type of polymer nanocomposites is the magnetic nanocomposites, which are composed of a multi-layered structure that have nanosized magnetic materials. Magnetic nanocomposites are characterized by their high separation capability, high surface area, and surface functional groups that can interact with heavy metal ions (Hu et al, 2008; Zhu et al., 2013). For instance, magnetite oxide composite was synthesized with a magnetite particle size of 10–15 nm for the removal of cobalt (II) ions from aqueous solutions (Liu et al., 2011).

Landfill Leachate Treatment

Leachate is the contaminated water that originated from waste internal water, waste biochemical reactions, and precipitation (Mandal et al., 2017; Reshadi et al., 2020). Leachate is mainly produced in landfills and contains hazardous chemicals such as heavy metals, ammonia, and organic compounds, which might cause serious diseases and contaminate the environment. Common leachate treatment methods include physical, chemical, and biological technologies. Most physical methods such as precipitation and adsorption, however, lack the capability to eliminate the adverse impacts on the environment. Similarly, biological, and chemical methods such as activated sludge process, aerobic stabilization pond, and membrane technologies are proved to be poor in the efficiency of compounds removal from old landfill leachate and have high operating cost (Chen et al., 2021; Hu et al., 2018; Tałałaj et al., 2019; Usman et al., 2020; Wang et al., 2018).

The application of nanotechnology provides alternative technologies for water pretreatment and purification of wastewater that is based on using nanoparticles. More specifically, iron nanoparticles and titanium dioxide are the most employed nanoparticles in landfill leachate treatment; these will be further discussed below:

Iron nanoparticles. Highly effective in removing organic materials and chlorinated solvents from leachate due to their small size and high surface area. Iron based nanoparticles could be used both in *ex-situ* for the treatment of contaminated soil and other solid wastes, and *in-situ* remediations to degrade the chlorinated organics such as trichloroethylene into eco-friendly compounds. In favorite conditions, iron nanoparticles can remove about 48% of leachate pollutants. Iron nanoparticles are also used in municipal solid waste landfill plants for the pretreatment and post-treatment of the leachate (Chen et al., 2021; Singh, 2022; Zaidi et al., 2015).

Titanium dioxide. Extensively used for the removal of heavy metals present in landfill leachate. For instance, titanium dioxide can remove 96% of lead and chromium particles present in leachate. Due to its chemical stability and nontoxic nature, it is considered a stable adsorbent and photocatalytic (Kumar et al., 2021).

Plastic Waste Remediation

Plastics are non-biodegradable materials that have been used in daily life and industries for decades due to their low-cost, lightness, stability, and durability. However, the increasing demand for plastics and the high consumption rates have raised the issue of the improper management of plastic waste and the associated environmental and health risks (Ganesh Saratale et al., 2021; Zheng et al., 2022). Traditional methods of plastic waste management include landfilling, incineration, and pyrolysis. However, these treatment methods produce toxic volatile substances (by pyrolysis), dust and irritant gases (by incineration), and leachate (by landfilling).

The problems with traditional plastic management methods can be overcome by applying nanotechnology approaches such as recycling tagging and tracking. In this approach, nanoparticle tags will be incorporated into the product at the manufacturing stage, which will allow tracking the product position during its entire life cycle. There are many advantages to using nanoparticle tags that are attributed to the small size of nanoparticles. One advantage is that no changes will occur to the basic structure of the product(Kumar et al., 2021; Zheng et al., 2022). In addition, nanoparticle tags will aid in product recycling based on its production method. Furthermore, the nanoparticle tags will not be removed from the product and, therefore, will be easily identified by optical, electronic, or magnetic detection mechanisms. Lastly, a clear result of using nano-sized tags is the reduction of the quantity of plastic waste materials disposal into landfills (Ganesh Saratale et al., 2021).

In another approach to plastic waste remediation, nanotechnology is used to enhance the physiochemical properties and mechanical properties of biodegradable plastics, which improves their decomposition and recycling. For this purpose, nanotechnologies such as nanofillers and photodegradation are commonly used; a brief description of these nanotechnologies is provided below:

Nanofillers. Nanoscale additives that can improve the performance of plastic in terms of stability, barrier property, and gas permeability. Nanofillers are classified into four types: organic, inorganic, nanoclay, and carbon nanostructures. Organic nanofillers include chitosan and cellulose, while inorganic nanofillers include metals, such as silver, and metal oxides, such as titanium dioxide. Carbon nanostructures include graphene, nanofibers, and nanotubes. Nanoclay is a compound made from elements such as silicon and oxygen (Kumar et al., 2020; Othman & Fadzil, 2021).

Photodegradation. The use of light-induced effect of plastic under the sun that causes the break of chemical bonds, destroy the structure of the plastic, and the oxidization to water and carbon dioxide. Additive nanomaterials can also be added as photocatalysts to improve the degradation efficiency and reduce the cost. Numerous recent photocatalyst plastic waste research proved the notable significance of using photodegradation in plastic waste management (Ariza-Tarazona et al., 2020; Tofa et al., 2019; Zheng et al., 2022).

Finally, the benefits of the application of nanotechnology in plastics waste remediation include, but not limited to, reducing the amount of solid waste in landfills, manufacturing eco-friendly products, and better recycling of the product using the tracking tag

Pollutants Detection

Monitoring of Environmental pollution is a major challenge as an increase in living standards have increased earth resources with diverse contaminant materials such as oil spills, pesticides, and other non-biodegradable materials. Manmade activities have also caused pollution of water bodies with bacteria, and other micro-pollutants (Patel et al., 2020). Therefore, regular detection and monitoring of environmental pollutants should be performed by the analysis of collected environmental samples. The conventional method that is used to achieve this goal is based on the exact analysis of physical or chemical properties of the pollutant, which is, however, proved to be time consuming and costly since it requires specific equipment and well-equipped for analysis of pollutants. Nanosensors provide an alternative approach for environmental monitoring that is simple, rapid, and sensible for analysis of pollutants, particularly the limit of detection and sensitivity (Singh & Singh, 2021).

A nanosensor is a nanoscale device that measures a physical quantity and converts it to a signal that can be detected and analyzed. A nanosensor is composed of three components that characterize it. These are a nanomaterial, a recognition element that provides specificity, and a signal transduction method that indicates the detection of a single or multiple analytes. Based on the signal transduction type, nanosensors can be classified into three types: optical (e.g., Optical fiber), electrochemical (e.g., impedance and amperometric), piezoelectric (e.g., crystal microbalance), and magnetic nanosensors (Mohamed, 2020). Different types of analytes can be detected and monitored using nanosensors such as hazardous chemicals and pesticides (Willner & Vikesland, 2018). Environmentally hazardous chemicals such as hydrazine and ammonia can be detected using nanosensors. For example, gold-based nanosensors were used to detect hydrazine using the cyclic voltammetry analytical technique. These nanosensors proved a good selectivity and sensitivity for the detection of hydrazine (Karthik et al., 2016; Liu et al., 2013; Mandal et al., 2018). Furthermore, gold-based nanosensor were proved efficient in detecting ammonia compound based on the optical characteristics of gold nanoparticles with a lowest detection of 1ppb (Pandey et al., 2013). In addition, silver-based nanosensors are used for the detection of hydrogen peroxide. The nanosensor revealed excellent catalytic activity in the reduction of H_2O_2 with a response limit of 8 μM (Liu et al., 2013). Cellulose-coated silver-based nanosensors are also used for the detection of hydrogen peroxide (Subramanian et al., 2017).

Detection of pesticides is of prime importance due to their toxicity and potential bioaccumulation. There are more than 800 ingredients present in commercial pesticides (Willner & Vikesland, 2018). The main classes of pesticides that have been the focus of nanosensor detection are Organophosphorus (OP), carbamates, neonicotinoids, and triazines (Liu et al., 2013; Willner & Vikesland, 2018). Fiber-optic biosensor based on sol-gel is used to detect dichlorvos at a concentration of 5.2 ppb (Andreou & Clonis, 2002).

A fiber-optic biosensor based on the enzyme glutathione S-transferase I is also used for detection of atrazine. Furthermore, a co-immobilizing acetylcholinesterase and glutathione S-transferase in a sol-gel film biosensor is utilized for the detection of OP compounds at concentrations from 0 to 2 ppm (Choi et al., 2003).

Lithium-ion Batteries (LIBs) Recycling

LIBs are dominating the battery market in the fields of electronics and electric vehicles. This is attributed to the many advantages LIBs have over traditional batteries in terms of energy density, lifespan, and reduced cost (Awan, 2020). Recently, nanotechnology has been extensively employed in manufacturing

LIBs to enhance their performance based on the unique physical and chemical properties of NMs (Rarotra et al., 2020). Spent LIBs, however, impose potential sources of pollution to the ecosystem if disposed improperly. LIBs contain heavy metals such as Co and Ni, and organic electrolytes that might pollute groundwater and the soil in case of leaking. Hence, recycling LIBs contributes to the sustainability of resources in terms of recovery of valuable metals such as iron, chromium, nickel, and manganese, and the proper management and removal of toxic metals contained in LIBs such as cadmium, copper, zinc, and mercury (Fan et al., 2021; Harper et al., 2019; Olapiriyakul & Caudill, 2008).

The recycling processes that are applied to recover valuable metals can be divided into four main types: pre-treatment, pyrometallurgy, hydrometallurgy and biometallurgy processes, and often recovery methods utilize a combination of these processes. These processes, however, might result in hazardous fumes and require large amounts of water and neutralizing solutions to treat the acid waste used in hydrometallurgy. Nanofibers and graphene provide valid alternatives that are capable of recovering metals due to their unique properties (Fan et al., 2021; Pavoski et al., 2022; Rarotra et al., 2020).

Electric and Electronic Waste Recycling

The demand for electrical and electronic equipment has been increasing rapidly due to urbanization, industrialization, and increasing population. The generated waste consists of both harmful and valuable components and is termed as Electric and Electronic Waste (EEW). It has been reported that about 54 Mt e-waste was generated in 2019 out of which only 17.4% was collected and recycled (Forti et al., 2020; Parthasarathy, 2021). The precious and rare earth metals present in EEW are highly relevant for nanotechnological applications, and improvements in EEW recycling could secure the amounts of raw materials that are needed to maintain the sustainability of nanotechnology.

EEW recycling is complex under many folds; the valuable metals are found in very small quantities within the bulk material of the equipment. In addition, sorting and dismantling electronics processes are time consuming and labor-intensive. Finally, the overall cost is high due to the need to provide adequate equipment that is capable of handling the toxic gases. Hydro- and pyrometallurgy are the current methods employed for the recovery of metals from EEW. These methods, however, consume high energy and generate significant amounts of toxic liquid and gases. The alternative environmentally friendly solutions represented by bioleaching, the use of supercritical carbon dioxide as extraction solvent, and electrochemistry. For example, electrochemistry is used to recover rare earth metals such as europium, neodymium and gadolinium using carbon nanotubes via their selective electrodeposition at different applied potentials (O'Connor et al., 2018; Pavoski et al., 2022).

Nuclear Waste Decontamination

The increasing demand for energy sources that are alternative to fossil fuel, and the need to cope with the rapid developments in technological and industrial sectors that demand more energy, made nuclear energy a valid renewable energy source that can be used for diverse purposes. The byproduct of nuclear power generation is a hazardous radioactive waste that poses a serious threat to all forms of life as well as the environment and thus, should be strictly regulated. Two different disposal strategies are applied in nuclear waste management, these are: (1) the geological disposal of the spent fuel after a few decades of wet or dry storages, and (2) the spent fuel storage that is reprocessed with aqueous dissolution, and conditioned in borosilicate glass before the geological disposal (Nath, 2018; Zhang & Liu, 2020).

However, decontamination of radioactive waste using current treatment and disposal procedures, such as membrane separation, ion exchange and chemical precipitation, is complicated due to the extremely long half-life of radionuclides, and presence of certain radionuclides in substantially lower concentrations than those of other components (Shi et al., 2012). Consequently, high-efficiency selective removal technological innovation in nuclear waste management is required. Nanotechnology and more specifically, NMs, can be used for safe disposal, classification, storage, and tracking of radioactive nuclear waste. It is possible now to design and produce a nanosized specialized functional materials. The unique physicochemical properties, such as the nano-size, high chemical reactivity, and large specific surface area, have made the use of these materials an efficient method for the disposal of radioactive nuclear waste. Various types of NMs are used to decontaminate nuclear waste and remove radionuclides. In the following, a brief description of the basic nano-based techniques that are used for nuclear and radioactive waste are given:

Carbon nanotubes. They have the potential to be stabilized with other atoms, and when used as an adsorbent, they are capable of isolating radionuclides from nuclear waste streams. The spectrum of possible applications of carbon nanotubes includes treatment of liquid effluents from nuclear plants, environmental remediation of contaminated sites, waste immobilization for final disposal. Carbon nanotubes are also used as an ion exchanger to remove radioactive cesium ions from nuclear wastewater (Belloni et al., 2009; Shi et al., 2012; Zhang & Liu, 2020).

Graphene oxide. It has a large surface area and a high concentration of oxygen- containing functional groups on the surface. This property allows the radionuclides to be easily added to the graphene oxide (Shi et al., 2012). To reduce the self-aggregation of graphene oxide and increase its dispersion in nuclear wastewater, graphene oxide is usually functionalized or combined with other functional materials. The graphene oxide adsorption capability can be further enhanced by grafting nano-polymers to its surface (Belloni et al., 2009; Shi et al., 2012).

Nanoparticle modified microrobots. These were tested for the purpose of removing radioactive components from the nuclear wastewater. This is supported by the fact that the microrobots have an extreme adsorption ability as well as rapid autonomous mobility in liquids. In a simulated radioactive wastewater, microrobots recovered 96% of uranium-loaded rods using a magnet within one hour (Ying et al., 2019). This could be a good indicator on the effective employment of microrobots in the management and remediation of radioactive waste.

DISCUSSION

The rapid developments in nanotechnology and NMs sciences have brought significant improvements into environmental sustainability procedures in general and into waste management practices in particular. In this context, NMs are being successfully employed in various environmental applications such as detecting and monitoring various pollutants and waste disposal using techniques such as nano-sorbents, nanomembrane separation, and environmental nano-catalysis.

However, many challenges in the field of environmental nanotechnology applications still persist and need to be addressed in a manner that maintains the balance among activity, reactivity, and stability (Sun, 2019). The challenges have several aspects which can be classified into four categories (Louie et al., 2016): physicochemical transformations of NMs; metrology; toxicity assessment; risk assessment exposure modeling. Table 2 provides detailed description of each challenge.

Table 2. Categories of current challenges that face the application of nanotechnology procedures for environmental sustainability and waste management

Challenge	Process	Reference
Metrology	Detection of selective detection methods of NMs might be enforced by environmental regulations. Physicochemical characterization is required for identifying the mechanisms that control the behavior of nanoparticles and their transformations.	(Louie et al., 2016)
Nanoparticles physicochemical transformations	Transformation of individual nanoparticles depends on their origin, composition, and amount. The transformation takes place in several chemical and physical modes that include aggregation, dissolution, and redox reactions.	(Maurer-Jones et al., 2013)
Toxicity assessment	Dosing of the nanomaterial can be problematic due to size heterogeneity of the sample or aggregation, which can change the exposure concentration due to settling and nanoparticle diffusion through biological coatings or mucin layers.	(Hull et al., 2013; Louie et al., 2016; Maurer-Jones et al., 2013)
Risk assessment exposure modeling	Exposure modeling is hindered by the incomplete understanding of the physicochemical transformations of nanoparticles and practical limitations of data availability.	(Louie et al., 2016; Lowry et al., 2012)

In particular, there is a necessity to find efficient methods to deal with the negative environmental impacts of nano-waste, which is a byproduct of nanotechnology (Sun, 2019). Furthermore, an international classification of nano-waste must be established to facilitate waste treatment procedures. Nano-waste management, in fact, is a promising field of environmental research that deserves the allocation of generous economical resources by the governments. The issue is of global concern and made part of the Agenda 2030 of the G20 forum that held in 2017 in Germany (Faunce & Kolodziejczyk, 2017). Examples of potential future research fields include: better understanding of the extent and rate of physiochemical transformations of NMs, detection and characterizing NMs in complex media, the development of environmentally relevant ecotoxicology assays, and the design of sufficiently detailed fate and transport models (Louie et al., 2016).

CONCLUSION

Waste management constitutes an increasing challenge due to the increasing rate of waste production and the introduction of new types of solid waste materials such as e-waste materials. Overcoming the issue involves using modern technologies such as nanotechnology, which has proved efficient in many other fields. Nanotechnology is the science of manipulating and controlling matter at atomic or molecular scale. It uses nanomaterials, which are materials that have at least one dimension less than 100 nm. Due to their small size, these materials increase reactivity in processes such as adsorption and oxidation/reduction. Nanomaterials can be biological (proteins and liposomes), carbon compounds (2D graphene, 1D carbon nanotubes and 0D carbon dots), oxides (copper oxide, silicon oxide, zinc oxide, titanium oxide), metal nanoparticles (silver, gold, platinum, etc.), nanostructured catalysts and polymers.

The application of nanotechnology is significant in waste management and disposal, particularly, in the production of new materials to replace current raw materials, and in providing novel solutions and facilities for recycling and disposal. In addition, nanofiltration is effective in remediation of heavy metals, toxic waste and nonbiodegradable materials of leachate. Furthermore, nano-filters such as nano-membrane filtration can be used to remove a wide range of pollutants such as biological fouling, COD and ammonium. To be emphasized, however, that the application of nanotechnology in waste management is still in its primary phases and, thus, enormous research is expected to be conducted, including nanomaterials risks to health and the environment.

REFERENCES

Abid, N., Khan, A. M., Shujait, S., Chaudhary, K., Ikram, M., Imran, M., Haider, J., Khan, M., Khan, Q., & Maqbool, M. (2022). Synthesis of nanomaterials using various top-down and bottom-up approaches, influencing factors, advantages, and disadvantages: A review. *Advances in Colloid and Interface Science, 300*, 102597–102597. doi:10.1016/j.cis.2021.102597 PMID:34979471

Adrados, A., de Marco, I., Caballero, B. M., López, A., Laresgoiti, M. F., & Torres, A. (2012). Pyrolysis of plastic packaging waste: A comparison of plastic residuals from material recovery facilities with simulated plastic waste. *Waste Management (New York, N.Y.), 32*(5), 826–832. doi:10.1016/j.wasman.2011.06.016 PMID:21795037

Ajayan, P. M., Schadler, L. S., & Braun, P. V. (2006). *Nanocomposite science and technology*. John Wiley & Sons.

Ali, I., Alharbi, O. M. L., Alothman, Z. A., Al-Mohaimeed, A. M., & Alwarthan, A. (2019). Modeling of fenuron pesticide adsorption on CNTs for mechanistic insight and removal in water. *Environmental Research, 170*, 389–397. doi:10.1016/j.envres.2018.12.066 PMID:30623886

Anand, A., & Singh, S. (2014). Membrane technique for leachate treatment: A literature review. *International Journal of Environmental Research and Development, 4*(1), 33–36.

Andreou, V. G., & Clonis, Y. D. (2002). Novel fiber-optic biosensor based on immobilized glutathione S-transferase and sol–gel entrapped bromcresol green for the determination of atrazine. *Analytica Chimica Acta, 460*(2), 151–161. doi:10.1016/S0003-2670(02)00250-7

Andreou, V. G., & Clonis, Y. D. (2002). A portable fiber-optic pesticide biosensor based on immobilized cholinesterase and sol–gel entrapped bromcresol purple for in-field use. *Biosensors & Bioelectronics, 17*(1-2), 61–69. doi:10.1016/S0956-5663(01)00261-5 PMID:11742736

Ariza-Tarazona, M. C., Villarreal-Chiu, J. F., Hernández-López, J. M., Rivera De la Rosa, J., Barbieri, V., Siligardi, C., & Cedillo-González, E. I. (2020). Microplastic pollution reduction by a carbon and nitrogen-doped TiO2: Effect of pH and temperature in the photocatalytic degradation process. *Journal of Hazardous Materials, 395*, 122632–122632. doi:10.1016/j.jhazmat.2020.122632 PMID:32315794

Arni, S. S. A., & Elwaheidi, M. M. (2020). *Concise Handbook of Waste Treatment Technologies*. CRC Press.

Awan, A. (2020). Batteries: An essential technology to electrify road transport. In Global EV Outlook 2020: Entering the decade of electric drive? (pp. 185-219): International Energy Agency.

Azeh Engwa, G., Udoka Ferdinand, P., Nweke Nwalo, F., & Unachukwu, N. M. (2019). Mechanism and Health Effects of Heavy Metal Toxicity in Humans. In Poisoning in the Modern World - New Tricks for an Old Dog?: IntechOpen.

Baldé, C. P., Wang, F., Kuehr, R., & Huisman, J. (2015). *The global e-waste monitor–2014, United Nations University, IAS–SCYCLE*. ISBN Print.

Barceló, D., & Farré, M. (2012). *Analysis and risk of nanomaterials in environmental and food samples* (Vol. 59). Newnes. doi:10.1016/B978-0-444-56328-6.09999-0

Bayda, S., Adeel, M., Tuccinardi, T., Cordani, M., & Rizzolio, F. (2019). The History of Nanoscience and Nanotechnology: From Chemical–Physical Applications to Nanomedicine. *Molecules (Basel, Switzerland)*, *25*(1), 112–112. doi:10.3390/molecules25010112 PMID:31892180

Belloni, F., Kütahyali, C., Rondinella, V. V., Carbol, P., Wiss, T., & Mangione, A. (2009). Can carbon nanotubes play a role in the field of nuclear waste management? *Environmental Science & Technology*, *43*(5), 1250–1255. doi:10.1021/es802764g PMID:19350887

Bhunia, P., Chatterjee, S., Rudra, P., & De, S. (2018). Chelating polyacrylonitrile beads for removal of lead and cadmium from wastewater. *Separation and Purification Technology*, *193*, 202–213. doi:10.1016/j.seppur.2017.11.001

Borbón, L. O. P. (2011). *Computational studies of transition metal nanoalloys*. Springer Science & Business Media. doi:10.1007/978-3-642-18012-5

Chen, G., Wu, G., Li, N., Lu, X., Zhao, J., He, M., Yan, B., Zhang, H., Duan, X., & Wang, S. (2021). Landfill leachate treatment by persulphate related advanced oxidation technologies. *Journal of Hazardous Materials*, *418*, 126355–126355. doi:10.1016/j.jhazmat.2021.126355 PMID:34329014

Chen, W., Gu, Z., Ran, G., & Li, Q. (2021). Application of membrane separation technology in the treatment of leachate in China: A review. *Waste Management (New York, N.Y.)*, *121*, 127–140. doi:10.1016/j.wasman.2020.12.002 PMID:33360812

Choi, J.-W., Kim, Y.-K., Oh, B.-K., Song, S.-Y., & Lee, W. H. (2003). Optical biosensor for simultaneous detection of captan and organophosphorus compounds. *Biosensors & Bioelectronics*, *18*(5-6), 591–597. doi:10.1016/S0956-5663(03)00016-2 PMID:12706567

Cloete, T. E. (2010). *Nanotechnology in water treatment applications*. Caister Academic Press.

Dermatas, D., Mpouras, T., & Panagiotakis, I. (2018). Application of Nanotechnology for waste management: Challenges and limitations. *Waste Management & Research*, *36*(3), 197–199. doi:10.1177/0734242X18758820 PMID:29466938

Dhawan, A., Singh, S., Kumar, A., & Shanker, R. (2018). *Nanobiotechnology: Human Health and the Environment*. CRC Press. doi:10.1201/9781351031585

Duffus, J. H. (2002). Heavy metals" a meaningless term?(IUPAC Technical Report). *Pure and Applied Chemistry*, *74*(5), 793–807. doi:10.1351/pac200274050793

EPA. U. (2020). Advancing sustainable materials management: 2018 tables and figures. Assessing Trends in Materials Generation and Management in the United States. United States Environmental Protection Agency.

Erdogan, O., & Kara, M. (2019). Analytical approach to the waste management of nanomaterials in developing countries. *Front. Drug Chem. Clin. Res*, *2*, 1–5.

Fan, M., Chang, X., Meng, Q., Wan, L. J., & Guo, Y. G. (2021). Progress in the sustainable recycling of spent lithium-ion batteries. *SusMat*, *1*(2), 241–254. doi:10.1002us2.16

Faunce, T., & Kolodziejczyk, B. (2017). Nanowaste: Need for disposal and recycling standards. *G20 Insights, 148*, 202-213.

Forti, V., Baldé, C. P., Kuehr, R., & Bel, G. (2020). *The global e-waste monitor 2020. United Nations University (UNU), International Telecommunication Union (ITU) & International Solid Waste Association*. ISWA.

Ganesh Saratale, R., Cho, S.-K., Dattatraya Saratale, G., Kadam, A. A., Ghodake, G. S., Kumar, M., Naresh Bharagava, R., Kumar, G., Su Kim, D., Mulla, S. I., & Seung Shin, H. (2021). A comprehensive overview and recent advances on polyhydroxyalkanoates (PHA) production using various organic waste streams. *Bioresource Technology*, *325*, 124685–124685. doi:10.1016/j.biortech.2021.124685 PMID:33508681

Gunatilake, S. (2015). Methods of removing heavy metals from industrial wastewater. *Methods (San Diego, Calif.)*, *1*(1), 14.

Harper, G., Sommerville, R., Kendrick, E., Driscoll, L., Slater, P., Stolkin, R., . . . Lambert, S. (2019). Recycling lithium-ion batteries from electric vehicles. *Nature, 575*(7781), 75-86.

He, Z. L., Yang, X. E., & Stoffella, P. J. (2005). Trace elements in agroecosystems and impacts on the environment. *Journal of Trace Elements in Medicine and Biology*, *19*(2-3), 125–140. doi:10.1016/j.jtemb.2005.02.010 PMID:16325528

Hillie, T., & Hlophe, M. (2007). Nanotechnology and the challenge of clean water. *Nature Nanotechnology*, *2*(11), 663–664. doi:10.1038/nnano.2007.350 PMID:18654395

Hornak, J. (2021). Synthesis, Properties, and Selected Technical Applications of Magnesium Oxide Nanoparticles: A Review. *International Journal of Molecular Sciences*, *22*(23), 12752–12752. doi:10.3390/ijms222312752 PMID:34884556

Hu, J. S., Zhong, L. S., Song, W. G., & Wan, L. J. (2008). Synthesis of hierarchically structured metal oxides and their application in heavy metal ion removal. *Advanced Materials*, *20*(15), 2977–2982. doi:10.1002/adma.200800623

Hu, W., Zhou, Y., Min, X., Liu, J., Li, X., Luo, L., Zhang, J., Mao, Q., Chai, L., & Zhou, Y. (2018). The study of a pilot-scale aerobic/Fenton/anoxic/aerobic process system for the treatment of landfill leachate. *Environmental Technology*, *39*(15), 1926–1936. doi:10.1080/09593330.2017.1344325 PMID:28625109

Hull, M. S., Vikesland, P. J., & Schultz, I. R. (2013). Uptake and retention of metallic nanoparticles in the Mediterranean mussel (Mytilus galloprovincialis). *Aquatic Toxicology (Amsterdam, Netherlands)*, *140-141*, 89–97. doi:10.1016/j.aquatox.2013.05.005 PMID:23765031

Imani, M. M., & Safaei, M. (2019). Optimized Synthesis of Magnesium Oxide Nanoparticles as Bactericidal Agents. *Journal of Nanotechnology*, *2019*, 1–6. doi:10.1155/2019/6063832

Karthik, R., Chen, S.-M., Elangovan, A., Muthukrishnan, P., Shanmugam, R., & Lou, B.-S. (2016). Phyto mediated biogenic synthesis of gold nanoparticles using Cerasus serrulata and its utility in detecting hydrazine, microbial activity and DFT studies. *Journal of Colloid and Interface Science*, *468*, 163–175. doi:10.1016/j.jcis.2016.01.046 PMID:26835586

Kaufman, S. M., & Themelis, N. J. (2009). Using a Direct Method to Characterize and Measure Flows of Municipal Solid Waste in the United States. *Journal of the Air & Waste Management Association*, *59*(12), 1386–1390. doi:10.3155/1047-3289.59.12.1386 PMID:20066903

Kaza, S., Yao, L. C., Bhada-Tata, P., & Van Woerden, F. (2018). *What a Waste 2.0: A Global Snapshot of Solid Waste Management to 2050*. World Bank. doi:10.1596/978-1-4648-1329-0

Kosuge, T., Imato, K., Goseki, R., & Otsuka, H. (2016). Polymer–inorganic composites with dynamic covalent mechanochromophore. *Macromolecules*, *49*(16), 5903–5911. doi:10.1021/acs.macromol.6b01333

Kumar, A. (2018). Recent Advances in Solid Waste Minimization using Nanotechnology. *International Journal for Research in Applied Science and Engineering Technology*, *6*(6), 759–764. doi:10.22214/ijraset.2018.6118

Kumar, R., Kumar, R., & Kaur, G. (2021). *New frontiers of nanomaterials in environmental science*. Springer. doi:10.1007/978-981-15-9239-3

Kumar, S., Mukherjee, A., & Dutta, J. (2020). Chitosan based nanocomposite films and coatings: Emerging antimicrobial food packaging alternatives. *Trends in Food Science & Technology*, *97*, 196–209. doi:10.1016/j.tifs.2020.01.002

Kyung, H., Lee, J., & Choi, W. (2005). Simultaneous and Synergistic Conversion of Dyes and Heavy Metal Ions in Aqueous TiO ₂ Suspensions under Visible-Light Illumination. *Environmental Science & Technology*, *39*(7), 2376–2382. doi:10.1021/es0492788 PMID:15871279

Li, J., Zhao, L., Qin, L., Tian, X., Wang, A., Zhou, Y., Meng, L., & Chen, Y. (2016). Removal of refractory organics in nanofiltration concentrates of municipal solid waste leachate treatment plants by combined Fenton oxidative-coagulation with photo – Fenton processes. *Chemosphere*, *146*, 442–449. doi:10.1016/j.chemosphere.2015.12.069 PMID:26741550

Liao, C.-H., Huang, C.-W., & Wu, J. C. S. (2012). Hydrogen Production from Semiconductor-based Photocatalysis via Water Splitting. *Catalysts*, *2*(4), 490–516. doi:10.3390/catal2040490

Liu, M., Chen, C., Hu, J., Wu, X., & Wang, X. (2011). Synthesis of magnetite/graphene oxide composite and application for cobalt (II) removal. *The Journal of Physical Chemistry C*, *115*(51), 25234–25240. doi:10.1021/jp208575m

Liu, R., Wei, Y., Zheng, J., Zhang, H., & Sheng, Q. (2013). A Hydrogen Peroxide Sensor Based on Silver Nanoparticles Biosynthesized by <i>Bacillus subtilis</i>. *Chinese Journal of Chemistry*, *31*(12), 1519–1525. doi:10.1002/cjoc.201300487

Liu, S., Zheng, Z., & Li, X. (2013). Advances in pesticide biosensors: Current status, challenges, and future perspectives. *Analytical and Bioanalytical Chemistry*, *405*(1), 63–90. doi:10.100700216-012-6299-6 PMID:22892800

Louie, S. M., Dale, A. L., Casman, E. A., & Lowry, G. V. (2016). Challenges facing the environmental nanotechnology research enterprise. *Engineered Nanoparticles and the Environment: Biophysicochemical Processes and Toxicity: Biophysicochemical Processes and Toxicity*, 1-19.

Lowry, G. V., Gregory, K. B., Apte, S. C., & Lead, J. R. (2012). Transformations of Nanomaterials in the Environment. *Environmental Science & Technology*, *46*(13), 6893–6899. doi:10.1021/es300839e PMID:22582927

Mahapatra, I., Clark, J. R. A., Dobson, P. J., Owen, R., Lynch, I., & Lead, J. R. (2018). Expert perspectives on potential environmental risks from nanomedicines and adequacy of the current guideline on environmental risk assessment. *Environmental Science. Nano*, *5*(8), 1873–1889. doi:10.1039/C8EN00053K

Mallikarjunaiah, S., Pattabhiramaiah, M., & Metikurki, B. (2020). Application of nanotechnology in the bioremediation of heavy metals and wastewater management. *Nanotechnology for food, agriculture, and environment*, 297-321. Springer.

Mandal, D., Mishra, S., & Singh, R. K. (2018). Green Synthesized Nanoparticles as Potential Nanosensors. In (pp. 137-164). doi:10.1007/978-981-10-7751-7_7

Mandal, P., Dubey, B. K., & Gupta, A. K. (2017). Review on landfill leachate treatment by electrochemical oxidation: Drawbacks, challenges and future scope. *Waste Management (New York, N.Y.)*, *69*, 250–273. doi:10.1016/j.wasman.2017.08.034 PMID:28865908

Maurer-Jones, M. A., Gunsolus, I. L., Murphy, C. J., & Haynes, C. L. (2013). Toxicity of Engineered Nanoparticles in the Environment. *Analytical Chemistry*, *85*(6), 3036–3049. doi:10.1021/ac303636s PMID:23427995

Mohamed, H. M. (2020). Sensors and biosensors for environment contaminants. *Nanosensor technologies for environmental monitoring*, 109-134. NIH.

Narasaiah, B., & Mandal, B. (2019). Bio-fabricated CuO NPs as green catalyst towards remediation of environmental pollutants. *Lett. Appl. NanoBioSci*, *8*, 597–603.

Nath, D. (2018). Nanomaterial for the Management of Radioactive Waste. In *Handbook of Ecomaterials* (pp. 1–18). Springer International Publishing. doi:10.1007/978-3-319-48281-1_49-1

Ngo, H.-H., Guo, W., & Xing, W. (2009). Applied technologies in municipal solid waste landfill leachate treatment. *Waste Water Treat Technol*, *2*, 199.

Njuguna, J., Pielichowski, K., & Zhu, H. (2021). *Health and environmental safety of nanomaterials: polymer nanocomposites and other materials containing nanoparticles*. Woodhead Publishing.

O'Connor, M. P., Coulthard, R. M., & Plata, D. L. (2018). Electrochemical deposition for the separation and recovery of metals using carbon nanotube-enabled filters. *Environmental Science. Water Research & Technology*, *4*(1), 58–66. doi:10.1039/C7EW00187H

Olapiriyakul, S., & Caudill, R. J. (2008, 19-22 May 2008). A framework for risk management and end-of-life (EOL) analysis for nanotechnology products: A case study in lithium-ion batteries. Paper presented at the *2008 IEEE International Symposium on Electronics and the Environment*. IEEE. 10.1109/ISEE.2008.4562877

Othman, S. A., & Fadzil, N. F. (2021). Implementation of Nanofiller for Bioplastic Reinforcement: A Review. *International Journal of Advanced Research in Technology and Innovation*, *3*(2), 19–27.

Pandey, S., Goswami, G. K., & Nanda, K. K. (2013). Green synthesis of polysaccharide/gold nanoparticle nanocomposite: An efficient ammonia sensor. *Carbohydrate Polymers*, *94*(1), 229–234. doi:10.1016/j.carbpol.2013.01.009 PMID:23544533

Part, F., Berge, N., Baran, P., Stringfellow, A., Sun, W., Bartelt-Hunt, S., Mitrano, D., Li, L., Hennebert, P., Quicker, P., Bolyard, S. C., & Huber-Humer, M. (2018). A review of the fate of engineered nanomaterials in municipal solid waste streams. *Waste Management (New York, N.Y.)*, *75*, 427–449. doi:10.1016/j.wasman.2018.02.012 PMID:29477652

Parthasarathy, M. (2021). Challenges and emerging trends in toner waste recycling: A review. *Recycling*, *6*(3), 57. doi:10.3390/recycling6030057

Patel, N., Pathak, P., Rai, D., & Chaudhary, V. K. (2020). Biosensors used for monitoring of environmental contaminants. *Nanosensor technologies for environmental monitoring*, 69-83. NIH.

Pavoski, G., Junior, A. B. B., Chaves, R. M., Maraschin, T., Oviedo, L. R., Martins, T. A. G., & Espinosa, D. C. R. (2022). Nanotechnology and recycling, remanufacturing, and reusing battery. In Nano Technology for Battery Recycling, Remanufacturing, and Reusing (pp. 53-78): Elsevier. doi:10.1016/B978-0-323-91134-4.00022-4

Pushparaj, K., Liu, W.-C., Meyyazhagan, A., Orlacchio, A., Pappusamy, M., Vadivalagan, C., Robert, A. A., Arumugam, V. A., Kamyab, H., Klemeš, J. J., Khademi, T., Mesbah, M., Chelliapan, S., & Balasubramanian, B. (2022). Nano- from nature to nurture: A comprehensive review on facets, trends, perspectives and sustainability of nanotechnology in the food sector. *Energy*, *240*, 122732–122732. doi:10.1016/j.energy.2021.122732

Qu, X., Alvarez, P. J. J., & Li, Q. (2013). Applications of nanotechnology in water and wastewater treatment. *Water Research*, *47*(12), 3931–3946. doi:10.1016/j.watres.2012.09.058 PMID:23571110

Rajkovic, S., Bornhöft, N. A., van der Weijden, R., Nowack, B., & Adam, V. (2020). Dynamic probabilistic material flow analysis of engineered nanomaterials in European waste treatment systems. *Waste Management (New York, N.Y.)*, *113*, 118–131. doi:10.1016/j.wasman.2020.05.032 PMID:32531660

Rarotra, S., Sahu, S., Kumar, P., Kim, K. H., Tsang, Y. F., Kumar, V., Kumar, P., Srinivasan, M., Veksha, A., & Lisak, G. (2020). Progress and challenges on battery waste management: A critical review. *ChemistrySelect*, *5*(20), 6182–6193. doi:10.1002lct.202000618

Reddy, P. J. (2011). Municipal solid waste management. The Netherlands: CRC Press/Balkema..

Reshadi, M. A. M., Bazargan, A., & McKay, G. (2020). A review of the application of adsorbents for landfill leachate treatment: Focus on magnetic adsorption. *The Science of the Total Environment*, *731*, 138863–138863. doi:10.1016/j.scitotenv.2020.138863 PMID:32446150

Robinson, T. (2007). Membrane bioreactors: Nanotechnology improves landfill leachate quality. *Filtration & Separation*, *44*(9), 38–39. doi:10.1016/S0015-1882(07)70288-4

Rosenzweig, C., Solecki, W. D., Romero-Lankao, P., Mehrotra, S., Dhakal, S., & Ibrahim, S. A. (2018). *Climate change and cities: Second assessment report of the urban climate change research network.* Cambridge University Press. doi:10.1017/9781316563878

Safaei, M., Taran, M., & Imani, M. M. (2019). Preparation, structural characterization, thermal properties and antifungal activity of alginate-CuO bionanocomposite. *Materials Science and Engineering C*, *101*, 323–329. doi:10.1016/j.msec.2019.03.108 PMID:31029325

Sen, M. (2020). Nanocomposite materials. *Nanotechnology and the Environment*, 1-12. InTech Open.

Sheet, I., Kabbani, A., & Holail, H. (2014). Removal of heavy metals using nanostructured graphite oxide, silica nanoparticles and silica/graphite oxide composite. *Energy Procedia*, *50*, 130–138. doi:10.1016/j.egypro.2014.06.016

Shekdar, A. V. (2009). Sustainable solid waste management: An integrated approach for Asian countries. *Waste Management (New York, N.Y.)*, *29*(4), 1438–1448. doi:10.1016/j.wasman.2008.08.025 PMID:19081236

Shi, W. Q., Yuan, L. Y., Li, Z. J., Lan, J. H., Zhao, Y. L., & Chai, Z. F. (2012). Nanomaterials and nanotechnologies in nuclear energy chemistry. *Radiochimica Acta*, *100*(8-9), 727–736. doi:10.1524/ract.2012.1961

Singh, K. R., & Singh, R. P. (2021). *Utility of nanobiosensors in environmental analysis and monitoring*. Research Gate.

Singh, V. V. (2022). Green nanotechnology for environmental remediation. In *Sustainable Nanotechnology for Environmental Remediation* (pp. 31–61). Elsevier. doi:10.1016/B978-0-12-824547-7.00017-5

Sitharaman, B. (2016). *Nanobiomaterials handbook*. CRC Press. doi:10.1201/b10970

Skoulidas, A. I., Sholl, D. S., & Johnson, J. K. (2006). Adsorption and diffusion of carbon dioxide and nitrogen through single-walled carbon nanotube membranes. *The Journal of Chemical Physics*, *124*(5), 054708–054708. doi:10.1063/1.2151173 PMID:16468902

Solano, R., Patiño-Ruiz, D., Tejeda-Benitez, L., & Herrera, A. (2021). Metal- and metal/oxide-based engineered nanoparticles and nanostructures: A review on the applications, nanotoxicological effects, and risk control strategies. *Environmental Science and Pollution Research International*, *28*(14), 16962–16981. doi:10.100711356-021-12996-6 PMID:33638785

Spellman, F. R. (2018). *The Science of Water*. CRC Press. doi:10.1201/b17484

Subramanian, L., Thomas, S., & Koshy, O. (2017). Green synthesis of silver nanoparticles using aqueous plant extracts and its application as optical sensor. *International Journal of Biosensors & Bioelectronics*, *2*(3), 82–85.

Sun, H. (2019). Grand Challenges in Environmental Nanotechnology. *Frontiers in Nanotechnology*, *1*, 2. doi:10.3389/fnano.2019.00002

Tałałaj, I. A., Biedka, P., & Bartkowska, I. (2019). Treatment of landfill leachates with biological pretreatments and reverse osmosis. *Environmental Chemistry Letters*, *17*(3), 1177–1193. doi:10.100710311-019-00860-6

Taran, M., Safaei, M., Karimi, N., & Almasi, A. (2021). Benefits and application of nanotechnology in environmental science: An overview. *Biointerface Research in Applied Chemistry*, *11*(1), 7860–7870.

Tchounwou, P. B., Yedjou, C. G., Patlolla, A. K., & Sutton, D. J. (2012). Heavy metal toxicity and the environment. *Molecular, clinical and environmental toxicology: volume 3: environmental toxicology*, 133-164. NIH.

Tiruvalam, R. C., Pritchard, J. C., Dimitratos, N., Lopez-Sanchez, J. A., Edwards, J. K., Carley, A. F., Hutchings, G. J., & Kiely, C. J. (2011). Aberration corrected analytical electron microscopy studies of sol-immobilized Au + Pd, Au{Pd} and Pd{Au} catalysts used for benzyl alcohol oxidation and hydrogen peroxide production. *Faraday Discussions*, *152*, 63–63. doi:10.1039/c1fd00020a PMID:22455039

Tofa, T. S., Kunjali, K. L., Paul, S., & Dutta, J. (2019). Visible light photocatalytic degradation of microplastic residues with zinc oxide nanorods. *Environmental Chemistry Letters*, *17*(3), 1341–1346. doi:10.100710311-019-00859-z

Tomar, R. S., Jyoti, A., & Kaushik, S. (2020). Nanobiotechnology: Concepts and Applications in Health. *Agriculture, and Environment*.

Tonelli, F. M. P., Goulart, V. A. M., Gomes, K. N., Ladeira, M. S., Santos, A. K., Lorençon, E., Ladeira, L. O., & Resende, R. R. (2015). Graphene-based nanomaterials: Biological and medical applications and toxicity. *Nanomedicine (London)*, *10*(15), 2423–2450. doi:10.2217/nnm.15.65 PMID:26244905

Topmiller, J. L., & Dunn, K. H. (2013). *Current strategies for engineering controls in nanomaterial production and downstream handling processes*. Retrieved from Usman, M., Cheema, S. A., & Farooq, M. (2020). Heterogeneous Fenton and persulfate oxidation for treatment of landfill leachate: A review supplement. *Journal of Cleaner Production*, *256*, 120448–120448. doi:10.1016/j.jclepro.2020.120448

Walser, T., Limbach, L. K., Brogioli, R., Erismann, E., Flamigni, L., Hattendorf, B., Juchli, M., Krumeich, F., Ludwig, C., Prikopsky, K., Rossier, M., Saner, D., Sigg, A., Hellweg, S., Günther, D., & Stark, W. J. (2012). Persistence of engineered nanoparticles in a municipal solid-waste incineration plant. *Nature Nanotechnology*, *7*(8), 520–524. doi:10.1038/nnano.2012.64 PMID:22609690

Wang, K., Li, L., Tan, F., & Wu, D. (2018). Treatment of Landfill Leachate Using Activated Sludge Technology: A Review. *Archaea (Vancouver, B.C.)*, *2018*, 1–10. doi:10.1155/2018/1039453 PMID:30254508

Wang, X., Huang, K., Chen, Y., Liu, J., Chen, S., Cao, J., Mei, S., Zhou, Y., & Jing, T. (2018). Preparation of dumbbell manganese dioxide/gelatin composites and their application in the removal of lead and cadmium ions. *Journal of Hazardous Materials*, *350*, 46–54. doi:10.1016/j.jhazmat.2018.02.020 PMID:29448213

Willner, M. R., & Vikesland, P. J. (2018). Nanomaterial enabled sensors for environmental contaminants. *Journal of Nanobiotechnology*, *16*(1), 95–95. doi:10.118612951-018-0419-1 PMID:30466465

Ying, Y., Pourrahimi, A. M., Sofer, Z., Matějková, S., & Pumera, M. (2019). Radioactive Uranium Preconcentration <i>via</i> Self-Propelled Autonomous Microrobots Based on Metal–Organic Frameworks. *ACS Nano*, *13*(10), 11477–11487. doi:10.1021/acsnano.9b04960 PMID:31592633

Zaidi Ab, G., Yusoff, M. S., & Andas, J. (2015). Review on Applications of Nanoparticles in Landfill Leachate Treatment. *Applied Mechanics and Materials*, *802*, 525–530. doi:10.4028/www.scientific.net/AMM.802.525

Zhang, X., & Liu, Y. (2020). Nanomaterials for radioactive wastewater decontamination. *Environmental Science. Nano*, *7*(4), 1008–1040. doi:10.1039/C9EN01341E

Zhao, G., Huang, X., Tang, Z., Huang, Q., Niu, F., & Wang, X. (2018). Polymer-based nanocomposites for heavy metal ions removal from aqueous solution: A review. *Polymer Chemistry*, *9*(26), 3562–3582. doi:10.1039/C8PY00484F

Zheng, Y., Mao, C., Zhang, Z., Liu, R., & Yan, B. (2022). Nanotechnology for the Remediation of Plastic Wastes. In *Sustainable Nanotechnology* (pp. 117–143). Royal Society of Chemistry. doi:10.1039/9781839165771-00117

Zhu, J., Wei, S., Chen, M., Gu, H., Rapole, S. B., Pallavkar, S., Ho, T. C., Hopper, J., & Guo, Z. (2013). Magnetic nanocomposites for environmental remediation. *Advanced Powder Technology*, *24*(2), 459–467. doi:10.1016/j.apt.2012.10.012

Chapter 2
An Analytical Analysis of the Present-Day Procedures for Disposing of Waste

Binay Kumar Pandey
https://orcid.org/0000-0002-4041-1213
Department of Information Technology, College of Technology, Govind Ballabh Pant University of Agriculture and Technology, Pantnagar, India

Vinay Kumar Nassa
Rajarambapu Institute of Technology, India

Digvijay Pandey
https://orcid.org/0000-0003-0353-174X
Department of Technical Education, IET, Dr. A.P.J. Abdul Kalam Technical University, India

A. Shaji George
https://orcid.org/0000-0002-8677-3682
Almarai Company, Riyadh, Saudi Arabia

Monika Gupta
https://orcid.org/0000-0002-2853-0201
Chitkara Business School, Chitkara University, India

Pankaj Dadheech
https://orcid.org/0000-0001-5783-1989
Swami Keshvanand Institute of Technology, Management, and Gramothan, India

A. S. Hovan George
Tbilisi State Medical University, Georgia

ABSTRACT

Population growth, urbanisation, industry, modernization, and digitalization increase residential, industrial, commercial, mining, radioactive, agricultural, hospital, and electronic wastes in the 21st century. Waste management is becoming the biggest global challenge. Waste management includes collecting, transporting, sorting, destroying, processing, recycling, controlling, monitoring, and regulating garbage, sewage, and other waste. Waste management preserves the environment, prevents pollution, and protects health. Global waste management is modern. Biological reprocessing, recycling, composting, waste-to-energy, bioremediation, incineration, pyrolysis, plasma gasification, ocean/sea disposal, etc. Waste management enhances life. This ensures future peace and wellness. Global health depends on waste management. This optimises waste management. This document discusses worldwide garbage management. It also offers the best waste management approach by critically reviewing previous researchers' findings.

DOI: 10.4018/978-1-6684-8117-2.ch002

INTRODUCTION

In the 21st century, a rapid increase in population, urbanisation, industrialization, modernization, and digitalization all contribute to an increase in wastes. These wastes include domestic garbage, industrial waste, commercial waste, mining waste, radioactive waste, agricultural waste, hospital waste, and electronic waste, amongst other types of waste. It won't be long until the correct management of these wastes is seen as one of the most important problems confronting the international community (Singh et al., 2022). Waste management is defined as the process of collecting, transporting, sorting, discarding, destroying, processing, recycling, controlling, monitoring, and regulating garbage, sewage, and other types of waste materials. The fundamental objectives of waste management are the mitigation of negative impacts on the surrounding natural environment and the avoidance of polluting activities. Protecting human health from the potentially negative effects that garbage can have is a secondary goal of waste management. The management of these wastes is currently being approached from a variety of innovative perspectives in many different parts of the world. Incineration, pyrolysis, composting, waste to energy, bioremediation, pyrolysis, plasma gasification, biological reprocessing, recovering materials through recycling, dumping rubbish in sanitary landfills, composting, and dumping waste in the ocean or sea are all examples of these types of waste disposal methods. These approaches to waste management contribute to an improvement in the overall quality of the environment, which in turn makes it simpler for living organisms to thrive in that setting. Because of the steps that are being taken right now, this paves the way for future generations to live in an atmosphere that is peaceful and conducive to good health. Finding and implementing the most efficient method of waste management that can be discovered is not only a pressing necessity in the present moment (Verma et al., 2022), but it is also essential to the health and happiness of people living in every region of the world. As a direct consequence of this change, the process of managing rubbish will become substantially more successful and effective. The purpose of this work is to make an attempt to describe the various waste management systems that have been used in various parts of the world at various times in history. In addition to this, it conducts an analysis of the discussions and findings of the research carried out by other researchers in an effort to determine the waste management systems that are the most successful (Dogiwal et al., 2022).

We are completely reliant on the environment around us in order to achieve the goals we have set for ourselves. Food, materials for clothing, electronic equipment, buildings, and transportation are just some of the things that we get from our surrounds and use in our day-to-day lives. Our surroundings also provide us with a lot of the things that we need. The condition of the material eventually deteriorates to the point where it is regarded as undesirable, defective, or otherwise useless. For the sake of leading a more refined way of life, we made a lot of things, used them for a while, and then threw them away as garbage in the dirt. Materials that have neither been wanted nor used appropriately are referred to be wastes. It is conceivable for it to exist in any state, including the solid, liquid, or gaseous states, depending on the circumstances. It is possible that it will be quite low in magnitude, or it may be rather high (Gupta, M., (2022). The rise in the number of people produces an increase in garbage that is precisely proportional to that rise. As a result of the buildup of wastes in the environment, the state of the environment has evolved into a major cause for concern for humankind. It lowered the standard of the air we breathe, the water we drink, and the soil we live on. Our earth and our species both began to be put in jeopardy as a direct result of the destructive effects it produced. Protecting our home planet Earth from the many pollutants has developed into a difficult and time-consuming endeavor for humans. It is imperative that we have effective waste management in place in order to guarantee the long-term survival

Figure 1. Steps to take for efficient and effective waste disposal

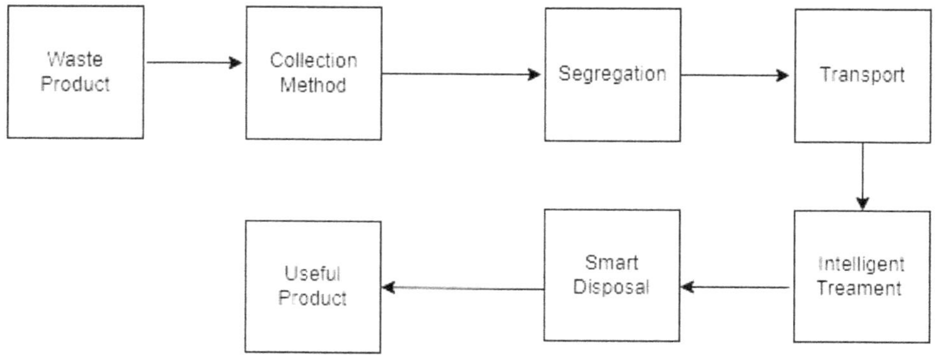

of our natural resources. That every living thing that now resides on our planet is able to do so without putting itself in harm's way and enjoys the benefits of this endeavor (Scanlon, 2007). The process of managing garbage is referred to as "waste management," and the name "waste management" describes the process. The process of managing waste involves a series of steps, including the collection of waste, the sorting of waste according to the type of waste, the transportation of waste with care and labeling of waste, the application of multiple treatments to waste in order to lessen the harmful effects of waste, and the disposal of waste through burning, burying, recycling, and other methods, amongst other disposal options, amongst others (Shah et al., 2001). These steps put a stop to the spread of pollution, decrease the damage it causes to the ecosystem, and maintain a clean and healthy environment for future generations. Because of this, the environment changed into a habitat that is conducive to the survival of all types of life (Radwan, 2023). The management of garbage in the modern world is becoming more technologically advanced and forward-thinking all the time. This study is being carried out with the goals of both enhancing the current level of environmental protection and locating the method of waste management that is both the most efficient and the most effective.

Figure 1 explain the process of managing waste involves a series of steps, including the collection of waste, the sorting of waste according to the type of waste, the transportation of waste with care and labelling of waste, the application of multiple treatments to waste in order to lessen the harmful effects of waste, and the disposal of waste through burning, burying, recycling, and other methods, among others. These measures stop the spread of pollution, lessen the environmental damage it causes, and preserve a clean and healthy environment. As a result, the environment transformed into a suitable habitat for all forms of life. In today's world, waste management has become increasingly technological and forward-thinking. This study is being conducted in order to discover the waste management strategy that is both the most efficient and effective, as well as to improve the protection of the environment.

Waste Product

The waste that is produced from medical procedures can be categorised in a few different ways. The quantity of trash that is generated during this stage, as well as the manner in which it is processed, is of the utmost importance in minimising the risk of injury to the workers who are exposed to it. The amount of garbage that is produced by medical facilities can be decreased in order to reduce the accumulation of waste. The reduction can be accomplished in a variety of ways, including the reduction of waste at its point of origin, the recycling of materials, and the management of stockpiles. Reusing goods that do not pose a risk to the people who use them, such as washable tablecloths, dinnerware, and refill containers for cleaning supplies, is one way to cut down on waste at its origin. By recycling materials like plastics and metals and composting organic waste, recycling can help reduce the amount of trash produced. Last but not least, stock management will assist in the establishment of an organised system for the medications inventory. This will help to prevent the purchase of superfluous products and duplicates, both of which have the potential to be wasted.

Waste Segregation

The various forms of garbage and the methods by which they can be collected independently are catalogued and categorized during the segregation process. The primary method for achieving segregation is to place waste from various types of MW into distinct coloured containers or bags that have been designated for each category. Therefore, this causes a problem when collecting from various sources due to the lack of standardization of the colours associated with each waste category (Menon, 2022). As a result, this results in an increase in the required amount of time as well as the financial cost of labour and equipment to separate the waste and direct it to the appropriate waste stream, which in turn results in a problem. The segregation should take place at the point source the majority of the time, with the exception of waste that is undergoing the same treatment, which should be separated within the treatment facility . Always ensure that the pointed items are kept distinct at the point of origin . The separation is done by medical workers, who must have had proper training to appropriately dispose of waste in order to prevent the spread of illnesses. It is important to remember that if a mistake is made during sorting waste, it must not be fixed in order to avoid spreading contamination to the other debris. It is important to properly store medical waste to prevent illnesses from spreading due to unauthorised human contact with the trash.

Waste Collection and Transport

It is important to prevent the accumulation of waste, which can contribute to the development of illnesses, by collecting MW at a rate of at least once each day. In addition, the employees responsible for collection should be outfitted with safety gear to prevent contaminations and illnesses, which should then be disposed of in a secure manner following Flores-Albornoz et al. (2023). The trash is collected from the health care entity and brought to the treatment facility through secondary transportation for the purposes of disposal, recycling, and treatment processes. The treatment facilities might be found either on the premises of the health care facility or in a different place.

Smart Waste Treatment

Treatment of medical waste is a procedure that is performed prior to the disposal of MW in order to mitigate the potentially negative impacts of this category of waste on both human health and the natural environment. The absence of appropriate therapy can have a number of consequences, including the following:

Poisoning caused by poisonous substances, bacterial and fungal infections, and other infectious diseases releasing toxic chemicals into the atmosphere, allowing them to leach into the ground and pollute the aquifers underneath, Bioaccumulation, Leaving a trace on the environment, the destruction of habitats for animals and plants (Paredes, 2023).

It is essential, throughout the manufacturing phase of any piece of medical equipment, to take into account the impact that these pieces of equipment will have by carrying out a life cycle assessment (LCA) and adhering to the appropriate treatment procedures. However, the processes and procedures that are utilised for treatment have a negligible impact on the amount of carbon emissions that are discharged into the surrounding air. For instance, during the disposal phase, a single intravitreal injection results in the release of 0.05 kg carbon dioxide.

Requirement for the Intelligent Waste Disposal System

The rapid increase in population seen all over the world has led to an increase in the amount of rubbish deposited in all locations. This results in an increase in the amount of undesired materials, which in turn leads to the production of various types of pollution and the spread of diseases. The result is that the populace suffers from a wide variety of ailments and continues to be unhealthy. Because of all of these factors, the environment has suddenly become unsuited for human habitation. Therefore, it is of the utmost importance to implement an effective waste management system in order to control the various forms of pollution, put a halt to the spread of diseases, restore the environment along with all of its resources, and so on. The management of waste requires a variety of approaches, each of which entails a significant number of stages and actions. In today's world, more complex technologies are being utilized in an effort to lessen the amount of garbage produced. This article will continue to be beneficial as long as we need to be aware of the various waste management ways and put those methods into practise in our day-to-day lives (David et al., 2023). Therefore, an article of this nature is required during this period of time. Based on the outcomes of previous research, this article makes an effort to determine which waste management strategies are the most effective and accurate.

Motivation of Study

- **Study waste and Smart Waste Management**
- **Different type of Waste**
- **Useful product after safe disposal of waste**

Waste

Materials that are undesirable or inapplicable are referred to as waste (or wastes). Any substance that is thrown after its primary use or that is worthless, faulty, or has no purpose at all is considered waste. In contrast, a by-product is a joint output that has relatively little value from an economic perspective. Through the application of an innovation that increases the value of a waste product over zero, a waste product may be converted into a by-product, joint product, or resource.

Problems in Its Waste

Despite the fact that there is widespread agreement that waste management services are necessary services that must be given in every community, very little information is known regarding the particular elements that make up waste. Recognizing that the idea of trash is very relative, given that the garbage of one person might serve as a resource for another. As a result, it is essential to have a detailed guide that specifies what kinds of things could be considered waste. In light of this, the current study investigates the idea of waste management and wastes themselves with the purpose of elucidating what waste is, how wastes are categorized, and how wastes are handled.

Waste Management

The term "waste management" refers to the process of collecting, removing, processing, and disposing of things that are deemed to be waste. The materials that are considered garbage might be solid, gaseous, liquid, or even dangerous, and they are typically formed as a result of human activity and as per Figure 2 different stages of waste managements are defined.

Figure 2. Different stages of waste management

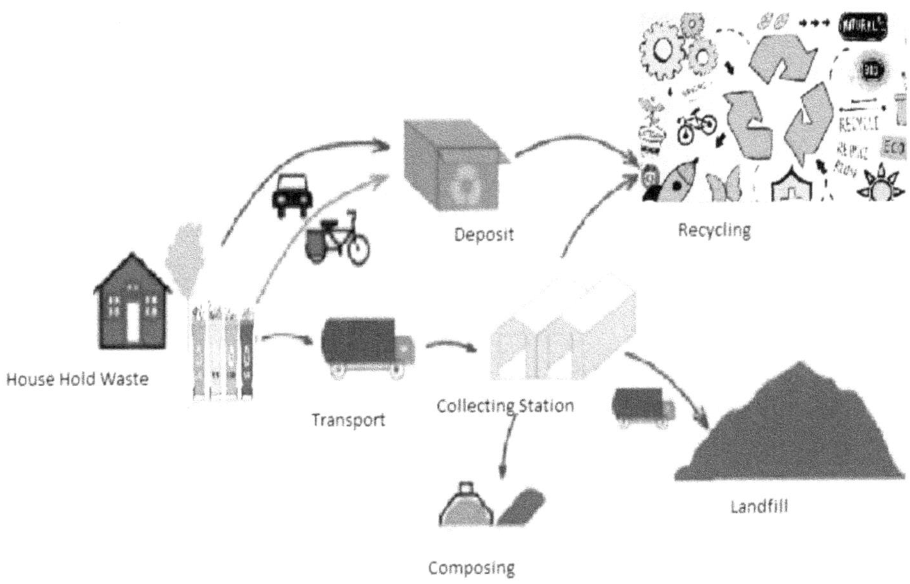

Methodology

His work, which is a literature review, relies mainly on secondary data, as is the case with the vast majority of desktopstudies, in which preexisting information is utilised for the purpose of conducting research and drawing important findings. Books, journal articles, unpublished studies, government reports, organisational and private webpages are some of the specific sources of data that were used for the study. When a large amount of work has already been done on a research issue and when the purpose of the study is to answer particular questions based on past works, this form of research methodology is chosen as a method of investigation. For all of these reasons, the approach that will be discussed in this study to investigate what many academics have claimed regarding wastes, their classification, and their management will be utilised.

Different Type of Waste

Commercial Wastes

Its include bulky wastes from shops, offices, hotels, non-government markets, stores, tyres, electronics, plastic bags, bottles, buckets, packaging materials, paper fibres, thermocol, discarded electric waste, and so on. Commercial trash also includes wastes that have been disposed electrically.

Industrial Wastes

Paper and pulp wastes, oil refineries, tanneries, distilleries, thermal power plants, chemical industries, metal smelters, coal, ash, acids, chemicals, textiles, plastics, nuclear wastes, unused metal sheets, metal scraps, rubber, leather, toxic effluents, fibres (residues), heavy metals, solvents, resins, sludge, abrasive, and so on are all examples of industrial waste.

Agricultural Wastes

It include things like farm trash, animal yards, agricultural leftovers, bagasse from sugarcane, outdated pesticides and fertilizers, manure, weedicides, fungicides, slaughter waste, plastics and containers, organics, and other such things.

Hotel and Restaurants Waste

Hotels and restaurants: According to Scandlon (2007), the hotel industry has the potential to have a negative effect on the environment as a result of the enormous volume of solid waste that it produces. This is especially true of restaurants. Hotels also produce a lot of garbage. According to the findings of a survey that analyzed the procedures that were followed by small hotels for the disposal of solid waste, the vast majority of hotel owners and managers in the United Kingdom do not care about the state of the environment. According to the findings of the survey, the vast majority of hotels do not recycle their waste. As a direct consequence of this, humans are to blame for the deterioration of the natural world as a direct result of the large quantity of solid trash that they generate on a daily basis (Khan et al., 2023). In addition, the findings of the study indicated that hotels do not have a desire to limit the amount of solid waste that they produce. This may be one of the contributing factors to the lack of information regarding the effect that the generation of solid waste by hotels has on the environment (Iyyana et al., 2023).

COVID-19 and Medical Waste

The initial few months of the pandemic saw a reduction in the amount of air pollution and greenhouse gas emissions throughout the majority of countries. This was due to the fact that many human activities were either scaled back or stopped totally during this time. The MW from used plastic containers, test kits, gowns, and syringes increased significantly after the COVID-19 epidemic and continued to do so throughout the process of manufacturing the vaccine.

During the COVID-19 outbreak that occurred in the year 2020, China's MW generation reached a peak of roughly 6000 tonnes, which is equivalent to an increase of 24%(Bommareddy, S., et al., (2022). According to the findings of a second study, the incidence of MW decreased by up to 30 percent in medium and large cities in China while the COVID-19 pandemic was taking place there. On the other hand, the prevalence of MW increased by 400 percent in the Hubei region. When the epidemic was over, Wuhan's daily production of MW had climbed to more over 240 tonnes, which was 190 tonnes more than it had been before.

It is thought that abandoned medical supplies and personal protective equipment contributed to the spread of illness during the COVID-19 pandemic. The MW rates in Wuhan, which acts as the epicenter of the pandemic, have grown from 0.6 kg/bed/day to 2.5 kg/bed/day as a direct result of the Corona virus (Pandey, D.,(2022). When COVID-19 started to spread throughout Bangladesh, the country's daily production of biological and MW waste grew to more than roughly 14,000 tones (Pandey, D., (2023), up from just 200 tonnes before the virus began to spread. After the COVID-19 pandemic, researchers conducted investigations on trash, including biological waste, and discovered that due to the high number of cases that were recorded, all garbage should be considered infectious. This conclusion was reached as a result of the studies. Cross-contamination and sickness transmission became more common as a result of the high infectious potential and widespread diffusion of the virus. [Cross-contamination] and [sickness transmission] are both terms that refer to the same thing (Goyal, S., et al., (2022).

Burning, chemical disinfection, steam disinfection, microwave disinfection, and reverse polymerization were the five ways that Voudri as explored in his search for the most successful approach to the treatment of medical waste. He found that chemical disinfection was the most effective option(Ngulube, P.,et al., (2022). Using multicriteria analysis, the goal was to select the tactic that would be most successful in its implementation. When it comes to evaluating technology, the analytic hierarchy approach places a higher priority on environmental, economic, technological, and social concerns, as the author of the approach explains(Dadheech, P., et al., (2022). The author's environmental sub-criteria included the reduction of volume, the inactivation of microorganisms, the emission of greenhouse gases, and air emissions, as well as solid wastes and liquid residues, patterns of energy and water consumption, and solid waste and liquid residue emissions (Janu, N., (2021). For the technical criteria, the author proposed examining treatment effectiveness, automation, and qualified operators(Pandey, D.,et al (2022). For the economic factors, the author suggested considering capital, operational, maintenance, and disposal costs. In addition to this, it was decided that the rate of technological advancement and the costs should be sub-factors of the social criterion. In order to deploy these technologies in a scenario that is more typical of the actual world, the author proposed doing an analysis of other site-specific variables, such as the maximum load that can be supplied by the selected technology and the local regulatory constraints.

CONCLUSION

75% of medical waste (MW) is non-hazardous. MW must be handled and treated properly since it contains dangerous pollutants that can cause illness and spread infections. With local laws and regulations, incineration, the most widely used method for treating MW, can be better managed to reduce cross-contamination and pollution. Due of their environmental hazards and low degradability, personal protective equipment (PPE) like masks, gowns, and vaccine needles and syringes increased dramatically during the COVID-19 pandemic.

Several factors must be considered to improve MW management and treatment, minimise production, disposal, and treatment costs, and reduce environmental damage: (i) to reduce the amount of waste generated by regulating the use of materials and disposable equipment, (ii) to segregate waste according to regulations with more strictness and attention, (iii) to limit the use of incineration, (iv) to follow stricter technological measures for the incineration of MW, such as filtration and treatment of emissions from incinerators, (v) to invest in new eco-friendly technologies for disinfection and treatment .

REFERENCES

Bommareddy, S., Khan, J. A., & Pandey, D. (2022). IoT implementation and challenges in healthcare industries. In *Networking Technologies in Smart Healthcare* (pp. 211–230). CRC Press.

Dadheech, P., Dogiwal, S. R., Kumar, A., Raja, L., Varshney, N., & Janu, N. (2022, May). A Compressive Review on Internet of Things in Healthcare: Opportunity, Application and Challenges. In *Emerging Technologies in Computer Engineering: Cognitive Computing and Intelligent IoT: 5th International Conference, ICETCE 2022*, (pp. 226-235). Cham: Springer International Publishing.

David, S., Duraipandian, K., Chandrasekaran, D., Pandey, D., Sindhwani, N., & Pandey, B. K. (2023). Impact of blockchain in healthcare system. In *Unleashing the Potentials of Blockchain Technology for Healthcare Industries* (pp. 37–57). Academic Press. doi:10.1016/B978-0-323-99481-1.00004-3

Dogiwal, S. R., Dadheech, P., Kumar, A., Raja, L., Kumar Singh, M., & Janu, N. (2022, May). Internet of Things Based Real-Time Monitoring System for Grid Data. In *Emerging Technologies in Computer Engineering: Cognitive Computing and Intelligent IoT: 5th International Conference, ICETCE 2022,* (pp. 236-247). Cham: Springer International Publishing. 10.1007/978-3-031-07012-9_21

Flores-Albornoz, J., Asis, E. H., Asnate-Salazar, E. A. S., Espinoza-Rodríguez, H., Ramirez-Asis, N., Pandey, D., & Dadheech, P. (2023). Metaverse: The Development of Human Culture in the Future and Cyberspace. In Applications of Neuromarketing in the Metaverse (pp. 52-61). IGI Global.

Goyal, S., Pandey, D., Singh, H., Singh, J., Kakkar, R., & Srinivasu, P. N. (2022). Mathematical modelling for prediction of spread of corona virus and artificial intelligence/machine learning-based technique to detect COVID-19 via smartphone sensors. *International Journal of Modelling Identification and Control, 41*(1-2), 43–52. doi:10.1504/IJMIC.2022.127096

Gupta, M., Sharma, S., & Bansal, S. (2022, April). Neuromarketing: An Emerging Domain in the Formal Education System. In *2022 3rd International Conference on Intelligent Engineering and Management (ICIEM)* (pp. 53-58). IEEE.

Iyyanar, P., Anand, R., Shanthi, T., Nassa, V. K., Pandey, B. K., George, A. S., & Pandey, D. (2023). A Real-Time Smart Sewage Cleaning UAV Assistance System Using IoT. In *Handbook of Research on Data-Driven Mathematical Modeling in Smart Cities* (pp. 24–39). IGI Global.

Janu, N., & Kumar, A. (2021). Automated fruit grading system using image fusion. In Smart Agricultural Services Using Deep Learning, Big Data, and IoT (pp. 32-45). IGI Global. doi:10.4018/978-1-7998-5003-8.ch002

Janu, N., Kumar, A., Dadheech, P., Sharma, G., Kumar, A., & Raja, L. (2021, April). Multiple Watermarking Scheme for Video & Image for Authentication & Copyright Protection. []. IOP Publishing.]. *IOP Conference Series. Materials Science and Engineering*, *1131*(1), 012020. doi:10.1088/1757-899X/1131/1/012020

Khan, H., Veeraiah, V., Jain, V., Rajkumar, A., Gupta, A., & Pandey, D. (2023). Integrating Deep Learning in an IoT Model to Build Smart Applications for Sustainable Cities. In *Handbook of Research on Data-Driven Mathematical Modeling in Smart Cities* (pp. 238–261). IGI Global. doi:10.4018/978-1-6684-6408-3.ch013

Menon, V., Pandey, D., Khosla, D., Kaur, M., Vashishtha, H. K., George, A. S., & Pandey, B. K. (2022). A Study on COVID–19, Its Origin, Phenomenon, Variants, and IoT-Based Framework to Detect the Presence of Coronavirus. In *IoT Based Smart Applications* (pp. 1–13). Springer International Publishing.

Ngulube, P., Pandey, B. K., Pandey, D., Nassa, V. K., George, A. S., Gupta, M. K., & Dadheech, P. (2022). Evaluation of the SARS-COV-2 Omicron variant's features, prospective consequences, and upcoming prospects. *An Interdisciplinary Approach in the Post-COVID-19 Pandemic Era*, 63-76. Nova Science Publishers.

Pandey, D., Hasan, A., Pandey, B. K., Lelisho, M. E., George, A. H., & Shahul, A. (2023). COVID-19 epidemic anxiety, mental stress, and sleep disorders in developing country university students. *CSI Transactions on ICT*, 1-9. NIH.

Pandey, D., Lelisho, M. E., Hasan, A., Rajput, N., Kumar, B., George, A. H., & Shahul, A. (2023). Factors Associated With Sleep Disorders Among Indian and Ethiopian Students in the Digital World in the Midst of the COVID-19 Pandemic. In Applications of Neuromarketing in the Metaverse (pp. 137-154). IGI Global.

Pandey, D., Wairya, S., Pradhan, B., & Wangmo. (2022). Understanding COVID-19 response by twitter users: A text analysis approach. *Heliyon*, *8*(8), e09994. doi:10.1016/j.heliyon.2022.e09994 PMID:35873536

Paredes, E. M. A., Murillo, M. F. M., Ramírez, V. B. S., & Pandey, D. (2023). Lighting Evaluation and Implementation of an Energy Efficiency Plan in the Company Mantransve. In *Applications of Neuromarketing in the Metaverse* (pp. 62–77). IGI Global. doi:10.4018/978-1-6684-8150-9.ch005

Radwan, E., Alattar, E., Radwan, A., Radwan, W., Alajez, M., Pandey, D., & Jebreen, K. (2023). Knowledge, attitudes, and practices of palestinian students toward COVID-19: a cross-sectional study during the first wave of the COVID-19 pandemic. *Human Arenas*, 1-22. NIH.

Scanlon, N. L. (2007). An analysis and assessment of environmental operating practices in hotel and resortproperties. *International Journal of Hospitality Management*, *26*(3), 711–723. doi:10.1016/j.ijhm.2006.07.003

Shah, S., Mehta, M., & Mukherjee, M. D. (2001). *Occupational Health Hazards Encountered at Health Care Facility and Medical College in India*. American Industrial Hygiene Association. doi:10.3320/1.2765786

Silva, C. E., Hoppe, A. E., Ravanello, M. M., & Mello, N. (2005). Medical waste management in the south of Brazil. *Waste Management (New York, N.Y.)*, *25*(6), 600–605. doi:10.1016/j.wasman.2004.03.002 PMID:15993344

Singh, D., Chaturvedi, P., Dhawan, G., Singh, A., & Aggarwal, R. (2022, February). WBMS: Waste Bin Management System for densely populated urban areas. In *2022 2nd International Conference on Innovative Practices in Technology and Management (ICIPTM)* (*Vol. 2*, pp. 770-774). IEEE.

Singh, I.B., and Sarma, R.K. (1996) Hospital Waste Disposal System and Technology. *Journal of Academy of Hospital Administration, 8*(2), pp. 44-48.

Verma, T., Aggarwal, R., Bhardwaj, G., Verma, V., & Agarwal, S. (2022, February). Geyser Time Controller for Effective Electricity and Water Management. In *2022 2nd International Conference on Innovative Practices in Technology and Management (ICIPTM)* (*Vol. 2*, pp. 378-382). IEEE. 10.1109/ICIPTM54933.2022.9753816

Chapter 3
Situation and Impact of Solid Waste Mismanagement Perception of Households in Rajshahi City of Bangladesh

Renuka Sharmin
Varendra University, Bangladesh

Md Ataul Gani Osmani
https://orcid.org/0000-0003-0207-2596
Alma Mater Studiorum Università di Bologna, Italy

Firdous Ahmad Malik
https://orcid.org/0000-0002-7815-0143
National Institute of Public Finance and Policy, India

Laeeq Janjua
WSB University, Poland

Maryum Raja
https://orcid.org/0009-0004-2556-3827
Southwest-Jiaotong University, China

ABSTRACT

The study empirically investigates the situation and effect of household solid waste mismanagement in Rajshahi city of Bangladesh. For this purpose, the study collects data from 120 households in Rajshahi city through personal in-depth interviews to investigate the situation and impacts of household waste management. The present study reveals that the city produces mostly food waste, collected by the scavengers employed through city corporations but poorly managed due to a lack of categorization and recycling technologies. According to the perception and responses of the households, waste odor and disease infections are the most common impacts of household waste mismanagement in Rajshahi city in Bangladesh. The conclusion reveals that lack of proper knowledge and use of technology with regards to

DOI: 10.4018/978-1-6684-8117-2.ch003

waste management and its recycling is one of the significant reasons for bad impacts through unplanned open dumping and burning system practiced by the city authority. It is recommended to reorganize the waste management system in the City by installing proper dustbins.

1. INTRODUCTION

Considering the fact, that now a day's improper waste management is a global issue. In poor nations, waste treatment and disposal systems generally consist of open dumping and open burning (Ferronato and Torretta, 2019). According to Haider and Riaz (2021), proper solid waste management benefits both the environment and its population. However, research that addresses all aspects of solid waste management is lacking. This comprises regular operating methods, existing problems, and proposed solutions. Urbanization and population growth are the direct reasons of the rapid increase of solid waste, which poses a significant problem to Municipal Corporations worldwide (Alam and Ahmade, 2013). Moreover, to live a sophisticated life in the world, people produce a lot of things, used it and then threw it as a waste in the earth. It can be in any form, either in solid or liquid or gaseous. It may range from a small to large size (Barthel et al. 2010; Chen et al. 2013).

In Bangladesh, waste disposal and its management is one of the most critical functions of the city corporations. The government spends a lot of money on trash collection in metropolitan areas, but it's never enough to keep up with the amount of garbage people generate every day, especially when there are also problems with physical design, government laws, and administrative and managerial structures. (Hasan, 1998). Waste management is a tricky task where mismanagement sometimes poses serious problem in urban areas. The general people and government always bypass their responsibilities to each other. But specifically local authorities like city corporations in Bangladesh are bound to keep their area clean (Yousuf, 2014). In Bangladesh, household waste (HW) management is a major issue, considering the fact as it involves huge costs. Since more space is required to store garbage, this problem is only becoming worse. Household and other waste is thrown in open fields in majority of Bangladesh's cities, including Rajshahi City Corporation (RCC).It has an immediate and serious environmental effects on the city people. Observed that the impacts of solid waste disposal on environment in Rajshahi City are aesthetic problem, water pollution, unhygienic, spreading of diseases, CO2 and methane emissions. Furthermore, it was discovered that the contamination of the city's drain water by waste is a major contributor to the spread of diseases including malaria and respiratory issues. There is a widespread lack of finances for the establishment of common waste management facilities, and the legal structure does not permit swift enforcement proceedings. This report provides commentary on each of these factors. Rajshahi is an underprivileged municipal corporation compared to others, and home trash management is one of the city's major concerns. Every day, the city's industries, factories, markets, hospital, clinic, and families generate enormous amounts of garbage. Due to a lack of personnel and technology, the Rajshahi City Corporation is unable to properly dispose of garbage or store it in the right location(Rahman and Jakia, 2013).

In Rajshahi City, approximately 350 tons of trash are generated daily, with the number increasing to 400 tons in the summer, and 12 tons of waste are classified as hazardous. Only 210 tonnes of the entire collected wastes are put on the landfill, while the remaining 140 tonnes are dumped directly into sewers and drains, aquatic bodies, and open spaces (Atik, 2013). Waste management or waste disposal refers to all of the activities and steps necessary to manage waste from its inception through its final disposal.

Figure 1. Waste and its causes in Rajshahi
Source: (Habib et al., 2021)

Management of waste begins with its resources. The chemicals leach into the soil and can become hazardous to the local population (Habib et al., 2021), (see figure 1).

There are problems of inadequate landfill disposal. Knowing the costs of waste disposal activities is necessary for governments to make decisions about their programs with an eye toward cost reduction and improved future planning. Additionally, there are concerns associated with the management of hospital and clinic waste. Indirect health concerns posed by the breeding of disease vectors, especially flies and rats, provide the greatest threat to the general public. (Alam, 2013). People are becoming increasingly concerned about the issue of household waste as uncollected garbage piles up along highways, in trash cans, and on the ground, posing a threat to the environment and harming public health. (Khan and Pathan, 2018).Therefore, this study based on two main objective, first one to describe the methods and strategies of Solid Waste Management in Rajshahi city and second one to find out the impacts of Solid Waste mismanagement in Rajshahi city.

The term "solid waste" refers to materials from urban, agricultural, and industrial sources that have been discarded, abandoned, or deemed unneeded in the modern day (Pop et al. 2022; Khan et al. 2022). The two primary types of solid waste are hazardous and nonhazardous. (Gabay, M. 2012). For example, liquid, semi-solid, wet waste and e-waste are the physical varieties of solid waste. Municipal solid waste (MSW) constitutes product labels, domestic waste, electronic waste, paper and wooden (Gundupalli et al. 2017). Furthermore, the complexity of the e-waste phenomenon, the high growth rates of both total quantity generated and quantity per capita, the negative externalities generated, especially on children and vulnerable people, and waste export practices in countries with permissive or nonexistent regulations have all prompted studies to find solutions for the most responsible management of this waste (Zhang et al., 2012). The need of consumer education to decrease the amount of e-waste and raise the collection rate, as well as the importance of legal requirements to increase the recycling rate, are all topics explored by the researchers. Furthermore, if residents separate their electrical and electronic waste from their regular garbage, they can greatly speed up the collection process. Authorities can increase the rate at which e-waste is collected if they designate drop-off locations.

The pace of urbanization is threatened by the rapid increase of MSW as the current and expected waste generation rate is quite high across the globe (Kaza et al. 2018). But a great percentage of this generated waste is not handled in an ecologically sustainable way and it is even higher across developing countries (Jerin et al. 2022). Habib et al. (2021) pointed out that in recent years, it is observed the highest growth of MSW in South Asia and East Asia (Pacific). This increasing rate of solid waste in South Asia is causing difficulty in waste disposal sustainably as there are heterogeneity and complexity of MSW compositions (D'Adamo et al. 2020). Moreover, there are lacks of cooperation among stakeholders and different institutions and thereby, the solid waste is not managed efficiently in developing countries (Rada et al. 2013). The main challenges of waste management are low collection and scattered coverage, use of old technologies, open dumping and burning as final disposal, lack of laws and review, and limited public awareness (Guerrero, et al. 2013). Specifically in South Asia, this sort of open dumping and burning is the most common which has a significant impact on the economic prosperity and environmental sustainability social, environmental, and economic impact, and thus the situation requires urgent action (Kumar and Agrawal, 2020). There is not nearly enough defined space to accommodate open-air waste dumping. Land scarcity, a lack of technological skills to handle massive amounts of rubbish, and the absence of clear legal regulations all exacerbate an already severe situation. (Rana et al. 2014).

Similarly, Ferronato and Torretta (2019) said that environmental pollution is becoming one of the major problems due to waste mismanagement. In the lower income country they dump or burnt their waste in the open place. It also highlighted the major health risks associated with waste disposal. It is found that the impact of waste mismanagement on the environment is widespread worldwide. That is, the effects of harmful chemical waste on air, soil and water and also shows the direct impact of harmful chemical waste on the health of waste pickers. Bahauddin and Uddin (2012) explain a model, called Environment Management Measure (EMM) model where they used three instruments such as economic, regulatory and suasive. They showed how the three instruments can minimize the polluting characteristics of individual and industries and how to recover the cost of the pollution in city areas. They recommended economic, regulatory and suasive instruments combination for solid waste management.

On the other hand, Farah, (2019) told that, because of poorly managed solid waste city areas face a lot of problems. But there are challenges related with solid waste management and the factors which are influenced its effectiveness. In that paper, they found political reason have a significant effect to collect and manage waste. This study induced that financial restrain has a significant effect on solid waste collection and transportation. The recommendation of this study based on findings strict enforcement of law and policy, need more budget allocation, retain staff with good salary and waste reduction, reuse and recycling is mandatory to manage waste. Sichiweza et al. (2017) reported irregular collection of household waste and indiscriminate dumping of waste, their health falling in danger. Their study uses the concept of circular economy to investigate how household participation in household solid waste management can contribute sustainability. In their paper they use 3R (Reduce, Reuse and Recycle) concept to manage household solid waste in the circular economy. They analyse the existing situation and relate it with circular economy concept and they also give some recommendation for this problem.

Gavrilita et al. (2006) use ORWARE model for system analysis. They used this model to find out the environmental impact of municipal solid waste management in the Republic of Moldova. Besides this model they also use some sub-model in their paper and this model includes material recycling like plastic, paper etc. Municipal solid waste management is a growing national problem in China. In China urban population is growing rapidly which create a large amount of municipal solid waste and it imposed a negative effect in the environment. Usually, landfills are used to handling solid waste but in small city they have no gas control system also don't have any collection system to manage their waste also there is a lack of equipment (Hoornweg & Thomas, 1999).

Likewise, Maskey (2018) found municipal solid waste management is one of the biggest problem for the local and national government in Nepal. There is no proper and actual waste collection system in Nepal and in the open place dumping activities are practised all over the country. It focuses on awareness, behaviour and attitude of the household who contribute 75% of waste generation. In this study it is estimated that municipality of household is generated 1621.14 ton waste every year in the open dumpsite and recyclable potential waste is also there which is uncollected. He suggests that segregation of waste at source and separate collection of waste is the fundamental step to manage waste.

Similarly, the management of municipal solid waste in Nepal is a serious concern for both the provincial and federal governments (Maskey, 2018). Unfortunately, Nepal lacks a dependable waste collection system, so individuals frequently dump rubbish in public locations. Individual households are responsible for seventy-five percent of all waste, hence the program targets them. According to this study, 1621 additional households are added to their respective towns annually. The open landfill receives 14 tons of waste every year, some of which could be repurposed but is instead left to decay. He argues that the first step in waste management is "source segregation," which means sorting garbage at the residence prior to transporting it to the curb. Shafy and Mansour (2018) said in their paper that, today municipal solid waste management is one of the major problems for every country mainly developing country. Their main research area is esteem of organic food waste. Agricultural trash, domestic food waste, human and animal waste, etc., make up the majority of organic solid waste, which is often processed as animal feed and disposed of in landfills. This organic waste is rich in proteins, minerals, and carbohydrates that could be utilised as raw materials in other processes. Moreover, Mohsin and Chinyama (2016) studied about safe and secure solid waste disposal which is a common issue of most of the developing country in the world. This study studied the causes and effects of solid waste management procedures on the environment and public health in the surrounding area. They collected information at random regarding solid waste generation, collection, and disposal, as well as its impact on human health and the environment. They analysed their data by applying chi-square test and descriptive statistical method. In the result, they found poor waste management practices like land degradation, create of breeding ground for disease carrying vectors like rat, mosquito etc. Yosufi (2019) told that solid waste management is not appropriate and each part of the system, different challenges are exists. Most of the people have no knowledge about hazardous solid waste impact on health and environment and they don't know about recycle or how to separate it. Metal component of municipal solid waste it contaminated flood from drain occlusion. From public side more behavioural changes to be need and better waste management system need from the local government side. The main problem is lack of infrastructural planning, technical leadership and lack of treatment program.

Considering the fact the Bangladesh appear to be the 8th high populated country of the world and millions of ton trash everyday disposed by the people. Give below are the few challenges faced by Bangladesh in terms of their waste management.

A. Lack of sustainable plan for waste management system.
B. Lack of people involvement and education regarding waste management system.
C. Lack of new technology used in dumping and waste segregation.
D. High level of corruption in local lower Municipal Corporations.
E. Lack of finance for local government for investment in waste management system.
F. Lack of infrastructure for waste and trash management.
G. Lack of public awareness regarding trash management.

Therefore the aim of writing this chapter is to empirically analyse the waste management of Rajshahi city. Rajshahi city is one of the most populous cities of Bangladesh with approximate 1.2 million people. Numerous other authors also highlighted put forward their research work as chapter theme using survey analysis such as (Pargal et al., 2002 ; Judd, 2010; Bratby,2016).

2. EMPIRICAL ANALYSIS

2.1 Data Source

The study is mainly based on primary data. To perform this field-level study, Rajshahi City Corporation is selected as the study area. The study collects in-depth information from randomly selected 120 households of the City Corporation areas. There are 30 words in Rajshahi City Corporation. By using simple random sampling technique, the study selects 12 words randomly. The word numbers are 6, 7, 9, 10, 11, 12, 15, 23, 24, 25, 27 and 28. And 10 households are surveyed randomly with a questionnaire schedule from each word, this constitutes 120 households.

2.2 Collection Method and Analysis

Collected quantitative data have been analysed using descriptive statistical techniques involving logical reasoning processes. Firstly, the raw data are edited in SPSS 23 software. Thereby, the researcher has sorted out important variables and analysis according to study objectives. To describe the household waste management conditions and patterns, the study uses descriptive Statistics such as, mean, percentage, minimum value, maximum value, standard deviation, and frequency. The analysed data has been presented in tabular and graphical form.

3. RESULTS AND DISCUSSION

3.1 Main Sources of Waste Generations

Waste is produced from various sources such as residential, commercial, industrial and street sweeping. Table 1 shows the distribution of major sources of waste generations in Rajshahi city. It is found that about 87.5% waste is generated from residential sources, whereas 3.3% is commercial waste and 9.2% waste is generated on the street. Therefore, the top most waste generated from household residential sources.

Table 1. Main sources of waste generations

Attributes	Frequency	Percent
Residential	105	87.5
Commercial	4	3.3
Street	11	9.2
Total	120	100.0

3.2 Composition of the Waste in Rajshahi City

Both the composition and volume of solid waste are evolving as a result of modernization (EEA, 2002). Similarly, air pollution, carbon and sand particles, residual of bio mass burning, paper and packing of all kinds, and many other types of rubbish are all examples of the solid wastes generated by urban households, streets, businesses, and industries in Bangladesh. The results of the field study reveal the presence of various organic chemicals in the RCC solid waste. Rajshahi City's average percent fraction of total generated solid garbage collected from various sources across the city (see figure 2)

Figure 2. Solid waste composition
Source: Author own based on data

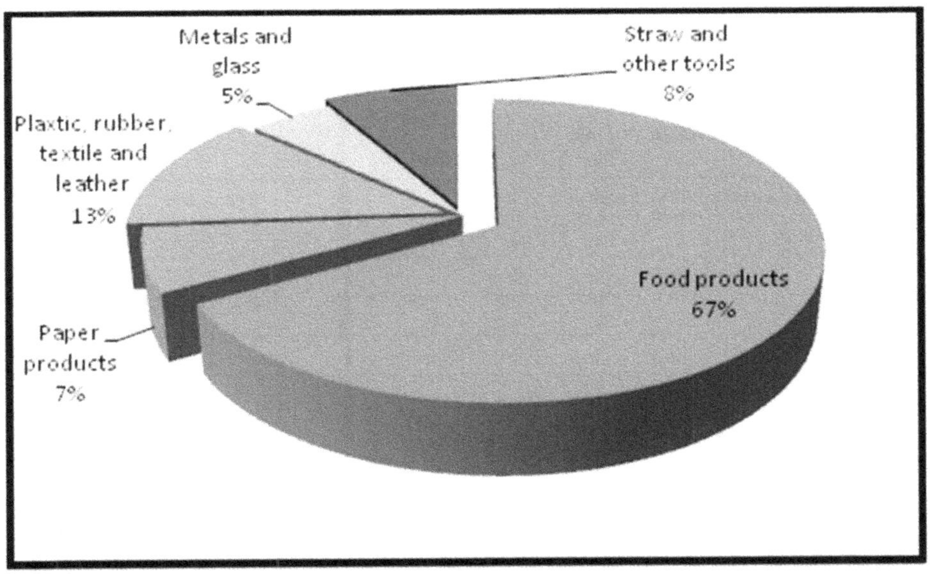

3.3 Methods of Waste Collection

The development of residential and commercial-industrial solid waste occurs in every house, apartment building, commercial and industrial facility, as well as in streets, parks, and vacant areas, making it difficult and complicated to collect solid waste in an urban area. The Rajshahi City Municipality consists of

Table 2. Method of waste collection

Sources	Frequency	Percent
Dustbin	9	7.5
Waste Van	103	85.8
Open Place	8	6.7
Total	120	100

thirty wards. 66 percent of the city is serviced with door-to-door waste pickup. Table 2 indicates that 85.8 percent of trash is placed in garbage trucks, 7.5 percent in trash cans, and 6.8 percent in other containers.

3.4 Waste Transportation and Disposal

Most garbage, solid waste and electronic waste in Rajshahi city is dumped improperly, on open land. Since the RCC does not provide enough infrastructure maintenance, many residents of the region dump their garbage in open locations such as drains. At now, low-lying areas are where the most bulk of garbage is dumped in landfills. Every principle of responsible garbage disposal has been broken at this dump. There is a complete lack of organization at this dump, with trash strewn about everywhere. Due to a lack of resources, the landfill is not being maintained adequately. The wind has carried the nasty odor from the dumping area.There are several responsible parties of waste disposal activities. The main concerned authorities are City Corporation. According to table 3, 5% are concerned people of households, 5.8% are scavenger, 89.2% waste disposal works are handled by City Corporation.

Table 3. Engagement in waste disposal works

Waste Disposal Worker	Frequency	Percent
Concern people of households	6	5
Scavenger	7	5.8
City corporation workers	107	89.2
Total	120	100

3.5 Impacts of Waste Mismanagement in Rajshahi City

City people have realistic expectations for healthy living, their expectation include a cleaner city, and a healthier environment. However, city people are negatively affected by inadequate waste management. Numerous regions of the city are contaminated due to ineffective garbage management. Therefore, good waste management is vital for minimizing the harmful environmental effects of garbage. The drainage system plays an important role in rubbish removal. Nonetheless, Rajshahi lacks a uniformly constructed drainage infrastructure. Consequently, urban pollution is characterized by its variability. The pollution in the city is not evenly distributed. This is especially prevalent in regions with inadequate waste management and drainage infrastructure. According to table 4, 50% respondent perceives that the environment

Table 4. Impacts of solid waste mismanagement

Attributes	Frequency	Percent
Waste Odour	60	50
Waste Fluids	10	8.33
Disease Infection	45	37.5
Others	5	4.17
Total	120	100.0

of Rajshahi city is affected by waste odour and 37.5% talk about diseases infection. This means that solid waste in the form of liquid from households is spreading serious infectious diseases like, diarrhea, pneumonia, asthma, and bronchitis

Inadequate waste collection causes ecological and maritime pollution as well as choked water drains. Due to flooding and other sources of standing water in trash, cholera and vector-borne diseases like malaria and dengue flourish. Garbage is a breeding ground for parasites that are easily transmitted to people by skin contact and inhalation. Hepatitis, influenza, and whooping cough are triggered by microorganisms released into the air by biological waste. Meningitis is an additional danger of transmission via body fluids, which contain microorganisms that inflame the membranes surrounding the brain and spinal cord. According this table 31.6 percent people said yes wastes are harmful for health and it spread disease but in this century it's a very sad matter that 63.2 percent people don't know about this that disease spread by wastes. Even germs as dangerous as anthrax are capable of causing skin infections through medical waste. Though anthrax is extremely uncommon, laboratories that produce immunizations for it may produce potentially dangerous medical waste if they aren't cleaned properly up.

3.6 Reasons of Waste Mismanagement

According to table 5, 48.3% thinks that people have no knowledge about waste management, 25% think that unskilled labour force related with waste management activities, 14.2% think unconcerned people and 10% think lack of waste management planning.

Table 5. Reasons of waste mismanagement

Reason	Frequency	Percent
No knowledge about waste management policy	58	48.3
Not enough budget for waste management	3	2.5
Unskilled labour force	30	25
Unconcerned people	17	14.2
Lack of waste management planning	12	10
Total	120	100.0

4. CONCLUSION AND POLICY RECOMMENDATIONS

This study provides an overview of how people of Rajshahi, Bangladesh, manage their own solid waste. As a result of the rapid acceleration of social and economic growth in Rajshahi city, the rate of solid household waste production has increased in both quality and quantity. The majority of Rajshahi City's low-income residents produce the most organic household waste, which is a big problem. Prior to garbage management, sorting and separating solid waste at home is essential. Raising public awareness through educational institutions and integrated, long-term media efforts can aid in reducing the quantity and quality of household rubbish. The organic components of municipal solid waste can also be used as natural fertilizers. Paper scraps in Rajshahi City can be utilized by packaging paper mills.

The Bangladeshi city of Rajshahi is a major population hub. The city's constraints and problems with managing solid waste are typical of urban places across the nation. This city creates approximately 350 tons of solid trash every day. RCC collects a maximum of 230 tons each day. The remainder is discarded informally in open dumps, storm sewers, and along the side of roads. It is currently inhabited by Rajshahi residents. Composting, recycling, incineration, burial, and sanitary landfill are not viable solutions for the parts that RCC collects. It has been dumped next to a livestock market, contaminating the groundwater and bringing disease and stench to adjacent neighborhoods.

None of the households surveyed had any kind of waste separation system in place. Therefore, in order to properly manage municipal waste, it is necessary to implement source separation of trash. From the numbers, we may infer that roughly two-thirds of people are open to contributing to the cost of a better waste management system. Many respondents reported experiencing illnesses during the recent year, including diarrhoea, pneumonia, asthma, and bronchitis, which they attribute to poor solid waste management. Therefore, this study recommends raising public consciousness, organizing waste disposal into more efficient time frames, setting up more disposal and collection points, increasing the number of staff and vehicles, instituting a "door to door" waste collection system, and separating out industrial and clinical wastes.

Based on our research which is presented in this chapter below are our proposed policy recommendations.

I. Mass and social media should promote waste disposal and recycling programs in Rajshahi city
II. Increase solid waste budgets and its related infrastructure in Rajshahi city.
III. Implement additional solid waste management projects with NGOs, donors, and international organizations.
IV. Every urban community should have a waste disposal site that the city corporation or municipal body collects daily.
V. The environmental conservation legislation should firmly enforce extended producer responsibility (EPR) for industrial and e-waste.
VI. Build city sewage and industrial waste treatment plants.
VII. Rajshahi city government should introduce more accountability in local city council which ultimately reduces corruption.

Availability of Supporting Data

This study is based on primary data; authors are responsible for providing supporting data upon request.

REFERENCES

Abdel-Shafy, H. I., & Mansour, M. S. (2018). Solid waste issue: Sources, composition, disposal, recycling, and valorization. *Egyptian journal of petroleum, 27*(4), 1275-1290.

Alam, P., & Ahmade, K. (2013). Impact of solid waste on health and the environment. [IJSDGE]. *International Journal of Sustainable Development and Green Economics*, *2*(1), 165–168.

Ashikuzzaman, M., & Howlader, H. M. (2019). *Sustainable Solid Waste Management in Bangladesh: Issues and Challenges*. IGI Global., doi:10.4018/978-1-7998-0198-6.ch002

Bahauddin, K. M., & Uddin, M. H. (2012). Prospect of Solid Waste Situation and an Approach of Environmental Management Measure (EMM) Model for Sustainable Solid Waste Management: Case Study of Dhaka City. *Journal of Environmental Science & Natural Resources*, *5*(1), 99–111. doi:10.3329/jesnr.v5i1.11601

Berthel, S., Folke, C., Cloding, J. (2010). *Social–ecological memory in urban gardens Retaining the capacity for management of ecosystem services.* Project: SUPER Sustainable Urban Planning for Ecosystem Services and Resilience. . doi:10.1016/j.gloenvcha.2010.01.001

Bratby, J. (2016). *Coagulation and flocculation in water and wastewater treatment.* IWA publishing. doi:10.2166/9781780407500

Chen, W. T., Wardill, J. T., Sun, Y., Pulver, S. R., Renninger, L. S., Baohan, A., Schreiter, E. R., Kerr, R. A., Orger, M. B., Jayaraman, V., Looger, L. L., Svoboda, K., & Kim, D. S. (2013). Ultrasensitive fluorescent proteins for imaging neuronal activity. *Nature*, *499*(7458), 295–300. doi:10.1038/nature12354 PMID:23868258

Cheng, H., & Hu, Y. (2010). Analysis of the Spatial Distribution of Lead Concentration in the Soil of Anka, Zamfara State, Nigeria. *Journal of Environment Pollution and Human Health.*, *4*(1), 16–23. doi:10.12691/jephh-4-1-3

D'Adamo, I., Gastaldi, M., & Rosa, P. (2020). Recycling of end-of-life vehicles: Assessing trends and performances in Europe. *Technological Forecasting and Social Change*, *152*, 119887. doi:10.1016/j.techfore.2019.119887

Farah, M. J. (2019). Challenges of Solid Waste Management and factors influencing its effectiveness: A case study in Burao Municipality.

Ferronato, N., & Torretta, V. (2019). Waste Mismanagement in Developing Countries: A Review of Global Issues. *International Journal of Environmental Research and Public Health*, *16*(6), 1060. doi:10.3390/ijerph16061060 PMID:30909625

Ferronto, N., & Torretta, V. (2019). Waste Mismanagement in Developing Countries: A Review in Global Issues (30909625). *International Journal of Environmental Research and Public Health*, *16*(6), 1060. doi:10.3390/ijerph16061060

Gabay, M. (2012). Resource Conservation and Recovery Act: Management of Pharmaceutical Waste. *Hospital Pharmacy*, *47*(10), 769–770. doi:10.1310/hpj4710-769

Gavrilita, P. (2006). *Environmental Systems Analysis of Municipal Solid Waste Management in Chisinau.*

Guerrero, L. A., Maas, G., & Hogland, W. (2013). Solid waste management challenges for cities in developing countries. *Waste Management (New York, N.Y.)*, *33*(1), 220–232. doi:10.1016/j.wasman.2012.09.008 PMID:23098815

Gundupalli, S. P., Hait, S., & Thakur, A. (2017). A review on automated sorting of source-separated municipal solid waste for recycling. *Waste Management (New York, N.Y.)*, *60*, 56–74. doi:10.1016/j.wasman.2016.09.015 PMID:27663707

Habib, M. A., Ahmed, M. M., Aziz, M., Beg, M. R. A., & Hoque, M. E. (2021). Municipal solid waste management and waste-to-energy potential from rajshahi city corporation in Bangladesh. *Applied Sciences (Basel, Switzerland)*, *11*(9), 3744. doi:10.3390/app11093744

Habib, M. A., Ahmed, M. M., Aziz, M., Beg, M. R. A., & Hoque, M. E. (2021). Municipal Solid Waste Management and Waste-to-Energy Potential from Rajshahi City Corporation in Bangladesh. [CrossRef]. *Applied Sciences (Basel, Switzerland)*, *11*(9), 3744. doi:10.3390/app11093744

Haider, M. Z., & Riaz, M. R. (2021). Municipal Solid Waste Management In Bangladesh: A Study Of X Municipality. *Khulna University Studies*, *18*(1), 37–43. doi:10.53808/KUS.2021.18.01.2101-S

Hasan, S. (1998). *Problems of municipal waste management in Bangladesh: An inquiry into its nature.* Habitat International. . doi:10.1016/S0197-3975(97)00039-8

Hoornweg, D., & Thomas, L. (1999). *What a waste: solid waste management in Asia*. The World Bank.

Jerin, D. T., Sara, H. H., Radia, M. A., Hema, P. S., Hasan, S., Urme, S. A., Audia, C., Hasan, M. T., & Quayyum, Z. (2022). An overview of progress towards implementation of solid waste management policies in Dhaka, Bangladesh. *Heliyon*, *8*(2), e08918. doi:10.1016/j.heliyon.2022.e08918 PMID:35243053

Judd, S. (2010). *The MBR book: principles and applications of membrane bioreactors for water and wastewater treatment*. Elsevier.

Kaza, S., Yao, L., Bhada-Tata, P., & Van Woerden, F. (2018). *What a Waste 2.0: A Global Snapshot of Solid Waste Management to 2050; Urban Development*. The World Bank. doi:10.1596/978-1-4648-1329-0

Khan, A. H., Sharholy, M., Alam, P., Al-Mansour, A. I., Ahmad, K., Kamal, M. A., Alam, S., Pervez, M. N., & Naddeo, V. (2022). Evaluation of cost benefit analysis of municipal solid waste management systems. *Journal of King Saud University. Science*, *34*(4), 101997. doi:10.1016/j.jksus.2022.101997

Kumar, A., & Agrawal, A. (2020). Recent trends in solid waste management status, challenges, and potential for the future Indian cities—A review. *Current Research in Environmental Sustainability*, *2020*(2), 100011. doi:10.1016/j.crsust.2020.100011

Maskey, B. (2018). *Municipal solid waste management in Nepal: A case study of Gorkha Municipality.* Sl: sn.

Mohsin, M., & Chinyama, A. (2016). Impacts of solid waste management practices on environment and public health: A case of Bahawalpur city, Pakistan. *Journal of Environmental and Agricultural Sciences*, *9*, 69–79.

Mohsin,M.; Chinyama,A. (2016) Impact of Solid Waste Management practices on Environment and Public Health: A Case of Bahawalpur City, Pakistan. *Journal of Environmental and Agricultural Sciences.*

Nowakowski, P. (2016). The influence of residents' behaviour on waste electrical and electronic equipment collection effectiveness. Waste Management &amp. *Waste Management & Research, 34*(11), 1126–1135. doi:10.1177/0734242X16669997 PMID:27705882

Pargal, S., Gilligan, D., & Huq, M. (2002). Does social capital increase participation in voluntary solid waste management? Evidence from Dhaka, Bangladesh. *The role of social capital in development*, 188-212. Research Gate.

Pop, M., Bucur, I., Zoldan, D., Imre, K., Nichita, I., Cristina, G., Tîrziu, A., & Tîrziu, E. (2022). Chemical and Microbiological Air Quality in a Municipal Solid Waste Landfill and Its Surroundings, in South-Eastern Romania. *Sustainability (Basel), 14*(1), 156. doi:10.3390u14010156

Quazi, B., Hassan, K. M., & Haque, E. M. (2012). *Solid waste recycling in Rajshahi city of Bangladesh.* ELSIEVER Journal., doi:10.1016/j.wasman.2012.05.036

Rada, E. C., Ragazzi, M., & Fedrizzi, P. (2013). Web-GIS oriented systems viability for municipal solid waste selective collection optimization in developed and transient economies. *Waste Management (New York, N.Y.), 33*(4), 785–792. doi:10.1016/j.wasman.2013.01.002 PMID:23402896

Rahman, M. M., & Jakia, T. (2015). *Solid Waste Management of Rajshahi City in Bangladesh and Its Impacts on Human Health and Environment.* NTUA.

Rana, R., Ganguly, R., & Gupta, A. (2014). *Solid Waste Management in Chandigarh – A Case Study.* Research Gate.

Sichiweza, E. (2017). Participation of households in solid waste management and circular economy towards sustainability: a case study of Kabwe Town, Central Province of Zambia [Master's thesis, University of Twente].

Staley, B. F., & Barlaz, M. A. (2009). Composition of Municipal Solid Waste in the United States and Implications for Carbon Sequestration and Methane Yield. *Journal of Environmental Engineering, 135*(10), 901–909. doi:10.1061/(ASCE)EE.1943-7870.0000032

Yosufi, A. (2019). Poor Municipal Waste Management and Its Health Implication: A Case Study of Kabul City. *Education Quarterly Eeview, 2*(2). doi:10.31014/aior.1993.02.02.75

Yousuf, T. B. (2014). 3R (Reduce, Reuse and Recycle) in Bangladesh. In *Municipal Solid Waste Management in Asia and the Pacific Islands* (pp. 61–75). Springer. doi:10.1007/978-981-4451-73-4_4

Zahur, M. (2007). *Solid waste management of Dhaka city: public private community partnership.* Research Gate.

Zhang, K., Schnoor, J. L., & Zeng, E. Y. (2012). E-waste recycling: Where does it go from here? Environmental Science &amp. *Environmental Science & Technology, 46*(20), 10861–10867. doi:10.1021/es303166s PMID:22998401

Chapter 4
Globalization and Emerging Opportunities and Challenges in Sustainable Environment in Industry 4.0

H. R. Swapna
https://orcid.org/0000-0003-0362-409X
Jain University (Deemed), India

Sanjeet Singh
Chandigarh University, India

Geetika Madaan
Chandigarh University, India

Sonali Mishra
Llyod Law College, India

Digvijay Pandey
https://orcid.org/0000-0003-0353-174X
Department of Technical Education, IET, Dr. A.P.J. Abdul Kalam Technical University, India

Uday Kumar Kanike
https://orcid.org/0000-0002-8792-5721
Georgia State University, USA

ABSTRACT

The rapid technological advancement of the manufacturing sector over the past few decades inevitably led to the rise of Industry 4.0. It has the potential to significantly alter how globalisation is practiced in the production and consumption of products and services across international markets. In this chapter, the authors take a closer look at the rise of Industry 4.0 and the new technical architecture that underpins it, as well as the benefits it's expected to bring. In addition, how technology can be used to fortify a company's competitiveness and shield it from the perils of this transformation. The new global division of labour, the worldwide supply chain, and the global value chain will all be profoundly affected by this multifaceted technology, as evidenced by a thorough examination of the relevant literature. It will alter the competitive landscape by shifting the advantage away from large corporations and toward small and medium-sized enterprises (SMEs) in emerging marketsand developed markets. As human and technological skills advance quickly, businesses may be able to profit.

DOI: 10.4018/978-1-6684-8117-2.ch004

INTRODUCTION

The rapid technological advancement of the manufacturing sector over the past few decades inevitably led to the rise of Industry 4.0. It has the potential to significantly alter how globalisation is practiced in the production and consumption of products and services across international markets. In this chapter, we'll take a closer look at the rise of Industry 4.0 and the new technical architecture that underpins it, as well as the benefits it's expected to bring.In addition, how we may use technology to fortify a company's competitiveness and shield it from the perils of this transformation. The new global division of labour, the worldwide supply chain, and the global value chain will all be profoundly affected by this multifaceted technology, as evidenced by a thorough examination of the relevant literature. It will alter the competitive landscape by shifting the advantage away from large corporations and toward small and medium-sized enterprises (SMEs) in emerging marketsand developed markets. As human and technological skills advance quickly, businesses may be able to profit from this innovation. However, there are a number of obstacles that must be conquered in order to make a profit from this new technical development, including a lack of qualified workers, technological difficulties, questionable business practices, and questionable values.

With an emphasis on globalisation, international trade, and foreign direct investments, Industry 4.0 has enhanced the relevance of digital transformation as well as its potential applications in manufacturing(Ghadge et al. 2020). "Artificial intelligence, the Internet of Things, additive manufacturing, robotics, cloud computing, and other digital and physical technologies" are all part of Industry 4.0, which seeks to increase firms' flexibility, responsiveness, and connectivity so that more informed judgments can be made(Ghadge et al. 2020; Chalmeta and Santos-dele 2020). When the research and development, design, production, marketing, distribution, and customer service phases of the value chain are all computerised, production efficiency and the worldwide market both rise(Handfield, Jeong, and Choi 2019). The German-conceived Industry 4.0 is currently trending in the international manufacturing sector(Bag et al. 2020). Managers and legislators in developed, emerging, and developing nations are debating how to take advantage of the Fourth Industrial Revolution while shielding their nations from its potential ill impacts(Preindl, Nikolopoulos, and Litsiou 2020). This is an emerging topic that is sparking heated debate amongst experts across disciplines and industries(Stock and Seliger 2016).

The purpose of this chapter is to add to the ongoing conversation about Industry 4.0 and its potential impact on global trade and the supply chain over the next few years(Tjahjono et al. 2017). What is Industry 4.0 in today's interconnected world, and how might it help international corporations and organisations advance their own globalisation agendas?Industry 4.0 is so advanced that it has surpassed all of the previous industrial revolutions that humanity has seen over the course of the last few centuries. The rise of cyber-physical systems and other developments that blur the line between the digital and physical worlds usher in what some have called the "fourth industrial revolution."The rise of cyber-physical systems and other developments that blur the line between the digital and physical worlds usher in what some have called the "fourth industrial revolution.", but self-sufficient, self-learning, and self-acting ones that don't need human intervention(Taylor and Conexxus 2018). The term "Industry 4.0" encompasses a wide range of practices, from ERP software and robotic process automation to AI and "smart" manufacturing. Increases in productivity are already being seen across the business world thanks to cutting-edge, high-value technologies whose horizons we cannot yet see.

To introduce our methodology, we shall first do a brief literature study. Then, we'll dive deep into the components of Industry 4.0 and discuss how they're impacting the way organisations operate in today's digital economy. While discussing the benefits of the fourth industrial revolution, we will also discuss the difficulties that have arisen as a result of it, in order to provide a balanced and realistic perspective. Finally, we will demonstrate the key approaches that will allow the company to take advantage of Industry 4.0's benefits.

2. INDUSTRY 4.0: ITS OPPORTUNITIES AND STRENGTHS

Connectivity and interaction between machines, especially between machines and humans, are greatly aided by the virtualization of a physical manufacturing environment that is made possible by Industry 4.0(Castagnoli et al. 2022). Automation, networked systems, and machine learning all work together to streamline the business's operations and boost its productivity(Murata et al. 2020). The following sections detail the presentation of a few of them.

Figure 1. Opportunities and strengths

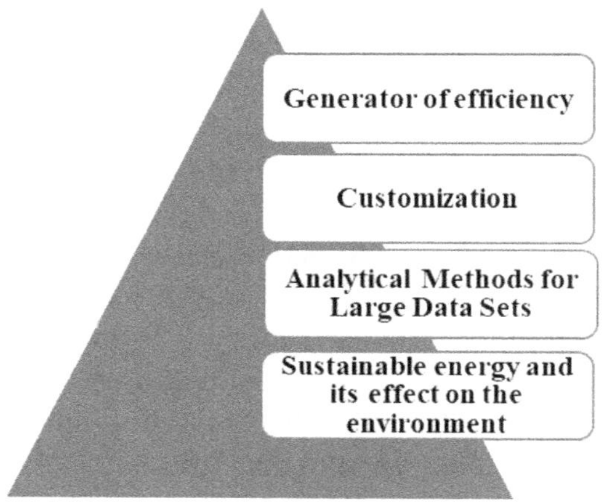

2.1. Generator of Efficiency

There will be increased productivity thanks to Industry 4.0. The extensive application of technologies across the many stages of the supply chain does, in fact, contribute to the latter's optimization(Gabriel and Pessl 2016). Therefore, it is efficient in terms of both money and time, and it boosts productivity, adaptability, and product quality across the board(Lin et al. 2017). This improves the firm's general dependability, which in turn increases the value of the business and its ability to compete. When we talk about Industry 4.0, we're not just talking about speeding up production using high-tech machinery. Indeed, there is a great deal more to it(Bienhaus and Haddud 2018). Through its impact on logistics, management, networks, and, more broadly, the entire company structure in the network, Industry 4.0 can help streamline the value chain(Müller and Voigt 2018).

First, businesses may progress their planning and managing using timely, relevant data. It's true that a company's ability to collect and make good use of data improves its internal and external communications and organisation, even on a global scale, from its headquarters in Quebec City, Canada, to a plant in Dhaka City, Bangladesh(Ashodian 2016). All workers who require it have quick and easy access to the planning, which may also be modified quickly and easily as needed. Due to increased accessibility of information at all times and in all locations, increased management of the organisation is a natural by-product of enhanced communication(Kiel 2017).

All the advancements in AI and the understanding of the chances that may be taken in enterprises are founded on data(Gillpatrick 2019). To better meet the needs of their clients, businesses can benefit from data and communication tools that, when used properly, also improve the precision of their projections. Industry data can illuminate inefficient practices in manufacturing and point the way toward solutions(Samans, 2019). This allows for improved opportunity costs by facilitating the identification of bottleneck products(Albayrakoglu, 2020; Handfield et al., 2019; Samans, 2019). In-house software customizations can provide for improved management and command of the business. Production time and costs can be analysed with data, as can comparisons of supplier and material portfolios for optimal performance. As a result, adjustments and corrections can be made quickly(Jin and Shin 2020). The stock level can be adjusted quickly, either from the production or the purchasing side of things. In conclusion, data is what gives AI its foundation. Machine learning makes it possible to use data to gain insight from both failure and success, a key ingredient for achieving exponential growth.

Data allows for the development of programmes that enhance corporate networks like those connecting suppliers to industries or customers, or between customers and workers in response to a request for a shift in the ordering process. Everything flows more easily, and decisions can be made more quickly. In addition to making the workplace safer, technology also improves health and welfare on the job. Better utilisation of resources and the ability to connect smart production systems vertically are hallmarks of smart production facilities, which boost the effectiveness of cutting-edge technology and promote long-term viability for organisations. "A powerful needs-oriented, customised, and customer-specific production operation is at the heart of smart production systems, which can be defined as Smart factories and smart goods, and the networking of smart logistics, production, and marketing, and smart services."The vast majority of businesses believe that the improvements brought forth by Industry 4.0 are beneficial to

Figure 2. Advantages of Industry 4.0

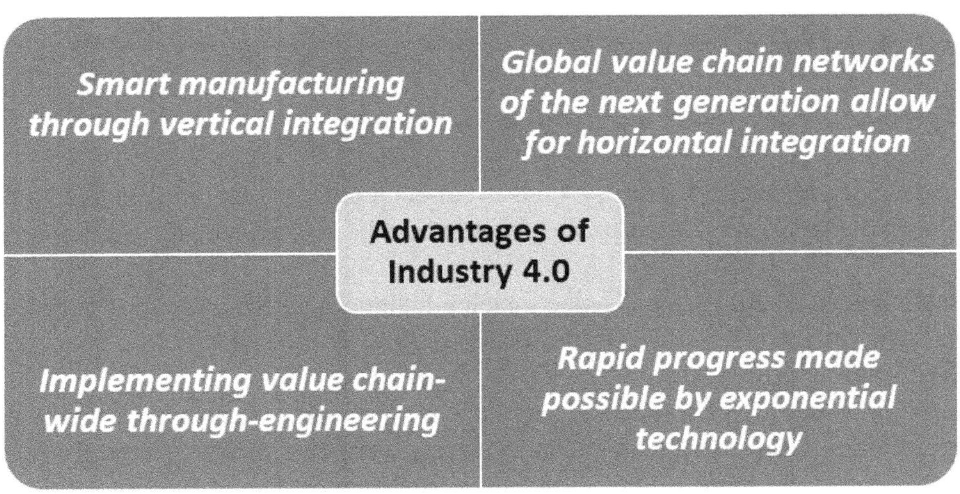

their bottom line(Zheng et al. 2020). When it comes to staying ahead of the competition, Industry 4.0 is the way to go. The effect on a business' bottom line is substantial. Part of the reason for this is that we are able to save money by making better use of what we have. What's more, the advantages of Industry 4.0 may be seen in its four primary parts:

- ***Smart manufacturing through vertical integration***

Integrating vertically to quickly adjust output to meet the needs of individual customers in response to fluctuations in demand and supply, "smart" factories use data that has been fully integrated. "Cyber-production systems (CPS) and smart sensor technologies (SST)" automate the organisation, assembly it self-directed, and keep tabs on the production in real-time. So, manufacturing system management and upkeep are improved. Everything from the beginning of a process until the end, including the resources used, is recorded, allowing for constant updates to the fluctuations and rapid adaptation to changes. This vertically integrated system seeks to maximise resource effectiveness while also meeting the varied and particular needs of its clientele. These cutting-edge systems permit businesses to have constant two-way communication with all links in their global supply chain, from manufacturers to customers to wholesalers.

- ***Global value chain networks of the next generation allow for horizontal integration***

"The horizontal integration is a novel form of the value-creation network, consisting of efficient real-time networks" that allow for unified visibility and responsiveness to issues and defects in real time. Horizontal integration, like the preceding integration approach, employs CPS to record and address all logistics issues in real time. Therefore, processes are more adaptable across the board. It's possible to make changes and adjustments at any stage to meet the needs of the customer.

- ***Implementing value chain-wide through-engineering***

The third defining feature of Industry 4.0 is the integration of several disciplines throughout the product and customer life cycles and the value chain as a whole. Integrating and coordinating production systems with the product life cycle optimises how products are made in relation to the rest of the manufacturing process. Similarly, this aspect of Industry 4.0 relies heavily on the comprehensive utilisation of data throughout the entire value chain.

- ***Rapid progress made possible by exponential technology***

The fourth major advantage of Industry 4.0 is the effect it has on "flexibility, individualization, and cost reductions" in relation to industrial processes thanks to its ever-expanding range of new solutions.

a. **Self-sufficiency:** Artificial intelligence enables highly cognitive and autonomous automation systems with the help of sophisticated robotics and sensor technology to optimise every step of the manufacturing process(Li 2020).
b. **Protectiveness and high standards:** Due to the advancements in nanotechnology and the development of nano-sensors, production may be closely monitored, leading to more accurate quality assessments and safer human-robot partnerships in the future.

c. **Fully Automated logistics processes:** Drones equipped with artificial intelligence (AI) for inventorying, and "driverless vehicles in factories and warehouses for delivering components, are particularly effective due to their capacity to work around the clock and in adverse weather." By utilising these smart factory logistics technologies, we can save money, time, and ensure product quality.
d. **3D printing:** By delaying product diversification, additive manufacturing (3D printing) enables smaller stockpiles and supply chain risk pooling between facilities, reducing both costs and the environmental impact of manufacturing. However, there are still obstacles to be solved before additive manufacturing may have an impact on manufacturing sectors, such as the high unit cost of production and the lack of adherence to economy of scale.

2.2 Customization

One of the most obvious results of efficiency in this technological revolution is increased happiness among consumers. In fact, thanks to cutting-edge technologies, client demand is in lockstep with the market's requirements and preferences while still yielding substantial profits for the business.

Customization is a major concern in the international manufacturing sector, and its importance is only likely to grow. Customers want to have a hand in shaping the direction of product design, and they may do so by providing feedback at any point in the process, from concept to final assembly. Due to this shift, manufacturers must shift their focus from making better products to learning more about each customer's unique demands and developing tailor-made solutions for each market(Stankovic, Gupta, and Figueroa 2017). That is to say, a dramatic transition from a scale economy to a customisation economy, independent of the geographic location of individual manufacturing sites along the global value chain. As a direct result of the ever-evolving wants and needs of consumers, system designers and assemblers are being faced with ever-greater challenges. Not only that, but the commissioning process isn't fully compensated by manufacturers' efforts to standardise and modularize. "End-to-end product data modelling from engineering to commissioning, on the other hand, paves the way for effective production in locations with a global presence and supply chain" and provides an effective method of dealing with the growing complexity of products due to the demand for individualised system solutions.Assembly line production methods are one method for personalization in manufacturing. With the advancement of assembly line technology, it is now possible to tailor the final product to the specific tastes and preferences of each individual consumer. These production lines will benefit greatly from the emerging technologies of Industry 4.0. In addition, this innovation allows for the introduction of business models encompassing product customisation, in which clients can alter product qualities even after production has begun. For this model to operate, manufacturing equipment must be able to make adjustments and communicate them to the client in real time, all while taking into account the current volume of work being processed by the factory. Being able to make alterations in real time online also mitigates the negative effects of the vast distance between the manufacturer and the client that arises from engaging in international commercial activities. There will be ripple effects throughout the production chain, including the assembly line. Late customization (i.e., after the order has been placed) of mass-produced goods has also been predicted in one study, thanks to the availability of up-to-the-moment data on the manufacturing process. Like mass-customized goods, this means customers will have a hand in determining the product's identity and features before it's released for the first time in a "late customization" state. "In contrast, mass-modified manufacturing facilities often create huge quantities of products that share a similar core but may be customised to some degree, resulting in production process portions with repeated bigger batches (e.g., automobile press shop) and areas of considerable product variety (e.g., final automotive assembly)."

"Mass-customized production facilities' clients typically personalise their purchases via a standardised set of configuration options that can be integrated into a modular design."The automotive industry and Tier-1 suppliers are among the most common places to find mass-customized production type plants. This is also true for the truck, bus, agricultural, and construction equipment industries. The factories of industrial equipment manufacturers are examples of mass-customized production since they produce in big quantities and have to accommodate a growing number of product variants. Because of this, the strategy of offering distinct products results in a paradigm shift in factory planning, which creates significant difficulties for the industrial sector as a whole. In order to appeal to the modern consumer, businesses have adapted more nimble models, playing to their individual strengths in order to win in the marketplace. In order to compete with the volatility of the market and provide customers with the same low prices as mass-production businesses, these production methods emphasise rapid organisational responsiveness. Since product differentiation reduces economies of scale, it is difficult to manage the manufacture of large quantities of customised products.

Nonetheless, it will be difficult to coordinate the assembly of customised or customised products due to their unique specifications. In conclusion, the production sequence in execution and the availability of an application to re-sequence to incorporate the late customizations determine whether or not the customer's late customizations can be approved. Not only that, but the benefits of late customisation procedures can only be realised if the system is self-sufficient and able to maintain the fabrication process. Customers require constant updates on the status of their customised product. It may be inferred, in particular for manufacturers, the broad use of the underlying technologies has the potential to cause profound changes in the location and structure of manufacturing facilities all over the world.

2.3 Analytical Methods for Large Data Sets

Companies formerly relied on limited-scope data sources like production records, internal accounting, and market research reports before the advent of Industry 4.0. There is a shift happening in the way information is gathered. Increases in data production can be seen from previously unanticipated places, such as search engine and social media site data and sensor data from smart devices. As a result of this technical development, MNCs can have access to previously unavailable sources of globally relevant business information. Computing power is improving and the price of data storage is falling thanks to technological advancements. As a result, the field of big data analytics has emerged.

Big data analytics is a powerful driver of innovation, but the way we think about time has shifted significantly with the advent of Industry 4.0. Future-oriented, big data analytics works to create both new and expanded data pools. In the past, IT has mostly served as a retrospective tool, keeping tabs on internal operations and alerting the proper authorities to any discrepancies it may have detected. Those businesses that have embraced big data analytics have seen gains in efficiency and profitability. Management can benefit from analysing big data in a number of ways, including improved resource utilisation and predictive maintenance, earlier detection of production flaws, faults, and deficiencies, and optimization of automation processes and trend analysis.

There are a number of ways in which big data analytics could affect international trade. In particular, businesses will be able to optimise their supply, production, and distribution activities on a global scale with less wasted effort and fewer wasted resources without having to invest much in local marketing affiliates in those foreign markets. There are, however, two significant downsides. The first is that high-quality big data may be a source of value for businesses, but that doing anything useful with it will require a wide variety of technological and governance capabilities(Garay-Rondero et al. 2019).Second, extensive use of big data applications poses a risk to people's privacy. Just as Facebook can deduce our tastes, so can Google our habits, and Twitter our thoughts, social media platforms can increasingly gauge our interests and behaviours.

To protect the privacy of individuals and businesses, as well as to establish guidelines for the collection, storage, and transmission of data at the local, state, national, and international levels, new data protection laws and/or stricter industry self-regulation will need to be enacted. Products involving consumer-generated intellectual property raise questions about who owns what, and who is responsible for what, and how these concerns will be resolved in international contexts. The tension between the previously outlined advantages of big data analytics and data privacy will require some kind of (transnational) governance structure to manage. At the end of the day, this may make or break a company's ability to fully exploit the commercial value of big data analytics.

2.4 Sustainable Energy and its Effect on the Environment

Industry 4.0's ability to optimise manufacturing and logistics could have a significant effect on how the environmental crisis is handled. Conservation of natural resources and biodiversity would benefit just as much from the processes of Industry 4.0 that enable energy savings and waste reductions to lower costs for enterprises. One of the United Nations' Sustainable Development Goals is to "increase energy efficiency and better manage waste." Since government environmental rules and regulations are expected to get much harsher in the next decades, the advancements brought about by Industry 4.0 would be adequate with them.

Improved production management is made possible by Industry 4.0's ability to track and analyse factory output. Opportunities for increased productivity, less waste, and better dependability abound thanks to constant production efficiency monitoring and intelligent quality management. As a result, businesses would be more efficient in their use of energy. Because of the data-mining opportunities provided by sensors and AI, businesses can make better decisions thanks to Industry 4.0(Jin and Shin 2020). To help businesses make informed decisions, relevant data on "manufacturing efficiency, product life cycle, and energy consumption" will be made available to them. Due to the exchange of information made available by smart data communication technologies, connecting the consumer to their energy usage is another key feature of Industry 4.0. Users will be able to track their energy consumption, which will encourage more independent behaviour.

However, "these benefits may be outweighed by the fact that Industry 4.0 necessitates the collecting and storage of vast amounts of data in data centres, which contribute greatly to global warming."But this is not an insurmountable problem, and businesses shouldn't be deterred by it, because technological advancements have made it possible to recycle the heat generated by data centres. Typically, this tactic is employed in Scandinavian countries like Finland and Sweden.

3. EVOLUTION OF INDUSTRY 4.0 AND IMPLICATIONS FOR GLOBAL BUSINESS

There has been a plethora of research on Industry 4.0, and it is constantly developing. Several studies have examined how Industry 4.0 (I4.0) has impacted international trade and commerce. Specifically, I4.0's primary motivators highlight the sector's participation in improving present-day business procedures by standardising production and supply-chain networks. Those advancements in Industry 4.0 have significant ramifications for the existing state of business on a national and international scale. One of the studies has discussed the inner workings of Industry 4.0 and provided a wide-ranging explanation of the concept by conducting a thorough literature review. One report summarised how businesses might improve their operations by implementing Industry 4.0.An additional research project looked into customization, detailing how Industry 4.0 has adapted the manufacturing sector to meet consumer needs, and providing examples of how businesses might benefit from these emerging technologies(Li 2020). "Impact analysis and examples of how Industry 4.0 might promote environmental sustainability have been described in a UNIDO report."Industry 4.0 has the potential to help businesses achieve triple bottom line (TBL) sustainability, as demonstrated by the work of Preindl, Nikolopoulos, and Litsiou (2020). With the help of the resources provided by Pandey et al. (2020), businesses who are interested in making the switch to Industry 4.0 can evaluate their readiness for the change.

3.1 A World of Industry 4.0 and International Trade

The possible gainsand difficulties for business and international trade are considerable with the advent of Industry 4.0. It not only aids in modernising the manufacturing process and autonomous execution, but it also enables managers to take on global production management via a malleable global supply chain(Zheng et al. 2020). The primary components of the fourth industrial revolution and their impact on the corporate world are transforming the way many industries operate. "By combining the Internet of Things (IoT), big data, cloud computing, artificial intelligence (AI), and communication technologies to build smart factories, Industry 4.0 is reshaping the global value chain and bringing about a new era of production."

"Digitalization, optimization, and customisation of production, automation, and adaptability, human and machine interactions and collaborations, high value-added offers, and automatic exchanges of data and communication" are the five primary drivers of Industry 4.0, as stated by Fatorachian & Kazemi, (2020). Factors such as human capital, monetary systems, management, organisational architecture, and manufacturing processesare just a few examples of areas where the fourth industrial revolution has had a profound influence on the international corporate landscape. "Cyber-physical systems, the Internet of Things, smart factories, smart technologies, the cloud, and big data" are all examples of the technologies that are fuelling this transformation(Preindl, Nikolopoulos, and Litsiou 2020). The newest design for industrialisation has diverse goals and is confronted by new kinds of problems as it expands on a global scale. With the help of Industry 4.0, high-cost countries (HCCs) can reorganise their manufacturing sectors, while developed nations can keep their robust industrial bases. With manufacturing in industrialised countries on the decline, this trend may present a significant opportunity. Strengthened competitiveness, adaptability, and speed in the face of regionalization of production and fluctuations in global end-market demandare three of the biggest questions that Industry 4.0 sets out to answer. In conclusion, Industry 4.0 can be defined as a collection of initiatives aimed at reshaping the global manufacturing sector in order to increase competitive advantage and market share, even in the face of uncertainty. This objective is realisable by the use of Industry 4.0'ssuggested institutional shifts, such as the use of smart technologies and factories that improve competitiveness and flexible management and enable rapid responses to the market's ever-changing conditions.

3.2 International Business and the Implications of Industry 4.0

A highly integrated network is at the heart of the fourth industrial revolution. Many formerly excluded small businesses are now able to take part in international trade thanks to the proliferation of the internet(Tripathi and Gupta 2020). "It paves the way for a greater variety of goods to be exported to more markets, frequently by more nimble and younger businesses. Findings show that a ten percent increase in internet use in the exporting country leads to a 0.4 percent rise in the volume of goods transferred between the two countries. When two countries both increase their internet usage by the same percentage, their average bilateral trade value per product rises by 0.6%. The fourth industrial revolution, or Industry 4.0, may need fundamental modifications to current company paradigms, opening up new opportunities for value creation. As a result of these shifts, conventional value chains are anticipated to evolve. Manufacturing small and medium-sized businesses (SMEs) will be impacted in three ways by Industry 4.0: value creation, value capture, and value offer."

The techniques and goals of global value chains will be drastically altered by the Fourth Industrial Revolution. Increasing customer involvement in the manufacturing process is one example of a growing trend that isn't motivated solely by financial gain. Other motivations include the pursuit of efficiency, the development and maintenance of global competitive advantages, the discovery of novel production methods, the emergence of innovations, the promotion of automation and learning, and so on.

The many new possibilities made possible by Industry 4.0 are changing the way businesses operate on all levels. In the age of Industry 4.0, a wide variety of business models are thriving. A common theme throughout the rapidly developing models of them is the anticipation. The merged model conveys the anticipation that a company would acquire expertise in the development of its overall offering. As a result of this emerging market need, one business either developed advisory services (relating to products or processes) or a new platform-based business model. To "platformize" a product is to utilise a cloud-based platform to generate digital goods that provide solutions to client problems. The term "platformization" refers to the incorporation of smart factories' ideas and operational procedures onto a digital platform to increase efficiency and productivity. Adding value through an interconnected ecosystem of digital goods and services powered by IT. We can see how the complexity of implementing Industry 4.0 within a company and its potential effect on the status quo of business from this forecast. The primary factors behind Industry 4.0 will alter the nature of commercial operations (Janu,Neha et al., (2022)).

Despite the evolution of I4.0, IoT still serves as a central tenet of the new paradigm. This technology gives users access to the web via deep learning methods. This technology makes machines into "smart objects" that can monitor and respond to their environment in real time, allowing for improvements in areas like as wear detection, manufacturing process control, capacity planning, and inventory management(Patil 2020). With cloud computing, computers are able to communicate with one another over the web. It has the potential to address a wide range of problems, including Big Data storage and the associated costs and capabilities (Dadheech, Pankaj, et al.(2022)).

What we call "cyber-physical systems" (CPS) are the technologies that make it possible for computers, apps, and physical objects to communicate and collaborate. It suggests a systemic level of interconnection. This method's primary goal is to facilitate the dissemination of vital data for the purpose of improved strategic decision-making, and it does so by means of a virtual connection. It forges tight ties between the physical world, machines, and the digital realm, all of which operate and interact in real-time owing to computation and the internet. When hardware and software are in sync, manufacturing efficiency can be monitored and tweaked in real time, facilitating sound strategic decision-making. CPS also improves human-machine collaboration and the interoperability of autonomous machines.

Deep learning abilities are included into the development of autonomous robots. New artificially intelligent robots can use deep learning technology to learn from their surroundings, make improvements for the better, and respond appropriately to problems they observe. Given that it has the potential to make Human labour obsolete while simultaneously increasing a company's bottom line, this is a crucial component of I4.0 that presents significant difficulties.

The enterprise resource planning system is "the backbone of the Industry 4.0. An example of an ERP system is SAP, which stands for Systems, Applications, and Products in Data Processing. SAP is a piece of software that helps businesses improve their operations by integrating their processes and boosting the flow of information and cooperation within the business and with its external partners.ERP systems aid businesses in a wide variety of ways, including increased departmental communication and collaboration, enhanced workflow and supply chain management, real-time data, process, and technology integration across internal and external value chains, standardisation of various business practices, and enhanced order management and accurate accounting for inventory."

The integration of these several "drivers" (IoT, CPS, IoS, ERP) into a single production model will result in the development of "smart factories," which use all the various "smart tools" and "production models" to maximise efficiency. This unified platform will promote globalisation of manufacturing and supply chain growth(Reza, 2020). This unified platform will aid in the globalisation of production and the development of supply chains around the world. Some manufacturing may be brought back home, but the new smart technology will also enable companies to reorganise their production network without regard to borders or physical proximity. That's why we're still in the early conceptualizations of Industry 4.0. consequences It's a never-ending process that alters the ways we do business and think about industrial processes. Therefore, the production model of smart factories becomes sustainable, low-cost, and adaptable to market shifts. This represents the maximum possible level of efficiency. This advanced technology will alter the factors that drive GPNs, lessen the significance of physical distance, and reorganise the global value chain (GVC). It is expected that the NGDL would arise and re-distribute manufacturing activity by incorporating several hubs in mature, emerging, and developing markets.

4. CHALLENGS STEMMING FROM INDUSTRY 4.0

Industry 4.0 will undoubtedly provide significant obstacles to the established norms of "production, consumption, and the global value chain." For the long-term growth and profit of "the fourth industrial revolution, businesses and governments" must explore sufficient methods for better implementing the Industry 4.0 systems, re-organizing company activities in light of the new context, and adopting additional precautions(Matt, Modrák, and Zsifkovits 2020). In this section, we will describe and examine some of the issues that have been brought up by Industry 4.0, beginning with implementation issues, moving on to HR management issues, then data security threats, then challenges related to Big Data, and lastly environmental issues.

Figure 3. Challenging stemming from Industry 4.0

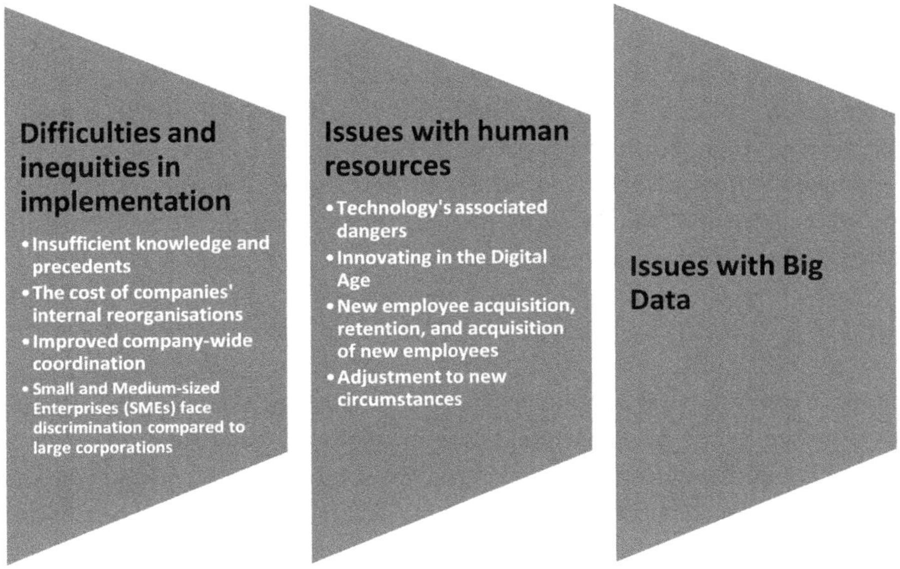

4.1 Difficulties and Inequities in Implementation

We've proven that the advent of Industry 4.0 ushers in a plethora of possibilities for companies to grow and improve throughput and productivity. "However, there are many obstacles to implementing Industry 4.0, the most significant of which are shortages of skill-sets that are necessary in the industry 4.0 phenomenon.In addition, there is a deficiency of knowledge and insight into the transformation, the need for an organisational overhaul, and the synchronisation of data resources. The disparity between small and medium-sized businesses and multinational corporations in terms of resources, attention, and strategies throughout the transition to Industry 4.0 exacerbates these problems." In this section, we will discuss the top three challenges encountered by organisations when they adopt Industry 4.0.

4.1.1 Insufficient Knowledge and Precedents

One of the main obstacles that prevents organisations from making the jump to the industry 4.0 age is a lack of relevant experience(Reza 2020b). However, there is no definitive business case available for enterprises to use as a point of reference, and the expense to undergo transformation is high. When there is no context, no pattern to follow, and no successful instance to replicate, it is exceedingly challenging to make the case for such investments. As a result, the majority of businesses either fail to successfully transition (60% of all transitions fail) or lack the intestinal fortitude to go through with the necessary changes to their organisation and the corresponding investments of time and money. The good news is that promising studies have recently emerged, laying the groundwork for the development of initial instruments with which to confront this dilemma. Pandey et al. (2020) suggest that before implementing any Industry 4.0 initiatives, businesses should conduct a maturity assessment of the technology within their own unique environment or supply chain.

With this analysis in hand, they can better understand where they stand and how they can best develop. Even though numerous consulting firms conduct research in this area, only a small amount of academic study has been done so far. Two consulting firms, "IMPULS management consulting GmbH (IMPULS) and PWC", have developed Industry 4.0 evaluation models in 2016 to help evaluate businesses' level of industry maturity. While the PWC model facilitates self-assessment of organisational maturity on the basis of a number of criteria, the IMPULS model is more concerned with providing improvement guidance in light of the organization's current degree of readiness. More of these models are in the works, indicating that "the difficulty of a lack of experience will be surmounted as Industry 4.0 extends throughout the production world."

4.1.2 The Cost of Companies' Internal Reorganisations

With more and more routines being roboticized and more and more decisions being predetermined by code, a company must undergo a complete reorganisation in order to make the shift to Industry 4.0. Companies who are embracing the industry 4.0 paradigm. We need to develop novel economic models and map out an alternative course of action. To make this change, many things must be shaken up, including the structure and workflow of the company as a whole, the departments within it, the equipment they use, the roles its employees play, the people they hire to fill those jobs, and so on. Large-scale shifts like these are expensive for businesses in many ways, including, but not limited to, time. To avoid a failed Industry 4.0 implementation, businesses must exercise extreme caution while making investments in these transformations.

4.1.3 Improved Company-Wide Coordination

A further difficulty in implementing Industry 4.0 is the efficient linking of all data. Every aspect of a business, from production to research and development to information technology to sales and finance, must work together to implement Industry 4.0. The adjustment is already challenging, and the barriers that separate these departments don't help(Bag et al. 2020). However, this issue is not just related to the coordination of duties: A crucial part of implementing Industry 4.0 is learning how to organise the massive amounts of information being produced by a business's many systems (Dogiwal, Sanwta Ram, et al. (2022)). Data from client information systems, for instance, must be processed and linked with production data. Thus, data integration is a challenging endeavour, and businesses must have access to skilled data scientists who can interpret and model this information.

4.1.4 Small and Medium-sized Enterprises (SMEs) Face Discrimination Compared to Large Corporations

In developed economies, 98% of all businesses are classified as SMEs, and they are becoming more and more crucial to international trade and the flow of materials around the world. Companies of different sizes and with different resources will face different challenges when adopting the new Industry 4.0 paradigm. Naturally, large firms don't suffer the same challenges that small and medium-sized enterprises (SMEs) do. Careful preparation is essential before implementing Industry 4.0 as part of a company's economic strategy, especially for a small or medium-sized business.

Due to their greater resources, large corporations can more easily and quickly catch up to the higher maturity level in the technology sector than SMEs. To be sure, they can devote greater resources to this endeavour in terms of both money and manpower. Because of this, I4.0 is spreading rapidly throughout significant firms that are investing heavily and working tirelessly to introduce and enable the necessary technologies. But SMEs lag behind because they don't have the means to investigate the dangers and rewards of I4.0 implementation. These challenges widen the chasm that already existed between SMEs and big conglomerates. Small and medium-sized enterprises (SMEs) have an edge over giant corporations due to the latter's more convoluted production and management processes. Therefore, alterations to the structure and ethos of SMEs are much less of a challenge to undertake.

4.2 Issues With Human Resources

Stankovic, Gupta, and Figueroa (2017) found that there are a wide variety of business process and model-related obstacles to implementing Industry 4.0 across 176 SMEs and 71 large businesses.The most significant obstacles include, but are not limited to, the myriad problems that arise when developing an Industry 4.0 strategy and the ever-increasing demand for professionals with specialised knowledge and training, especially in the areas of advanced technologies, smart engineering, automation, digitalization, etc.

In spite of the widespread development of technical and future-oriented themes applicable to industries, which is present in many different types of businesses, human resources continue to pose a significant barrier to the widespread adoption of Industry 4.0 ideas, methods, models, and equipment. An essential human resources facet of implementing Industry 4.0, as brought up by Mubarak and Petraite (2020), is the idea of digital trust. Between traditional trust and the adoption of Industry 4.0, digital trust raises personnel concerns. The following are the most pressing human resource concerns associated with Industry 4.0:

4.2.1 Technology's Associated Dangers

A potential risk of the digital trust method is the rise of technocentric businesses, which fail to recognise the importance of human interaction and instead focus solely on technological solutions. As a matter of fact, in this model, stakeholders will put more faith in machines, AI, and automation than in humans to manage their businesses and organisations. The Deloitte studies did raise concerns about HR's level of readiness for the global value chain and Industry 4.0. Only 12% of businesses are prepared for the challenges posed by the industry 4.0 ecosystem, despite the fact that 32% of businesses overall are prepared to confront technological challenges.This number shows the shift in emphasis away from hiring and keeping employees through retention policies and toward efficiency with the most advanced technological implementation.

4.2.2 Innovating in the Digital Age

Indeed, ISHR technologies make it possible to automate and standardise with the help of smart technologies a wide range of HR processes, including the following: Human resource activities such as orientation and development programmes, payroll processing, talent acquisition, and employee advancement(Ghadge et al. 2020). Enterprise resource planning (ERP) and automation of robotic processes (ARP) were among the several methods used to re-evaluate HR's role in the organisation. As a result of these many trends, human resources professionals are being challenged to develop cutting-edge skill sets in light of the smart technologies that are reshaping their workplace and posing novel dangers to their everyday operations. There is the challenge of keeping up with the fast-paced evolution of the human resources (HR) function, which comes with its own set of hazards associated with the safety of data storage and transmission, which could ultimately benefit the competition. Remember that ISHR systems like the one SAP recommends can store sensitive information like "salaries, positions, personal data of each employee, curriculum vitae, yearly reports, information posted on the various job boards, information about the company's important partners, etc."

4.2.3 New Employee Acquisition, Retention

The implementation, development, use, and maintenance of the innovations of the fourth revolution called for a highly skilled labour force. Therefore, many different sectors require some degree of machine-human collaboration. If this is what Industry 4.0 is all about, then it's no surprise that GAFAM (Google, Apple, Facebook, Amazon, and Microsoft) are increasingly expanding into previously unconquerable traditional industries where their expertise and resources make them unrivalled rivals(Matt, Modrák, and Zsifkovits 2020).For instance, "we can remember the creation of the Google autonomous car, the collaboration between the Volkswagen group with Amazon web services to collect and analyze its industrial data and becoming a major global leader of the automotive industry, and even the collaboration between Apple and General Electric (GE) to create new applications for the internet of things (IoT) platforms that benefit to GE industries purposes thanks to IOS (exploitation system owned by Apple) opportunities."Competition is heightened in traditional sectors and marketplaces (but not only those) as a result of Industry 4.0, which appears to be swamped by the rapid development of the fourth industrial revolution. Recruiting highly skilled personnel and incorporating ongoing reskilling, learning, and training into HR strategies are necessary for businesses to meet these problems.Additionally, the introduction of retention policies that will limit the spread of sensitive business information and boost efficiency and profitability while maintaining a competitive edge is one of the most pressing HR challenges. It's easy to see how an increase in the need for highly trained personnel could be risky if it weren't paired with human resources strategies aimed at preparing the workforce for the smart, autonomous workplace of the future.

4.2.4 Adjustment to New Circumstances

The resistance to change is a serious problem that the HR department must address immediately. Deloitte found that only 17% of those polled felt prepared to manage workplaces that included humans, AI, and other machines working together, while the remaining 60%-70% were doomed to fail because of poor change management. Castagnoli et al., (2022)also corroborated this issue. They demonstrated that some workers are averse to change because they worry about being laid off and replaced by robots. Standardized tasks have been pushed aside by smart and autonomous systems, as also agreed upon by Bienhaus & Haddud, (2018). Therefore, HR managers are under intense pressure to hire and train a highly skilled workforce while simultaneously laying off individuals whose jobs will be automated by smart technology so that their companies may become more efficient and competitive on the global market(Min et al. 2019).

Ahead to the inauguration of the Davos meeting in 2016, the World Economic Forum (WEF) announced that about 5 million jobs would be eliminated in the key global economies during the following five years. The global HR sector must respond to the exponential demands of the Fourth Industrial Revolution by accounting for its impact on the need for human resources around the world and its potential consequences for economic growth and international competitiveness.

4.3 Issues With Big Data

Challenges in areas like Big Data management and cyber security are just the beginning of what the Fourth Industrial Revolution will bring to the business world. There are still many insurmountable roadblocks to the full implementation of Industry 4.0: Digitization of the sector brings with it several limitations due to the severe difficulties of standardisation and cybersecurity(De Propris and Bailey 2020).

According to their findings, the widespread use of Big Data poses a significant issue in the workplace and highlights the growing need for a regulatory framework to safeguard individuals' privacy and personal information. The Deloitte research highlights the top 10 cyber dangers in Industry 4.0 and the associated serious privacy concerns. There will be a variety of problems even if laws are tightened to deal with big data challenges (Gupta, Sunita, et al. (2022)).

Furthermore, the European Union has enacted the General Data Protection Regulation (GDPR) to reduce the various problems associated with Big Data management. The General Data Protection Regulations (GDPR) seek to "impose a legal framework on the handling of personal data". There is still work to be done until a comprehensive Industry 4.0 framework is in place, despite efforts to establish a global regulatory framework to reduce anxieties about the security of big data.

Big data analytics may be the key to a company's future success, but they may also cause management headaches on the inside. Data expansion difficulties, or "what do I do with this much information?" must be taken into account, since the collecting of such a large amount of data must be valuable for something, but there is a lack of good knowledge of big data and, thus, proper exploitation of the latter(Min et al. 2019). So, what is it exactly? Confusion over which Big Data technology to use is another problem. For organisations to collect and analyse reports and data relevant to their operations, there is a growing selection of software options.

When one lacks this self-awareness, though, it might be difficult to make a precise and straightforward decision. Finally, there is the problem of a general deficiency of specialists in the Big Data field. Data scientists, engineers, and analysts are in high demand, and many businesses are actively seeking them out. Given the scarcity of certain skills and the law of demand and supply, it can be expensive for a company to hire new employees, but doing so is essential if it wants to expand, particularly in today's increasingly international economy.

5. STRATEGIES FOR EXECUTING THE FOURTH INDUSTRIAL REVOLUTION

Key references in terms of "Smart Manufacturing architectures" may be found in the following models: "Reference Architecture Model or Industry 4.0 (RAMI 4.0), Smart manufacturing ecosystem (SME), and Intelligent manufacturing system architecture (IMSA)." Smart industrial architecture is the planning and implementation of an established organisational layout and network. To get a general idea of what Smart Manufacturing Systems are, one can go to the Smart manufacturing ecosystem (SME) (SMS). The product, the manufacturing process, and the economics are the three facets. It facilitates the sharing of data for the purpose of tracking a product from its inception to its final disposal. It examines the three dimensions and their relationship to the full value chain. The cyber-physical production system, manufacturing operations management (MOM), and enterprise resource planning (ERP) all play a role in the vertical integration of these three dimensions (CPPS). Intelligent Manufacturing System Architecture (IMSA) is a three-dimensional framework for intelligent manufacturing systems that includes life cycles, a hierarchy of systems, and intelligent functions. It incorporates the norms and characteristics of all IMS into a framework with key dimensions (Pandey 2020): the Lifecycle, to represent the integration of the supply chain; the System Hierarchy, to represent the different levels of control, the workshop, the enterprise, and the cooperation between them; and the Intelligent Functions, to represent the various resources, connections, information fusions, and new business patterns(Murata et al. 2020).

The German industrial sector is expected to see significant productivity increases thanks to the widespread adoption of Industry 4.0, as stated in a report by BGC (90 to 150 billion EUR). Between fifteen and twenty-five percent will be added to output. Productivity gains of 5–8% are achievable, after factoring in the price of materials. The magnitude of each effect is industry-dependent. It is estimated that the automobile industry can expect a 10- to 20-percent increase in productivity, while the industrial component manufacturing sector would witness improvements of 20-30 percent. To add insult to injury, Industry 4.0 will have an effect on Germany's manufacturing sector's ability to bring in additional money. Sales are expected to increase by roughly 30 billion euros year, or about 1 percent of German GDP, due to rising demand for a wider variety of ever-more-tailored products and the need of manufacturers for enhanced devices and new data applications(Ivanov, Dolgui, and Sokolov 2019).

The aforesaid expansion and productivity improvement are expected to have a favourable effect on employment growth of 6% over the next decade, according to the BCG report. Additionally, the need for mechanical engineers may rise by as much as 10% throughout the same time frame. Some low-skilled individuals performing routine or tedious activities will be replaced by automation in the next years. The demand for workers with ICT skills, however, will rise as big data analytics becomes increasingly popular. One of the greatest threats to future growth and innovation is the inability to adapt to this shift in required capabilities. Manufacturers will lose roughly 1% to 1.5% of their sales in absolute value over the next decade if they don't take the necessary steps to compete in the industry 4.0 economy. Last but not least, it is certain that Industry 4.0 will alter the global value chain, labour market, and logistics.

6. CONCLUSION

As a result of geopolitical unpredictability and natural disasters, global transactions are experiencing a great deal of volatility(Machado, Winroth, and Ribeiro da Silva 2020). In light of the current climate, companies need to formulate short-, medium-, and long-term strategic objectives for growth, risk management, and the global supply chain (Pandey 2022). Companies in the manufacturing sector need to adopt multifaceted and interdisciplinary strategies to deal with the problems caused by present and future supply chain interruptions. Using cutting-edge tools like "artificial intelligence and business analytics," companies may better analyse vast volumes of data, keep tabs on supply chain operations in real time, and develop the agility to quickly reorganise their Supply chain(Salam 2019). When competing on a global scale, a company's success will hinge on the agility with which its industry responds to shifting market conditions.

In conclusion, it appears that global business structure has been greatly improved thanks to Industry 4.0. We outlined how Industry 4.0 could improve the global competitiveness of enterprises operating on a global scale. Via an examination of the developmental stages through which the historical backdrop had reached its current, highly developed state, We discussed the ways in which Industry 4.0 had benefited both small and large businesses. Therefore, we defined the industry and its main subsets and emphasised their importance to the business. The challenges and opportunities of the modern industrial period give a learning opportunity for managers. We analysed its effectiveness and impact on the environment, as well as the possibility it offers of catering to a diverse set of requirements on account of its adaptability and the power of Big Data Analytics (Pandey 2021). Of course, there are also some drawbacks to consider with the many advantages. As a result of the profound social impact that robotics and automation will have, the field of human resources, in particular, stands to benefit greatly from the innovations and structural shifts that Industry 4.0 presents. Industry 4.0-based technologies will present significant competition to manufacturing production locations in developing and emerging nations that are part of the global value chain, such as a cheap and plentiful labour supply. We need to invest in people and the digital transformation if we're going to be able to face the challenges of the industry 4.0 era.

Finally, we discovered that, despite Big Data's potential benefits to business, it is still a field in need of special attention because of how hard it may be to utilise. Still, this revolutionary period of the modern era is not the last of its kind. "As a matter of fact, the rise of Industry 5.0 has reversed the script because it is a provider of solutions for both people and the environment. Indeed, the new revolution of Industry 5.0 places a premium on worker safety and incorporates all the value-added components of Industry 4.0 in the hopes of economic growth in the new industry as a path to a better future."

REFERENCES

Reza, M. (2020b). Relationship with Industry 4.0 and Supply Chain Management System of FMCG. *Journal of Business Management Studies.*

Albayrakoglu, M. (2020). *Supply Chain Management In The Textile Industry : A Supplier.* Semantic Scholar.

Ashodian, J. (2016). *Industry 4. 0 Is Transforming Supply Chains.* SICK. https://scg-mmh.s3.amazonaws.com/pdfs/sick_wp_industry4.0_10
0616.pdf

Bag, S., Wood, L. C., Mangla, S. K., & Luthra, S. (2019, September). Procurement 4.0 and Its Implications on Business Process Performance in a Circular Economy. *Resources, Conservation and Recycling, 152*, 104502. doi:10.1016/j.resconrec.2019.104502

Bienhaus, F., & Haddud, A. (2018). Procurement 4.0: Factors Influencing the Digitisation of Procurement and Supply Chains. *Business Process Management Journal, 24*(4), 965–984. doi:10.1108/BPMJ-06-2017-0139

Castagnoli, R., Büchi, G., Coeurderoy, R., & Cugno, M. (2022). Evolution of Industry 4.0 and International Business: A Systematic Literature Review and a Research Agenda. *European Management Journal, 40*(4), 572–589. doi:10.1016/j.emj.2021.09.002

Chalmeta, R. & Santos-dele, N. (2020). *Sustainable Supply Chain in the Era of Industry 4 . 0 and Big Data : A Systematic Analysis of Literature and Research.* MDPI.

Dadheech, P. (2022). A Compressive Review on Internet of Things in Healthcare: Opportunity, Application and Challenges. *Emerging Technologies in Computer Engineering: Cognitive Computing and Intelligent IoT: 5th International Conference, ICETCE 2022*. Cham: Springer International Publishing. 10.1007/978-3-031-07012-9_20

Dogiwal, S. R. (2022). Internet of Things Based Real-Time Monitoring System for Grid Data. *Emerging Technologies in Computer Engineering: Cognitive Computing and Intelligent IoT: 5th International Conference, ICETCE 2022*. Cham: Springer International Publishing. 10.1007/978-3-031-07012-9_21

Fatorachian, H., & Kazemi, H. (2020). Impact of Industry 4.0 on Supply Chain Performance. *Production Planning and Control*, *32*(1), 63–81. doi:10.1080/09537287.2020.1712487

Gabriel, M., & Pessl, E. (2016). Industry 4.0 and Sustainability Impacts: Critical Discussion of Sustainability Aspects With a Special Focus on Future of Work and Ecological Consequences. *Annals of the Faculty of Engineering Hunedoara*, *14*(2), 131.

Garay-Rondero, C. L., Martinez-Flores, J. L., Smith, N. R., Santiago, O. C. M., & Aldrette-Malacara, A. (2019). Digital Supply Chain Model in Industry 4.0. *Journal of Manufacturing Technology Management*, *31*(5), 887–933. doi:10.1108/JMTM-08-2018-0280

Ghadge, A., Kara, M. E., Moradlou, H., & Goswami, M. (2020). The Impact of Industry 4.0 Implementation on Supply Chains. *Journal of Manufacturing Technology Management*, *31*(4), 669–686. doi:10.1108/JMTM-10-2019-0368

Gillpatrick, T. (2019). Understanding The Role Of Consumer Behavior In Forecasting The Impact Of Industry 4 . 0 And The Wave Of Digital. *Accenture.*

Gupta, S., Nawal, M., Janu, N., & Goyal, D. (2022). IoT, Enabling Technologies, and Sensor Node Deployment Pattern in WSN. *ECS Transactions*, *107*(1), 7441–7455. doi:10.1149/10701.7441ecst

Handfield, R., Jeong, S., & Choi, T. (2019). Emerging Procurement Technology: Data Analytics and Cognitive Analytics. *International Journal of Physical Distribution & Logistics Management*, *49*(10), 972–1002. doi:10.1108/IJPDLM-11-2017-0348

Ivanov, D., Dolgui, A., & Sokolov, B. (2019). The Impact of Digital Technology and Industry 4.0 on the Ripple Effect and Supply Chain Risk Analytics. *International Journal of Production Research*, *57*(3), 829–846. doi:10.1080/00207543.2018.1488086

Janu, N. (2022). *Cross-lingual Word Sense Disambiguation Using Multilingual Co-occurrence Graphs*. Graph Learning and Network Science for Natural Language Processing. doi:10.1201/9781003272649-9

Janu, N. (2022). Query-Based Image Retrieval Using SVM. *Emerging Technologies in Computer Engineering: Cognitive Computing and Intelligent IoT: 5th International Conference, ICETCE 2022*. Cham: Springer International Publishing, 2022.

Jin & Shin. (2020). ScienceDirect Changing the Game to Compete : Innovations in the Fashion Retail Industry from the Disruptive Business Model. *Business Horizons*, *7*(2), 23–36. . doi:10.1016/j.bushor.2020.01.004

Kiel, D. (2017). What Do We Know about Industry 4.0" so Far? *International Association for Management of Technology 3*, 9–22. https://www.researchgate.net/publication/316684847

Li, L. (2020). Education Supply Chain in the Era of Industry 4.0. *Systems Research and Behavioral Science*, *37*(4), 579–592. doi:10.1002res.2702

Lin, D., Lee, C., Lau, H., & Yang, Y. (2017). Industrial Management & Data Systems Strategic Response to Industry 4.0: An Empirical Investigation on The Chinese Automotive Industry Article Information. *Industrial Management & Data Systems 118* (3), 0–18.

Machado, C. G., Winroth, M. P., & da Silva, E. H. D. R. (2020). Sustainable Manufacturing in Industry 4.0: An Emerging Research Agenda. *International Journal of Production Research*, *58*(5), 1462–1484. doi:10.1080/00207543.2019.1652777

Matt, D. T., Modrák, V., & Zsifkovits, H. (2020). *Industry 4.0 for Smes: Challenges, Opportunities and Requirements. Industry 4.0 for SMEs: Challenges*. Opportunities and Requirements. doi:10.1007/978-3-030-25425-4

Min, J., Kim, Y., Lee, S., Tae-won, J., Kim, I., & Song, J. (2019). The Fourth Industrial Revolution and Its Impact on Occupational Health and Safety, Worker ' s Compensation and Labor Conditions. *Safety and Health at Work*, *10*(4), 400–408. doi:10.1016/j.shaw.2019.09.005 PMID:31890322

Mubarak, M. F., & Petraite, M. (2020). Industry 4.0 technologies, digital trust and technological orientation: What matters in open innovation? *Technological Forecasting and Social Change*, *161*, 120332. doi:10.1016/j.techfore.2020.120332

Müller, J. M., & Voigt, K. (2018). Sustainable Industrial Value Creation in SMEs: A Comparison between Industry 4.0 and Made in China 2025." International Journal of Precision Engineering and Manufacturing - Green Technology 5 (5): 659–70. . doi:10.1007/s40684-018-0056-z

Murata, C., Pant, N., Lyer, S., Veitch, J., & Campbell, M. (2020). *Seizing the Trade 4. 0 Opportunity About the Authors*. Deloitte Insights.

Pandey, D., Pandey, B. K., Noibi, T. O., Babu, S., Patra, P. M., Kassaw, C., & Canete, J. J. O. (2020). Covid-19: Unlock 1.0 risk, test, transmission, incubation and infectious periods and reproduction of novel Covid-19 pandemic. *Asian Journal of Advances in Medical Science*, 23-28.

Pandey, D., Babu, S. Z. D., Noibi, T. O., Pandey, B. K., Dey, P. T. D., Kassaw, C., & Canete, J. J. O. (2020). A study on novel covid-19 epidemic threat for universal physical situation in 2020. *Asian Journal of Advances in Research*, 27-33.

Pandey, D., Aswari, A., Taufiqurakman, M., Khalim, A., & Azahrah, F. F. (2021). System Of Education Changes Due to Covid-19 Pandemic. *Asian Journal of Advances in Research*, 10-15.

Pandey, D., Islam, T., & Malik, M. A. (2021). Novel coronavirus disease (SARS-COV-2): an overview. *Asian Journal of Advances in Medical Science*, 8-12.

Pandey, B. K., Pandey, D., Wariya, S., Aggarwal, G., & Rastogi, R. (2021). Deep learning and particle swarm optimisation-based techniques for visually impaired humans' text recognition and identification. *Augmented Human Research*, *6*(1), 1–14. doi:10.100741133-021-00051-5

Pandey, D., Islam, T., Magray, J. A., Gulzar, A., & Zargar, S. A. (2021). Use of statistical analysis to monitor novel coronavirus-19 cases in Jammu and Kashmir, India. *European Journal of Biological Research, 11*(3), 274–282.

Pandey, D., & Pandey, B. K. (2022). An Efficient Deep Neural Network with Adaptive Galactic Swarm Optimization for Complex Image Text Extraction. In *Process Mining Techniques for Pattern Recognition* (pp. 121–137). CRC Press. doi:10.1201/9781003169550-10

Pandey, D., Wairya, S., Sharma, M., Gupta, A. K., Kakkar, R., & Pandey, B. K. (2022). An approach for object tracking, categorization, and autopilot guidance for passive homing missiles. *Aerospace Systems*, 1-14.

Patil, D. A. (2020). The study of industry 4 . 0 and its impact on supply chain management. *International Research Journal of Engineering and Technology.*

Preindl, R., Nikolopoulos, K., & Litsiou, K. (2020). Transformation Strategies for the Supply Chain: The Impact of Industry 4.0 and Digital Transformation. *Supply Chain Forum, 21* (1), 26–34. 10.1080/16258312.2020.1716633

Propris, L., & Bailey, D. (2020). *Industry 4.0 and Regional Transformations.* Routledge. . doi:10.4324/9780429057984

Reza, M. (2020a). Relationship with Industry 4.0 and Supply Chain Management System of FMCG. *Journal of Business and Management Studies, 2*(2), 24–31.

Salam, M. A. (2019). Analyzing Manufacturing Strategies and Industry 4.0 Supplier Performance Relationships from a Resource-Based Perspective. *Benchmarking*. doi:10.1108/BIJ-12-2018-0428

Samans, R. (2019). *Globalization 4.0: Shaping a New Global Architecture in the Age of the Fourth Industrial Revolution*. World Economic Forum. https://www3.weforum.org/docs/WEF_Globalization_4.0_Call_for_Engagement.pdf

Stankovic, M., Gupta, R., & Figueroa, J. (2017). *Industry 4.0 - Opportunities behind the Challenge. UNIDO Background Paper*, 1–56. United Nations Industrial Development Organization. https://www.unido.org/sites/default/files/files/2017-11/UNIDO Background Paper on Industry 4.0_27112017.pdf

Stock, T., & Seliger, G. (2016). Opportunities of Sustainable Manufacturing in Industry 4.0. *Procedia CIRP 40* (Icc): 536–41. . doi:10.1016/j.procir.2016.01.129

Tjahjono, B., Esplugues, C., Ares, E., & Pelaez, G. (2017). What Does Industry 4.0 Mean to Supply Chain? *Procedia Manufacturing, 13*, 1175–1182. doi:10.1016/j.promfg.2017.09.191

Tripathi, S., & Gupta, M. (2020). A Framework for Procurement Process Re-Engineering in Industry 4.0. *Business Process Management Journal, 27*(2), 439–458. doi:10.1108/BPMJ-07-2020-0321

Zheng, T., Ardolino, M., Bacchetti, A., & Perona, M. (2020). The Applications of Industry 4.0 Technologies in Manufacturing Context: A Systematic Literature Review. *International Journal of Production Research, 0*(0), 1–33. doi:10.1080/00207543.2020.1824085

Chapter 5
Review on Smart Sewage Cleaning UAV Assistance for Sustainable Development

Binay Kumar Pandey
https://orcid.org/0000-0002-4041-1213
Department of Information Technology, College of Technology, Govind Ballabh Pant University of Agriculture and Technology, Pantnagar, India

Digvijay Pandey
https://orcid.org/0000-0003-0353-174X
Department of Technical Education, IET, Dr. A.P.J. Abdul Kalam Technical University, India

Pankaj Dadheech
https://orcid.org/0000-0001-5783-1989
Swami Keshvanand Institute of Technology, Management, and Gramothan, India

Darshan A. Mahajan
NICMAR University, Pune, India

A. Shaji George
https://orcid.org/0000-0002-8677-3682
Almarai Company, Riyadh, Saudi Arabia

A. Shahul Hameed
Consolidated Techniques Co. Ltd., Riyadh, Saudi Arabia

ABSTRACT

Autonomous systems that cannot adapt to real-world environments have proliferated due to rapid technological improvement. These programmes will free people from repetitive, inefficient chores. The monotonous, dusty, and dangerous conditions of the unmanned aerial vehicle (UAV) pose a severe threat to human health and safety. Autonomous systems improve supply chain, tracking, and hazardous climate control. This chapter proposes merging compression and track architecture to improve UAV performance. The UAV and pipe increase trackwheel friction. The UAV doesn't move in the tube. Cleaning commences when sensors detect the mouth. Dirt sensors prevent the washing process from working. This chapter focuses on drain cleaning automation. Device automation addresses mobility and space. This study supports this method for garbage disposal and filtering. The technology removes manual cleaning and relies on human control of system movement.

DOI: 10.4018/978-1-6684-8117-2.ch005

INTRODUCTION

Municipal corporations are responsible for the cleaning, maintenance, and protection of cities. The drainage systems of many communities are buried beneath. When the drainage system is not maintained properly, potable water can become tainted with drainage water, which can lead to the spread of infectious diseases (Pandey, D., et al. (2021)). Drainage systems become clogged during rainy seasons, which makes day-to-day life more difficult. It's possible that the atmosphere, the traffic, and the people will all aggravate you. Most of the time, clogged drains will cause sewage and other kinds of wastewater to overflow into your property. Consider constructing a facility that would provide authorities or other interested parties with instant access to information regarding the quantity of blockages as well as their location. Therefore, sensors watch for any lapses in maintenance. The sensor notifies the transmitter whenever a drain becomes clogged or water overflows its intended path. It is difficult for humans to maintain the maintenance holes due to the deteriorating environment and the difficulties of monitoring the entire status. It is impossible to know whether someone purposefully entered the maintenance hole or whether they were there because of an accident. Drainage systems are essential to urban populations in order to prevent flooding. Many people are harmed by inadequate drainage system upkeep. Road conditions are negatively impacted by inadequate drainage monitoring systems in many places. Roads are designed to handle both automobile traffic and pedestrian traffic. Insufficient monitoring of drainage systems is the root cause of water contamination and the spread of diseases that are transmitted through water in many cities and towns across the globe. Highways frequently have problems with water pooling up. Flooding results in traffic congestion, which wastes time, money, and employment opportunities. The contamination of groundwater can be very challenging to eradicate. For its economy to expand, India requires a reliable drainage system. Smart cities depend on drainage system design. More employees are required in order to keep a drainage system in good working order. As a result of the unstructured nature of the cleaning environment, engineers have made significant progress in the areas of artificial intelligence, unmanned aerial vehicles, and robotics (George et al., 2021). However, this continues to be a challenge. Pandey et al. (2022) provide a home wastewater purifying unmanned aerial vehicle (UAV) in this work. The treatment of wastewater is the primary focus of the majority of mobile UAV experiments. Documentation regarding drainage UAVs is limited. It would appear that their expertise lies in the coverage algorithm. This article explains how to construct a fully autonomous UAV mobile for the purpose of cleaning drainage systems. The planning of tracks on a global scale needs to end at a specific point in artificial areas such as home hallways, for example at the intersection of two corridors. The UAV plans out local detours to get past obstacles like walls. Therefore, the navigation system requires a framework in order to describe and find these building placements on a world map that ultimately leads somewhere. The entry hallway has been designed with an open floor plan. According to Yadeta et al. (2020), the IIoT is essential to improving industrial efficiency, quality, cost, and resource reduction. IIoT-based project implementations, on the other hand, have very little documentation. The technology of the IIoT makes it difficult for typical businesses to upgrade their antiquated automation devices. This article takes a look at the automation of sewage treatment plants that relies on technical skills. The core controller of the STM32 processor located in Geneva, manufactured by STMicroelectronics, makes older automation equipment better. Tencent's We Chat Applet, which is located in Shenzhen, serves as the enlarged host computer. Buildings are being outfitted with IIoT-based remote monitoring and control systems for sewage treatment. The use of unmanned aerial vehicles (UAVs) has witnessed phenomenal growth in popularity over the past few years, with applications ranging from agriculture and urban services to the arts and recreational activities. UAVs have been researched for serious and commercial applications, specifically as mobile sensing, Smart sewage cleanup, and actuation platforms (that is, as a delivery platform for an ever-expanding array of sensors and actuators). Both The NEC Research and TNO are doing research into the potential applications of unmanned aerial vehicles, often known as UAVs, and swarms of UAVs in circumstances where the employment of traditional methods would be ineffective due to the requirement for high resolution, the presence of prohibitive surroundings, or the presence of severe time limitations. This review paper presents a brief outline of application areas centred on security and protection that we identified as essential objectives for industrial and commercial activities. The background for this article is intelligent self-driving robots and independent or semi-autonomously operating groups. The deployment of unmanned aerial vehicles (UAVs) presents a number of challenges, which are explored in this section, both in general and in connection to the specific deployments contained within the specified application areas. This article's purpose is not only to provide a summary of the relevant research and the present state of the art, but also to offer a glimpse into the (possible) future work that we may perform and the challenges that we may face. Both of these objectives will be accomplished by reading this article.

LITERATURE REVIEW

The authors of proposed that mechanical drain purification should be utilised instead of human drain cleaning. AkshayDeshmukh et al. (2021) are responsible for making this recommendation. Although automatic drainage is required for industrial applications (Pandey, B. K., et al. (2021)), it is challenging to dispose of waste from production and from firms in an efficient manner. It is possible to use the drainage pipelines, but if the obstruction is allowed to wash away, then there will be fatalities among the population. They built a mechanised and semi-automatic drainage water cleaner so that they could tackle this problem and stop the additional loss of human life that was occurring. In addition to the elimination of waste and trash, they planned to make use of this ineffective technology to manage daily waste filtration. The Cleaner Drainage System emphasises the treatment of waste before it enters the drainage system in order to preserve the natural world. This is done in order to safeguard it. Burning this waste in residential areas will contribute to climate change and will obstruct drainage systems, which will lead to flooding if it is not properly disposed of.

Manual scavenging was reported to be unlawful in the year 2013, according to Bakker, T., et al. (2010). Despite the fact that private contractors working for the local government continue to utilise portable scavengers, manual scavenging was reported to be banned in the year 2013. The current situation can be improved with the help of the people complement that the informatics system provides. A computer-controlled robotic automation system is used to check and clean the wastewater in order to accomplish both goals. (Meslie, Y., et al. (2021)), the automated stitching unmanned aerial vehicle (UAV) BarytaArtana is tasked for examining and removing garbage pipelines. Second, the unmanned aerial vehicle does an inspection of the drainage pipes to check for things like fractures, corrosion inhibitors, and other similar elements. When compared to pipeline feeds, the amount of time that UAV cameras stream for is substantially greater. These views display the operator in order to facilitate comparisons between different programmer registrations. In addition to that, the unmanned aerial vehicle (UAV) has a proximity sensor installed in its front body. After the unmanned aerial vehicle (UAV) has approached the obstruction, the rotor will start to move it out of the way. Using this method puts an end to any and all challenges connected to the washing of sewage.

According to the findings of Patil, G. S., et al. (2018), the presence of water is an unavoidable prerequisite for the development of life on earth. The reason for this is because more than 70 percent of the water on Earth is not appropriate for consumption by humans. It is not only expensive to drink, bathe in, or otherwise consume, prepare, or distribute freshwater; it is also expensive to do so. The presence of contaminants in water can cause a variety of health concerns, some of which could be fatal. The difference between wastewater and river water that originates from houses, businesses, endeavours, company operations, and foundations that are subjected to treatment facilities can be distinguished by using a funnel system that has been suitably planned and constructed.

According to Ramdas, B. Y., and et al. (2014), it is crucial for today's society for factories to be automated. Managing hazardous trash is another difficult undertaking that must be undertaken. The garbage pipelines posed little danger, but blockages in the sewage pipes can be extremely dangerous to people's health. We are currently working on creating what we refer to as a "Automatic Sewage Washing Machine" for use in wetlands and for the disposal of waste.

These methods either result in harmful soils being released into the wastewater or in their disappearance, depending on which occurs. In recent years, advancements in the treatment of wastewater have made it possible to remove toxins without resorting to their annihilation. This sewage's causes are unique. The generation of between 200 and 500 litters is caused by each wastewater frame that is processed. The wastewater treatment facility monitors and controls this source of contamination until it can be released into the environment. The treatment of wastewater is something that the majority of institutions, especially those that run around the clock, can benefit from. The fundamental goal of wastewater treatment plants is to reduce or eliminate the amount of pollutants that are released into natural water sources. There are a few different treatment facilities that are in desperate need of maintenance and new equipment right now.

The requirement for human labour in potentially hazardous sewer pipe conditions has prompted a non-governmental organisation to call for the development of a pipe-mounted Unmanned Aerial Vehicle (UAV) for Sewer Garbage Collection (SGC UAV) (Pandey, B. K., et al. (2022).). This NGO has lobbied for the establishment of the SGC UAV. According to the findings of the research, it would be possible to collect the item, place it in a basket in the maintenance hole, and then deliver it on the ground if a moving UAV was used in conjunction with other equipment inside the garbage pipe. The unmanned aerial vehicle (UAV) serves to insulate the engines from the wastewater environment, makes use of the sewage curved field to maximise attraction, generates a low-inertia manipulator to boost efficiency, and provides a waste management system that is able to modify its shape in order to comply with environmental regulations. The concept and the fundamental notions offer resounding support for the utilisation of the SGCR system. Carry out a simulated flight using the unmanned aerial vehicle.

GENERAL METHOD FOR SMART SEWAGE UAV ASSISTANCE FOR SEWER WORKERS

Because of the unique nature of this technology, unmanned aerial vehicles (UAVs), a normal explanation will not do. However, there are a few fundamental features that a UAV should have, which can make it easier to define what a UAV is and what it is not, as well as which components should be present in a device before it can be deemed a UAV. These qualities include a remote control system, a camera, and an

Figure 1. Smart sewage UAV assistance system

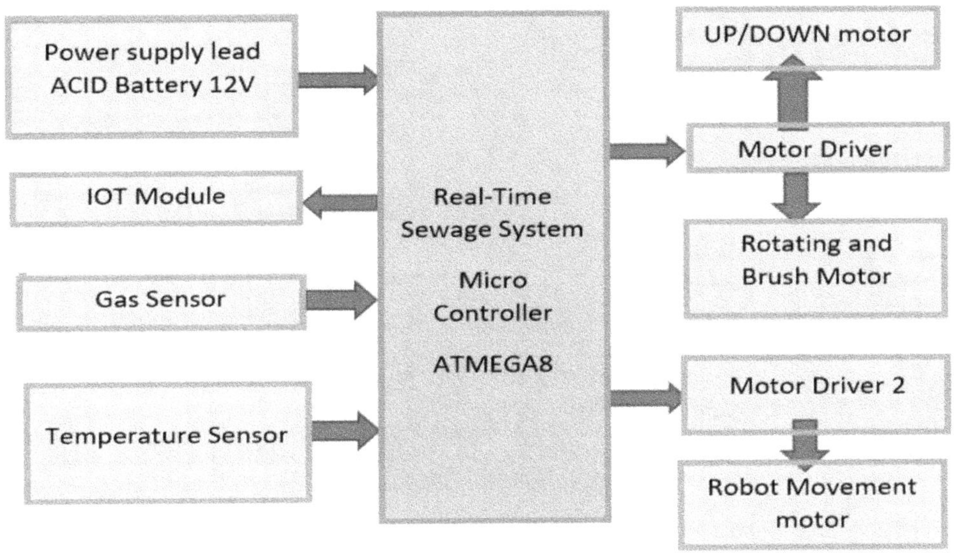

onboard computer. Because of the cleaner, it's possible that automatic sanitation of the UAV underground drainage system will be achievable. UAV cleaners initially went down the subsurface drains to clean them when the incident first began. Both the front wheel and the back wheel of the UAV cleaner each have their own independent drive systems (Khan, B., et al. (2021).). The quantity of other production is maintained to a minimum, and infrasound sensors are placed on the left, right, and front of the vehicle to detect any obstructions. Logical outputs are generated by a comparator LM324, and the amount of other production is minimised.

Infrasound sensors fabricated with the use of an LM324 (Suresh, D. S., et al. (2013)). The micro-controller ATmega8A and the H-bridge driver L293 are responsible for controlling the engine's speed as well as the direction it rotates in. Figure 1 illustrates the comprehensive intelligent UAV assistance system for sewage operations.

An unmanned aerial vehicle (UAV) that can detect obstacles and manoeuvre itself around them is an example of a closed-loop method that is not only effective but also relatively straightforward. You will have the opportunity to show the UAV how to avoid the barriers when they appear. A demonstration of the intelligent UAV support equipment for the sewage system is shown in Figure 1. Figure 2 provides an overview of the real-time smart sewage robotic aid that may be used by sewer employees thanks to the Internet of Things.

Gas and Temperature Sensors

The gas sensor in this system is responsible for identifying the presence of potentially hazardous gases that are released into the environment by sewage treatment facilities. In addition to that, this system includes a live sewage level detection and monitoring system as an integrated component. An alert is sent to the observer who is situated in a different location from the action that is being observed once a threshold value has been reached that was previously determined. Following that, several measurements

Figure 2. Overview sewage robot system

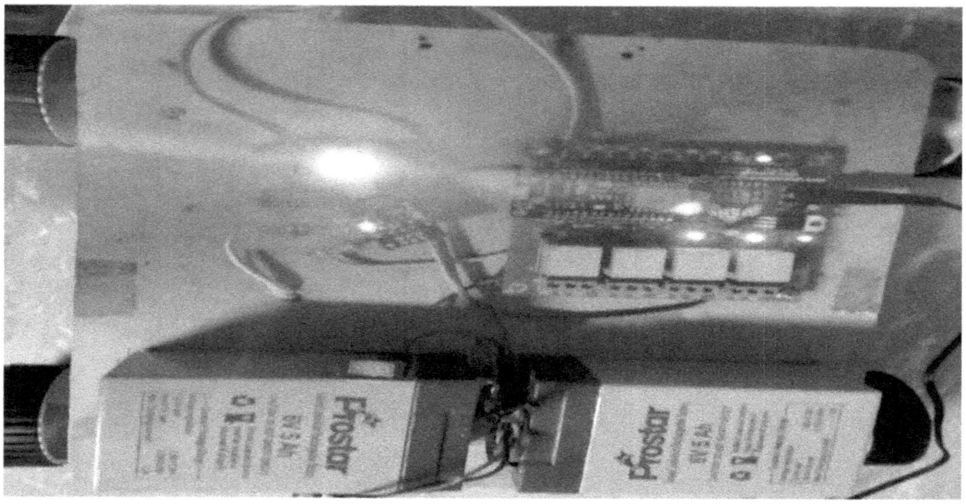

of the concentration of the gases in the air are provided to the personnel so that they may evaluate for themselves whether or not it is safe for them to continue working or cleaning in that setting. The architecture of the suggested system makes use of monitoring gear for the Internet of Things that is remotely integrated, as well as a platform for the Internet of Things. It is crucial for industrial applications to calibrate gas sensors (Pandey, J. K., et al. (2022)), and it is also essential to locate the proper threshold values for septic plants and facilities. It is essential to give a warning to the sewage personnel before they arrive at their destination in order to prevent possibly hazardous circumstances. This is of the utmost significance whenever there is an increase in the concentration of gaseous pollutants. The multiple sensors are utilised to monitor various characteristics of the sewage, such as the amounts of gas that are present, the temperature, and so on. Specifically, the sensors detect the quantities of gas that are there. The sewage worker will receive a warning from this system in the form of an SMS message and a phone call anytime the threshold value falls below the detected values. This serves to protect sewage workers as well as the general public from contracting potentially harmful infections. In figure 3, the gas sensor is shown in its full in its completeness.

The ThingSpeak analysis tool presents example sensor data taken from a system that is still in the process of being developed. Both carbon monoxide and methane sensors measured levels that were above the threshold, with carbon monoxide at 2.3 ppm and methane measuring 60 ppm. As a result, a GSM module (Iyyanar, P., et al. (2023).) transmitted an alert to a phone number that had been entered using the sensor's code.

Temperature sensors sense the heat energy that is surrounding them. IoT sensors are used in a variety of sectors, including industry, agriculture, and sewage treatment in real time. The temperature sensor for the safety of the sewage system is displayed in Figure 4.

Figure 3. Gas Sensor used in the sewage system

Figure 4. Temperature Sensor used in the sewage system

DRONES AS UAV IN WASTE MANAGEMENT

The data and knowledge that may be gathered by low-cost uncrewed aerial vehicles, which are often sometimes referred to as "drones," can be useful to the environment. These vehicles are also known as unmanned aerial vehicles (UAVs). Operators of unmanned aerial vehicles have access to a wide variety of different types of aircraft that were developed specifically for the task of cleaning up waste. On the other hand, selecting the appropriate drone is not always a simple process to complete. The extent of the survey, the amount of money that is available, the kind of camera that is required for surveying applications, and the question of whether or not RTK-enabled instruments, such as the Phantom 4 RTK drone, are required are all things that need to be taken into consideration (Menon, V., et al. (2022)). In spite of the fact that the potential contributions of such data to coastal management activities have been described in a number of recent studies that span a variety of disciplines (including ecology, engineering, and various branches of physical and human geography), the full extent to which such data can be used is still largely uncharted at this point in time. According to research conducted by Singh et al. (2022, July), the utilisation of drones in the construction of high-resolution 3D models of landfills could help the management team cut costs and save time. The increase in the dumping of solid waste over time is an important statistic, and these models can be used to track that rise (Khan, H., et al. (2023)).

The utilisation of drones, which have been shown to be the most cost-effective, helpful, efficient, and precise equipment available, has unlocked the enormous upside potential of the waste management company (Vinodhini, V., et al. (2022)). Drones have been shown to be the most helpful, efficient, and precise equipment available. The use of drones has been demonstrated to be the most effective method for realising this immense upside potential. Buildings may be inspected from a safe distance while workers stay on the ground to complete the inspections. Drones are currently being utilised in every industry, from the forensics department of law enforcement to the production of documentaries to the broadcasting of sports. There are a great many applications for these systems, some of which include monitoring the flow of traffic, the management of pipelines, the safety of ports, environmental tracking, imaging of soil moisture, waste collection, crop observation, and a great many others (Juneja, V., et al. (2023)). Babu, S. Z. D., (2022) lists some of the numerous uses for these systems.

CONCLUSION

The water services provided by the city are beneficial to both the surroundings and human health. Sewage equipment is at the centre of water challenges in both developing countries and nations with greater resources. In intelligent cities, human garbage has been cleaned up for decades, which has made the subject matter less appealing. Sewerage services were already in place in early settlements across the world before toilets were even invented. The automation of intelligent sewer systems was introduced by this effort. UAVs that perform automated drainage utilise this technology. The drain will be cleaned by an unmanned aerial vehicle (UAV) that is capable of cleaning sewage autonomously underground. The autonomous tube-cleaning UAV has access to four other pathways in addition to a track. Foldable links allow for the unmanned aerial vehicle to move either horizontally or vertically within the tubing. Therefore, UAVs are used to clean wire brushes. In this paper, intelligent and real-time drainage monitoring for metropolitan cities based on the internet of things is discussed. Using sensors for gas, water, and blockages, the system is able to identify any issues with the drainage system.

FUTURE WORK

Sensor networks are crucial enablers for the Internet of Things. The Internet of Things is utilised by this system to monitor wastewater. Using sensors such as flow indications, water levels, and block identification, we are able to identify problems that occur inside the drainage system in real time. We will be better able to tackle the issues if we increase the number of early blocking warnings. This can create a drainage system that is both intelligent and efficient, allowing for control and the diagnosis of problems.

REFERENCES

Akshay, D., Gaikwad, S., & Mendhe, A. (2021). AUTOMATIC DRAINAGE CLEANING MACHINE. *International Journal of Innovations in Engineering Research and Technology,* 1–3. https://repo.ijiert.org/index.php/ijiert/article/view/1730

Babu, S. Z. D., Pandey, D., Naidu, G. T., Sumathi, S., Gupta, A., Bader Alazzam, M., & Pandey, B. K. (2022, July). Analysation of Big Data in Smart Healthcare. In *Artificial Intelligence on Medical Data: Proceedings of International Symposium, ISCMM 2021* (pp. 243-251). Singapore: Springer Nature Singapore.

Bakker, T., Van Asselt, K., Bontsema, J., Müller, J., & van Straten, G. (2010). A path following algorithm for mobile robots. *Autonomous Robots*, *29*(1), 85–97. doi:10.100710514-010-9182-3

George, A. H., Fernando, M., George, A. S., Baskar, T., & Pandey, D. (2021). Metaverse: The next stage of human culture and the internet. [IJARTET]. *International Journal of Advanced Research Trends in Engineering and Technology*, *8*(12), 1–10.

Iyyanar, P., Anand, R., Shanthi, T., Nassa, V. K., Pandey, B. K., George, A. S., & Pandey, D. (2023). A Real-Time Smart Sewage Cleaning UAV Assistance System Using IoT. In *Handbook of Research on Data-Driven Mathematical Modeling in Smart Cities* (pp. 24–39). IGI Global.

Juneja, V., Singh, S., Jain, V., Pandey, K. K., Dhabliya, D., Gupta, A., & Pandey, D. (2023). Optimization-Based Data Science for an IoT Service Applicable in Smart Cities. In *Handbook of Research on Data-Driven Mathematical Modeling in Smart Cities* (pp. 300–321). IGI Global. doi:10.4018/978-1-6684-6408-3.ch016

Kassaw, C., & Pandey, D. (2021). COVID-19 pandemic related to anxiety disorder among communities using public transport at Addis Ababa, Ethiopia, March 2020: cross-sectional study design. *Human Arenas*, 1-10.

Khan, B., Hasan, A., Pandey, D., Ventayen, R. J. M., Pandey, B. K., & Gowwrii, G. (2021). Fusion of datamining and artificial intelligence in prediction of hazardous road accidents. In *Machine learning and iot for intelligent systems and smart applications* (pp. 201–223). CRC Press. doi:10.1201/9781003194415-12

Khan, H., Veeraiah, V., Jain, V., Rajkumar, A., Gupta, A., & Pandey, D. (2023). Integrating Deep Learning in an IoT Model to Build Smart Applications for Sustainable Cities. In Handbook of Research on Data-Driven Mathematical Modeling in Smart Cities (pp. 238-261). IGI Global.22.Bansod, S. J., &Thakre, S. (2014). Near Infrared Spectroscopy based Soil Nitrogen measurement-A Review. *International Journal of Current Engineering and Technology*, *4*(1), 268–272.

Maxwell, J. C. (1873). *A treatise on electricity and magnetism* (Vol. 1). Clarendon Press.

Menon, V., Pandey, D., Khosla, D., Kaur, M., Vashishtha, H. K., George, A. S., & Pandey, B. K. (2022). A Study on COVID–19, Its Origin, Phenomenon, Variants, and IoT-Based Framework to Detect the Presence of Coronavirus. In *IoT Based Smart Applications* (pp. 1–13). Springer International Publishing.

Meslie, Y., Enbeyle, W., Pandey, B. K., Pramanik, S., Pandey, D., Dadeech, P., & Saini, A. (2021). Machine intelligence-based trend analysis of COVID-19 for total daily confirmed cases in Asia and Africa. In *Methodologies and Applications of Computational Statistics for Machine Intelligence* (pp. 164–185). IGI Global. doi:10.4018/978-1-7998-7701-1.ch009

Pandey, B. K., Pandey, D., & Agarwal, A. (2022). Encrypted Information Transmission by Enhanced Steganography and Image Transformation. [IJDAI]. *International Journal of Distributed Artificial Intelligence, 14*(1), 1–14. doi:10.4018/IJDAI.297110

Pandey, B. K., Pandey, D., Wariya, S., Aggarwal, G., & Rastogi, R. (2021). Deep learning and particle swarm optimisation-based techniques for visually impaired humans' text recognition and identification. *Augmented Human Research, 6*(1), 1–14. doi:10.100741133-021-00051-5

Pandey, D., Pandey, B. K., & Wairya, S. (2021). Hybrid deep neural network with adaptive galactic swarm optimization for text extraction from scene images. *Soft Computing, 25*(2), 1563–1580. doi:10.100700500-020-05245-4

Pandey, D., & Wairya, S. (2022). Perfomance Analysis of Text Extraction from Complex Degraded Image Using Fusion of DNN, Steganography, and AGSO. In *VLSI, Microwave and Wireless Technologies: Select Proceedings of ICVMWT 2021* (pp. 195-203). Singapore: Springer Nature Singapore.

Pandey, D., Wairya, S., Sharma, M., Gupta, A. K., Kakkar, R., & Pandey, B. K. (2022). An approach for object tracking, categorization, and autopilot guidance for passive homing missiles. *Aerospace Systems*, 1-14.

Pandey, J. K., Jain, R., Dilip, R., Kumbhkar, M., Jaiswal, S., Pandey, B. K., & Pandey, D. (2022). Investigating Role of IoT in the Development of Smart Application for Security Enhancement. In *IoT Based Smart Applications* (pp. 219-243). Cham: Springer International Publishing.

Patil, G. S., Pawar, R. A., Borole, M. D., Ahire, S. G., Krishnani, A. L., & Karwande, A. H. (2018). *Review Paper on Drainage Water Cleaner Machine. International Research Journal of Engineering and Technology* (Vol. 5). IRJET.

Prema, K., Kumar, N. S., & Sunitha, K. A. (2009, June). Online temperature control based on virtual instrumentation. In *2009 International Conference on Control, Automation, Communication and Energy Conservation* (pp. 1-4). IEEE.

Ramdas, B. Y., & Galande, S. G. (2014). Green Growth Management by Using Arm Controller. *International Journal of Engineering Research and Applications*, *4*(3), 360–363.

Singh, H., Pandey, B. K., George, S., Pandey, D., Anand, R., Sindhwani, N., & Dadheech, P. (2022, July). Effective Overview of Different ML Models Used for Prediction of COVID-19 Patients. In *Artificial Intelligence on Medical Data: Proceedings of International Symposium, ISCMM 2021* (pp. 185-192). Singapore: Springer Nature Singapore.

Suresh, D. S., Jyothi Prakash, K. V., & Rajendra, C. J. (2013). Automated soil testing device. *ITSI Transactions on Electrical & Electronics Engineering,* (pp. 2320-8945). Semantic Scholar.

Vaani, I., Sushil, S. J., Kunjamma, U. V., Ramachandran, A., Bai, V. T., & Thyla, B. (2017, May). BhrtyArtana (A pipe cleaning and inspection robot). In *2017 Third International Conference on Sensing, Signal Processing and Security (ICSSS)* (pp. 422-425). IEEE. 10.1109/SSPS.2017.8071632

Vinodhini, V., Kumar, M. S., Sankar, S., Pandey, D., Pandey, B. K., & Nassa, V. K. (2022). IoT-based early forest fire detection using MLP and AROC method. *International Journal of Global Warming*, *27*(1), 55–70. doi:10.1504/IJGW.2022.122794

Yadeta, A. E., & Pandey, D. (2020). Analysis of the global impact of the pandemic (COVID-19) on construction industry: Possible scenarios. *Current Trends in Civil & Structural Engineering*, *6*(4), 1–8. doi:10.33552/CTCSE.2020.06.000641

Chapter 6
E-Waste Management Transition Towards Circular Economy

Kumar Shalender
Chitkara Business School, Chitkara University, India

Babita Singla
https://orcid.org/0000-0002-8861-6859
Chitkara Business School, Chitkara University, India

Sandhir Sharma
https://orcid.org/0000-0002-3940-8236
Chitkara Business School, Chitkara University, India

ABSTRACT

Electronic waste is one of the most challenging problems faced by the globe today. In order to solve it in an efficient and effective manner, this chapter highlights the role of the circular economy and its elements in effectively recycling and managing electronic waste. With the help of extant literature, the research digs deep into the concept of electronic waste and circular economy and how the elements of this economy can be combined to effectively manage electronic waste. This is followed by the development of a conceptual framework that encompasses all important stakeholders including suppliers, manufacturers, and customers for effective management of electronic waste. The research also emphasizes the adoption of a holistic and all-encompassing policy framework for effectively managing electronic waste and also details significant implications for manufacturers and policymakers.

1. INTRODUCTION

Electronic waste (E-waste) can be easily counted among the most prominent problems the world is facing today. There is no doubt about the utility and benefits the digital revolution has brought to our doorsteps

DOI: 10.4018/978-1-6684-8117-2.ch006

but at the same time, the rising number of electronic gadgets, equipment, and devices is assuming the proportions of a huge problem. Countries and Nations across the globe are currently contending with the issue to solve the problem of electronic waste and scientists are working overnight to find a credible solution that will help to discard electronic waste without significantly impacting our ecosystem. The Covid-19 has only exacerbated this problem as the net sales and use of electronic equipment during and after the pandemic have risen significantly (Kulshrestha et al., 2022). Complicating the situation further is the fast rate of changing technology and enhanced innovation capabilities of the companies that have resulted in the launch of new products and devices in the market at breakneck speed. The new iterations that used to take at least a couple of years to come into the market in the past are getting launched today every other day (Awasthi et al. 2018). This entices the customers to buy the new model and dispense with older ones, thereby aggravating the issue of electronic waste.

Another prominent part of the problem is generation of electronic waste throughout the value chain of manufacturing of items. E-waste not only include end products such as our mobile devices, tablets, systems, and PCs but the entire manufacturing cycle starting right from the sourcing of ores and components required to produce them. In addition, production lines, assembly points, and other touch points also lead to generation of electronic waste. At each and every step of this value chain, the electronic waste gets generated and that complicates the situation in more than one way. The impact of electronic waste is comprehensively damaging to our ecosystem and environment. The disposal of the waste itself is associated with the emission of strong radiation emanating from the electronic material, leading to compromised health for the factory workers as well as other allied stakeholders involved in the product manufacturing and disposal of electronic items. In fact, for the entire ecosystem of developing, manufacturing, and disposing of electronic waste has a detrimental impact on the health and environment (Cui and Zhang 2008). Looking closely at the definition of electronic waste, one can easily understand that this waste includes both broken as well as functional electronic equipment. The problem of electronic waste can be addressed with the help of a circular economy. The concept has come into prominence thanks to the enhanced awareness of people, nations, and environment activists related to electronic waste. The following section of the chapter talks in detail about the concept of circular economy in navigating through the issue of electronic waste and how these waste management practices can lead to a more sustainable environment with minimum damage and help in the effective management of electronic waste.

2. CONCEPT OF CIRCULAR ECONOMY

As the name suggests, the concept of circular economy can be described as a model that reduces the consumption of materials for producing and manufacturing goods through a sustained efforts on the reducing, reusing, and recycling. In the specific context of e-waste management, the circular economy can help manufacturers and producers to reduce their requirement for raw materials by effectively recycling the material and components required for the production of items. In simple words, the recycling of the disposed electronic products will go into the manufacturing and production of new devices and help to reduce the overall electronic waste that is generated across the globe (Ganguly 2016). It also pertains to giving a second life to electronic products so that the net quantity of electronic waste that usually goes into the landfills can be reduced to significant levels. Of course, achieving zero electronic waste is not an easy task but if the implementation of the circular economy takes place with the desired efficiency and effectiveness then this ambitious dream can be realised effectively and efficiently. The process of prolonging the life of electronic devices also works in the favour of manufacturers and customers and bring enormous benefits to the economy of the country (Vyas et al., 2023) . Especially the amount the companies need to spend in order to manage waste of electronic products can be curtailed and using recycling practices, the entire value chain of electronic products can be managed in a far more efficient and effective manner. The benefits of circular economy are multifaceted as a concept is advantageous not only from the commercial viewpoint but also has strong ecological benefits for the entire ecosystem and society at large. The resources can be effectively utilised, and the overall optimisation of their usage will lead to the creation of higher-order economic benefits for stakeholders throughout the system (Garg and Adhana 2019). This, in turn, will lead to more sustainable economic development in which a balance can be struck between both environmental and commercial aspects of doing business. The concept of circular economy can also lead to enormous advantages for the recycling industry which is home to a large number of entrepreneurs and new-age business ventures effectively contributing to the world economy including India. Electronic waste management will encourage companies to venture into the recycling industry and along with managing electronic waste, other industries which are having damaging impact on the environment (such as automobile, coal power plants) can also be inspired to take the same steps for effectively reducing their carbon footprints.

The creation of a successful circular economy with a focus on recycling electronic waste can lead to a series of innovative steps that will help new inventions to manifest in the field and contribute to both economy and the ecological preservation of our environment. Therefore, it comes as no Surprise that governments across the globe are putting the focus on the circular economy, encouraging stakeholders and new business ventures to come into the field, and offering open policy frameworks to make sure that the desired results on the front of electronic waste and other practices can be obtained within the constraints of resources, time, and money (Herat 2021).

3. E-WASTE MANAGEMENT: CIRCULAR ECONOMY FRAMEWORK

The development of the conceptual framework for circular economy in the case of E-waste management comprises the sourcing of raw materials, manufacturing of products, usage by customers, and life extensions of the gadgets and equipment (insert figure 1).

The very first part of this framework focuses on the design and development of durable and reliable products that can be used for an extended period of time (Ravindra and Mor 2019). It is extremely important for the suppliers to focus on and source raw materials that can be used for extended use even on the pattern of rough usage. The design of the product must also focus on its recyclable properties and should be conceptualized in a manner that after its life is over, the material and components used in making it can be recycled and again used for the production process. The second phase of the process is related to manufacturing where the producing unit must have production facilities that can accommodate the scrap material from the recyclables products in manufacturing the new items (Tesfaye et al. 2017). This accommodation might require some crucial and critical changes in the production and assembly lines and need some investment on the part of manufacturers. The third part of the conceptual framework is focused on the use of digital systems and equipment by consumers. The idea here is to encourage extended use and provide life extensions to the products by offering superior repair and service facilities to consumers. The extended life can be easily infused to many electronic gadgets and if the customers are provided with reliable services related to the repair and maintenance of the product,

Figure 1. Circular economy for e-waste management
Source: Authors' conceptualization

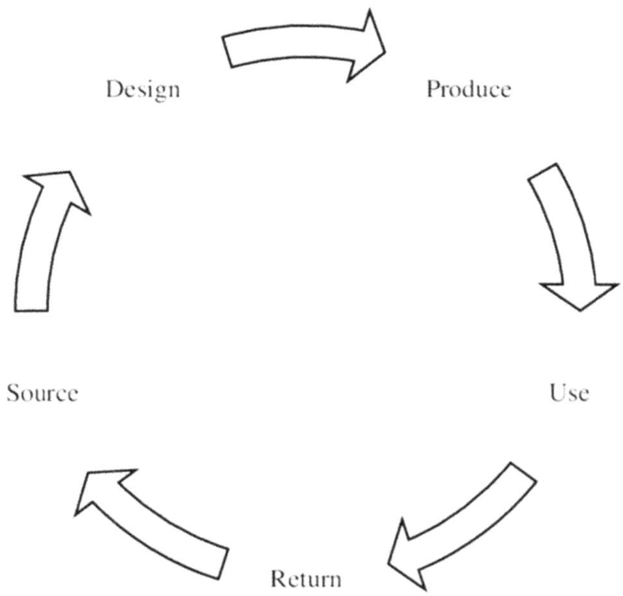

the desired results can be obtained on this front. Towards the end of their life, it is the responsibility of the manufacturers, in collaboration with policymakers, to offer attractive discounts and financial incentives for the consumers who are willing to return the old gadgets to the manufacturer so as to help with the overall process of recycling (Wang 2019). Of course, this step has financial implications, and that's why the financial support by the government becomes crucial to encourage the adoption of a circular economy in the industry. The last step of the economy encompasses all tools and technologies that are required to effectively capture the recyclable material from old gadgets and equipment. This is particularly true for enhancing the overall adoption of recyclable practices and helping the process to achieve desired outcomes as far as the recycling practices are confirmed. The success of this process will also determine the net amount of electronic waste and can help the nation curb the menace of this problem of humongous proportions (Shalender and Yadav 2019). Therefore, the concept of circular economy in the case of E-waste management is not significantly different from conventional value chain models which is a positive thing to start with. This similarity also inspires and encourage companies to take credible steps to reduce electronic waste and add to their brand equity by emerging as credible protector of ecosystem for the future generations.

4. CONCLUSION AND DISCUSSION

Nobody can contest the utility and advantages of a circular economy for electronic waste management. It is the pressing need of the hour to limit our electronic waste otherwise it can wreak havoc on the entire ecosystem and spoil it for our upcoming generations. All the stakeholders involved in the ecosystem of this particular industry need to play a very critical and crucial role here. Suppliers need to become more particular about using recyclable material and manufacturers also need to ensure that a massive part of their manufacturing accommodates the recycled material for making new products and equipment. The major burden of responsibility lies on the shoulders of customers who must ensure that they use electronic gadgets and equipment in a responsible and sustainable manner. They must also put in efforts to search for reliable service partners so that the longevity and the entire life cycle of electronic gadgets can be enhanced in the long run. This is the responsibility that must be shared by the manufacturers as offering reliable services and maintenance is key to the longevity of devices in the future. Customers can also contribute by returning the damaged products to the designated actors and outlets which can be then utilised for recycling and allied processes.

The critical and crucial role of policymakers couldn't be overemphasised in the process of successfully managing electronic waste across the globe. In fact, it's the primary responsibility of all officials, regulators, and councilpersons to come up with an open framework that will encourage each and every stakeholder in the ecosystem to adopt practical steps towards minimising electronic waste. The governments must also offer attractive financial benefits to both manufacturers and customers so that these stakeholders feel encouraged and inspired to come forward and take proactive steps to reduce the harmful impact of electronic waste. It is also the responsibility of global multilateral organisations to spread awareness among the masses to responsibly use electronic gadgets so that synergy can be derived from the efforts of all stakeholders involved in the process. It is important to understand that to effectively manage electronic waste, each stakeholder including society must come forward and play a constructive role in managing the entire system in an effective and efficient manner. The desired results of effective e-waste management can be achieved only when everybody recognises the pressing need to address this issue and put in collaborative efforts in the direction that will lead to the success of the process in the long term. It is only through the joint efforts of every stakeholder the process of e-waste management will meet its desired goals and become a benchmark for all others to follow.

REFERENCES

Awasthi, A. K., Wang, M., Wang, Z., Awasthi, M. K., & Li, J. (2018). E-waste management in India: A mini-review. *Waste Management & Research*, *36*(5), 408–414. doi:10.1177/0734242X18767038 PMID:29732961

Cui, J., & Zhang, L. (2008). Metallurgical recovery of metals from electronic waste: A review. *Journal of Hazardous Materials*, *158*(2-3), 228–256. doi:10.1016/j.jhazmat.2008.02.001 PMID:18359555

Ganguly, R. (2016). E-waste management in India-An overview. *International Journal of Earth Sciences and Engineering*, *9*(2), 574–588.

Garg, N., & Adhana, D. (2019). E-waste management in India: A study of the current scenario, International Journal of Management. *Technology and Engineering*, *9*(1), 2791–2803.

Herat, S. (2021). E-waste management in Asia Pacific region: Review of issues, challenges and solutions. *Nature Environment and Pollution Technology*, *20*(1), 45–53. doi:10.46488/NEPT.2021.v20i01.005

Kulshrestha, D., Tiwari, M.K., Shalender, K., & Sharma, S. (2022). Consumer Acatalepsy Towards Buying Behaviour for Need-Based Goods for Sustainability During the COVID-19 Pandemic. *Indian Journal of Marketing*, *52*(10), 50–63, doi: 10.17010/ijom/2022/v52/i10/172347

Ravindra, K., & Mor, S. (2019). E-waste generation and management practices in Chandigarh, India and economic evaluation for sustainable recycling. *Journal of Cleaner Production*, *221*, 286–294. doi:10.1016/j.jclepro.2019.02.158

Shalender, K., & Yadav, R. K. (2019). Strategic Flexibility, Manager Personality, and Firm Performance: The Case of Indian Automobile Industry. *Global Journal of Flexible Systems Managment*, *20*(1), 77–90. doi:10.100740171-018-0204-x

Tesfaye, F., Lindberg, D., Hamuyuni, J., Taskinen, P., & Hupa, L. (2017). Improving urban mining practices for optimal recovery of resources from E-waste. *Minerals Engineering*, *111*, 209–221. doi:10.1016/j.mineng.2017.06.018

Vyas, V., Mehta, K., & Sharma, R. (2023). The nexus between toxic-air pollution, health expenditure, and economic growth: An empirical study using ARDL. *International Review of Economics and Finance*, *84*, 154–166. doi 10.1016/j.iref.2022.11.017

Wang, X. (2019). System Simulation Optimization of Resource and Environmental Effects of Circular Economy. *Nature Environment & Pollution Technology*, *18*(5), 1573–1578.

Chapter 7
Waste-to-Materials/Energy: A Need for Waste Management to Achieve Sustainable Development Goals

Rakesh Ashok Afre
https://orcid.org/0000-0002-7412-6437
Assam down town University, Guwahati, India

Gopal K. Upadhyay
Assam down town University, Guwahati, India

Sunandan Baruah
Assam down town University, Guwahati, India

Milind M. Kulkarni
Ajeenkya D. Y. Patil University, Pune, India

ABSTRACT

The increasing urbanization in India has led to the development of an environment that is characterized by the lack of proper management of solid waste. This issue contributes to the development of health hazards and the degradation of the urban environment. This chapter aims to provide a comprehensive understanding of the various parameters of solid waste management to achieve the sustainable development goals. Apart from this it aims to provide a comprehensive analysis of the current status of solid waste management and its conversions to various useful things. It also explores the various opportunities that can be achieved through the establishment of effective and efficient waste processing units.

INTRODUCTION

The management of municipal solid waste is a major concern for many countries around the world. This issue is why it is important to study it to find out what can be done better. The management of solid waste is a major concern for many countries around the world. Due to the large amount of waste that is

DOI: 10.4018/978-1-6684-8117-2.ch007

generated each day, it is very important that the government and private sectors work together to reduce the amount of waste. Some of the harmful substances that can be used in the treatment of municipal solid waste include pesticides, toxic chemicals, and medicines. In various countries, such as India, the management of solid waste is a major challenge due to the increasing population and the improving lifestyle. Despite the various development activities that have occurred in the country, the management of solid waste systems in India has remained the same (World Bank, 2018). This is because the lack of proper management of the waste has been identified as a threat to the environment and human health. According to the World Health Organization, the lack of proper disposal of solid waste can lead to various health issues (Malinauskaite et al., 2017). Studies have shown that proper utilization can be both desirable and not an option. In India, there is a problem with the segregation of construction and e-waste, as well as the proper disposal of plastic, industrial refuse, demolition debris, and commercial waste. In total, over 12 million tonnes of construction and demolition debris, as well as various other types of solid waste, are generated each year in India. It is very important that we properly utilize the waste that comes from our society. This is because it has a huge potential to transform into valuable resources and energy. There are various techniques that can be utilized to recover this resource, such as energy-from-waste conversion (Nanda and Berruti 2020). Various types of waste-to-energy technologies, such as biological and thermochemical conversion, can be utilized to transform municipal solid waste into various fuels. Incineration is the most common method of converting municipal solid waste into energy. However, it can also significantly reduce the volume of the waste.

In order to achieve the Sustainable Development Goals, it is very important that the management of solid waste (MSW) is carried out in a way that is eco-friendly (Rodić, L., & Wilson, D., 2017). There are various ways that can be used to convert the waste into useful products such as biofuels and nanomaterials. However, the major issue that needs to be resolved is the handling of biomedical waste, plastic waste, and e-waste (MSW rules, 2016). In wet/green waste, the abundant carbon atoms can be transformed into nanomaterials because of the presence of hydrocarbons. For instance, after heating a corn cobs and sugarcane bagasse in an inert gas atmosphere, the former gets converted to porous carbon, which can be utilized in the desorption or adsorption of greenhouse gases (Iwuozor et. al 2022). The management of solid waste (MSW) is expected to undergo significant changes due to various factors. Some of these include the increasing emphasis on the role that consumers play in the creation and treatment of waste, as well as the development of new technologies that treat it as a resource. Despite the various factors that have been identified as contributing to the increasing complexity of the waste management process, the challenges that remain are still significant. Some of these include the integration of the informal sector into the planning process for solid waste management in developing countries, the use of standardized definitions for waste, and the reduction of greenhouse gas emissions from solid waste. In less-industrialized cities, the informal sector has a wide variety of characteristics. As cities look to develop their solid waste management plans, they need to consider the various trade-offs that need to be made between the environmental and public health benefits of utilizing the services of the informal sector. One of the most effective ways to reduce the generation of solid waste is through the reduction of materials at the source. This can be done through the establishment of effective reuse opportunities at the household or community level (Shazwin and Nakagoshi, 2010, Yadav and Devi, 2009).

The chapter will focus on possible conversion of MSW and practical approach of utilization to day-to-day activities to achieve the sustainable development goals. This will help to students, teachers, researchers and industries as a reference guide and do implementation of various possible conversion/ utilization of MSW.

Table 1. Types of waste and its contents

Type of waste	Content of solid waste
Biodegradable	Kitchen waste, green waste (unused parts of vegetables and fruits), garden waste (leaves, shoots etc), wooden furniture
Recyclable	Paper, cardboard, cans, bottles, glass, plastics etc.
Composite waste	Tetra packs, cloths, hard plastics
Harmful household & toxic waste	shoe polish, medicines, paints, spray cans, shoe polish, chemicals, diapers
Inert waste	Construction and demolition waste, metals bars, wood, concrete, discarded furniture, Stones, silt, ash etc.
Electronic waste	Light bulbs, fluorescent tubes, batteries, computers, mobiles, wires,
Medical waste	Syringe, bandages, medicines, saline bottles, cotton swabs, etc.
Liquid Waste	water used for different industries, household, tanneries, distilleries, thermal power plants

METHODOLOGY

The literature search was conducted using the Google Scholar, Web of Science and through google search. These platforms store the largest citation databases, and the dates of the searches varied from 2000 to 2022 and relevant literature were sorted out to study and understand the qualitative approach in this MSW domain.

OUTLOOK FOR MSW GENERATION AND ITS IMPACT

Municipal solid waste is waste that contains domestic waste predominantly from household in addition of wastes from commercial, sanitation, construction, debris from demolition and streets waste in and around cities. It is in the form of solid/semisolid form with excluding harmful industrial waste and it has been categorised into various categories as mentioned below:

Municipal or household wastes can be generated from different sources depending on the activities of people. Most studies indicate that the solid waste generated from developing countries is mainly from households. The later refers to the variable quantity of solid waste that can be generated from various sources such as factories, streets, institutions, and other similar facilities. Due to their heterogeneous nature, the solid waste produced from these sources has varying chemical and physical characteristics. Some of these include food waste, yard waste, leather, metal, glass, batteries, textiles, paper, and wood. Due to the nature of the solid waste produced from these facilities, it is very important that the various components of the waste are sorted and separated before they are used for any purpose. This method is one of the most important steps in the solid waste management process.

The success of solid waste segregation depends on the public's involvement and the proper implementation of principles and procedures related to its sorting and separation. The generation of solid waste (SWG) is a global issue that is considered to be a major concern for developing countries. Due to the large amount of SWG that are produced in urban areas, the resulting pollution can severely affect the environment and the basic services provided in these regions. Studies have shown that solid waste management is a major issue in urban areas and cities. Due to the low level of achievement in this field,

Figure 1. UN 2015-the sustainable development goals (SDGs) agenda by 2030
(adopted from the UN 2015)

many countries and regions, including East and North Africa, are experiencing serious problems with this issue. Due to the limited resources available, many developing countries are not able to effectively address the issue of solid waste management. This is considered one of the biggest problems faced by cities all around the world. (Okot-Okumu, J., & Nyenje, R. (2011; Al-Khatib el. al., 2010; Valkenburg et al., 2008; Nabegu, A. B. 2010)

MSW decay and direct combustion in open space contribute to various air and environmental pollution. They also cause biological and chemical contamination in the groundwater and soil. Due to the decomposition of the organic waste, methane gas can contribute to the global warming and environmental change. High organic content in the waste can increase the growth of microbes and contribute to the development of chronic and infectious diseases in people who live near the site of solid waste. These include waste pickers and residents of nearby areas. In addition, burning of the waste can cause various respiratory and skin conditions. In 2013, a study conducted by Jayakrishnan et al. revealed that over 70% of the waste collectors in Kolkata and 25% in Mumbai were suffering from respiratory ailments. Complications such as cancer and chemical poisoning can also be caused by the improper disposal of electronic and hazardous waste. The mixing of biomedical products and household waste with solid waste can increase the risk of getting HIV and Hepatitis B (Cremiato et. al., 2018; Jayakrishnan et. al., 2013; Pathak et. al. 2017).

Sustainable Development Goals Pertaining to Solid Waste Management

One of the most challenging issues that countries face when it comes to achieving the 17 Sustainable Development Goals (SDGs) is the implementation of regulations that address the negative impacts of urbanization on the environment and society as mentioned in the fig.1. The objective of this study is to promote the implementation of solid waste management plans and programs that are designed to meet the SDGs. Through an analysis of the various factors that affect the implementation of these plans and programs, it was able to identify the main factors that contribute to achieving these goals. In addition to analyzing the multiple targets and strategies that are related to these plans and programs, it was also able to identify the most and least affected groups of people by these plans and programs. According to the experts, the main goals of sustainable communities and cities and good health and well-Being are the ones that are most likely to be achieved through the implementation of SWM programs and plans. However, other goals such as quality education and peace and justice are also likely to be affected by these plans and projects (Elsheekh et. al. 2021; Jerath, 2021).

The SDGs are a set of goals that aim to improve the well-being of people and the environment. They also address the various ways in which these changes can be induced. These goals are additionally designed to enable a secure and peaceful environment. The different sectors of development can make a significant contribution to the achievement of the SDGs. One of these is solid waste management. This sector can play a vital role in helping to achieve the goals of the Sustainable Development Agenda mentioned in Table 2 below (Rodić, L., & Wilson, D. 2017).

CURRENT CHALLENGES AND IMPLICATIONS IN MSW MANAGEMENT

MSW collection and segregation is one of the major challenges. Non-biodegradable materials such as glass and plastics can affect the treatment process of municipal solid waste. They can also increase the operating and labour costs associated with sorting and segregation at the facility. These materials can be used for various recycling processes, such as gasification, liquefaction, or pyrolysis. Without the proper segregation of mixed organic, plastic, glass, paper, textiles, and metal waste streams, the waste management systems of the future will not be feasible. Due to the varying types of solid waste streams, industries, households, and institutions should be responsible for segregating them. The residents should be aware of the various types of waste that can be classified into different categories. These include non-recyclable, recyclable, electronic waste, paper, glass, plastics, composites, and pharmaceutical wastes. Unfortunately, the collection methods of large municipal solid waste trucks are not always able to handle these types of wastes. This is why it is important that the various agencies handling solid waste provide residents with clear and legible containers and bags for their garbage collection (Narayana, 2009; Priti, & Mandal, K., 2019).

In order to improve their operations and expand their market share, the recycling industry's major players should develop new technologies and improve their customer relationship management. They should also expand their product standards and develop effective marketing strategies. For instance, many solid waste management companies in developed countries provide environmental and management solutions for both residential and industrial wastes. Some of the major companies in the solid waste management industry operate various facilities such as transfer stations, collection operations, and landfills. They also provide customers with up-to date information about the various activities of their operations (Kumar et. al., 2009) [20].

WASTE TO ENERGY/MATERIALS GENERATION

One of the most critical factors that contributes to the development of a society is energy/materials generation. Today, there are various barriers that prevent the efficient utilization of waste. One of these is the lack of awareness among the stakeholders in the supply chain. Various methods can be utilized to recover energy from biowastes, such as through the treatment of these materials in anaerobic digestion.

Table 2. Specific targets of the sustainable development goals related to MSW management (Source: Adapted from Rodić, L., & Wilson, D. (2017) & THE 17 GOALS | Sustainable Development (un.org).

Sustainable Development Goals (SDGs)	Specific Targets
SDG 3: Good health and well being	3.2 End preventable deaths of children under 5 years of age. 3.3 End the epidemics of AIDS, tuberculosis, malaria and neglected tropical diseases and combat hepatitis, water-borne diseases and other communicable diseases. 3.9 Reduce the number of deaths and illnesses from hazardous chemicals and air, water and soil pollution and contamination
SDG 6: Clean water and sanitation	6.3 Improve water quality by reducing pollution, eliminating dumping and minimizing release of hazardous chemicals and materials, halving the proportion of untreated wastewater and substantially increasing recycling and safe reuse globally
SDG 7: Affordable and clean energy	7.2 Increase substantially the share of renewable energy in the global energy mix
SDG 11: Sustainable cities and communities	11.1 Ensure access for all to adequate, safe and affordable housing and basic services and upgrade slums 11.6 Reduce environmental impact of cities, including by paying special attention to air quality and municipal and other waste management
SDG 12: Responsible consumption and production	12.1 Sustainable consumption and production national action plans or targets incorporated into national policies 12.2 Sustainable management and efficient use of natural resources. 12.3 Halve per capita global food waste at the retail and consumer levels and reduce food loss during production, supply chains, and post-harvest steps 12.4 Environmentally sound management of chemicals and waste to minimize their adverse impacts on human health and the environment 12.5 Promotion of 3Rs to substantially reduce waste generation
SDG 13: Climate action	13.2 Integrate climate change measures into policy and planning 13.3 Build knowledge and capacity to meet climate change
SDG 14: Life below water	14.1 Prevent marine pollution, particularly from land-based activities

It is important that the public encourages the use of biomass instead of using petroleum products. In 2013, a study conducted by a team of researchers revealed how to maximize the biomass supply. Bio-energy can be used in sectors where there is high social demand. In 2017, a study conducted by Garcia and Hora noted that although wood wastes are typically burned directly to produce energy, the use of wood waste in European countries is expected to increase due to the recently enacted renewable energy act (Keegan, 2013; Garcia, & Hora, 2017).

Processing Mechanism and Technologies of Waste to Energy/Materials: An Indian Scenario

Technologies Available

Through the use of waste-to-energy technologies, such as electricity and biogas, we can recover the energy from our waste:

Biomethanation

Anaerobic digestion is a process that involves the fermentation of organic materials without the need for free oxygen. This process yields a mixture of methane, carbon dioxide, and other gases. The process can be utilized in a decentralized manner to biodegrade various types of wet wastes, such as those from restaurants, slaughter houses, and institutions. Biogas from this process can be burned either in a gas boiler or a burner to produce electricity or heat for certain thermal applications. The organic materials can be cleaned and transformed into BioCNG by removing the harmful substances and carbon dioxide. This can then be used as a fuel for vehicles or injected into the national grid. Using 20 kg of cattle dung can yield around 1m3 of biomethane, which is equivalent to about 0.4kgs of BioCNG.

Incineration

In addition to burning solid waste, incineration technology involves recovering the heat from the combustion to produce steam, which then powers a turbine. In order to treat the gases produced by the combustion, an air pollution control system has to be installed. The resulting ash can then be used for construction or used as a building material. In India, this technology has been successfully used to treat various types of solid waste, such as industrial and municipal solid waste. It can also generate electricity.

Gasification

The process of gasification involves using high temperatures to break down solid waste materials. It produces synthetic gas, which is a mixture of hydrogen and carbon monoxide, through the decomposition of various materials. Some of the components used in the gasifier include biomass, agro-residue, and segregated municipal solid waste. Gasification of solid waste can help increase the efficiency of the power generation process and reduce the emissions. It is an ideal alternative to thermal treatment of this type of waste.

Pyrolysis

The process of pyrolysis involves breaking down combustible materials using heat. This produces a mixture of gases, such as carbon monoxide, hydrogen, methane, and complex hydrocarbons. The products of this process are usually liquids or solid residue. Pyrolysis can be used to generate heat in a boiler, while the process of decarbonization involves breaking down solid waste. This process can also be used to clean up and enhance combustion turbine generators' output.

(Current Status I Ministry of New and Renewable Energy, Government of India (mnre.gov.in)

Technological Innovation and Best Practices Adopted at State Level

Some of the technological initiatives taken at city level of states like Andhra Pradesh, Ladakh, Karnataka, and Odisha-

New technologies are needed in order to make waste management more sustainable. They can help improve the efficiency of the collection, transportation, and treatment of waste. They can also help prevent waste from developing harmful substances and improve the environment. In India, the development of new technologies related to waste treatment and management is still in its early stages. In order to recover the wealth from the waste, the waste management sector of India should adopt new technology. Some of the cities that have adopted new technology include those that have made use of various types of wireless communication and navigation systems, such as global positioning system. They also use various web-based and mobile applications to improve the efficiency of their waste management processes.

Innovative Models

Some of the innovative models developed at city level of states like Odisha), Maharashtra and Kerala.

In the city of Paradeep, a community-driven model has been adopted, which includes establishing micro-composing facilities and material recovery units. Through the active participation of various groups, such as women's associations and third-gender groups, the city has shown a new approach to waste management. The concept of the project is based on the economic sustainability of the community. The revenue generated from the sale of recycled products and the user fees collected from the users has been higher than the budget. During the period of February to July 2021, the total expenditure of the waste management department was around Rs.20.5 lakh. On the other hand, the revenue generated by the project was around Rs.23.3 lakh. The primary source of revenue for the city is the collection of user fees and fines, as well as the sale of recyclables. The city is able to process about 100 percent of its waste.

E-Waste Management

Jamshedpur (Jharkhand) The rapid growth of electronic equipment has led to an increase in the amount of e-waste. The UN has referred to this phenomenon as a tsunami. Despite its negative effects, e-waste can be molded into various useful products. It is also filled with valuable metals and materials. Although e-waste does not usually fall under the jurisdiction of local solid waste management agencies, it is now time for the government to step up its efforts in addressing the issue. Although most cities do not have plans to address the issue of e-waste, some have already started implementing measures to reduce the amount of this waste.

Biodegradable Waste Management

Best practices adopted for biodegradable waste management at some of the cities of states like Andhra Pradesh, Karnataka and Maharashtra

Cities in India spend a lot of their budget on solid waste transportation and collection. Since we are spending our taxes on fuel to transport water, we should be responsible for managing our organic waste in a way that is close to its source. This is what the cities in India have been doing.

The introduction of zero-waste management facilities in Mysuru made a huge difference in the city's efforts to reduce its solid waste. Since the city has a comprehensive infrastructure for processing biodegradable waste, it is regarded as the leading authority in this field. In order to minimize the amount of waste, Vengurla has adopted a two-bin system that consists of a green bag for non-biodegradable waste and a blue bag for biodegradable waste. The collection of domestic hazardous waste and sanitary waste was carried out separately. The council has also adopted various technologies and systems that can help minimize the amount of waste that it generates. These include the use of biomethanation and vermicomposting, as well as the briquetting of green foliage and biomass waste.

Material Processing

Best practices adopted for material processing at some of the cities of states like Madhya Pradesh, Gujarat, Jharkhand and Odisha

We use various types of materials in our daily lives when it comes to municipal solid waste. Some of these include plastics, paper, metals, glass, and more. The value of the various materials that we use is inversely related to the volume of solid waste. According to the rule, solid waste should be placed in three different types of containers: domestic hazardous waste, non–biodegradable waste, and biodegradable waste. However, at a secondary sorting facility, over 50 different materials have been identified that need to be transported to various facilities for processing.

The city of Surat faced a daunting task when it came to managing its solid waste, as its population grew from 1981 to 1991. To meet the increasing demand for its services, it launched a comprehensive approach that involved implementing a variety of measures such as establishing a 100 percent source segregation facility and improving the efficiency of its operations. In an effort to implement decentralised waste management in the state of Odia, the government has made it mandatory for every local body to adopt this practice. The Durban Development Department has now come up with a standard operating procedure that will help municipal agencies develop their own waste management units. Through the training of local self-help groups, the city councils were able to increase the capacity of their citizens to manage their waste. This resulted in a behavioral change among the residents. By implementing a political will, the municipality was able to achieve a material processing rate of 100 percent.

Plastic Waste Management

Best practices adopted for plastic waste management at some of the cities of states like Goa, Sikkim and Tamil Nadu

Plastics have become an integral part of our lives due to their low cost, lightweight, and versatile properties. However, despite their numerous advantages, these materials have become a cause for concern among urban local governments. Due to the lack of awareness regarding the various aspects of plastic waste management, it has become a daunting task for local authorities and consumers. This is why it is important that they are able to identify the type of plastic that they are dealing with and how it is recycled. Some of India's cities have already taken the lead in addressing this issue.

In response to the growing concerns about plastic waste pollution, the city of Sikkim adopted an alternative strategy. In June 1998, the state became the first in India to ban the use of plastic bags. In 2016, the city also prohibited the use of bottled water in government offices. Besides the ban on plastic utensils and plates, the city also enforced other measures to reduce the use of plastic in the city. These included regular enforcement activities and an awareness campaign. In order to create a plastic-free city, Kumbakonam has mandated source segregation. After the state government banned the use of plastic in 2019, the city immediately established a resource recovery facility. While recyclable materials were being recycled, non-recyclables were converted into fuel and used for cement manufacturing. To achieve a garbage-free and bin-free town, the city has not resorted to shortcuts. Instead, it has focused on a comprehensive waste management system.

Sanitary Waste Management

Best practices adopted for sanitary waste management at some of the cities of states like Maharashtra

One of the most debated and least explored aspects of solid waste management is the channelization and handling of sanitary waste. In 2016, the solid waste management rules were changed requiring that both dry and sanitary waste be handled and sent to the same bin. Initially, it was difficult to manage the city's sanitary waste due to various technical and social issues. Through the efforts of the city administration, it was able to achieve a collection rate of 100 percent. The sanitary waste is then separated and transported to a treatment facility operated by the CBWTF. Through the implementation of a comprehensive enforcement strategy and minimal investment in infrastructure, the city was able to achieve its goals. The Red Dot campaign, which was launched by the Pune city administration, was a unique initiative that brought together the various stakeholder groups to take responsibility for their waste. It was carried out through a well-designed and efficient collection and disposal system. The city is now exploring the use of technology to transform its sanitary waste into valuable products.

Zero-Landfill Model

Best practices adopted for zero landfill model at some of the cities of states like Chhattisgarh, Maharashtra and Kerala

A zero-landfill model is an economically, technically, and environmentally sustainable waste disposal method that is based on the principles of the circular economy. It advocates for the continuous effort to end the reliance on landfills. The goal of a zero-landfill environment is to ensure that the maximum amount of waste is treated and managed through scientific procedures and recycling. This method eliminates the need for new landfills and allows for the reuse and recycling of various materials. It is a multi-stakeholder process that ensures that the waste is segregated and utilized in a manner that is beneficial to the environment.

In 2015, the residents of the city of Ambikapur faced the usual issues of overflowing community bins, garbage mountain, and indiscriminately dumped waste along the roads. After the implementation of the Garbage Clinic Model by the local administration, the city was able to achieve a 100 percent segregation rate. After the waste is collected, it is taken to the Solid & Liquid Resource Management Centre, where it is subjected to secondary segregation. It then undergoes tertiary segregation to extract the remaining inorganic fractions. The site where the legacy waste dump used to be is now being used as a recycling center.

Until 2012, the town of Taliparamba was sending its waste to a landfill site that affected the environment and the local population. After 2012, the city implemented a decentralised waste management system that involved the help of women's groups. Today, almost all households in the town follow a door-to-door collection method. Through the city's acquisition of the dumpsite, the material recovery facility of the town has been established. Moreover, the city has provided over 9,500 households with bio-bins, which are designed to help them compost their waste. This practice has helped minimize the transportation costs and improve the environment.

Figure 2. Effect of MSW on environment

MSW-Effect on Environment

The improper disposal of waste has a significant negative impact on the environment, including the spread of diseases and pollution mentioned in Fig.2 (Di Maria, F., Mastrantonio, M., & Uccelli, R. 2021). Without proper treatment, the organic wastes that are landfilled can contaminate the soil, groundwater, and air. When these materials decompose, they produce methane gas, which is a type of greenhouse gas that can replicate oxygen in confined spaces. One of the most common problems with the management of landfills is the issue of leachate, which can be mixed with groundwater and surface waters. This can occur in older landfills that lack proper liners . In (Giang, Kochanek, Vu, & Duan, 2018) addition, poor landfill management can attract flies and mice that carry diseases. In developing countries, where dumping is common, landfills are often located in metropolitan areas. These facilities can be malodorous, decreasing the value of homes within the area. Although proper engineering and designs can be used to minimize the effects of landfilling, they are rarely utilized in all countries (Njoku, Edokpayi, & Odiyo, 2019).

Incineration, specifically thermal treatment, has been regarded as an unsuitable waste management technique due to concerns about its negative effects on the environment and human health (Assamoi & Lawryshyn, 2012). However, disregarding incineration when considering other methods to dispose of solid waste can lead to even worse problems. Despite the negative effects of incineration, it is still not safe to assume that it is a safe method to dispose of solid waste. Studies have shown that it can produce volatile gases that can negatively affect the environment (Liu, Ren, Lin, & Wang, 2015). (Current Status | Ministry of New and Renewable Energy, Government of India (mnre.gov.in)

Solid Waste to Carbon Materials

Activated Carbon

According to Ahmed, 2017, the production of activated carbon from agro-industrial wastes has been carried out to reduce the cost of treating wastewater. Currently, the products are made from coals, coconut shells, and wood. Porosity and large surface area are some of the desirable features of these materials that allow them to be utilized in the treatment of wastewater. The lower cost of treating wastewater is also beneficial since it is less lucrative compared to other industrial sectors. Over the past decade, there has been a significant increase in the number of studies on the properties of bio-waste materials to remove various contaminants from the wastewater. (Ahmed 2017; Wong et. Al., 2018).

Carbon Nanomaterials (CNMs)

In order to effectively address the issue of solid waste management, new technologies need to be developed. This can be done through the development of new materials and procedures. Another major challenge that the scientific community faces is the creation of a new energy storage device that can be used to address the global energy crisis. Various scientific groups have been addressing these global issues by conducting studies on the efficient management of solid waste. They have also been able to identify new materials and technologies that can be utilized in the energy storage industry. Among the various types of solid waste are waste plastic, industrial waste, agricultural waste, and paper waste. These materials have carbonic skeletons that serve as the structural framework for nano-science. The proficiency in performing the activation can modify the carbonic skeleton's morphology depending on the processing method. Some of these include carbon nanotube, graphene, and quantum dots. Carbon-based nanomaterials known as CNMs have various applications. Solid waste can be utilized as a feedstock for the production of other CNMs. Due to the properties of these materials, they have been able to be used as a feedstock for various applications. These include the development of supercapacitors, fuel cells, bioimaging devices, and solar cells. The numerous research articles that were published during 2014 to 2019 on the various aspects of the conversion of solid waste into CNMs clearly showed the scientific community's ongoing efforts in this field (Spitalsky et. Al., 2010; Sahoo et. Al., 2010; Xie et. Al., 2005) [25-27]. In order to create various carbon materials, Sharon and colleagues have been trying to use plant-based materials such as turpentine oil and camphor. As an extension of their work, they have been exploring the possibility of using oil seeds and plant fibers to make carbon materials. These materials could be used to study how they can store hydrogen. The researchers believe that the process of pyrolysis could be performed without the use of a catalyst. This could make the production of carbon fibers very economical. (Sharon et. al., 1998, 2002; Kumar et. al., 1999).

Graphene

In order to create graphene, which is a type of carbon, the researchers led by Essawy and colleagues crushed the plastic waste bottles that were filled with PET plastic. They then placed the sieved residue inside an autoclave reactor, which was placed inside an electric furnace for about 1 h. The resulting sample was cooled to a temperature of 800 C. They utilized an up-cyclization process to convert the plastic waste into graphene. The researchers were able to create high-quality graphene by introducing an organic modified catalyst. This process was performed inside a tube furnace at a temperature of 700 ^{o}C. The process of converting reeds into porous carbon nanosheets involves the synthesis of graphitic carbon. In the first case, the process of carbonization was carried out at high temperature. After that, the sodium metal ion was activated, which further modified the carbon with an organic molecule. On the other hand, the process of converting the cellulosic reeds into porous carbon nanosheets was done through a hydrothermal method (Essawy et. al., 2017; Gong et. al., 2014; Wang et. al., 2015).

DISCUSSION AND SUGGESTIONS ON MSW MANAGEMENT

Most people in the low-to-mid-income group generate organic waste, while those in the high-income category tend to produce more metal and glass. Municipal solid waste management involves various activities such as recycling, incineration, landfilling, and waste-to energy conversion. The process of solid waste disposal is carried out through the operation of a landfill site, which is an ecological reactor. Some of the factors that are considered when it comes to sustainable landfilling include the thickness of the cover, the collection of leachate, and the gas recovery facilities. There might be resistance from the public when it comes to implementing a new approach to waste management. This is why it is important that the public adopts new ideas and mindsets during any transition. This is because changing their mindsets can take time. The public's trust in the services that are delivered is very important to gain a positive understanding of the issues related to waste management. To improve the efficiency of the waste management process, the public should adopt a new mindset about garbage and how it is generated. This can be done by conceptualizing what garbage is and how it is collected. Aside from technological solutions, the public also needs to adopt new behaviors and mindsets in order to solve the problems related to waste management. The concept of waste avoidance can be shifted from trash generation to a more effective waste management strategy by implementing reduce, reuse and recycle process. A major campaign should be conducted at all levels to raise awareness about the value of waste. The implementation of a more strategic approach to waste management at the household, business, and municipal levels can help decrease the amount of garbage that accumulates in landfills.

In 2014, the government of India introduced the Waste-to-energy policy. It has been estimated that the country can achieve a total energy production of > 1500 MW by utilizing the waste generated by the municipal solid waste. The use of MSW as the feedstock for the production of bioenergy is one of the most promising pathways to achieving the more than 10 gigawatts of bioenergy by 2025. Life cycle analysis (LCA) can be used to estimate the economic and environmental costs associated with the treatment of waste. It is also believed that segregation of waste can be done in combination with the production of electricity and biogas. In India, the use of various methods such as composting, refuse derived fuel, material recovery, and sanitary landfilling can reduce the environmental damages caused by municipal solid waste. It is estimated that a single metric ton of organic waste can generate around

20 to 80 m^3 of bioenergy and 474 kilowatt hours of electricity. Following the assessment of its social, economic, and environmental impacts, the appropriate techniques should be employed (Pujara Y. et. al., 2019). Through the use of the LCA, the government can help address the various social and environmental issues related to the treatment of municipal solid waste. This process can also contribute to the sustainable development of the country. It includes issues such as climate change, sanitation, and public health. In addition, the use of LCA in integrated solid waste management can help minimize the impact of the waste on the environment. Out of the 17 SDGs, about 10 of them are directly or indirectly related to the management of solid waste. In order to minimize the impact of the waste on the environment, it is essential that India's primary information on the LCA is analyzed and collected. Apart from this, more focus should be given on waste to nanomaterials conversion using physical, chemical and biological methods. The derived nanomaterials can have various application in adsorption of gases, toughened surfaces, coatings, paints, concretes and so on.

MSW management systems can help address various SDGs in developing economies. Unfortunately, despite the widespread adoption of solid waste management systems, there is still a lack of awareness about their importance among the public. Unfortunately, the current policies and frameworks do not encourage private-public partnerships to help fund MSW management systems. Instead, they rely on public funding, making it difficult for them to operate efficiently. This is why it is important that policymakers involve all stakeholders in the planning process for MSW systems. In the long run, the implementation of institutionalized waste separation is a key component of a sustainable solid waste management system. It can help accelerate the shift from a linear system of consumption, production, and waste to a circular economy.

CONCLUSION

Municipal solid waste is a unique and challenging type of waste that can be difficult to manage. This issue is often presented to the relevant entities and municipalities. The assessment found that the cost of technology for the generation of energy and materials from this type of waste needs to be more competitive. There is a lack of proper waste infrastructure and the informal sector is contributing to the country's growing waste problem. There are also various issues related to the participation of the public in the waste management process. This is why it is important that the government and the private sectors work together to improve the waste management system in the country. The goal of sustainable waste management is to ensure that the maximum amount of resources are extracted from the waste. This can be done through the development of waste-to-energy/materials facilities and engineered landfills. However, India faces various challenges in this regard. Some of these include the lack of skilled and trained waste management personnel, the waste policy, and the selection of waste technology. In order to avoid further damage to the environment and public health, India needs to meet the various requirements for sustainable waste management, proper segregation, conversion MSW to well utilized forms such as energy, materials and re-building the things.

ACKNOWLEDGMENT

The authors would like to thanks Prof. N. C. Talukdar, Vice-Chancellor, Assam down town University for his constant support and encouragement.

REFERENCES

Ahmed, M. J. (2017). Adsorption of non-steroidal anti-inflammatory drugs from aqueous solution using activated carbons [Review]. *Journal of Environmental Management*, *190*, 274–282. doi:10.1016/j.jenvman.2016.12.073 PMID:28063293

Al-Khatib, I. A., Monou, M., Abu Zahra, A. S., Shaheen, H. Q., & Kassinos, D. (2010). Solid waste characterization, quantification and management practices in developing countries. A case study: Nablus District – Palestine. *Journal of Environmental Management*, *91*(5), 1131–1138. doi:10.1016/j.jenvman.2010.01.003 PMID:20116162

Assamoi, B., & Lawryshyn, Y. (2012). The environmental comparison of landfilling vs. incineration of MSW accounting for waste diversion. *Waste Management (New York, N.Y.)*, *32*(5), 1019–1030. doi:10.1016/j.wasman.2011.10.023 PMID:22099926

Cremiato, R., Mastellone, M. L., Tagliaferri, C., Zaccariello, L., & Lettieri, P. (2018). Environmental impact of municipal solid waste management using life cycle assessment: The effect of anaerobic digestion, materials recovery and secondary fuels production. *Renewable Energy*, *124*, 180–188. doi:10.1016/j.renene.2017.06.033

Di Maria, F., Mastrantonio, M., & Uccelli, R. (2021). The Life Cycle Approach for assessing the impact of municipal solid waste incineration on the environment and on human health. *The Science of the Total Environment*, *776*, 145785. doi:10.1016/j.scitotenv.2021.145785 PMID:33647657

El Essawy, N. A., Ali, S. M., Farag, H. A., Konsowa, A. H., Elnouby, M., & Hamad, H. A. (2017). Green synthesis of graphene from recycled PET bottle wastes for use in the adsorption of dyes in aqueous solution. *Ecotoxicology and Environmental Safety*, *145*, 57–68. doi:10.1016/j.ecoenv.2017.07.014 PMID:28708982

Elsheekh, K. M., Kamel, R. R., Elsherif, D. M., & Shalaby, A. M. (2021). Achieving Sustainable Development Goals from the perspective of solid waste management plans. *Journal of Engineering and Applied Sciences (Asian Research Publishing Network)*, *68*(1), 9. Advance online publication. doi:10.118644147-021-00009-9

Garcia, C., & Hora, G. (2017). State-of-the-art of waste wood supply chain in Germany and selected European countries. *Waste Management (New York, N.Y.)*, *70*, 189–197. doi:10.1016/j.wasman.2017.09.025 PMID:28951150

Giang, N. V., Kochanek, K., Vu, N. T., & Duan, N. B. (2018). Landfill leachate assessment by Hydrological and geophysical data: Case study namson, Hanoi, Vietnam. *Journal of Material Cycles and Waste Management*, *20*(3), 1648–1662. doi:10.100710163-018-0732-7

Gong, J., Liu, J., Wen, X., Jiang, Z., Chen, X., Mijowska, E., & Tang, T. (2014). Upcycling waste polypropylene into graphene flakes on organically modified montmorillonite. *Industrial & Engineering Chemistry Research, 53*(11), 4173–4181. doi:10.1021/ie4043246

Iwuozor, K. O., Chizitere Emenike, E., Ighalo, J. O., Omoarukhe, F. O., Omuku, P. E., & George Adeniyi, A. (2022). A review on the thermochemical conversion of sugarcane bagasse into biochar. *Cleaner Materials, 6*, 100162. doi:10.1016/j.clema.2022.100162

Jayakrishnan, T., Jeeja, M. C., & Bhaskar, R. (2013). Occupational Health Problems of Municipal Solid Waste Management Workers in India. *International Journal of Environmental Health Engineering, 2*(1), 42. doi:10.4103/2277-9183.122430

Jerath, N. (2021). Awareness and attitude of students towards municipal solid waste management to achieve Sustainable Development Goals - A Case Study. *International Journal of Plant And Environment, 7*(01), 78–85. doi:10.18811/ijpen.v7i01.9

Keegan, D., Kretschmer, B., Elbersen, B., & Panoutsou, C. (2013). Cascading use: A systematic approach to biomass beyond the energy sector. *Biofuels, Bioproducts & Biorefining, 7*(2), 193–206. doi:10.1002/bbb.1351

Kumar, M., Kichambare, P. D., Sharon, M., Ando, Y., & Zhao, X. (1999). Synthesis of conducting fibers, nanotubes, and thin films of carbon from commercial kerosene. *Materials Research Bulletin, 34*(5), 791–801. doi:10.1016/S0025-5408(99)00066-5

Kumar, S., Bhattacharyya, J. K., Vaidya, A. N., Chakrabarti, T., Devotta, S., & Akolkar, A. B. (2009). Assessment of the status of Municipal Solid Waste Management in Metro cities, state capitals, class I cities, and class II towns in India: An insight. *Waste Management (New York, N.Y.), 29*(2), 883–895. doi:10.1016/j.wasman.2008.04.011 PMID:18595684

Liu, A., Ren, F., Lin, W. Y., & Wang, J.-Y. (2015). A review of municipal solid waste environmental standards with a focus on incinerator residues. *International Journal of Sustainable Built Environment, 4*(2), 165–188. doi:10.1016/j.ijsbe.2015.11.002

Malinauskaite, J., Jouhara, H., Czajczyńska, D., Stanchev, P., Katsou, E., Rostkowski, P., Thorne, R. J., Colón, J., Ponsá, S., Al-Mansour, F., Anguilano, L., Krzyżyńska, R., López, I. C., Vlasopoulos, A., & Spencer, N. (2017). Municipal solid waste management and waste-to-energy in the context of a circular economy and energy recycling in Europe. *Energy, 141*, 2013–2044. doi:10.1016/j.energy.2017.11.128

Municipal solid wastes (management and handling) rules, 2000. (n.d.). TSPCB. https://tspcb.cgg.gov.in/GOs/Municipal%20Solid%20Waste%20Rules.pdf

Nabegu, A. B. (2010). An analysis of Municipal Solid Waste in Kano Metropolis, Nigeria. *Journal of Human Ecology (Delhi, India), 31*(2), 111–119. doi:10.1080/09709274.2010.11906301

Nanda, S., & Berruti, F. (2020). Thermochemical conversion of plastic waste to fuels: A Review. *Environmental Chemistry Letters, 19*(1), 123–148. doi:10.100710311-020-01094-7

Narayana, T. (2009). Municipal solid waste management in India: From waste disposal to recovery of resources? *Waste Management (New York, N.Y.)*, *29*(3), 1163–1166. doi:10.1016/j.wasman.2008.06.038 PMID:18829290

Njoku, P. O., Edokpayi, J. N., & Odiyo, J. O. (2019). Health and environmental risks of residents living close to a landfill: A case study of thohoyandou landfill, Limpopo Province, South Africa. *International Journal of Environmental Research and Public Health*, *16*(12), 2125. doi:10.3390/ijerph16122125 PMID:31208082

Okot-Okumu, J., & Nyenje, R. (2011). Municipal solid waste management under decentralisation in Uganda. *Habitat International*, *35*(4), 537–543. doi:10.1016/j.habitatint.2011.03.003

Pathak, P., Srivastava, R. R., & Ojasvi. (2017). Assessment of legislation and practices for the sustainable management of Waste Electrical and electronic equipment in India. *Renewable & Sustainable Energy Reviews*, *78*, 220–232. doi:10.1016/j.rser.2017.04.062

Priti, & Mandal, K. (2019). Review on evolution of Municipal Solid Waste Management in India: Practices, challenges and policy implications. *Journal of Material Cycles and Waste Management*, *21*(6), 1263–1279. doi:10.1007/s10163-019-00880-y

Pujara, Y., Pathak, P., Sharma, A., & Govani, J. (2019). Review on Indian Municipal Solid Waste Management practices for reduction of environmental impacts to achieve sustainable development goals. *Journal of Environmental Management*, *248*, 109238. doi:10.1016/j.jenvman.2019.07.009 PMID:31319199

Rodić, L., & Wilson, D. (2017). Resolving governance issues to achieve priority sustainable development goals related to solid waste management in developing countries. *Sustainability (Basel)*, *9*(3), 404. doi:10.3390u9030404

Sahoo, N. G., Rana, S., Cho, J. W., Li, L., & Chan, S. H. (2010). Polymer nanocomposites based on functionalized carbon nanotubes. *Progress in Polymer Science*, *35*(7), 837–867. doi:10.1016/j.progpolymsci.2010.03.002

Sharon, M., Mukhopadhyay, K., Yase, K., Iijima, S., Ando, Y., & Zhao, X. (1998). Spongy carbon nanobeads—A new material. *Carbon*, *36*(5-6), 507–511. doi:10.1016/S0008-6223(98)00060-8

Sharon, M., Pradhan, D., Ando, Y., & Zhao, X. (2002). Electrical and optical properties of semiconducting Camphoric Carbon Films. *Current Applied Physics*, *2*(6), 445–450. doi:10.1016/S1567-1739(02)00154-2

Shazwin, T. M., & Nakagoshi, N. (2010). Sustainable waste management through international cooperation: Review of comprehensive waste management technique and training course. *J. Int. Dev. Coop.*, *16*(1), 23–33.

Spitalsky, Z., Tasis, D., Papagelis, K., & Galiotis, C. (2010). Carbon Nanotube–Polymer Composites: Chemistry, processing, mechanical and electrical properties. *Progress in Polymer Science*, *35*(3), 357–401. doi:10.1016/j.progpolymsci.2009.09.003

Valkenburg, C., Walton, C. W., Thompson, B. L., Gerber, M. A., Jones, S. B., & Stevens, D. J. (2008). *Municipal solid waste (MSW) to Liquid Fuels Synthesis, volume 1: Availability of feedstock and Technology*. OSTI. doi:10.2172/962858

Wang, H., Yi, H., Zhu, C., Wang, X., & Jin Fan, H. (2015). Functionalized highly porous graphitic carbon fibers for high-rate supercapacitive electrodes. *Nano Energy*, *13*, 658–669. doi:10.1016/j.nanoen.2015.03.033

Wong, S., Ngadi, N., Inuwa, I. M., & Hassan, O. (2018). Recent advances in applications of activated carbon from biowaste for wastewater treatment: A short review. *Journal of Cleaner Production*, *175*, 361–375. doi:10.1016/j.jclepro.2017.12.059

Xie, X., Mai, Y., & Zhou, X.XIE. (2005). Dispersion and alignment of carbon nanotubes in polymer matrix: A Review. *Materials Science and Engineering R Reports*, *49*(4), 89–112. doi:10.1016/j.mser.2005.04.002

Yadav, I. C., & Devi, N. L. (2009). Studies on municipal solid waste management in Mysore city — A case study. *Report and Opinion*, *1*(3), 15–21.

Chapter 8
Smart Waste Management: Issues and Adoption Challenges

Jashandeep Singh
Chitkara Business School, Chitkara University, India

Meenakshi Malhotra
Independent Researcher, Belgium

ABSTRACT

Waste has become a major concern in the current era. Most nations across the globe have also become more concerned about a safe and healthy environment, and they are working on the issue continuously. Therefore, waste management has emerged as a major challenge, and smart waste management is proving very helpful in dealing with waste, especially in developing countries. However, the implementation of smart waste management systems is not without its challenges. This chapter aims to identify and analyse the key issues and challenges in smart waste management.

INTRODUCTION

Smart waste management has gained increasing importance in recent years, as it promises to revolutionize the way waste is managed and reduce the negative impacts of waste on the environment. Smart waste management involves the use of technology and data to optimize waste collection, sorting, and disposal, resulting in more efficient and sustainable waste management practices. As we move towards sustainable development, a lot of innovations and fresh infrastructure is being developed to make conventional cities the smart cities. To fulfil this goal, proper management of all the resources needs to be ensured so that a better quality of life for everyone on this planet can be achieved (Albino et al., 2015). Waste has become a major concern in the current era. Most of the nations across the globe have also become more concerned about the safe and healthy environment, and they are working on the issue continuously. Therefore, the waste management has emerged as a major challenge and smart waste management is proving very helpful in dealing with the waste (Verma et al., 2020) especially in the developing countries (Islam, 2015).

DOI: 10.4018/978-1-6684-8117-2.ch008

Emerging economies have been successful in the implementation of the smart waste management, but the developing economies are still struggling with this concern as it is difficult to store and dispose-off the waste generated by the population. The transportation and disposal of the waste is costing heavily to the governments of these nations. Moreover, it is a matter of safe human existence (Tania, 2014). The waste needs to be managed in a way that it saves the cost of management with no or little impact on the existence (Kshirsagar et al., 2018). It is the high time that we handle the waste in a way so that the planet becomes safer to live.

The waste is of numerous types, such as domestic, recyclable, medicinal, toxic and radioactive. The waste can be generated due to the manufacturing of commercial and domestic goods. It is combination of multiple types of garbage and rubbish (Ihsanullah, 2022), which is generated in the form of plastic, glass, metal among other stuff. The annual waste production is around eight billion tons across the globe and the municipal waste alone accounts for two billions tons (Wilson and Velis, 2015).

A smart waste disposal idea has been presented by Saha et al. (2017) using solar powered sensors to monitor and control waste accumulation in a smart bin. It was also discussed that the accumulated waste can also be reduced by processing it multiple times. The remote sensors will share the information regarding the amount of accumulation and the waste will be processed via using wireless communications. The smart bin also functions as an internet access point and it can easily be fitted in to any size and shape of containers. These bins can be tracked using tracking devices and they are also capable to send real time data of the waste filled in them. It is claimed that this approach helps organisations to minimise the number of trash vans, reducing energy usage and increasing the collection time, while minimising the running costs.

Smart waste management is an emerging concept that involves the use of advanced technologies and data analysis techniques to optimize waste collection, processing, and disposal (Ilankoon et al., 2018). While this approach has the potential to improve the efficiency and sustainability of waste management systems, it also presents several challenges that must be addressed to achieve its full potential (Esmaeilian et al., 2018). Implementing smart waste management techniques and systems are still in the initial stage and it has a long way to go (Fuss et al., 2018).

Waste management depends upon the numerous factors and most importantly, it depends upon the economic condition of the country. This is further dependent upon the state governments as it is mostly a decentralised process (Nowakowski and Pamula, 2020). It is not just about the implementation by the centre and state governments but it is majorly dependent upon the citizen' understanding of the concept and their behaviours (Aljerf, 2018). A sustainable solution requires behavioural modifications and demands an overall renewal of the existing waste management systems (Zhang et al., 2019). Therefore, a lot of efforts are required before we actually achieve this goal of waste management. Nonetheless, the modern technologies and products seems efficient to deal with the problem. When it come to the smart waste management, the use of gadgets is a must and it also poses a threat to the data security of the citizens (Schafer, 2014).

Smart Waste Management Benefits

Smart waste management is a modern and innovative approach to dealing with waste. It involves the use of technology and advanced systems to manage waste in a more efficient, environmentally friendly, and cost-effective manner. Smart waste management has numerous benefits that make it an essential aspect of sustainable development.

Firstly, smart waste management helps to reduce the environmental impact of waste disposal. Traditional waste management methods such as landfilling and incineration can have significant negative effects on the environment. Landfills take up a lot of space and can contaminate soil and water sources. Incineration releases harmful pollutants into the air, contributing to air pollution and global warming. Smart waste management technologies such as waste-to-energy systems, recycling, and composting help to reduce the amount of waste that ends up in landfills and incinerators, thus reducing the environmental impact of waste disposal (Shukla, 2016).

Secondly, smart waste management helps to conserve natural resources. Recycling, for instance, helps to reduce the need for virgin materials to be extracted from the environment. By recycling, we can conserve natural resources such as timber, metals, and water. Waste-to-energy systems can also help to reduce the use of fossil fuels for energy production, thus reducing the carbon footprint of waste disposal (AbdWahab et al., 2014)

Thirdly, smart waste management can help to save costs. Traditional waste management methods can be expensive and inefficient. Landfills, for instance, require significant capital investments for construction and maintenance. Incineration is also expensive, and the cost of disposing of hazardous waste can be prohibitive. Smart waste management technologies such as recycling and composting can help to reduce the cost of waste disposal while generating revenue from the sale of recycled materials (Omar et al., 2016).

Fourthly, smart waste management can improve public health. Landfills and incinerators can emit harmful pollutants into the air, contributing to respiratory problems and other health issues. Smart waste management technologies such as waste-to-energy systems can help to reduce the emission of harmful pollutants, thus improving public health (Chen et al., 2018).

Fifthly, smart waste management can create job opportunities. Recycling and composting, for instance, require a significant workforce to collect, sort, and process waste. The development of waste-to-energy systems can also create job opportunities in the renewable energy sector (Patil and Rao, 2014).

Smart Waste Management Barriers

- Lack of awareness about smart waste management is the first hurdle in the execution. Smart waste management technologies are relatively new and requires knowledge and expertise to understand and implement it (Zhang et al., 2019).
- Lack of regulatory pressures also act as a barrier in the proper implementation of the smart waste management systems (Mathews and Tan, 2016). There must be a pressure on the waste management companies to implement the systems.
- Environmental regulations are not as strong as they need to be. Proper law enforcement must be there to tackle with the issue. There must be stronger laws for those who pollute the environment or dump the rubbish anywhere and everywhere (Kassinis and Vafeas, 2006).
- Organisations need enough capacity to innovate for the implementation of these programs. Adoption of technology is not an easy task and requires efforts (Malhotra and Singh, 2013). Technology needs a continuous update of the systems to keep it running. Therefore, waste management organisations must be having that much of capacity to run these systems successfully (Zhang et al., 2019).

- Not just the capacity, but organisations find it difficult to implement those technologies. It is not just one technology to work upon but there is an integration of technologies which makes it difficult for the system implementation (Hannan et al., 2015).
- Another barrier in smart waste management is the need to address social and cultural factors that influence waste behaviour. While technology can help optimize waste management processes, it cannot solve the underlying social and cultural factors that drive waste generation and disposal. Therefore, it is necessary to develop awareness campaigns and education programs that promote sustainable waste management practices and encourage citizen participation. Consumers must also take an active part in the implementation processes. A demand from the stakeholders can make a difference, as they are the ones who are going to benefit the most in the long run (Wolf, 2011; Morrissey and Browne, 2004). Organisations also need to be focused on the long-run benefits instead of the short-run profits (Giunipero et al., 2012).
- There must be some laid down standards and rules for the management of the waste of different kinds like plastic, food, electronics etc.

Smart Waste Management Issues

Many researchers have been talking about the dimensions of waste management which may be including the interdisciplinary studies, which includes engineering, urban sciences, ecology, social sciences etc. There are challenges in the implementation due to its interdisciplinary nature. As it requires an expertise in the multiple areas like waste management, public policy, environmental science, decision sciences, logistics, legislation, and technologies while implementing it. It gets further complex due to the numerous stakeholders including government and governmental agencies, technology providers, consumers, producers, and waste management companies (Zhang et al., 2019).

One of the main issues in smart waste management is the need for reliable and accurate data (Kanwar et al., 2022). To implement a smart waste management system, it is necessary to collect and analyze data on the quantity, composition, and location of waste generated in a given area. This data can be obtained through various sources, such as sensors, cameras, and citizen feedback. However, the quality and consistency of this data can vary depending on the technology used, the placement of sensors, and the level of citizen participation. Therefore, it is essential to develop standardized protocols for data collection and analysis to ensure the accuracy and reliability of the data (Fallavi et al., 2017).

Another issue in smart waste management is the integration of multiple technologies and systems. Smart waste management involves the use of various technologies, such as sensors, Internet of Things (IoT) devices, data analytics software, and communication networks. These technologies must be seamlessly integrated to provide a comprehensive waste management solution. However, the integration of multiple systems can be complex and requires careful planning and coordination (Yadav et al., 2023). Furthermore, the implementation of smart waste management systems requires significant investment in technology and infrastructure. For example, installing sensors and IoT devices to monitor waste generation and collection requires capital expenditure, as well as ongoing maintenance and support costs. Therefore, it is essential to assess the financial feasibility of these systems and identify sources of funding to support their implementation (Andeobu et al., 2022).

Smart waste management systems must be designed to accommodate the needs and preferences of diverse communities (Lu et al., 2015). Waste generation and disposal vary depending on factors such as socio-economic status, cultural practices, and geographic location. Therefore, it is necessary to develop tailored solutions that address the specific needs and preferences of different communities .

Challenges in Smart Waste Management

Smart waste management systems have the potential to revolutionize the way waste is managed, leading to more efficient and sustainable waste management practices. However, the implementation of smart waste management systems is not without its challenges. Some of the challenges are as follows:

Data Management and Analysis

Smart waste management systems rely heavily on data collection and analysis to optimize waste collection and disposal. However, managing and analyzing vast amounts of data generated by various sensors and devices can be challenging. Furthermore, data quality and reliability can also be an issue, as sensors and devices may malfunction or generate inaccurate data, leading to incorrect decisions (Folianto et al., 2015).

Infrastructure

Smart waste management systems require a robust and reliable infrastructure to function effectively. This includes a network of sensors and devices for data collection, as well as software and hardware systems for data processing and analysis. Furthermore, waste collection and disposal infrastructure must be modernized and upgraded to support smart waste management systems (Baldo et al., 2021).

Cost

The implementation of smart waste management systems can be costly, requiring significant investments in infrastructure, software, and hardware systems. Additionally, ongoing maintenance and updates can also be expensive, making it challenging for smaller municipalities and communities to adopt smart waste management systems (Omar et al., 2016).

Public Acceptance

Smart waste management systems often involve changes to waste collection and disposal practices, which may not be immediately accepted by the public. This can include changes in waste sorting and disposal procedures, as well as changes in the frequency and timing of waste collection. Public education and engagement are crucial to ensuring the successful implementation of smart waste management systems (Ajwang et al., 2021).

Privacy and Security

Smart waste management systems rely on the collection and analysis of large amounts of data, which can include personal and sensitive information. Ensuring the privacy and security of this data is crucial, as it can be vulnerable to hacking and other cyber threats. Furthermore, data privacy regulations and standards must be adhered to, which can add complexity and cost to the implementation of smart waste management systems (Sharma et al., 2020).

CONCLUSION

In conclusion, smart waste management has numerous benefits that make it an essential aspect of sustainable development. It helps to reduce the environmental impact of waste disposal, conserve natural resources, save costs, improve public health, and create job opportunities. Governments and other stakeholders should invest in smart waste management technologies to achieve sustainable waste management practices. However, smart waste management presents several challenges that must be addressed to achieve its full potential. The challenges include the need for reliable and accurate data, the integration of multiple technologies and systems, the financial feasibility of implementing these systems, the need to address social and cultural factors, and the need to accommodate diverse communities. Addressing these challenges, such as data management and analysis, infrastructure, cost, public acceptance, and privacy and security, is crucial to ensuring the successful implementation of smart waste management systems. It will further require collaboration between technology providers, waste management authorities, and communities to develop sustainable and effective waste management solutions.

REFERENCES

Abd Wahab, M. H., Kadir, A. A., Tomari, M. R., & Jabbar, M. H. (2014). Smart recycle bin: a conceptual approach of smart waste management with integrated web based system. In 2014 international conference on IT convergence and security (ICITCS),1-4. IEEE.

Ajwang, S. O., Abila, J. O., & Tajero-Dakay, I. (2021). Adoption conceptual model for intelligent waste management in smart cities: Theoretical review. *International Journal of Computer Science Research, 5*(1), 426–440. doi:10.25147/ijcsr.2017.001.1.50

Albino, V., Berardi, U., Dangelico, R.M. (2015). Smart cities: definitions, dimensions, performance, and initiatives. *Journal of Urban Technology, 22* (1), 3e21.

Aljerf, L. (2018). Data of thematic analysis of farmer's use behavior of recycled industrial wastewater. *Data in Brief, 21*, 240–250. doi:10.1016/j.dib.2018.09.125 PMID:30364627

Andeobu, L., Wibowo, S., & Grandhi, S. (2022). Artificial intelligence applications for sustainable solid waste management practices in Australia: A systematic review. *The Science of the Total Environment, 834*, 155389. doi:10.1016/j.scitotenv.2022.155389 PMID:35460765

Baldo, D., Mecocci, A., Parrino, S., Peruzzi, G., & Pozzebon, A. (2021). A multi-layer lorawan infrastructure for smart waste management. *Sensors (Basel), 21*(8), 2600. doi:10.339021082600 PMID:33917255

Chen, W. E., Wang, Y. H., Huang, P. C., Huang, Y. Y., & Tsai, M. Y. (2018). A smart IoT system for waste management. In *2018 1st International cognitive cities conference (IC3)*, 202-203. IEEE. 10.1109/IC3.2018.00-24

Esmaeilian, B., Wang, B., Lewis, K., Duarte, F., Ratti, C., & Behdad, S. (2018). The future of waste management in smart and sustainable cities: A review and concept paper. *Waste Management (New York, N.Y.)*, *81*, 177–195. doi:10.1016/j.wasman.2018.09.047 PMID:30527034

Fallavi, K. N., Kumar, V. R., & Chaithra, B. M. (2017). Smart waste management using Internet of Things: A survey. In *2017 International Conference on I-SMAC (IoT in Social, Mobile, Analytics and Cloud)(I-SMAC)*, (pp. 60-64). IEEE. 10.1109/I-SMAC.2017.8058247

Folianto, F., Low, Y. S., & Yeow, W. L. (2015). Smartbin: Smart waste management system. In *2015 IEEE Tenth International Conference on Intelligent Sensors, Sensor Networks and Information Processing (ISSNIP)*, (pp. 1-2). IEEE.

Fuss, M., Barros, R. T. V., & Poganietz, W. R. (2018). Designing a framework for municipal solid waste management towards sustainability in emerging economy countries-An application to a case study in Belo Horizonte (Brazil). *Journal of Cleaner Production*, *178*, 655–664. doi:10.1016/j.jclepro.2018.01.051

Giunipero, L. C., Hooker, R. E., & Denslow, D. (2012). Purchasing and supply management sustainability: Drivers and barriers. *Journal of Purchasing and Supply Management*, *18*(4), 258–269. doi:10.1016/j.pursup.2012.06.003

Hannan, M.A., Abdulla Al Mamun, M., Hussain, A., Basri, H., Begum, R.A. (2015). A review on technologies and their usage in solid waste monitoring and management systems: issues and challenges. *Waste Management*. *43*, 509e523.

Ihsanullah, I., Alam, G., Jamal, A., & Shaik, F. (2022). Recent advances in applications of artificial intelligence in solid waste management: A review. *Chemosphere*, *309*, 136631. doi:10.1016/j.chemosphere.2022.136631 PMID:36183887

Ilankoon, I. M. S. K., Ghorbani, Y., Chong, M. N., Herath, G., Moyo, T., & Petersen, J. (2018). E-waste in the international context–A review of trade flows, regulations, hazards, waste management strategies and technologies for value recovery. *Waste Management (New York, N.Y.)*, *82*, 258–275. doi:10.1016/j.wasman.2018.10.018 PMID:30509588

Islam, F. S. (2015). The engineers role in climate change mitigation. *Journal of Modern Science and Technology*, *3*(1), 117–124.

Kanwar, V. S., Shetty, A., & Mehta, T. (2022). Water audit of Chitkara University, Himachal Pradesh–A zero discharge campus. *Materials Today: Proceedings*, *69*, 383–386. doi:10.1016/j.matpr.2022.09.057

Kassinis, G., & Vafeas, N. (2006). Stakeholder pressures and environmental performance. *Academy of Management Journal*, *49*(1), 145–159. doi:10.5465/amj.2006.20785799

Kshirsagar, P. R., Chippalkatti, P. P., & Karve, S. M. (2018). Performance optimization of neural network using GA incorporated PSO. *Journal of Advanced Research in Dynamical and Control Systems*, *10*(4), 156–169.

Lu, J. W., Chang, N. B., Liao, L., & Liao, M. Y. (2015). Smart and green urban solid waste collection systems: Advances, challenges, and perspectives. *IEEE Systems Journal*, *11*(4), 2804–2817. doi:10.1109/JSYST.2015.2469544

Malhotra, M., & Singh, J. (2013). Factors affecting the adoption of online shopping in youngsters-an empirical study. *SSRN*, *2*(1), 44–54. doi:10.2139srn.2384552

Mathews, J. A., & Tan, H. (2016). Circular economy: Lessons from China. *Nature*, *531*(7595), 440–442. doi:10.1038/531440a PMID:27008954

Morrissey, A. J., & Browne, J. (2004). Waste management models and their application to sustainable waste management. *Waste Management (New York, N.Y.)*, *24*(3), 297–308. doi:10.1016/j.wasman.2003.09.005 PMID:15016418

Nowakowski, P., & Pamuła, T. (2020). Application of deep learning object classifier to improve e-waste collection planning. *Waste Management (New York, N.Y.)*, *109*, 1–9. doi:10.1016/j.wasman.2020.04.041 PMID:32361385

Omar, M. F., Termizi, A. A. A., Zainal, D., Wahap, N. A., Ismail, N. M., & Ahmad, N. (2016). Implementation of spatial smart waste management system in Malaysia. [IOP Publishing.]. *IOP Conference Series. Earth and Environmental Science*, *37*(1), 012059. doi:10.1088/1755-1315/37/1/012059

Patil, Y., & Rao, P. (2014). Industrial waste management in the era of climate change: a smart sustainable model based on utilization of passive biomass. *Patil, Yogesh and Rao, Prakash (2014) Industrial Waste Management in the Era of Climate Change: A Smart Sustainable Model Based on Utilization of Passive Biomass*. In W. L. Filho (Ed.), *Handbook of Climate Change Adaptation* (pp. 1–13). Springer-Verlag Berlin Heidelberg. doi:10.1007/978-3-642-40455-9_49-1

Saha, H. N., Auddy, S., Pal, S., Kumar, S., Pandey, S., Singh, R., Singh, A. K., Banerjee, S., Ghosh, D., & Saha, S. (2017). Waste management using Internet of Things (IoT). In *8th annual industrial automation and electromechanicalengineering conference*, Bangkok, Thailand. 10.1109/IEMECON.2017.8079623

Schafer, B. (2014). D-waste: Data disposal as challenge for waste management in the Internet of Things. *International Journal of Information Ethics*, *22*, 101–107. doi:10.29173/irie131

Sharma, M., Joshi, S., Kannan, D., Govindan, K., Singh, R., & Purohit, H. C. (2020). Internet of Things (IoT) adoption barriers of smart cities' waste management: An Indian context. *Journal of Cleaner Production*, *270*, 122047. doi:10.1016/j.jclepro.2020.122047

Shukla, R. (2016). Smart waste management. *Asian Journal of Pharm Educational Research*, *5*(2), 48–54.

Tania, F. (2014). Solid waste management of Dhaka City: A socio-economic analysis. *Banglavision*, *13*(1), 91–100.

Verma, V., Chauhan, H., Gupta, D., & Gupta, S. (2020). Different Convolution Neural Network Based Models for Garbage Detection: A Review. In *2020 IEEE International Women in Engineering (WIE) Conference on Electrical and Computer Engineering (WIECON-ECE)* (pp. 117-121). IEEE. 10.1109/WIECON-ECE52138.2020.9397961

Wilson, D. C., & Velis, C. A. (2015). Waste management–still a global challenge in the 21st century: An evidence-based call for action. *Waste Management & Research*, *33*(12), 1049–1051. doi:10.1177/0734242X15616055 PMID:26574579

Wolf, J. (2011). Sustainable supply chain management integration: A qualitative analysis of the German manufacturing industry. *Journal of Business Ethics*, *102*(2), 221–235. doi:10.100710551-011-0806-0

Yadav, H., Soni, U., & Kumar, G. (2023). Analysing challenges to smart waste management for a sustainable circular economy in developing countries: A fuzzy DEMATEL study. *Smart and Sustainable Built Environment*, *12*(2), 361–384. doi:10.1108/SASBE-06-2021-0097

Zhang, A., Venkatesh, V. G., Liu, Y., Wan, M., Qu, T., & Huisingh, D. (2019). Barriers to smart waste management for a circular economy in China. *Journal of Cleaner Production*, *240*, 118198. doi:10.1016/j.jclepro.2019.118198

Chapter 9
An Overview of E-Waste Management:
Health Obstacles and Prospects

Devanshri Bhatt
https://orcid.org/0009-0002-1183-5073
Jain University (Deemed), India

H. R. Swapna
Jain University (Deemed), India

Digvijay Pandey
https://orcid.org/0000-0003-0353-174X
Department of Technical Education, IET, Dr. A.P.J. Abdul Kalam Technical University, Lucknow, India

Nitesh Behare
https://orcid.org/0000-0002-9338-8563
Balaji Institute of International Business, Sri Balaji University, Pune, India

Rashmi Mahajan
https://orcid.org/0000-0001-9082-6874
Balaji Institute of International Business, Sri Balaji University, Pune, India

Priyadarshini Karthikeyan
AIMST University, Malaysia

ABSTRACT

The reuse and recirculation of products and materials are the basis of the concept of the circular economy. A circular economy entails markets that give incentives to reusing products, rather than scrapping them and then extracting new resources. The circular economy model has become highly relevant in recent years, with the electronics industry being one of the sectors that have considered its application. Electronic waste (e-waste), that is, waste arising from end-of-life electronic products similar to computers and mobile phones, is one of the swiftly growing waste aqueducts in this world moment. The developing world has come to the primary destination for the habituated electrical and electronic outfit (EEE) exported by the advanced world, making waste operations critical. Indeed, though numerous studies were conducted on waste operation, veritably many are conducted in developing countries that are significant donors of used EEE. India is among the top five-waste-producing countries in the world with an estimated periodic product of 2 million tons.

DOI: 10.4018/978-1-6684-8117-2.ch009

INTRODUCTION

Electronic waste, as known as e-waste, is generated when any electronic or electrical outfit becomes unfit for the intended use or if it has crossed its expiry date. Electrical and electronic equipment (EEE) has turned out to be the most vital accessory to our everyday life (Kazancoglu, Mangla, & Ram, 2020). Given the potential for growth, this concept is becoming increasingly relevant for e-products which is identified as one of the key focus areas by the recent Circular Economy (Isernia et al., 2019). The increasing population, excessive use of electrical and electronic products, and extreme demand for resources have compelled the linear economy to transform into Circular Economy (Isernia et al., 2019). A circular economy, by definition, covers the entire lifecycle of a product, which requires an optimal combination of business model and design strategy for slowing, narrowing, and closing resource loops (European Commission Communication, 2015). Due to rapid-fire technological advancements and the product of newer electronic outfits, the old bones get easily replaced with new models. This has particularly led to an exponential increase ine-waste in India. India is a big request for electronic bias. E-waste is an arising problem given the volumes ofe-waste being generated and the content of both toxic and precious paraphernalia in them. This presto-growing waste aqueduct is accelerating because the global request for particular computers (PC) is far from achromatism and the average life span of a PC is abating swiftly. Over the formerly three decades, the global request for electrical and electronic outfits (EEE) continues to grow exponentially, while the life span of those products becomes shorter and shorter. Predictably, the number of electrical biases will continue to increase on a global scale as the product of EEE is one of the nippy-growing global manufacturing exertions and microprocessors will be used in ever-adding numbers in quotidian objects. At present, general disquisition one- waste operation in developing countries is still in its childhood.

E-waste is posing a serious challenge in the disposal and recycling of both developed and developing countries. While having some of the world's most advanced high-tech software and attack development installations, India's recycling sector can be called medieval. The ditching of e-waste, particularly computer waste, into India from advanced countries and all this has made e-waste operation an issue of environmental and health concern. Compared to conventional external wastes, certain factors of electronic products contain toxic substances, which can induce trouble to the terrain as well as to mortal health. For case, television and computer spectators generally contain dangerous paraphernalia analogous to lead, mercury, and cadmium, while nickel, beryllium, and zinc can constantly be set up on circuit boards (Joon, Shahrawat, and Kapahi, 2017). In the last two decades, increasing environmental awareness has strongly influenced the relationship between the production and consumption of products, environmental protection, and sustainability (Kumar Awasthi et al., 2018). Due to the presence of these substances, recycling, and disposal ofe-waste becomes an important issue.

A number of people are ignorant of the implicit negative impact of the swiftly adding use of computers, spectators, and boxes. When these products are placed in tips or incinerated, they pose health risks due to the dangerous paraphernalia they contain. The infelicitous disposal of electronic products leads to the possibility of damaging the terrain. As further waste is placed in tips exposure to environmental venoms is likely to increase, performing to elevated risks of cancer and experimental and neurological conditions.CE is considered a new business model able to support more sustainable development by retaining, as much as possible, products, resources, energy, and materials (Bocken et al., 2016). A major auto mobilise of the growinge-waste problem is the short life span of utmost electronic products lower than two times of computers and cell phones. India, moment, is burdened with the enormous volume of e-waste, generated domestically or imported immorally. Over the last numerous decades, India, along with other Asian and African countries, has become a major destination for e-waste exports from OECD countries. In addition, Indians have been generating swiftly adding amounts of e-waste domestically. It's estimated that 75 percent of electronic particulars are disposed due to query about how to manage it.

This study provides a concise overview of India's current e-waste script, namely the magnitude of the problem, environmental and health hazards, and current disposal and recycling operations.

REVIEW OF LITERATURE

K Parajuly and H Wenzel (2017) comprehended a study on the current e-waste management system which suffers from losses in the form of materials as well as product and component functionality, which occur during the collection and processing of EoL products. Their study exclusively involved a product family approach pertaining to e-waste management. They identified the attributes upon which product families of e-products could be built for better EoL management. They also identified the three key improvements in the EoL management chain in order to avoid material and functionality losses like an improved collection system; a pre-sorting and testing platform; family-centric processing of EoL products. They suggested the establishment of a more robust system of categorizing e-products carries the potential of improving the EoL material recovery and facilitating reuse, refurbishment, and remanufacturing.

Sharma M, Joshi S, and Kumar A (2020) ascertained that excessive use of electrical and electronic products and extreme demand for resources have compelled the linear economy to transform into Circular Economy. They have conducted an extensive literature review and expert judgment to identify the most influencing key enablers of e-WM in circular economies.Their study has also highlighted that e-WM can be efficient if it focuses on producing eco-friendly products, developing strict legislation, building a green image, and supporting the producers to implement CE practices. They have found out that the 'Environmental management system' (EMS) is the most significant and important driving enabler to influence all the other existing enablers.

Abalansa S, El Mahrad, Icely J, and Newton A (2021) determined that there are numerous laws and policies for regulating the processing of e-waste at different governance scales such as the international Basel Convention, the regional Bamako Convention, and various national laws but many of these regulations are not fully implemented and there is substantial financial pressure to maintain the jobs created for processing e-waste. India along with many other nations has been selected for a comprehensive study of the transboundary movements of e-waste which includes a systematic review of existing literature, the application of the Driver, Pressure, State, Impact, Response (DPSIR) framework for analyzing complex problems associated with social-ecological systems, and the application of the Life Cycle Assessment (LCA) for evaluating the environmental impact of electronic devices from their manufacture through to their final disposal. In this study, they have suggested multiple techniques of e-waste management that are available to reduce the impact of the bad and the ugly, such as adopting the concept of a circular economy, urban mining, reducing loopholes, and improving existing policies and regulations, as well as reducing the disparity in income between the top and bottom of the management hierarchy for e-waste disposal.

LH Xavier, EC Giese, AC Ribeiro-Duthie, and FAF Lins (2021) expound sustainability as the reuse and recirculation of products and materials on the basis of the concept of the circulareconomy. The circular economy is an innovative proposal that can result in positive impacts such as reduced demand for raw materials, reduced consumption of basic resources, and jobcreation, as well as preventing negative impacts resulting from the exploitation and processing of natural resources. They suggest urban mining, a form of closed-loop supply chain management, that offers an attractive alternative to the management of waste electrical and electronic equipment (e-waste) and, at the same time, as a sustainable way to exploit mineral resources, reduces primary material intake and stimulates the circularity in the supply chain.

MT Islam, N Huda, A Baumber, and R Shumon (2021) comprehended a study involving a global assessment of consumer behavior towards e-waste and its implications for the circular economy. Studying consumer behavior allows the identification of appropriate approaches to achieving a circular economy. The study results show that consumers' disposal and recycling behaviors are the two main areas of research interest in the studies reviewed. In contrast, reuse and repair behavior were investigated to a lesser extent.

S Barapatre, and M Rastogi (2022) deduced the inappropriate handling techniques for electronic waste management and the high potency that causes environmental pollution as well as human health hazards resulting in a global predicament. Waste electrical and electronic equipment management in developed and developing countries (precisely, India) was explored along with befitting avenues concerned with recyclable components utilizing extended producer responsibility resulting in the development of a circular economy. They advocated that curtailing the existing electronic waste management issue would be possible through "sustainable consumption and production" which entails countries necessitating vital transformation for their societies to produce and consume goods. This would be accomplished by the collaborative participation of governments, international organizations, commercial enterprises, and individuals to revolutionize flawed consumption and production patterns.

V Murthy and S Ramakrishna (2022) analyzedThe trending need for smarter electrical and electronic equipment (EEE) is surging globally by the year and is giving rise to huge amounts of outdated EEE going into landfills. Their study was intended to bring out the global best practices in the field of e-waste management, and to shed light on the importance of policy implementation, technology requirements, and social awareness to arrive at a sustainable and circular economy. It examines the gap created among the stakeholders and their knowledge of the roles and responsibilities towards legalized E-waste management. It further illustrates the lack of awareness of extended producer responsibility (EPR) and producer responsibility schemes. They discover that like the agricultural and IT revolution, e-waste is on its way to creating a revolution with its unsustainable methods of processing. They suggest a need to create and enforce mandatory global best practices in developing nations accompanying a common global approach towards the legal framework, EPR, benchmark technologies, licenses or permissions, and transformation of the informal sector is the first and foremost step towards a sustainable future.

RESEARCH GAP

Research pertaining to e-waste management has been directed using the same keywords such as "e-waste", "circular economy", "public health" and "recycling" as formerly searched in this review in the light of Y Kazancoglu, YD Ozkan-Ozen and SK Mangla (2020) asserted that "using prospect theory", "two selves effect" and sustainability are dispensed with sustainability risk assessment point of view. With this in mind, the study explores critical constituents required in providing a better understanding towards e-waste management and circular economy. The goal, therefore, is to provide the most comprehensive elaborations on the connections between various factors and elements within the context of e-waste management.

E-WASTE RECYCLING PRACTICES IN INDIA

Electronic waste(e-waste) generally includes discarded computer observers, motherboards, mobile phones and dishes, compact discs, headphones, TV sets, air conditioners, and refrigerators. About 95 percent of India'se-waste is reclaimed in the informal sector and a formal manner.

Informale-waste recycling includes the disposal of obsolete electronics and other electronic waste(e-waste). This type of recycling is done by individuals or groups that collect, identify, and transport electronic waste in their communities. The unbridled recycling ofe-waste by the informal sector is the main concern in non-OECD countries like India, China, etc. In the utmost cases, the entiree-waste treatment is being carried out in a limited terrain in the informal sector, where there's no control over emigration. The dominance of the informal sector can be explained by the poisonous, yet effective styles used in recycling. Because of the important part of the refurbishment and resale request in India, only a certain quantum of e-waste was eventually reclaimed.

On the other hand, Formal recycling is when a company collects electronic biases and disposes of them duly as waste. Formal e-waste activities are authorized and have been officially documented by governments. These activities include the formal collection of e-waste from private companies or government agencies, involving controlled dismantling and recycling that consider aspects of human and environmental needs requiring the use of sophisticated machines and procedures.Formal recycling programs are generally done in a centralized position where particulars are sorted and put into a designated vessel. Developed countries generally use formal e-recycling installations, which tend to be larger and have better controls for health and environmental protection. These installations generally

Figure 1. Recycling an e-waste product

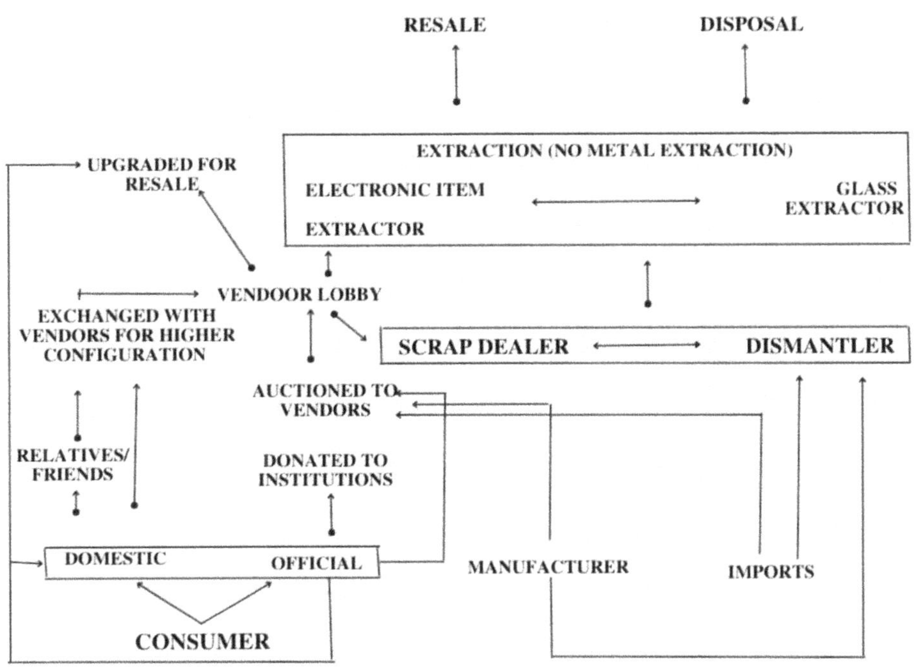

refurbish, resell, or strike and shred electronics into separate recyclable accouterments, similar to plastic and essence. The domination of informal recycling units is apparent in the e-waste recycling script of the country. Formalization is perceived then as an exchange sale between the government and enterprises.In the formal e-waste sector, e-waste generated in private and formal institutions tends to favorthe formal collection and processing of e-waste. Formal e-waste activities are expensive and capital intensive, thus leading to a lower number of formal recycling units in the e-waste sector.

DISCUSSION

Domestic mass production and management of e-waste are influenced by a country's economic situation. The management of electronic waste (e-waste) is challenged by rapid technological advancement with increasing amounts and diversity of electrical and electronic products (e-products). The idea of this study is to review thee-waste produced in India, its sources, current operation practices, and their environmental and health counteraccusations. As seen before the recycling of e-waste is a matter of huge concern. Only about 10 percent of e-waste is reclaimed encyclopaedically. The maturity of this e-waste is being handled by untrained workers in slum areas without any particular defensive outfit, posing a hazard to their health. numerous electronic particulars similar to computers and mobile phones contain dangerous substances like lead, mercury, and cadmium. Recycling or disposing of similar particulars can pose serious pitfalls to mortal health and the terrain. When e-waste is disposed into tips, these poisons can be released into the atmosphere or the girding land and lead to adverse environmental goods. Populations are decreasingly being exposed to dangerous chemicals. These dangerous chemicals are released during product

Figure 2. Chemicals found in e-waste and related health issues
(Grant et al., 2013; Perkins et al., 2014; Annamalai, 2015)

Constituent	Sources	Health effects
Cadmium	Switches, PCBs and solder joints rechargeable batteries, UV stabilizers in older PVC cables and CRTs	Carcinogen; pulmonary, kidneys and bone structure damage
Mercury	Switches and printed solders (as an alloy with tin) and as lead oxide in CRT, lead-acid batteries	Repeated exposure leads to irreversible effects on the nervous system, particularly in children
Lead	Solder in PCBs, glass panels and gaskets in computer monitors	Damage to central nervous system and kidneys, still births and miscarriages, children are more prone
PVC	Cable	Dioxins produced on burning are endocrine disrupters
Antimony	Trace component of metal solders	Possible human carcinogen
Phthalates	Used to soften plastics, especially PVC	Classified as "toxic to reproduction"
Chloro-benzenes	Formed during the combustion of the chlorinated plastic PVC	Persistent and bio-accumulative, hazardous impacts on the liver, thyroid and central nervous system
Barium	CRTs	Muscle weakness and damage to heart, liver, spleen
Beryllium	Circuit boards	Lung cancer and lung damage

manufacture and disposal. Some of these chemicals like heavy essence are poisonous indeed in small quantities and can bioaccumulate in organisms. Indeed, small quantities of these extremely dangerous poisons can negatively affect mortal health. Children are more vulnerable to the health pitfalls of e-waste exposure and, thus, need further protection. As they're still growing, children's input of air, water, and food in proportion to their weight is vastly increased compared to grown-ups, performing under a lesser threat of dangerous chemical immersion (Joon, Shahrawat, and Kapahi, 2017).

However, there are a number of ways to improve e-waste management in India. The government could adopt the following techniques in order to curb the harm caused by the e-waste produced in India:

- Assessment of e-waste: Proper assessment of e-waste (along with the significance), its types and associated health hazards should be carried out (Joon, Shahrawat, and Kapahi, 2017).
- Encourage product design by asking electronic manufacturers to
    a) Commit to product design and target dangerous chemicals: Product contrivers should design products in such a way that they use no/ negligible dangerous chemicals and can be fluently disassembled into the corridor that can be reused, reclaimed, or composted.
    b) Developing a commercial chemical policy program should include the preventative principle and an ongoing commitment to continually ameliorate products with the safest chemicals and accouterments.
    c) Demand responsibility throughout the force chain: Requires suppliers to expose the chemicals used in all factors.
- Carry out health threat assessment studies: Health threat assessment studies are needed for understanding the consequences of the unhappy operation of end-of-life electronic wastes in developing countries.
- Increase consumer mindfulness enterprise to make consumers apprehensive of 'post-life' running of obsolete bias including take-reverse programs enforced by authorities.
- Training of workers: Training of workers involved in e-waste handling operations to avoid occupational health hazards is a prerequisite for safe e-waste operation.

CONCLUSION

India is placed among the other global nations which have generated more e-waste in quantity and especially urban India needs an urgent approach to tackle this issue. After agricultural and IT revolutions, we can now see electronic waste making its way towarda revolution with its fleeting methods of processing. Practices such as illegal jilting of waste, virgin mining, and plasticizing the ocean have disturbed the earth's core, causing an imbalance in nature.It has been observed that the current management practices in India have the potential to adversely impact the health of the recycling workers and the environment as a whole. The recycling activities that are carried out offer more than a considerable amount of danger. As most of the E-waste in India is channelized to the non-formal sector, the formal sector is facing the problem of not having sufficient input materials, and the transformation of the informal sector is the need of the hour to achieve a structured process of e-waste operation.Hence, the evolution of the formal sector is still a leisurely process in the country.

There is a need to produce and apply obligatory global stylish practices in developing nations. A common global approach towards the legal frame, EPR, standard technologies, licenses or warrants, and metamorphosis of the informal sector is the first and foremost step towards a sustainable future. An integrated approach of administering a common legal frame would ensure strong mindfulness among people, including consumers regarding their behavioralpatterns with e-waste.The policy-level initiatives related to e-waste in India could minimize the level of risks with time.Technical and policy-level interventions, implementation and capacity building, and increasing public awareness can convert this challenge into an opportunity to show the world that India is ready to deal with future problems and can set global credible standards concerning environmental and occupational health.

REFERENCES

Annamalai, J. (2015). Occupational health hazards relate to informal recycling of e-waste in India: An overview. *Indian Journal of Occupational and Environmental Medicine*, *19*(1), 61–65. http://www.ijoem.com/article.asp?issn=00195278;year=2015;volume=19;issue=1;spage=61;epage=65;aulast=Annamalai. doi:10.4103/0019-5278.157013 PMID:26023273

Bocken, N. M. P., de Pauw, I., Bakker, C., & van der Grinten, B. (2016). Product design and business model strategies for a circular economy. *J. Ind. Prod. Eng.*, *2016*(33), 308–320. doi:10.1080/21681015.2016.1172124

European Commission Communication. COM (2015) 614/2: Closing the Loop–an EU Action Plan for the Circular Economy; European Commission Communication: Brussels, Belgium, 2015

Grant, K., Goldizen, F. C., Sly, P. D., Brune, M.-N., Neira, M., Berg, M. V. D., & Norman, R. E. (2013). Health consequences of exposure to e-waste: A systematic review. *The Lancet. Global Health*, *1*(6), 350–361. http://www.sciencedirect.com/ science/article/pii/S2214109X13701013. doi:10.1016/S2214-109X(13)70101-3 PMID:25104600

Isernia, R., Passaro, R., Quinto, I., & Thomas, A. (2019). The reverse supply chain of the e-waste management processes in a circular economy framework: Evidence from Italy. Sustainability (Switzerland) https://dx.doi.org/ doi:10.3390/su11082430

Joon, V., Shahrawat, R., & Kapahi, M. (2017). The emerging environmental and public health problem of electronic waste in India. *Journal of Health & Pollution*, *7*(15), 1–7. Advance online publication. doi:10.5696/2156-9614-7.15.1 PMID:30524825

Kazancoglu, Y., Mangla, S. K., & Ram, M. (2020). Risk assessment for sustainability in e-waste recycling in circular economy. *Clean Technologies and Environmental Policy*, *24*(4), 1145–1157. doi:10.100710098-020-01901-3

Kumar Awasthi, A., Wang, M., Wang, Z., Kumar Awasthi, M., & Li, J. (2018). E-waste management in India: A mini review. *Waste Management & Research*, *36*(5), 408–414. Advance online publication. doi:10.1177/0734242X18767038 PMID:29732961

Perkins, D. N., Drisse, M.-N. B., Nxele, T., & Sly, P. D. (2014). E-waste: A global hazard. *Annals of Global Health*, *80*(4). Advance online publication. http://www.annalsofglobalhealth.org/article/S2214-9996(14)00320-8/pdf. doi:10.1016/j.aogh.2014.10.001 PMID:25459330

ADDITIONAL READING

Abalansa, S., El Mahrad, B., Icely, J., & Newton, A. (2021). *Electronic waste, an environmental problem exported to developing countries: The good, the bad and the ugly*. Sustainability. doi:10.3390u13095302

Barapatre, S., & Rastogi, M. (2022). *e-Waste Management: A Transition Towards a Circular Economy*, 1499–1521. Springer. doi:10.1007/978-981-16-4230-2_68

Borthakur, A., & Sinha, K. (2013). Electronic waste management in India: A stakeholder's perspective. *Electronic Green Journal*, *1*(36). doi:10.5070/G313618041

Brigden, K., Labunska, I., Santillo, D., & Allsopp, M. (2005). *Recycling of electronic wastes in China and India: workplace and environmental contamination*. Greenpeace. http://www.greenpeace.org/ international/PageFiles/25134/recycling-of-electronicwaste.pdf

Chatterjee, S. (2012). Sustainable Electronic Waste Management and Recycling Process. *American Journal of Environmental Engineering*, *2*(1), 23–33. doi:10.5923/j.ajee.20120201.05

Greenpeace. (2005). *The e-waste problem*. Greenpeace International. https://www.greenpeace.org/international/campaigns/toxics/electronics/the-e-waste-problem#

Islam, T., Mahmud, M. P., Baumber, A., Shumon, M. R. H., Zaman, A., Ali, F., Hossain, R., & Sahajwalla, V. (2021). A global review of consumer behavior towards e-waste and implications for the circular economy. *Journal of Cleaner Production*, *316*, 128297. doi:10.1016/j.jclepro.2021.128297

Julander, A., Lundgren, L., Skare, L., Grander, M., Palm, B., Vahter, M., & Liden, C. (2014). Formal recycling of e-waste leads to increased exposure to toxic metals: an occupational exposure study from Sweden. *Environ Int. 73*, 243-51. https://www.sciencedirect.com/science/article/pii/ S0160412014002116

Kaushik, P. R., & Herat, S. (2020). Current state of e-waste management in India. *International Journal of Environment and Waste Management*, *25*(3), 322. doi:10.1504/IJEWM.2020.106293

Lieder, M., & Rashid, A. (2016). Towards circular economy implementation: A comprehensive review in context of manufacturing industry. *Journal of Cleaner Production*, *115*, 36–51. doi:10.1016/j.jclepro.2015.12.042

Manomaivibool, P. (2009). Extended producer responsibility in a non-OECD context:The management of waste electricaland electronic equipmentin India. *Resources, Conservation and Recycling*, *53*(3), 136–144. doi:10.1016/j.resconrec.2008.10.003

Maphosa, V., & Maphosa, M. (2020). E-waste management in Sub-Saharan Africa: A systematic literature review. *Cogent Business and Management.*, *7*(1), 1814503. doi:10.1080/23311975.2020.1814503

MoEF. (2008). *Guidelines for Environmentally Sound Management of E-waste*. MoEF.

Murthy, V., & Ramakrishna, S. (2022). A Review on Global E-Waste Management: Urban Mining towards a Sustainable Future and Circular Economy. *Sustainability (Basel)*, *14*(2), 647. doi:10.3390u14020647

Needhidasan, S., Samuel, M., & Chidambaram, R. (2014). Electronic waste: An emerging threat to the environment of urban India. *Journal of Environmental Health Science & Engineering*, *12*(1), 36. doi:10.1186/2052-336X-12-36 PMID:24444377

Parajuly, K., & Wenzel, H. (2017). *Product family approach in e-waste management: A conceptual framework for circular economy*. Sustainability. doi:10.3390u9050768

Robinson, B. H. (2009). E-waste: An assessment of global production and environmental impacts. *The Science of the Total Environment*, *408*(2), 183–191. doi:10.1016/j.scitotenv.2009.09.044 PMID:19846207

Sharma, M., Joshi, S., & Kumar, A. (2020). Assessing enablers of e-waste management in circular economy using DEMATEL method: An Indian perspective. *Environmental Science and Pollution Research International*, *27*(12), 13325–13338. Advance online publication. doi:10.100711356-020-07765-w PMID:32020449

Sinha, S. (2008). Dark shadows of digitization on Indian horizon. In R. Johri (Ed.), *E-waste: Implications, regulations, and management in India* (pp. 23–44). The Energy and Resource Institute.

Skinner, A., Dinter, Y., Lloyd, A., & Strothmann, P. (2010). The Challenges of E-Waste Management in India: Can India draw lessons from the EU and the USA? *ASIEN*, *117*, 7–26.

Turaga, R., Bhaskar, K., Sinha, S., Hinchliffe, D., Hemkhaus, M., Arora, R., Chatterjee, S., Khetriwal, D. S., Radulovic, V., Singhal, P., & Sharma, H. (2019). E-waste management in India: Issues and Strategies. *Vikalpa*, *44*(3), 127–162. doi:10.1177/0256090919880655

Wath, S., Vaidya, A. N., Dutt, P. S., & Chakrabarti, T. (2010). A roadmap for development of sustainable E-waste management system in India. *The Science of the Total Environment*, *409*(1), 19–32. doi:10.1016/j.scitotenv.2010.09.030 PMID:20951410

Watson, A., Brigden, K., Shinn, M., & Cobbing, M. (2010). *Toxic transformers: a review of the hazards of brominated and chlorinated substances in electrical and electronic equipment*. Greenpeace Research Laboratories. http://www.greenpeace.to/publications/ Toxic-Transformers-2010.pdf

Williams, E., Kahhat, R., Allenby, B., Kavazanjian, E., Kim, J., & Xu, M. (2008). Environmental, social, and economicimplicationsofglobal reuseand recyclingof personalcomputers. *Environmental Science & Technology*, *42*(17), 6446–6454. doi:10.1021/es702255z PMID:18800513

Xavier, L. H., Giese, E. C., Riberio-Duthie, A. C., & Freitas Lins, F. A. (2021). Sustainability and the circular economy: A theoretical approach focused on e-waste urban mining. *Resources Policy*, *74*, 101467. doi:10.1016/j.resourpol.2019.101467

Chapter 10
Biomining Method to Extract Metal Components Using Computer-Printed Circuit Board E-Waste

Sudheer Hanumanthakari
Faculty of Science and Technology, ICFAI Foundation for Higher Education, India

Murapaka Dhanalakshmi Bhavani
Aditya College of Engineering and Technology, India

M. D. Mohan Gift
Panimalar Engineering College, India

Kalyan Devappa Bamane
D. Y. Patil College of Engineering, Akurdi, India

K. V. Kanimozhi
Saveetha School of Engineering, Saveetha Institute of Medical and Technical Sciences, India

Sampath Boopathi
Muthayammal Engineering College, India

ABSTRACT

In the present scenario, the e-waste from the various electronic sectors has been increasing due to increased utilization of electric components. In this chapter, the bioleaching(biomining) process of a computer printed circuit board (CPCB) is illustrated to extract the metal components. Basic concepts for e-waste management, their impacts, and various e-waste treatment methods have been explained. The various existing conventional metal extraction methods for the wasted CPCB have also been explored. Definitions, types, cryogenic bioleaching (biomining), influencing factors, and procedures of the bioleaching process have been illustrated. The microbiological methods for the processing of e-waste, the selection of process parameters, and the optimization or maximization of metal extraction processes were demonstrated to promote the e-waste management processes.

DOI: 10.4018/978-1-6684-8117-2.ch010

INTRODUCTION

Equipment that is electrical or electronic is no longer intended for use. One of the constantly expanding solid waste sources is e-waste. Consumers are being compelled to buy newer or current electronic equipment due to factors including a high obsolescence rate, a short life span, the design for dump policy of manufacturers, constant developments in technical applications, and lowered pricing. It is estimated that the globe creates millions of metric tonnes of e-waste annually. Few are known to be adequately collected, processed, and recycled. Under unfavourable circumstances, the maximum is likely to be dumped into the residual garbage. Recycling e-waste can reduce the demand for primary resources, save landfill space, and create jobs. Printed circuit boards (PCBs) contain metals that are 10 times more pure than natural ores. Urban mining refers to the treatment of e-waste as an "urban mine" because of its high metal concentration. Because there are so many valuable metals present, recycling presents a business opportunity. On a global scale, both formal and unofficial industries recycle e-waste. In order to prevent exposure to dangerous metals and poisonous compounds, only small portions are processed by the formal sector, i.e. in a way that is safe for the environment and the employees(Yaashikaa et al., 2022).

The present methods for recovering metals from e-waste are expensive, energy-intensive, and dangerous (for both humans and the environment), and they produce a significant amount of secondary contaminants. In contrast, hydrometallurgical processes produce liquid waste, hazardous fumes, a lot of sludge, and spent acid. The most popular approach for removing metals from e-waste is bioleaching. Utilizing microorganisms' innate capacity to change metallic materials into their watery state is known as "bioleaching." The method has a number of benefits, including minimal chemical sludge production, low energy use, and simple metal recovery.

In this work, microbes are primarily used for the bioleaching of Cu, Au, and Ag from waste CPCBs. Metals like Cu, Zn, and Ni are dissolved in the reaction by ferric ions (Fe3+), which act as oxidising agents and react with the material. Additionally, bacteria use Fe2+ as an energy source and oxidise it to produce Fe3+. Most of the heterotrophic cyanogenic bacteria, including C. violaceum, B. megaterium, and Pseudomonas, have been used in the bioleaching of Au and Ag. Metals were extracted from used CPCBs by fungi in the current study(Yin et al., 2018).

E-waste is poisonous to microorganisms, and cell metabolism affects how much bioleaching occurs. Foreign microorganisms should not be used instead of local microorganisms from polluted locations. The native microorganisms have the physiological and metabolic apparatus to withstand and degrade the contaminant/pollutant. E-waste recycling facilities and mines both contain native microbes that were isolated and tested for bioleaching and toxin tolerance. Like other biological processes, bioleaching is influenced by abiotic and biotic factors that have an impact on the metabolic activity of the microorganisms and, as a result, the recovery of the metals. Temperature, pH, aeration rate, medium make-up, a lixiviant precursor, incubation period, particle size, and e-waste concentration are among the abiotic parameters. Statistical methods like the RSM and standard OFAT can be used to optimise the bioleaching process. The influence of pH and pulp density on gold dissolution from e-waste was investigated by Natarajan and Ting. The effects of pH and glycine content on the bioleaching of gold were assessed. Using the cyanogenic bacteria B. megaterium, they improved the procedure using a response surface approach. From battery wastes to metal-contaminated sediments and mining wastes, bioleaching can be used to remove metals, including precious metals. In order to recover metals profitably, there is an urgent need for intense bioleaching research that focuses on the discovery of novel microbial strains and, as a result, process parameter optimization. The goal of the current research was to investigate new microbial communities that could leach Cu, Au, and Ag from used CPCBs. The bacteria strains are from an abandoned gold mine and an e-waste recycling facility in Haryana, India, and researchers are using them to bioleach metals from used carbon monoxide cathode rays (CPCBs).

All electrical and electronic devices that have reached the end of their useful lives are referred to as "e-waste." E-waste is increasing dramatically in volume. It contains dangerous metals and non-metals, both of which, when treated incorrectly, can have serious effects on the environment and human health. The majority of the world's electronic trash gets recycled, either formally or informally, and the remainder ends up in landfills, endangering both the environment and human health. It offers chances for resource recovery and employability in relation to waste management. In addition to this, it is an "urban mining" source for valuable metals like gold. Traditional methods of recovering metals are used by both the formal and informal sectors. The use of bioleaching, a green method that makes use of the microorganism's innate capacity to convert metals into their soluble forms, is advised. The method is inexpensive and can recycle the smallest amount of metal that is inaccessible to conventional methods(Hubau et al., 2020).

Categories of e-waste: Lamps account for the least quantity of electronic garbage across all continents. The tiniest equipment was thrown away on a global scale. The categories of toys, entertainment, medical equipment, sporting goods, and automatic dispensers have all been combined into one.

Composition of E-waste: The variety of components found in the e-waste stream makes it is difficult to determine a general composition. The two materials that are most prevalent are iron and steel, with polymers coming in second. Many studies have divided e-waste into categories including ferrous and non-ferrous metals, plastics, glass, and other materials. Base metals including Cu, Zn, Ni, and Fe, as well as hazardous and precious metals like Pb, As, Hg, Cr, and Cd, are found in e-waste (Au, Ag, and Pd). Cu and Au are more abundant than the ores. Few electronic wastes are composed of glass, such as the LCD glass from mobile phones and the CRT glass used in televisions and monitor screens.

E-waste as "urban mine": The six metals or metal groups, such as steel, aluminium, copper, gold, silver, and platinum group metals, are primarily responsible for the volume and value of e-waste (PGMs). The most significant and valuable components are PCBs, which account for 40% of the value of the metals. Au and PGMs, which make up the majority of precious metals, are fixed in displays, monitors, and another small IT equipment. According to research, waste PCBs contain higher concentrations of Au, Ag, and Pd than main ores do. The price of extracting and processing the same amount of metal from ore is more than the price of the used cell phones. In poor nations, informal e-waste recycling is a rising industry that creates jobs. Metals with a life expectancy of 20 years or less, such as gallium and indium, can be recovered using it. Developing countries also need to build new treatment facilities to handle the massive volume of electronic trash produced.

E-WASTE RECYCLING AND ITS IMPACTS

Informal Recycling

Only 20% of e-waste is known to have undergone formal processing. The fate of the remaining 80% is unknown and is most likely to be recycled under subpar circumstances or added to residual garbage. The informal sector meets the majority of needs in developing countries such as India, China, and Africa. In contrast to the official industry's potential extraction efficiencies of over 95-99 percent, the informal sector can only recover 25 percent of metals. Polychlorinated dibenzo-p-dioxins (PCDDs), a dangerous toxin, are consequently discharged into the atmosphere(Yin et al., 2018).

Environmental and Health Impacts

Metals such as Cr, Cd, Pb, Cu, Au, and Ag are recovered illegally in the informal sector using the barest amount of safety measures. The recovery of metals under less than ideal circumstances not only contaminates the environment's soil, water, air, and plant life, but also creates a number of health risks. Serious diseases and problems are brought on by pollutants such as PBDEs, PCBs, and toxic metals that are inhaled as well as directly contacted by the skin through dust, soil, and water. CD poisoning or consumption causes infertility, severe lung damage, and, in rare cases, death.

Chronic health impacts in children, such as cancer of the lungs, bladder, kidney, and liver, are caused by high Pb and Cd levels in their blood. Lung cancer is brought on by Hg and Cr found in relays, switches, batteries, and printed wiring boards. Impact assessment studies carried out close to e-waste recycling operations revealed higher levels of PCB and PBDEs in addition to harmful metals.

Formal Recycling

Many businesses have embraced scientific, moral, and environmentally beneficial techniques for recycling e-waste as a result of the implementation of severe legislation. The central pollution control boards (CPCBs) in India have currently approved recycling or dismantling units. Given the demand in this industry, more businesses want to build cutting-edge facilities. It is possible to recycle electronic waste using automated, semi-automated, or manual operations. an increase in the amount of life-threatening metals like Ni, Pb, Zn, and Cd at the Philippines' official recycling facility and a higher concentration of Cu in the soil.

Air samples were analysed for heavy metal pollution using exposure indicators from official Swedish e-waste recycling facilities. Workers from these recycling plants have been found to have brominated flame retardants in their blood and serum samples in addition to metals. It presents a serious risk to those who work in the formal recycling of electronic waste, as well as to the environment, though to a lesser degree. The WHO advises routine health examinations, clean working environments, recurring environmental audits, and impact assessment studies(Hubau et al., 2020).

TREATMENT OF E-WASTES

Informal Treatment of E-Waste

The large and widespread informal sectors are made up of businesspeople who use crude methods to extract metals from e-waste. According to a study, people who work in the informal sector absorb a lot of toxic fumes, which shortens their lifespan.

Incineration: There is a significant risk of producing and distributing pollutants and harmful compounds during the burning of e-waste. This is particularly true when there has been no previous treatment or flue gas purification before burning. According to studies, when flame-retardants are burned, the copper (Cu), which is contained in cables, acts as a catalyst to cause the formation of dioxin. When

Figure 1. Formal metal extraction method

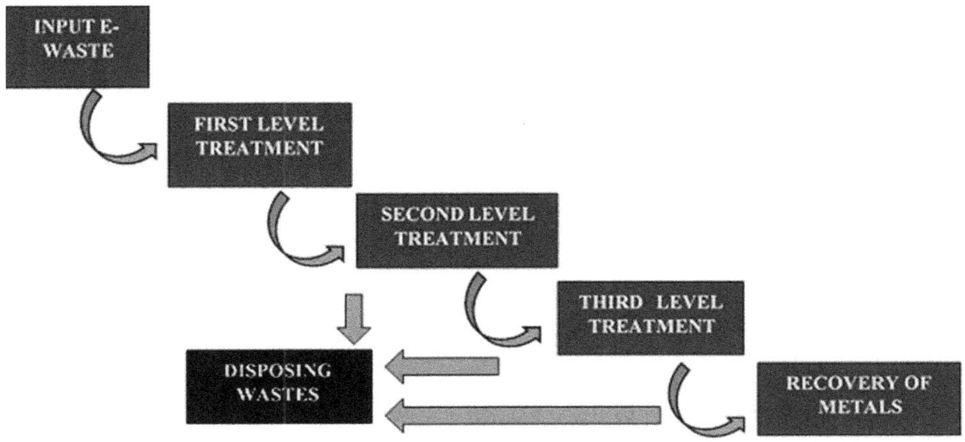

electronic debris is burned, highly dangerous dioxins and polybrominated flame retardants (PBDs and PBDFs), which are present in PVC, can be produced. Additionally, burning causes the loss of priceless trace components that may be recovered by sorting and handling them separately.

Open Burning: Asthma attacks, respiratory infections, and other issues including coughing, wheezing, chest pain, and eye irritation can all be brought on by inhaling emissions from open fires. Due to the low temperatures at which open burning takes place, many more pollutants are released than during a regulated incineration process in an plant. As an illustration, the burning of PVC releases hydrogen chloride, which, when inhaled into the lungs, reacts with water to form hydrochloric acid. Numerous respiratory disorders, as well as lung tissue deterioration, may result from this.

Acid dipping: The most common acids utilised were hydrochloric, nitric, and sulfuric acids. People rely heavily on aqua regia (a mixture of acids in varying ratios) for metal extraction. A small bit of metal is extracted, and the remaining acid is poured into the earth(Arya & Kumar, 2020).

Formal Treatment of E-Waste

Based on material movement, there are three levels of e-waste treatment. Each level of trash treatment consists of a set of unit activities, where the first level of waste is handled and the output becomes the second level of waste handling's input. The leftovers are disposed of or cremated following the third phase of treatment (Figure 1).

First level treatment: At the first level of e-waste treatment, there are three-unit operations. (1) Decontamination: Elimination of all gases and liquids, (2) Breaking down: manual or automated (3) Isolation.

Decontamination: E-waste must first be decontaminated in order to make it non-hazardous. Under negative pressure, all liquids and gases (if any) must be removed.

Dismantling: Sample safety precautions must be taken during operations, whether the dismantling procedure is human or mechanical. The used equipment is disassembled to extract the components from the decontaminated e-waste or those that don't require decontamination.

Segregation: Following disassembly, the parts are divided into hazardous and non-hazardous e-waste fractions and sent for second- and third-level treatment.

Second Level Treatment: Hammering, shredding, which reduces particle size, and special treatment procedures like electromagnetic separation and eddy current separation are all part of this step.

Third Level Treatment: Metal and plastic fractions are separated using specialised techniques, including electromagnetic separation and eddy current separation. The recovery rate of the metal and e-waste fractions for third level treatment depends on how effective this treatment is. The basic goal of this treatment is to recover valuable materials like polymers and ferrous and nonferrous metals.

METAL EXTRACTION

Metals can be extracted using a variety of techniques, such as pyrometallurgy. Some sophisticated nations and their businesses have a very secretive standard operating procedure for the recovery of metals. All of the methods, aside from cry metallurgy, are being studied. Pyrometallurgy's main drawback is the significant energy consumption required to burn the substrate at a high calorific value in order to extract metals. This process produces a significant amount of pollution, necessitating additional treatment tools like scrubbers to remove particulate matter(Işıldar et al., 2016; Liang et al., 2010).

Pyrometallurgy

Pyrometallurgy is the practise of extracting metals from their ores using high-temperature heat. To remove plastics during processing, crushed leftovers are burned in a furnace or molten bath, and the refractory oxides combine with some metal oxides to produce a slag phase. The traditional method for recovering metals from e-waste is pyrometallurgy. To recover the metals, the process entails melting, roasting, drossing, melting, and cremation. This approach has the benefit of accepting all types of electronic waste. Researchers have created a brand-new, cleaner pyrometallurgical method with ultrasonic assistance that has great recovery efficiency and less waste output. Despite the method's limitations, researchers tried a variety of improvements to the procedure to increase yield and lower associated pollutants.

E-waste was handled in a furnace at Umicore's metal smelter and refinery in Belgium to extract valuable metals. In order to recycle unwanted PCBs into a Cu-Ni-Si alloy, a mixed oxide (particularly Pb and Zn), and environmentally benign slag, the business has devised a process. While the remaining dust is processed separately to recover Pb, Sb, In, and Cd, PCBs are fed to a Cu converter in one smelter to extract metals like Ag, Au, Cu, Ni, Pd, Se, and Zn. Many harmful gases and chemicals are known to be released during the pyrometallurgical processes, which seriously pollute the environment.

Pyrolysis

Gases, oils, and chars are produced during the pyrolysis of polymers and can be used as fuels or as chemical feedstocks. The organic component of PCB wastes is broken down during pyrolysis, which makes it considerably simpler to separate the organic, metallic, and glass fibre portions of PCBs. The pyrolytic technique is appealing because it enables the recovery of valuable products from residue, gases, and oils. By using the right remedies, such as the addition of adequate scavengers or dehydrohalogenation, it is possible to stop the evolution of toxics. New technologies are suggested, like depolymerization in supercritical methanol or vacuum pyrolysis.

Vacuum pyrolysis: Solder has been recovered via vacuum pyrolysis, which has also made it easier to separate metals and glass fibres from PCB waste. Additionally, it lowers the temperature of decomposition, which lessens the frequency and strength of secondary reactions.

Depolymerization in supercritical fluids: Methanol has a lower critical temperature and pressure than water, which allows for softer conditions for the degradation of plastics and the removal of metals as a result.

Gasification and co-combustion: Gasification produces carbon monoxide and hydrogen from organic molecules (syngas). Syngas can be used in gas turbines to generate electricity or as a chemical manufacturing intermediary. Gasification is rivalled by combustion and co-combustion, both of which produce electricity.

Hydrometallurgy

Leaching is the process of using a solvent to remove a soluble component from a solid. Leaching of electronic waste necessitates the use of acid and/or halides.A practical method for eliminating base metals and releasing the surface of precious metals is acid leaching. Processes for recovering metal include crystallisation, chemical reduction, and electrorefining. Gold, silver, and palladium were recovered from the cyanide solution by adsorption on activated carbon, whereas silver was removed from the sulphate medium by precipitation with NaCl and palladium by cementation on aluminium.

In the hydrometallurgical process, the e-waste is mechanically crushed to release or expose the metals, which is followed by leaching with an appropriate lixiviant, purifying the leachate, and finally recovering the metals. These techniques benefit from high metal recovery efficiency and low capital costs. Thiosulfate, cyanide, sulfuric acid, aqua regia solution, nitric acid, and thiourea are among the chemical lixiviants used to dissolve the metallic portion of e-waste. Solvent extraction, cementation, ion exchange, and adsorption on activated carbon are used to recover au from leached or pregnant solutions. LCD powder was hydrometallurgical leached for indium (In) using H_2SO_4 at 90 °C for one hour while being stirred at 500 rpm. With NH_4OH and a pH of 7.4, 99.8% of the precipitation was achieved after leaching. When recycling e-waste, there are concerns about the production of toxic, extremely acidic or alkaline effluents, acid vapours, spent acidic medium, and cleaning solvents. This could put workers at risk for acute or chronic exposure hazards as well as environmental contamination(Mäkinen et al., 2015).

Bio Hydrometallurgy

One of the most promising technologies in metallurgical processing is biotechnology. Microbes have the capacity to transfer metal ions into the cell or bind them at the cell surface for a variety of intracellular processes. Depending on how this interaction works, metal recovery could be more or less selective. The method of bioleaching involves transferring metals from their ores into a lixiviant pool. In bioleaching, microbes such as fungi, bacteria, or algae may be employed. The most commonly used microorganisms are Thiobacillus species and Chromobacterium violaceum bacteria. To recover metals from e-waste and low-grade ores, a new method called “biohydrometallurgy” or “biometallurgy” as a bioremediation technique was developed. Bioleaching and biooxidation are steps in the process. Microorganisms work as biocatalysts to extract metals from metal-bearing fractions during the process. During gold mining operations, bio oxidation is frequently employed to purify and eliminate contaminants from the solid materials. Biooxidation is the microbial-assisted oxidation of host minerals that contain valuable metals. Unwanted metal impurities are eliminated as a result, while the valuable metal of interest is left in the solid residue.

The numerous studies concentrated on the commercial bioleaching of metals, including copper, gold, and silver, from low-grade ores and electronic trash. The process of converting or solubilizing metals from solid metallic fractions into their extractable form is known as bioleaching. Since they have been used since pre-Roman times, the Rio Tinto mines in south-western Spain are frequently regarded as the birthplace of biohydrometallurgy. In comparison to conventional approaches, the process of bioleaching has been shown to be a potential economic alternative for the extraction of metals. Similar to biogeochemical cycles, a process enables metal recycling, which lowers energy demand and conserves natural resources like ores and landfill space. Fly ash, industrial trash, and complicated materials like e-waste can all have metals extracted using bioleaching. The advantage of using bio metallurgy to process e-waste is that it can extract metals that can't be recovered using traditional techniques. In the last 30 to 40 years, 20 plants have been marketed, although bioleaching is still in its infancy(Liang et al., 2010).

BIOLEACHING

For many years, copper and precious metals have been recovered from ores using bioleaching. Gold may be biologically separated from PCBs using Chromobacterium violaceum, while copper can be effectively separated from PCB waste using a bacterial consortium enriched from acid mine drainage. The primary inorganic acid present in leaching situations is sulfuric acid. Thiobacilli and other sulfur-oxidizing microbes are responsible for its formation. Fungi are used less frequently than bacteria, yet they are nevertheless renowned for having odd features. The fastest-growing industrial waste in the twenty-first century has been e-waste. Acidophiles that are moderately thermophilic and mesophilic are among the microorganisms that are being studied. These microorganisms get their energy from oxidising ferrous iron and sulphur elements. Microbes can be used to bioleach metals from e-waste, including Cu, Zn, Ni, and Pb. Metals are known to be mobilised as a result of the bacteria that produce inorganic and organic acids (lixiviant). Researchers have suggested a two-step bioleaching procedure for the effective mobilisation of precious metals(Xiang et al., 2010). The various stages of bioleaching technique is illustrated in Figure 2.

Types

There are two kinds of bioleaching, or one and two stages, in which solid waste and microorganisms are both fed at the same time to a medium where microbial growth and metal dissolution are occurring. The presence of metal ions in solid waste causes environmental stress and toxicity, which inhibits microbial development and lowers the effectiveness of lixiviant secretion and metal dissolution. Higher pulp densities could not be tolerated by microbes, which restricts their use in large-scale industrial applications. Additionally, in the two-step bioleaching process, microorganisms are initially grown in a growth medium devoid of solid waste until they reach the ideal cell density. Microbial cells are grown in growth media in a single step until they produce the maximum number of metabolites. The second phase involves extracting metals from solid waste using a cell-free medium that contains biologically generated compounds(Yang et al., 2009).

Figure 2. Various stages of bioleaching technique

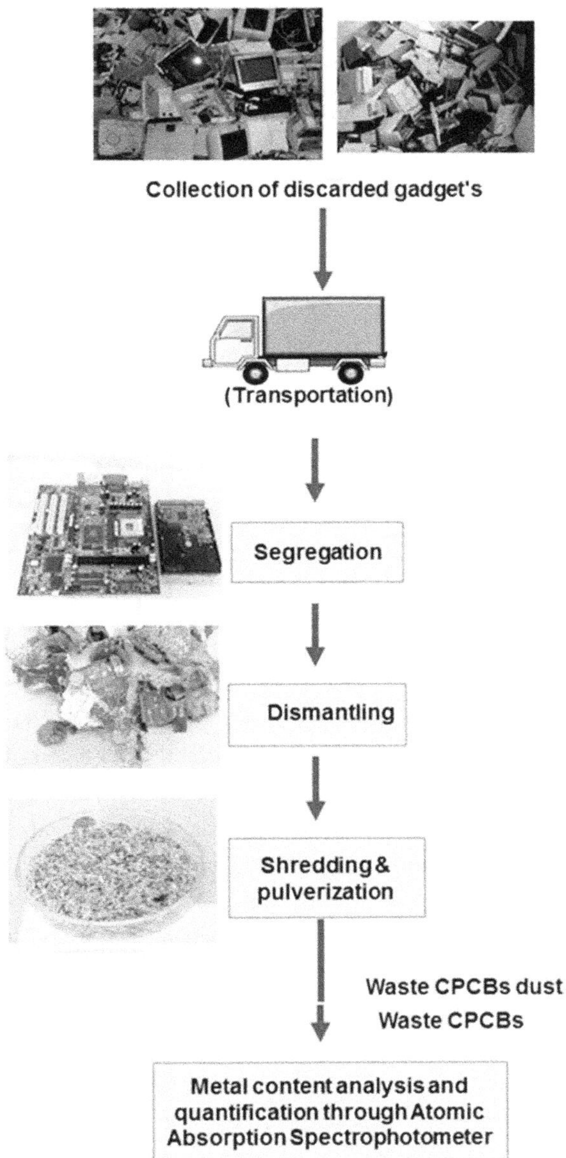

Direct leaching: It uses the organic acids created by the microorganisms to oxidise the hazardous metals that are insoluble, converting them into ions that are soluble.

Indirect leaching: The metal around the microbe is oxidised by bacteria that oxidise metal. Typically, the anion of the target metal compound is oxidized, releasing free metal ions into the aqueous medium. When compared to alternative treatment options, this technique is more time-consuming yet effective and less damaging. nevertheless, leads to bio-leachate (pool of metals). Based on the process, there are

Figure 3. Two stages of bioleaching technique

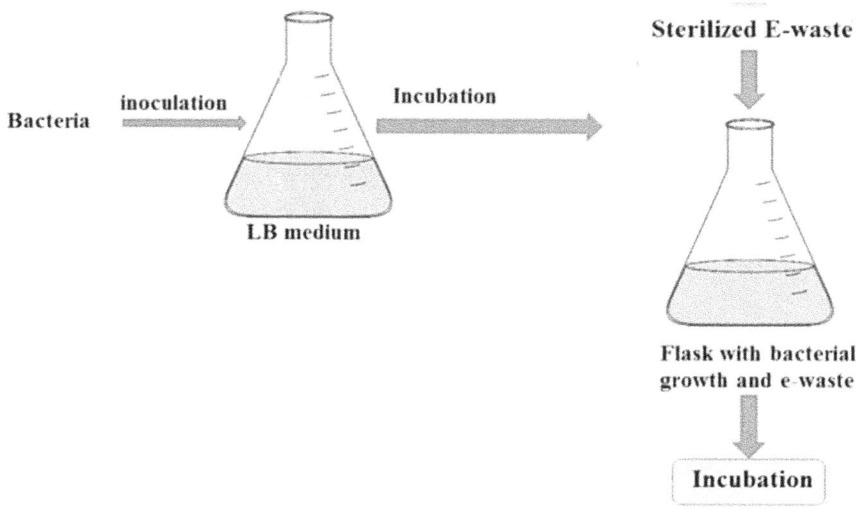

three different types of bioleaching(Awasthi et al., 2016; Işıldar et al., 2016). Two stages of bioleaching technique is represented in Figure 3.

One step bioleaching: Samples and microorganisms were combined, then incubated.

Two step bioleaching: Initially, microorganisms were injected. The sample was then introduced and incubated after that.

Spent medium Bioleaching: Inoculated microorganisms were cultivated until the early stationary phase. The spent medium was then combined with the sample after the microorganisms had been screened.

Biosorption: Bio sorbents are made from the biomass of algae, fungi, or bacteria that is naturally plentiful and/or waste. The biosorption-based technique has a number of benefits over traditional methods. Low operating expenses, a reduction in the amount of chemical and biological sludge that must be treated, and excellent detoxification efficiency. By selecting a proper biomass from the billions of microorganisms and their derivatives, it is possible to achieve the biosorption of precious metals like gold, silver, and platinum. Gold has received the majority of attention in biosorption studies, but silver and other metals should also be recovered from solutions with greater care(Liu et al., 2016; Priya & Hait, 2017).

Reactor: Only on a lab scale has the potential of microorganisms for metal extraction been used. By scaling up the reactor setup, steps should be taken to replace acid leaching and protect the informal recyclers. The bioleaching method has utilised a variety of reactor types. Due to their ease of construction, stirred tank reactors were frequently used in bioleaching research. If the reactor has a separate compartment for the Water Phosphorus Control Biosciences (WPCB), adding WPCB to the reactor would be simple.

Activated Carbon: A microporous inert carbon with a sizable interior surface is activated carbon. Organic molecules from gases or liquids can be adsorbed on the surface. Numerous chemical compounds and oxidising substances can be adsorbed by activated carbon. It can also remove certain heavy metals from water and other liquids. The main development of microspores during carbon activation produces a vast surface area for adsorption to take place. Activated carbons are well known for their numerous applications in removing contaminants in all three phases of matter—organic, inorganic, and inorganic. It has been tested on a wide basis as well as in pilot studies. It was known to adsorb anything with its active site, making it one of our prehistoric methods for cleaning utensils. The pores in activated carbon assist in adsorbate capture, resulting in pure retentate. Calcium, potassium, carbon, oxygen, magnesium, and potassium are abundant in orange peel. It is generally known that orange peel activated carbon may absorb nickel from the electroplating industry. Orange peels have negative zeta potential when exposed to metals and are effective in adsorption. By subjecting the orange peels to various acids, it is possible to improve the effectiveness of the adsorption capacity of particular metals. With a slight modification using magnetic nanoparticles on the surface since they were discovered to be rich in hydroxyl and carboxyl groups, it has also been used in the removal of cadmium. Activation of activated carbon can be done in a number of ways, including thermally, chemically, and with microwave assistance. These activations aid in boosting the carbon's adsorption capacity, allowing pollutants in the adsorbate to be absorbed. By putting the activated carbon in a microwave at a specific frequency, the pore size of the carbon was enhanced(Ji et al., 2022).

Cyanogenic Bioleaching of E-Waste

Au and Ag are frequently recovered from ore concentrations via cyanidation. At the moment, cyanide is used for gold recovery in about 90% of Au mining operations. Recently, a different strategy has been put into consideration where the microbe is used in place of the chemical cyanide. The word "bio-cyanidation" refers to the process of leaching metals using biogenic cyanide; it is also referred to as "alkaline bioleaching" or "heterotrophic bioleaching." The majority of the bacteria that make up the cyanogenic microorganism, which produces hydrogen cyanide (HCN) as a secondary metabolite, are found in soil microflora. In the early stationary phase of bacterial development, cyanide is normally produced. The HCN synthase enzyme converts glycine into HCN, but the amount that can be produced is constrained. In addition to producing cyanide, cyanogenic bacteria have the enzymatic resources needed to detoxify cyanide. The most common method for gold dissolution has been C. violaceum. It has been observed that cyanogenic bacteria can produce metal-cyanide complexes from metallic solid fractions like PCB trash. The authors demonstrated the mobilisation of both Ni and Au from electronic trash as tetracyano nickelate and dicyanoaurate, respectively(Xiang et al., 2010).

Factors Prompting Bioleaching Development

Only by increasing metal recovery efficiencies will metals bioleaching from e-waste find a potential industrial application. Like all other biological processes, bioleaching is regulated by a number of process variables, including pH, temperature, the type of microorganisms used, the compounds that act as precursors, and pulp density.

Choice of microorganism: Cyanogenic bacteria can recover precious metals like Au, Pd, and Ag in alkaline environments. In an acidic environment, an autotrophic group of bacteria has been observed to extract base metals. The presence of hazardous metals and the alkaline nature of electronic waste both affect the bioleaching of metals from e-waste.

Surface area: For quick metal dissolution, scientists suggest a particle size range of 40 to 200 m. The number of particles increased as the particle size decreased. This causes friction between germs and solid particles, which causes cell lysis or damage.

Pulp density: Increased pulp density during bioleaching is known to impede microbial development and, as a result, result in a lower rate of metal extraction. From an industrial standpoint, bioleaching operations with lower pulp densities are not cost-effective. Researchers conducted tests using various e-waste concentrations to determine the ideal pulp density for the best outcomes.

Precursor molecule: By oxidative decarboxylation, glycine is a direct precursor to cyanide, and HCN is produced from glycine. A glycine concentration greater than 10 g/L inhibited bacterial growth and cyanide generation. Glycine concentration needs to be optimised because it can improve metal recovery.

Temperature: To achieve successful metal dissolution, the temperature must be conducive to bacterial metabolism and support microbial development. The type of bacteria employed in the bioleaching will determine the best temperature range for a particular bacterium. While most cyanogenic bacteria do not function at this temperature, bioleaching investigations involving these bacteria are conducted at 30 °C.

pH: At an initial pH range of 7-8, both bacterial growth and cyanide production are at their highest levels. At this pH, cyanide can be lost through volatilization and is present as gaseous hydrogen cyanide (HCN). Experiments on metal bioleaching under alkaline conditions will cause the equilibrium to shift toward CN-ions.

MICROBIOLOGICAL METHODS

Microorganisms: Through the application of the enrichment culture technique, microorganisms were recovered from soil samples and industrial e-waste trash collected from Tamilnadu, India. At -20°C, glycerol stocks were used to preserve the bacterial cultures. Bacterial cultures were kept on nutrient agar slants and kept in storage at 4°C for routine usage.

Sterilization: Unless otherwise stated, all media, distilled water solutions, and e-waste were autoclaved for 20 minutes at 121°C and 15 pounds per square inch of pressure to sterilise them. Laminar air flow chamber was utilised to maintain aseptic conditions during microbiological investigation.

Bioleaching of Metals Waste CPCBs

Site characterization and sampling: The soil samples come from an abandoned gold mine in the Himachal Pradesh state of India. In a community outside of Salem, the abandoned Goldmine is located close to the banks of the Kaawari River. It was turned off years ago since there was a shortage of AU. The CPCB-containing electronic waste was gathered from a recycling centre, placed in zipper storage bags 37, and kept there at -20 °C for testing. After being collected from both sites, a representative sample of the soil was aseptically homogenised and used for microbiological isolation.

Mining of cultivable bacterial populations: Solan, HP, India's abandoned gold mine, provided the source of the bacteria, which were then cultured in growth medium, including sterilised e-waste. The test flasks were incubated in a shaker incubator for 6 days at 150 rpm and 30 °C. On nutrient agar plates, colony forming units (CFU) were counted and computed after 70 hours. For 70 hours, plates were incubated at 31°C, and colonies were watched for growth. Samples were taken from the flask holding 120 g/L of pulp density after one week of incubation, and they were serially diluted. The e-waste toxicity tolerance of bacteria isolates chosen during initial screening for the bioleaching of Cu, Au, and Ag was examined. In addition to isolated bacterial strains, their tolerance levels were also looked at. To assess each strain's tolerance for toxicity, the plate count method was used. The CFU per gramme of e-waste pulp density was used to calculate the amount of bacterial growth. After counting the number of bacterial cells using the spread plate method, bacteria were cultured on nutrient agar medium and serially diluted. The trials were carried out in flasks that had been inoculated with sterile e-waste and 1% inoculum at the appropriate pulp densities.

Dose-response Analysis: In a quantitative analysis known as dose-response, the maximum treatment concentration and the 50% effective concentration for metal bioleaching are estimated. The inhibitory response was calculated in this situation. Using the GraphPad Prism, data from toxicity assessment studies were plotted to produce a dose-response curve. For each bacterial strain, the EC50 values were calculated. The concentration of a chemical or pollutant at which 50% of the microbial population decreases within a specified time period is known as the "EC50." Log (agonist) vs. normalised response-variable slope was the equation used to produce dose-response curves.

Selection of the Process Parameter

Due to its poor recovery rates, lack of a favourable growth environment, and unoptimized process parameters, bioleaching has not been established as a successful method for extracting metals from metallic repositories, e-waste, or ores. Because even small changes can have a big impact on an industry's ability to succeed commercially or industrially, parameter optimization is a subject of paramount importance in every industrial production process. Both biotic and abiotic characteristics, as well as pH, temperature, aeration, and other environmental conditions, all affect how well such a process works. In order to accelerate the pace of reaction during the bioleaching process, important parameters must be optimised. The two-step bioleaching procedure was adhered to during optimization with consideration for prior work(Ji et al., 2022; Mäkinen et al., 2015).

Optimization of the Process for Maximum Recovery of Metals

This is a tried-and-true technique for analysing data to determine the best way to address issues in computer programming.

Effect of pulp density: An important factor in the bioleaching process is pulp density. The pulp density concentration affects the growth of microorganisms and the leaching of metals. Increased pulp density will undoubtedly raise the concentration of metals in the environment where leaching takes place. This might prevent microbial growth, which would lead to less metal mobilisation.

Effect of temperature: All bacterial strains were treated to two steps of bioleaching at a varied temperature (25–40 °C) while maintaining the same conditions in all other areas. All bacterial strains were exposed to a range of temperatures to determine the best range for mobilising metals.

Effect of glycine concentration: Glycine is a recognised immediate direct precursor molecule of cyanide, which undergoes oxidative decarboxylation to produce HCN and CO2. Glycine loses four hydrogen atoms to make HCN in the process, and then the HCN synthase enzyme produces cyanide. It was claimed that a particular concentration of glycine promoted the development of end products, such as cyanide.

Effect of pH: Because of its pKa of 9.3, cyanide has a strong pH dependence on concentration. It is possible to write down the cyanide gas and aqueous equilibrium. Because of the low pH below the pKa value, the equilibrium is shifted to the right and favours the synthesis of HCN rather than CN- in the equations. To avoid HCN loss by volatilization, commercial bioleaching trials are preferred at an alkaline pH > 10. Above pH 10, gold recovery is said to be effective, and pH above 7 is optimal for gold recovery. The majority of cyanogenic bacteria, however, solubilize gold at a physiological pH of 7-8 because higher pH levels are unfavourable for bacterial growth. Therefore, we kept other variables like glycine, rpm, incubation temperature, and pulp density constant while testing bacterial strains for their bioleaching potential at pH 7, 8, and 9.

Optimization Techniques: Systematic studies can be implemented to identify the best operating procedure using Taguchi, and RSM methods(Boopathi, 2019, 2021; Boopathi et al., 2012; Kalidas et al., 2012; Sampath et al., 2021; Saravanan et al., 2022). The best practises can be integrated with the IoT and machine learning techniques for the automations(Babu et al., 2022; Boopathi et al., 2022; Boopathi, Khare, et al., 2023; Boopathi, Siva Kumar, et al., 2023; Domakonda et al., 2022; Jeevanantham et al., 2022; Kumara et al., 2023; Palaniappan et al., 2023; S. et al., 2022; Samikannu et al., 2022; Sampath et al., 2022; Senthil et al., 2023; Vanitha et al., 2023; Vennila et al., 2022; Yupapin et al., 2022). The GA, TOPSIS, VIKOR, and NSGA-II multi-criteria optimization techniques can be applied to improve the best practises in e-waste management. The above techniques are important to improving e-waste processing(Boopathi, 2019; Boopathi et al., 2021; Boopathi & Sivakumar, 2013; Myilsamy & Sampath, 2021; Sampath & Myilsamy, 2021; Yupapin et al., 2022).

Bioleaching of Metals From Waste CPCBs

Pre-treatment of waste: In order to bioleach gold, 30 ml of nitric acid was combined with 10 g of waste polychlorinated biphenyls (PPCBs). A pre-treatment with nitric acid was carried out to get rid of base metal interference, particularly copper. The metal content of pre-treated waste CPCBs was analysed after they were dried, weighed to a consistent mass, and weighed again. The mixture was centrifuged at 5000 rpm for 10 minutes after 2 hours of shaking. Deionized water was used to wash the remaining trash, and the process was repeated until all traces of the blue-green colour were gone(Işıldar et al., 2016).

Bioleaching of untreated waste: Under previously optimised conditions, the bioleaching abilities of bacterial strains against pre-treated and untreated trash were examined. Only three of the five bacterial strains were utilised because of their ability to mobilise metal.

Disposal of the residual waste CPCBs: After bioleaching, the e-waste underwent additional compositional testing before being disposed of using an aqua-regium-based digestion method in accordance with protocol. The final disposal of the residual material was carried out in compliance with the regulations set forth by the Environment Protection Agency (EPA).

SUMMARY

A study has offered a viable, ecologically friendly solution—bioleaching for e-waste treatment and metals recovery—to a pressing global problem. The metal composition of several trash CPCB size fractions revealed significant size and weight variation. The metal concentration of CPCBs varied when digested for 1, 2, and 3 hours. Indigenous bacterial strains were identified and used to bioleach waste CPCBs from an abandoned gold mine and an e-waste recycling facility. Four well-known bacterial strains, P. balearica, B. megaterium, Bacillus sp., and L. sphaericus, recovered Cu, Au, and Ag from trash. Three bacterial strains have been shown to bioleach Cu, Au, and Ag from e-waste.

The study is the first to investigate and report on the e-waste toxicity evaluation of bacterial strains. To improve metal mobilisation, process variables including pH, pulp density, temperature, and the precursor molecule (glycine) were tuned. According to a study, the unprocessed waste of CPCBs with greater amounts of metal ions such as Cu, Au, and Ag demonstrated considerable metal mobilisation.

Using cyanogenic bacteria, the study combined chemical and biological techniques to recover metals from e-waste. In industrial-scale bioleaching processes, native bacterial strains may be a good contender for resource recovery from e-waste. When used to remediate trash with modest levels of metallic content, such as e-waste dust, bioleaching of metals from electronic waste is substantial. After resource recovery, the safe disposal of residual e-waste remains a significant problem that requires the scientific community's attention. However, recent studies provide strong support for the use of bioleaching to recover metals in large-scale operations. This might greatly advance the circular economy idea.

REFERENCES

Arya, S., & Kumar, S. (2020). Bioleaching: Urban mining option to curb the menace of E-waste challenge. *Bioengineered*, *11*(1), 640–660. doi:10.1080/21655979.2020.1775988 PMID:32538256

Awasthi, A. K., Zeng, X., & Li, J. (2016). Integrated bioleaching of copper metal from waste printed circuit board—A comprehensive review of approaches and challenges. *Environmental Science and Pollution Research International*, *23*(21), 21141–21156. doi:10.100711356-016-7529-9 PMID:27678000

Babu, B. S., Kamalakannan, J., Meenatchi, N., Karthik, S., & Boopathi, S. (2022). Economic impacts and reliability evaluation of battery by adopting Electric Vehicle. *IEEE Explore*, (pp. 1–6). IEEE.

Boopathi, S. (2019). Experimental investigation and parameter analysis of LPG refrigeration system using Taguchi method. *SN Applied Sciences*, *1*(8), 892. doi:10.100742452-019-0925-2

Boopathi, S. (2021). Improving of Green Sand-Mould Quality using Taguchi Technique. *Journal of Engineering Research*, 1–12. doi:10.36909/jer.14079

Boopathi, S., Arigela, S. H., Raman, R., Indhumathi, C., Kavitha, V., & Bhatt, B. C. (2022). Prominent Rule Control-based Internet of Things: Poultry Farm Management System. *IEEE Explore*, (pp. 1–6). IEEE.

Boopathi, S., & Khare, R. KG, J. C., Muni, T. V., & Khare, S. (2023). Additive Manufacturing Developments in the Medical Engineering Field. In Development, Properties, and Industrial Applications of 3D Printed Polymer Composites (pp. 86–106). IGI Global.

BoopathiS.MyilsamyS.SukkasamyS. (2021). Experimental Investigation and Multi-Objective Optimization of Cryogenically Cooled Near-Dry Wire-Cut EDM Using TOPSIS Technique. *Ijamt Preprint*. doi:10.21203/rs.3.rs-254117/v1

Boopathi, S., Siva Kumar, P. K., & Meena, R. S. J., S. I., P., S. K., & Sudhakar, M. (2023). Sustainable Developments of Modern Soil-Less Agro-Cultivation Systems. In Human Agro-Energy Optimization for Business and Industry (pp. 69–87). IGI Global. doi:10.4018/978-1-6684-4118-3.ch004

Boopathi, S., & Sivakumar, K. (2013). Experimental investigation and parameter optimization of near-dry wire-cut electrical discharge machining using multi-objective evolutionary algorithm. *International Journal of Advanced Manufacturing Technology*, *67*(9–12), 2639–2655. doi:10.100700170-012-4680-4

Boopathi, S., Sivakumar, K., & Kalidas, R. (2012). Parametric Study of Dry WEDM Using Taguchi Method. *International Journal of Engineering Research and Development*, *2*(4), 1–6.

Domakonda, V. K., Farooq, S., Chinthamreddy, S., Puviarasi, R., Sudhakar, M., & Boopathi, S. (2022). Sustainable Developments of Hybrid Floating Solar Power Plants: Photovoltaic System. In Human Agro-Energy Optimization for Business and Industry (pp. 148–167). IGI Global.

Hubau, A., Minier, M., Chagnes, A., Joulian, C., Silvente, C., & Guezennec, A. G. (2020). Recovery of metals in a double-stage continuous bioreactor for acidic bioleaching of printed circuit boards (PCBs). *Separation and Purification Technology*, *238*, 116481. doi:10.1016/j.seppur.2019.116481

Işıldar, A., van de Vossenberg, J., Rene, E. R., van Hullebusch, E. D., & Lens, P. N. L. (2016). Two-step bioleaching of copper and gold from discarded printed circuit boards (PCB). *Waste Management (New York, N.Y.)*, *57*, 149–157. doi:10.1016/j.wasman.2015.11.033 PMID:26704063

Jeevanantham, Y. A., Saravanan, A., Vanitha, V., Boopathi, S., & Kumar, D. P. (2022). Implementation of Internet-of Things (IoT) in Soil Irrigation System. *IEEE Explore*, 1–5.

Ji, X., Yang, M., Wan, A., Yu, S., & Yao, Z. (2022). Bioleaching of Typical Electronic Waste—Printed Circuit Boards (WPCBs): A Short Review. *International Journal of Environmental Research and Public Health*, *19*(12), 7508. doi:10.3390/ijerph19127508 PMID:35742757

Kalidas, R., Boopathi, S., Sivakumar, K., & Mohankumar, P. (2012). Optimization of Machining Parameters of WEDM Process Based On the Taguchi Method. *IJEST, 6*(1\&2), 91.

Kumara, V., Mohanaprakash, T. A., Fairooz, S., Jamal, K., Babu, T., & B., S. (2023). Experimental Study on a Reliable Smart Hydroponics System. In *Human Agro-Energy Optimization for Business and Industry* (pp. 27–45). IGI Global. doi:10.4018/978-1-6684-4118-3.ch002

Liang, G., Mo, Y., & Zhou, Q. (2010). Novel strategies of bioleaching metals from printed circuit boards (PCBs) in mixed cultivation of two acidophiles. *Enzyme and Microbial Technology*, *47*(7), 322–326. doi:10.1016/j.enzmictec.2010.08.002

Liu, R., Li, J., & Ge, Z. (2016). Review on Chromobacterium Violaceum for Gold Bioleaching from E-waste. *Procedia Environmental Sciences*, *31*, 947–953. doi:10.1016/j.proenv.2016.02.119

Mäkinen, J., Bachér, J., Kaartinen, T., Wahlström, M., & Salminen, J. (2015). The effect of flotation and parameters for bioleaching of printed circuit boards. *Minerals Engineering*, *75*, 26–31. doi:10.1016/j.mineng.2015.01.009

Myilsamy, S., & Sampath, B. (2021). Experimental comparison of near-dry and cryogenically cooled near-dry machining in wire-cut electrical discharge machining processes. *Surface Topography : Metrology and Properties*, *9*(3), 035015. doi:10.1088/2051-672X/ac15e0

Palaniappan, M., Tirlangi, S., Mohamed, M. J. S., Moorthy, R. M. S., Valeti, S. V., & Boopathi, S. (2023). Fused Deposition Modelling of Polylactic Acid (PLA)-Based Polymer Composites: A Case Study. In Development, Properties, and Industrial Applications of 3D Printed Polymer Composites (pp. 66–85). IGI Global.

Priya, A., & Hait, S. (2017). Comparative assessment of metallurgical recovery of metals from electronic waste with special emphasis on bioleaching. *Environmental Science and Pollution Research International*, *24*(8), 6989–7008. doi:10.100711356-016-8313-6 PMID:28091997

S., P. K., Sampath, B., R., S. K., Babu, B. H., & N., A. (2022). Hydroponics, Aeroponics, and Aquaponics Technologies in Modern Agricultural Cultivation. In *Trends, Paradigms, and Advances in Mechatronics Engineering* (pp. 223–241). IGI Global. doi:10.4018/978-1-6684-5887-7.ch012

Samikannu, R., Koshariya, A. K., Poornima, E., Ramesh, S., Kumar, A., & Boopathi, S. (2022). Sustainable Development in Modern Aquaponics Cultivation Systems Using IoT Technologies. In *Human Agro-Energy Optimization for Business and Industry* (pp. 105–127). IGI Global.

Sampath, B., & Myilsamy, S. (2021). Experimental investigation of a cryogenically cooled oxygen-mist near-dry wire-cut electrical discharge machining process. *Strojniski Vestnik. Jixie Gongcheng Xuebao*, *67*(6), 322–330. doi:10.5545v-jme.2021.7161

Sampath, B., Naveenkumar, N., Sampathkumar, P., Silambarasan, P., Venkadesh, A., & Sakthivel, M. (2021). Experimental comparative study of banana fiber composite with glass fiber composite material using Taguchi method. *Materials Today: Proceedings*, *49*(5), 1475–1480. doi:10.1016/j.matpr.2021.07.232

Sampath, B. C. S., & Myilsamy, S. (2022). Application of TOPSIS Optimization Technique in the Micro-Machining Process. In Trends, Paradigms, and Advances in Mechatronics Engineering (pp. 162–187). IGI Global. doi:10.4018/978-1-6684-5887-7.ch009

Saravanan, M., Vasanth, M., Boopathi, S., Sureshkumar, M., & Haribalaji, V. (2022). Optimization of Quench Polish Quench (QPQ) Coating Process Using Taguchi Method. *Key Engineering Materials*, *935*, 83–91. doi:10.4028/p-z569vy

Senthil, T. S., Puviyarasan, M., Babu, S. R., Surakasi, R., Sampath, B., & Associates. (2023). Industrial Robot-Integrated Fused Deposition Modelling for the 3D Printing Process. In Development, Properties, and Industrial Applications of 3D Printed Polymer Composites (pp. 188–210). IGI Global.

Vanitha, S. K. R., & Boopathi, S. (2023). Artificial Intelligence Techniques in Water Purification and Utilization. In *Human Agro-Energy Optimization for Business and Industry* (pp. 202–218). IGI Global. doi:10.4018/978-1-6684-4118-3.ch010

Vennila, T., Karuna, M. S., Srivastava, B. K., Venugopal, J., Surakasi, R., & Sampath, B. (2022). New Strategies in Treatment and Enzymatic Processes: Ethanol Production From Sugarcane Bagasse. In Human Agro-Energy Optimization for Business and Industry (pp. 219–240). IGI Global.

Xiang, Y., Wu, P., Zhu, N., Zhang, T., Liu, W., Wu, J., & Li, P. (2010). Bioleaching of copper from waste printed circuit boards by bacterial consortium enriched from acid mine drainage. *Journal of Hazardous Materials*, *184*(1–3), 812–818. doi:10.1016/j.jhazmat.2010.08.113 PMID:20869807

Yaashikaa, P. R., Priyanka, B., Senthil Kumar, P., Karishma, S., Jeevanantham, S., & Indraganti, S. (2022). A review on recent advancements in recovery of valuable and toxic metals from e-waste using bioleaching approach. *Chemosphere*, *287*, 132230. doi:10.1016/j.chemosphere.2021.132230 PMID:34826922

Yang, T., Xu, Z., Wen, J., & Yang, L. (2009). Factors influencing bioleaching copper from waste printed circuit boards by Acidithiobacillus ferrooxidans. *Hydrometallurgy*, *97*(1–2), 29–32. doi:10.1016/j.hydromet.2008.12.011

Yin, S., Wang, L., Kabwe, E., Chen, X., Yan, R., An, K., Zhang, L., & Wu, A. (2018). Copper bioleaching in China: Review and prospect. *Minerals (Basel)*, *8*(2), 32. doi:10.3390/min8020032

Yupapin, P., Trabelsi, Y., Nattappan, A., & Boopathi, S. (2022). Performance Improvement of Wire-Cut Electrical Discharge Machining Process Using Cryogenically Treated Super-Conductive State of Monel-K500 Alloy. *Iranian Journal of Science and Technology - Transactions of Mechanical Engineering*, 1–17.

APPENDIX: ABBREVIATIONS

CFU - Colony Forming Units
CPCB - Computer Printed Circuit Board
EPA - Environment Protection Agency
Fe^{3+} - Ferric Ions
HCN - Hydrogen Cyanide
LCD - Liquid Crystal Display
OFAT- One Factor at a Time
PCBs -Printed Circuit Boards
PCDDs - PolyChlorinated Dibenzo-p-Dioxins
PPCBs - PolyChlorinated Biphenyls
PVC - Polyvinyl Chloride
RSM – Response Surface Method
WHO – World Health Organization
WPCB- Water Phosphorus Control Biosciences

Chapter 11
Bibliometric Review on Healthcare Sustainability

Sanjeet Singh
Chandigarh University, India

Geetika Madaan
https://orcid.org/0000-0001-8141-9935
Chandigarh University, India

Jaskiran Kaur
https://orcid.org/0000-0002-4452-1807
Lovely Professional University, India

Swapna H. R.
Jain University (Deemed), India

Digvijay Pandey
https://orcid.org/0000-0003-0353-174X
Department of Technical Education, IET, Dr. A.P.J. Abdul Kalam Technical University, Lucknow, India

Amrinder Singh
Jain University (Deemed), India

Binay Kumar Pandey
https://orcid.org/0000-0002-4041-1213
Department of Information Technology, College of Technology, Govind Ballabh Pant University of Agriculture and Technology, Pantnagar, India

ABSTRACT

United Nations SDGs (sustainable development goals) have prompted health-care organisations in and around this world to implement health care initiatives that aim to improve their long-term sustainability. Analysis of 637 documents from the Scopus database was retrieved using bibliometric methods for the existing study. Research on health care sustainability has been documented over almost the past 12 years (2010-November, 2022) in this review. Results have indicated that the intellectual structure of this knowledge body comprises of five clusters. First, the authors saw that this review database had 22 writers listed as authors. This indicates that a high number of academicians are interested in this topic. But at the same time, it is also evident that majority of contributors produce only few publications on health care sustainability.

DOI: 10.4018/978-1-6684-8117-2.ch011

1. INTRODUCTION

United Nations SDGs (Sustainable Development Goals) have prompted health-care organisations in and around this world to implement health care initiatives that aim to improve their long-term sustainability. Analysis of 637 documents from the Scopus database was retrieved using bibliometric methods for the existing study. Research on health care sustainability has been documented almost over the past 12 years (2010-November, 2022) in this review. Results have indicated that, the intellectual structure of this knowledge body comprises of 5 clusters. First, we saw that our review database had 22 writers listed as authors. This indicates that a high number of academicians are interested in this topic. But at the same time, it is also evident that majority of contributors produce only few publications on health care sustainability. According to the statistics Buffoli M., Capolongo S. and Gola M. (136 Scopus citations), Torkayesh A.E. (113 Scopus citations), Di Noia M. (104 Scopus citations), Vagnoni E. (80 Scopus citations), and Cavicchi C. (74 Scopus citations) are the most influential researchers in the literature of health care sustainability. Recent themes in the field of health care sustainability show up in the density map. These themes shown up in study on health care sustainability depicts (1) health delivery policy for long-term development that takes into account the economic results (2) Organizations are managed by people who improve quality and lead them. It is important to have both "soft skills (leadership and changes) and hard skills (quality improvement systems)" in order to manage for sustainability in health care system. The most important aspect to take away from this review is that future studies need to take into account both social and environmental factors.

The "well-being" and "quality of life" of a country's natives is defined by health status of its population (Wan, 2007; Ruggeri K et al., 2020). As a result, citizens from all over the world called for higher levels of sustainable health care from their governments (Greenhalgh et al., 2012; Kumar and Preetha, 2012). It has become increasingly difficult to manage health care due to ongoing rise in costs and increase in awareness for issues in terms of health and wellness. Organizations in the healthcare industry have working to improve patient safety and care while stream line the process and making health services more accessible(Ament et al., 2012; Goh and Marimuthu, 2016).

Many academics, policymakers, and health care practitioners have been inspired by the publication of United Nations' MDGs (Millennium Development Goals) (Ramirez, West and Costell, 2013). Health care facilities can be built and operated according to the "Green Guide for Health Care's" guidelines as well as the way services are delivered. More impetus provided for sustainable health care with UN announcement in 2015 towards the attainment of sustainable Development Goals at global level (Punnakitikashem and Hallinger, 2020). Increased access to food and nutrition is a goal of SDG 1. Increased access to food and nutrition is a goal of SDG3. SDG 6 is the access to clean drinking water and hygiene to the people worldwide. SDG11 is to become more inclusive, robust, long lasting, and secure in terms of human settlements. These SDGs intended to achieve by the members by 2030 (Johnston, 2016).

In healthcare research, the "sustainability" is relatively new, but it is quickly emerging as a key topic (Ament et al., 2012). To achieve the sustainable development goals, "communities and institutions" will be heavily reliant on the adoption of new and more "sustainable" methods of managing health care services, organisations, and systems. There are those who believe that shifting health care organisations to focus on the "triple bottom line" of "social, environmental, and economic" results will lead to a more sustainable system (Ramirez, Oetjen and Malvey, 2011; Ramirez, West and Costell, 2013). In this study evaluation, the terms "health care sustainability" and "sustainable healthcare" are used interchangeably

(Greenhalgh et al., 2004; Scheirer, 2005; Gruen et al., 2008). Until recently, there were only a few systematic reviews published on relevant fields of health care research and long-term sustainability.

A scientific mapping review approach will be used to evaluate research on healthcare sustainability in the study. This review makes an attempt to address the following RQ (research questions):

RQ1: What are the most important research trends in healthcare management for long term sustainability?
RQ2: Which sustainable healthcare authors and papers have had the most impact on citations in the last few years?
RQ3: Are there any topics in the literature on sustainable healthcare that have received the most attention in research?

The study analysed a dataset of 637 Scopus-indexed texts using bibliometric methodologies. Trends in the growth and composition of sustainable healthcare analysed. Authorship, documents discovered by "citation, influential author citations, and keyword co-occurrence analysis."

2. REVIEW OF LITERATURE

Organization, societal, and system levels can all be used to promote sustainability(Scheirer, 2005);(Kenny and Priyadarshini, 2021). Proctor et al., (2015) tried to apply cluster mapping to investigate difficulty in sustainability research and offered suggestions to aid the transition to evidence-based health care. It was then that the scope of sustainability study in health-care management was enlarged by later literature (Vest and Gamm, 2009);(Anuar, Saad and Yusoff, 2018). Numerous research scholars and practitioners of management have emphasized that there is a need of "business transformation strategies," which focus on improvements in operational process to increase the efficiency and implementation of the sustainable organisations (Leape et al., 2009);(Vest and Gamm, 2009) . The "Toyota/Lean Production System", "Six Sigma (6σ)", and "Total Quality Management (TQM)" are some of the examples of these techniques. They all attempt to enhance organisational processes and performance results (Vest and Gamm, 2009). Despite this trend in healthcare administration, there is no evidence that continuous improvement initiatives produce long-term outcomes (Antonio et al., 2015). Furthermore, as opposed to environmental or social viewpoints, this area of study is the most likely associated with the "economic" perspective of sustainability.

Another area of SHM (Sustainable Health-care Management) research and practice has tried to focus on facilitating organizational transformation in this health-care environment which is highly dynamic in nature. This particular study of management theory and practice tried to look at how the organization's vision, mission, and strategy may be re-oriented to promote an agenda pivotal to sustainability. Different initiatives to "re-vision" healthcare organizations round the agenda have relied on the implementation of a "triple bottom-line" approach (Ramirez, West and Costell, 2013);(Ramirez, Oetjen and Malvey, 2011);(Shediac-Rizkallah and Bone, 1998). The importance of including stakeholders of health-care in development and attainment of sustainability goals that include environmental, social, and economic elements has been underlined in programs that have taken this approach (Ramirez, Oetjen and Malvey, 2011). Scholars and health-care executives have repeatedly used the theoretical design of diffusion, dissemination, and implementation of innovations to increase the inventive ability of health-care organisations in this 'change management approach.' (Greenhalgh et al., 2004).

Health-care management has also included sustainability by being more thoughtful for integrating management procedures with health-care facility architecture and design (Pinzone, Lettieri and Masella, 2012).

(1) ambient conditions (e.g., spatial layout and functioning);
(2) signs, symbols, and artefacts;
(3) materials; and
(4) technological systems are all architectural design components that promote sustainability.

Organizational culture, roles, structures, human resource practices, work processes, and leadership are all organizational and management elements that affect the social and environmental sustainability of health care organizations (Pinzone, Lettieri and Masella, 2012). Sheth et al., (2008) looked at health-facility design in the context to two new guidelines:

a. LEED (Leadership in Energy and Environmental Design) and
b. GGHC (Green Guide to Health Construction).

The redesigning of healthcare facilities was connected to the resolution of energy consumption and waste management issues (Janu, Neha, et al.(2022)). The critical need for advances in hospital waste management, as well as certain advancements in operations management, were underlined (Sheth et al., 2008). Recognizing the interdependence of sustainability concerns has lately led to an increased attention the "macro-level" health-care system's influence on sustainability, at the societal and organisational levels (Robinson et al., 2012). For example, owing to policy attempts by the NHS (National Health Service) in the UK (United Kingdom), "disinvestment" has become a hot subject. Investment in 'savings', 'substitution', 'contractual variation', 'limitation', and 'full withdrawal' are five forms of disinvestment that have been discovered and analysed (Daniels et al., 2013).

Given the current paper's use of bibliometric review methodologies, the researchers would like to acknowledge previous bibliometric reviews in this topic (Punnakitikashem and Hallinger, 2020). For instance, bibliometric evaluations on performance assessment in health care (Punnakitikashem and Hallinger, 2020), multi-criteria decision making in health-care (Ma, Diaby and Xiao, 2014), and health-care risk management have been published (Kenny and Priyadarshini, 2021). Unfortunately, no previous evaluations have looked at the literature on health care sustainability from the aspect of long-term sustainability. The current bibliometric review is framed by this gap in the literature (Gupta, Sunita, et al. (2022)).

3. RESEARCH METHODOLOGY

This section of the study, explains how they choose which documents to look at and how they analyse the data. Finding and extracting documents from the Scopus index was the first step in the process. It has been founded Scopus provides a wider range of sources in comparison to database of web of science particularly for Medical and Physical Science fields. To argue that web of science is more limited coverage produces a database with higher quality schools is subject to empirical verification in a specific field. Punnakitikashem and Hallinger, (2020) conducted a study on "knowledge base on healthcare manage-

Figure 1. Criteria for selection of articles through PRISMA flowchart
Moher et al., (2009)

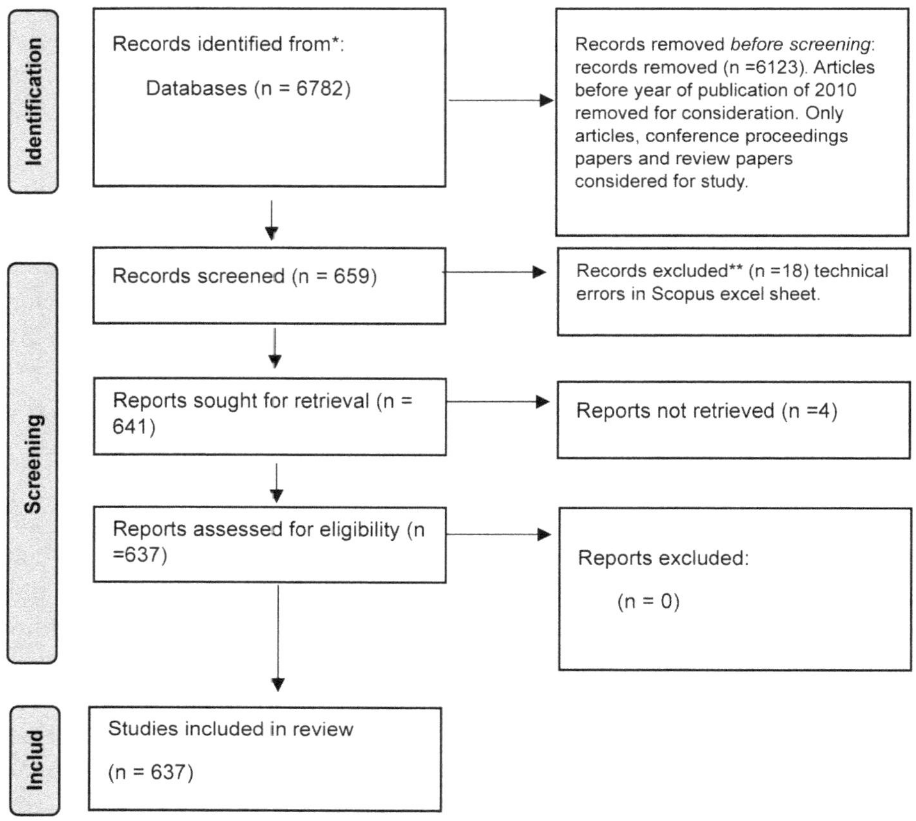

ment for sustainability", which we also mentioned in our research (Pandey, D., et al. (2023)). Studying distinct authors' articles and also citations from "Web of Science and Scopus", they discovered that the two databases were highly correlated (Punnakitikashem and Hallinger, 2020). As a result, we chose Scopus because of its comprehensive coverage (Zupic and Cater, 2015; Mongeon and Paul-Hus, 2016).

Until the end of 2020, this review is confined to the very first document that was identified by the authors, that was published in 2010. We only considered studies that looked at "healthcare" and "sustainability" and "sustainable development" from 2010-2022 (November) (e.g., hospitals, corporations, government non-profits, and educational institutions). Macro level issues like the long-term viability of healthcare systems are not included in the review of papers.

Searches for documents aided by "Preferred Reporting Items for Systematic Reviews (PRISMA) and Meta-Analysis" (Moher et al., 2009) (Figure1). In Scopus, a keyword search performed using the keywords 'sustainable development', 'sustainability' and 'healthcare'. 6782 documents were found in this research. 6145 documents were eliminated as a result of lack sufficient relevance (Figure 1). A total of 637 documents found in the final database.

Exporting and saving bibliographic data of 637 documents (authors, titles and affiliations, citations and etc.) was done so that the data could be used for later data analysis. Descriptive statistics and bibliometric analysis were used to look at the data (Iyyanar, P., et al. (2023).). These included a look at 'citation-analysis', and it means "visualization of similarities" by looking at 'author co-citation's and keywords that appear together (White and McCain, 1998; Moher et al., 2009; van Eck and Waltman, 2017). The Bibliometric analysis were also carried out using VOSviewer bibliometric software, along with Scopus analytical tools and Excel(Van Eck and Waltman, 2017).

4. RESULTS

In this section, the findings of bibliometric examination of the healthcare sustainability are presented. The following is the order in which 3 research questions were addressed:

4.1 Research Trends in Healthcare for Long-Term Sustainability

The earliest source considered for this review was, an article published in the year 2010(n=14). Interest of the scholars (figure 2) in this topic, on the other hand, did not take off until 2020 (n=108), when the number of publications in this field searched by 74.19%. Since 2019, curiosity in this area of 'sustainability research and practice' has grown dramatically, with around 50.86% of all articles coming in 2020-2022. While showing trend all 637 retrieved articles were taken into consideration.

Figure 2. Trends in healthcare for long-term sustainability, 2010–November, 2022 (n = 637)

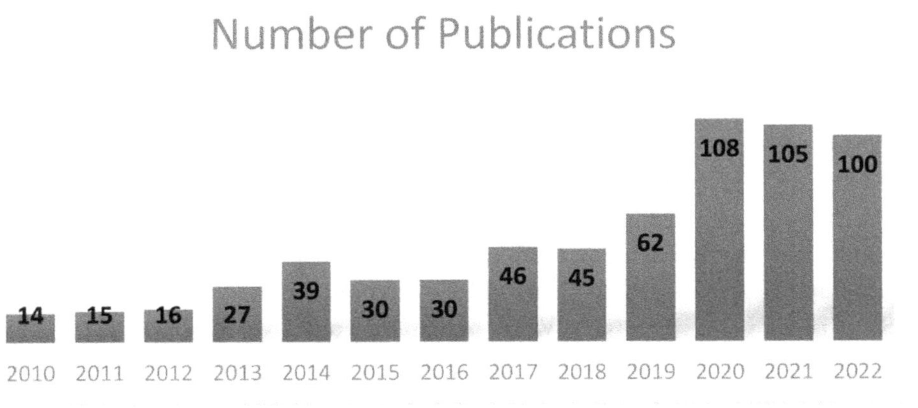

The geographical location of authors (figure 3) investigated in order to determine whether academic interest in sustainable health-care research is concentrated or not. This category of work has been written in nine different nations throughout the world, demonstrating a global interest in the subject (figure 3). However, we did see that authorship was concentrated at United States (142), Canada (28), Spain (35), and India (56); the United Kingdom (81), more than half of the health care sustainability articles collected for this research came from authors belonging to these mentioned 5 countries.

Figure 3. Concentrated global distribution of health care sustainability, 2010–April, 2022 (n = 637)

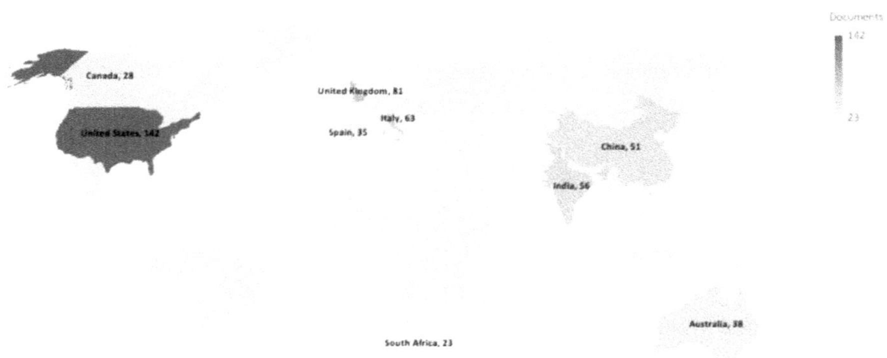

4.2 Authors and Articles Having the Most Impact (Influence) on Citations

Table 1 shows this Scopus citation analysis of authors. First, we saw that our review database had 52 writers listed as authors (not tabled). This indicates that a high number of academicians are interested in this topic. But at the same time, it is also evident that majority of contributors (Table 1) produce only few publications on healthcare sustainability. For instance, Buffoli M., Capolongo S., and Gola M. from Italy who are the most active academician found for this field, published 136 works. The authors' areas of interest range from quality, innovation, management, and policy to environmental science and other aspects of health care. These findings reveal the emergence of a new field of long-term health care management composed of authors with a diverse range of interests and viewpoints.

Italy (Cavicchi C., Di Noia M., Vagnoni E.), Turkey (Torkayesh A.E.) followed by Portugal (Mateus R., Bragança L). These scholars who have high citations, tried to focus on 'health-care innovation, health-care policy, health-care management, and environmental sciences to health-care delivery."

According to the statistics in Table 1, Buffoli M., Capolongo S. and Gola M. (136 Scopus citations), Torkayesh A.E. (113), Di Noia M. (104), Vagnoni E. (80), and Cavicchi C.

(74) are the most influential researchers in the literature of health care sustainability. Furthermore, the table omits numerous additionally highly cited writers who published a single impact document because we set a criterion of 'at least four publications' by each author for this research.

As shown in Table 2, the most frequently cited articles in the field of healthcare sustainability are listed here and determined by Scopus Citations. More than hundred citations were found in first six documents (Vest and Gamm, 2009; Antonio et al., 2015); (Cullen et al., 2005; Stirman et al., 2012). With regard to the health care sustainability's recent distribution, these citations considered to be of a moderate quality. "Scopus citations are always significantly lower for the same document or author because of its more selective coverage of literature, which should be made clear to readers used to assessing impact via Google Scholar Citations (Falagas et al., 2008)" As Table 2 illustrates, research reviews have had a significant impact on the advancement of this field of study(Gupta, Sunita, et al.(2022)).

Table 1. The most influential authors and articles on sustainable healthcare by Scopus citations and documents published (n=637)

Author	Country	Documents	Scopus citations	Citation per documents	Focus
Buffoli M.	Italy	5	136	27.2	Health care, Public administration and Sustainable development
Capolongo S.	Italy	5	136	27.2	health care facility
Gola M.	Italy	5	136	27.2	health care sustainability
Torkayesh A.E.	Turkey	4	113	28.25	health care management
Di Noia M.	Italy	4	104	26	Innovation in healthcare
Vagnoni E.	Italy	4	80	20	healthcare care policy
Cavicchi C.	Italy	5	74	14.8	healthcare care management
Braganca L.	Portugal	4	72	18	healthcare quality management
Mateus R.	Portugal	4	72	18	Environmental health
Scavarda A.	Brazil	4	49	12.25	Environmental health
Kumar S.	India	4	39	9.75	Environmental health
Godbole N.S.	India	5	34	6.8	health care management
Lamb J.	United States	5	34	6.8	healthcare policy
Svensson G.	Norway	5	31	6.2	healthcare programs
Zaman K.	Pakistan	4	29	7.25	e-healthcare
Overcash M.	United States	4	28	7	health care management
Twomey J.	United States	4	28	7	health care management
Rodriguez R.	Norway	4	22	5.5	health care management
Fang Y.	United States	8	12	1.5	Smart Health care Systems and sustainability
Zhou L.	United States	4	8	2	Smart Health care Systems and sustainability
Yang Z..	United States	8	6	0.75	Smart Health care Systems and sustainability
Yuan H.	United States	7	6	0.85	Smart Health care Systems and sustainability

Riboni et al., (2016), the most often referenced paper, used an experimental strategy to recognise fine grained anomalous behaviours in order to diagnose moderate cognitive impairment in its early phases. The focus of this study is on healthcare administration. Bains et al., (2012) conducted experimental study to assist cancer survivors in resuming their normal lives. It learned about the difficulties experienced by cancer patients when they return to work and resume the formula tasks (Kumar A, et al. (2021)). The findings of this study have shifted the conversion in the field of healthcare sustainability. "e-health" research conducted by Alami et al., (2017) sought to identify the factors that contributed to Norway's disparity in the use of telemedicine. This study employed plethora of research study methods with the pivot on e-health (medicine).

Table 2. The most influential scopus citations on healthcare sustainability, 2010-April,2022 (n = 637)

Authors	Title	Focus	Cited by	Type of Paper
(Punnakitikashem and Hallinger, 2020)	"Bibliometric review of the knowledge base on healthcare management for sustainability, 1994-2018"	Healthcare sustainability	5	Review
(Alami et al., 2017)	"Exploring factors associated with the uneven utilization of telemedicine in Norway: A mixed methods study"	e-health(medicine)	36	Exploratory study
(Graham et al., 2017)	"An integrated digital/ clinical approach to smoking cessation in lung cancer screening: Study protocol for a randomized controlled trial"	healthcare management	6	Experimentation approach
(Desmond et al., 2018)	"Health service planning and sustainable development: Considering what, where and how care is delivered through a pro-environmental lens"	Health service and sustainability	2	Conceptual
(Riboni et al., 2016)	"SmartFABER: Recognizing fine-grained abnormal behaviors for early detection of mild cognitive impairment"	Healthcare Management	62	Experimentation approach
(Bains et al., 2012)	"Helping cancer survivors return to work: What providers tell us about the challenges in assisting cancer patients with work questions"	Healthcare Management	41	Experimentation approach
(Kashyap, 2012)	"Building sustainable healthcare knowledge systems by harnessing efficiencies from biomedical linked open data."	Knowledge base (sustainable healthcare)	1	Conceptual
(Montour A et al., 2009)	"The changing nature of nursing work in rural and small community hospitals."	Healthcare Management	23	Exploratory study
(Lee et al., 2008)	"Expanding the reach of health sciences education and empowering others: The Open Course Ware initiative at Tufts University."	health sciences	31	Conceptual

The next highly cited document was conceptual research by (Lee et al., 2008). In this paper the nucleus of the study was on Health Sciences. Through this research, they made an attempt to look out

Figure 4. Representation of co-author analysis through network visualization (minimum number of documents of an author= 4 and minimum number of citations of an author=5; 22 meet the threshold out of 2309 authors)

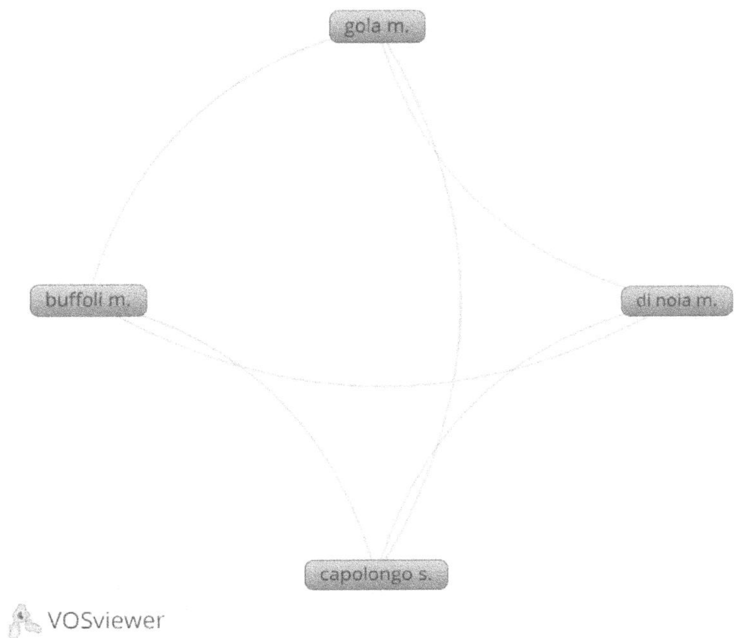

for ways to increase the number of people who have access to Health Science education and training. Tufts University in Greater Boston, Massachustts, USA is the site of the investigation.

In order to reveal the intellectual structure of healthcare sustainability, authors used VOSviewer software to analyse co-author analysis (figure4). It was determined how many times each of the 637 publications reference each other in their reference list by conducting an analysis on co-author analysis. It appears that researchers who are frequently cited by others share similar research interests, based on co-author analysis. VOS viewer software also makes a network map that "visualises commonalities" between the authors who are referenced in healthcare sustainability database by looking at the frequency of "author citations".

4.3 Topics in the Literature on Sustainable Healthcare That Have Received the Most Attention in Research

In order to study literature on healthcare sustainability ko word analysis was analysed. "Most frequent words" identified using VOS viewer software. Researchers in this field are currently focused on the topics represented by the "yellow/ lighter tinted bubbles", whilst previously popular issues are represented by "darker bubbles". in this map, the bubble size, colour, and position all play a role in determining how frequently they appear on the relationship with topics or themes. as indicated through different maps namely density map (figure 5), common network visualization (figure 4), overlay visualization (figure 6), "the most frequently occurring keywords" were Sustainable Development (occurrences = 8),

Figure 5. Keywords network visualisation map for healthcare sustainability literature, 2010–November,2022 (146 keywords meet the threshold out of 6249 keywords); minimum number of occurrences of a keyword=10

Figure 6. Keywords density visualisation map for healthcare sustainability literature, 2010–November,2022 (146 keywords meet the threshold out of 6249 keywords); minimum number of occurrences of a keyword=10

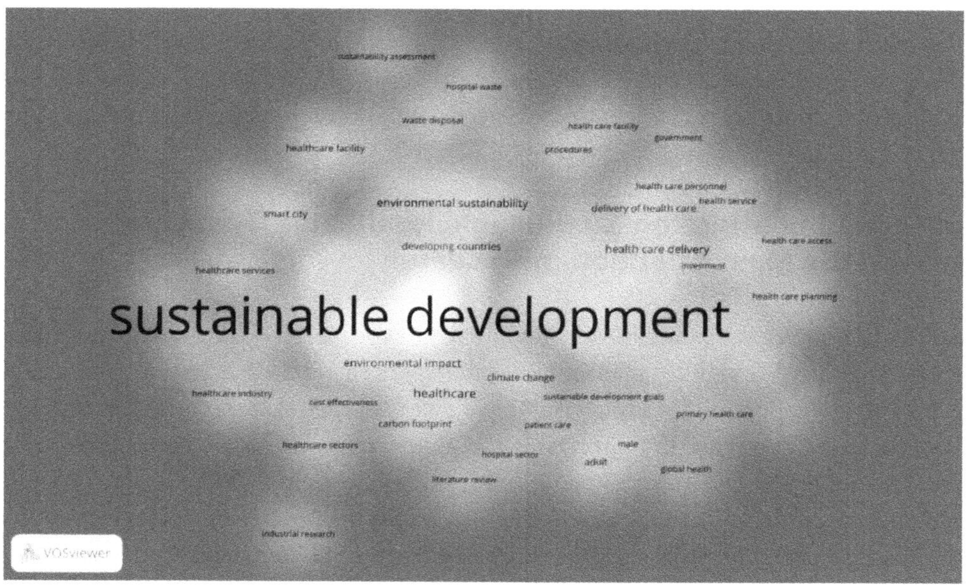

Figure 7. Keywords overlay visualisation map for healthcare sustainability literature, 2010–November, 2022 (146 keywords meet the threshold out of 6249 keywords); minimum number of occurrences of a keyword=10

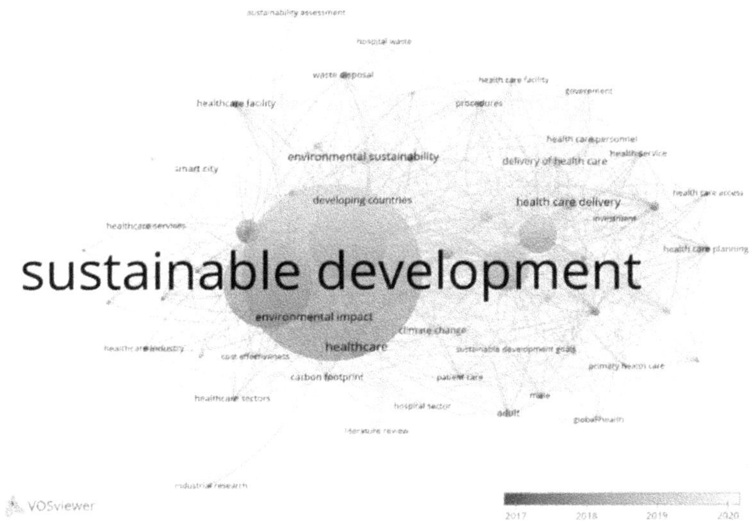

humans (occurrences = 7), healthcare (occurrences = 7) followed by sustainability (occurrences = 3); healthcare delivery (occurrences = 2); Overlay visualization map also depict the keywords occurrence in the articles published in the specific span from 2017 to 2020. Maps formed 5 cluster which visible network visualization (Lelisho, M. E., et al. (2023).). Cluster 1 (red in colour represented in network visualization map) having keywords comprising 25 items (adult, delivery of healthcare, global health, government, healthcare access, healthcare cost, healthcare delivery healthcare organization, healthcare personnel, healthcare planning, healthcare policy, healthcare quality, healthcare system, health insurance, health policy, health service, human, humans, investment, male, organization and management, patient care, primary health care, public health, and World Health Organization). Cluster 2 (green in colour represented in network visualization map) having keywords comprising 20 items (carbon footprint, cost effectiveness, developing countries, health, health care, health care providers, health risks, health-care system, healthcare industry, healthcare organizations, healthcare sectors, healthcare services, hospitals, information and communication, information dissemination, innovation, smart city, sustainability, sustainable development)(Ngulube, P., et al. (2022)). Cluster 3 (blue in colour represented in network visualization map) having keywords comprising 10 items (economics, environmental management, environmental protection, environmental sustainability, healthcare facility, health care facility, healthcare waste management, hospital waste, procedures, and waste disposal). Cluster 4 (purple in colour represented in network visualization map) having keywords comprising 8 items (climate change, environmental impact, health services, hospital sector, industrial research, literature review, sustainable development, Sustainable development). Cluster 5 (yellow in colour represented in network visualization map) having keywords comprising 2 items (economic sustainability, sustainability assessment).This pattern of VOS viewer strengthens that the results from keywords occurrence analysis highlighted the dominance of keyword of sustainable development, humans, sustainability and healthcare in the 637 selected articles.

DISCUSSION

In this section, the authors present the findings and discuss the reviews limitations. Studies first shortcoming is due to the fact that it relied on a quantitative approach to review documents in the healthcare sustainability area. Instead of examining research findings, the review examined associated bibliographic information. As a result, rather than synthesising research findings the reviews only have broad implications for the field's future growth.

There is a second restriction that is brought about by the definition of healthcare sustainability itself. Healthcare sustainability is still a new field with a lot of uncharted territory when it comes to conceptual definitions. Our search for sustainability turned up documents more concerned with programs and innovation sustainability than sustainability in the terms of environment and society. When we looked at bibliometric data, we were able to see what's going on with healthcare sustainability right now in terms of "health care delivery". A benchmark for future reviews of this field can be established by our findings.

CONCLUSION

The literature on healthcare management for sustainability was found to be very limited in this review of research. Most of this information has been gathered over the last 12 years, despite the fact that the first relevant academic work was published in 2019. "Sustainable development is becoming more widely accepted across a wide range of professional fields, thanks in part to the UN zephyr to publicise the SDGs.

Healthcare sustainability literature has a truly global reach, as evidenced by our examination of where it has appeared in different parts of the world. According to our research on sustainable healthcare management, however, people from developing countries haven't been giving as much as they should. A lack of research on healthcare sustainability in developing countries is shown by this imbalance in the distribution of research. This is because healthcare sustainability is very important for the social and economic development of these countries. This is why we believe it is important for researchers to focus on developing countries' healthcare systems' long-term viability. "Stakeholder theory (Salem, Shawtari and Shamsudin, 2016); (Hussain et al., 2018), corporate social responsibility(Lee and Hu, 2018), corporate shared value (Ray and Ray, 2018), green supply chain management (Lo et al., 2018), sufficiency economy philosophy (Chen et al., 2012), a sustainable development philosophy initiated by King Rama IX of Thailand (Kantabutra, 2020), and learning theory (Hovlid et al., 2012)" all of the ideas that can be used in this research are included .

The analysis of citations identified influential co-authors who have written about this topic. This group was made up of people including Buffoli M., Capolongo S. and Gola M. (136 Scopus citations), Torkayesh A.E. (113), Di Noia M. (104), Vagnoni E. (80), and Cavicchi C. (74). "Health care policy, healthcare innovation, healthcare management, and the application of environmental sciences to healthcare delivery" are among the areas of specialisation for these highly cited researchers.

This review is enriched by an empirical investigation of the intellectual structure of this new literature. This research adds to the findings of the previous reviews of research by identifying the schools of thought that make up interdisciplinary literature. According to a co-author citation analysis, there are three distinct approaches to sustainable healthcare: (1) Transformation in Healthcare services towards attainment of sustainable goals (Harris, Green, et al., 2017), (2) new approaches to managing health care facilities (Harris, Allen, et al., 2017), and (3) optimising resources for long-term viability (Harris,

Allen, et al., 2017). "The science map of the healthcare sustainability literature revealed that both micro level issues in healthcare management in organisations and macro level issues in healthcare systems."

It was argued in the first school of thought that the ability of staff to adapt to shifting demands and conditions was facilitated by factors such as "clearly articulated vision, strong program leadership, a strong corporate culture and ongoing capacity building." According to the second line of reasoning, "innovations in managing healthcare operations", innovation have helped people get better care and improve the way healthcare organizations are run. Our study will help future studies to think about how to improve operations management techniques so that they can help the environment. The most recent school of thought, which emphasised long-term sustainability and prioritised and allocated resources, emphasised the significance of health policy. This is how you say it: the best way to use limited healthcare resources, especially money to use them to make the most difference for both accessibility and quality.

Recent themes in the field of healthcare management show up in the density map. These themes show up in research on "healthcare management and health policy for long term development" that takes into account the economic results organizations are managed by people who improve quality and lead them. Leadership and change management as well as quality improvement systems are necessary for sustainable management. This is true, but we also pointed out that these trends are more focused on the economic side of sustainability (quality and efficiency). It looks like the social and environmental aspects haven't been given as much attention as they should have. Thus, the most important thing to take away from this review is that future studies need to take into account both social and environmental factors. We did find some studies that look at how to provide healthcare to older people, which is a big social problem. Even so this is only a small part of the many social effects of healthcare that have an impact on how long it lasts. We want to know how sustainable waste and supply chain management is being utilised to its fullest potential in health care organisations in order to become more environmentally friendly. How are concepts, namely "life cycle assessment and the circular economy" shifting the way health care organizations run? Again, the lack of information in the health care sustainability literature shows how new this field is and how Sustainable issues have received scant attention in this regard.

REFERENCES

Alami, H., Gagnon, M. P., Wootton, R., Fortin, J. P., & Zanaboni, P. (2017). 'Exploring factors associated with the uneven utilization of telemedicine in Norway : A mixed methods study', *BMC Medical Informatics and Decision Making. BMC Medical Informatics and Decision Making*, *17*(180), 1–15. doi:10.118612911-017-0576-4 PMID:29282048

Ament, S. M. C., Gillissen, F., Maessen, J. M. C., Dirksen, C. D., van der Weijden, T., & von Meyenfeldt, M. F. (2012). Sustainability of healthcare innovations (SUSHI): Long term effects of two implemented surgical care programmes (protocol). *BMC Health Services Research*, *12*(1), 423. Advance online publication. doi:10.1186/1472-6963-12-423 PMID:23174024

Antonio, A. (2015). 'Lean in Healthcare: A comprehensive review', *Health policy. Health Policy (Amsterdam)*, *9*(1), 197–209. doi:10.1016/j.healthpol.2015.02.002

Anuar, A., Saad, R., & Yusoff, R. Z. (2018). 'Sustainability through Lean Healthcare and Operational Performance in the Private Hospitals : A Proposed Framework', *Int. J Sup. Chain. Mgt*, *7*(5), 221–227.

Bains, M., Yarker, J., Amir, Z., Wynn, P., & Munir, F. (2012). Helping Cancer Survivors Return to Work : What Providers Tell Us About the Challenges in Assisting Cancer Patients with Work Questions. *Journal of Occupational Rehabilitation*, *22*(1), 71–77. doi:10.100710926-011-9330-4 PMID:21870065

Cao, S. (2020). Evaluation of Smart Humanity Systems and Novel UV-Oriented Solution for Integration, Resilience, Inclusiveness and Sustainability. *5th International Conference on Universal Village, UV 2020*, (21). 10.1109/UV50937.2020.9426226

Capolongo, S., Bottero, M. C., & Lettieri, E. (2015) Chapter 1 Healthcare Sustainability Challenge. *International Academic Conference*, *6*(3), 0–9. 10.1007/978-3-319-14036-0

Capolongo, S., Buffoli, M., & Noia, M. (2015) 'Chapter 2', (March 2016), pp. 10–22. doi:10.1007/978-3-319-14036-0

Castro, M. de F., Mateus, R., & Bragança, L. (2017). Healthcare Building Sustainability Assessment tool - Sustainable Effective Design criteria in the Portuguese context. *Environmental Impact Assessment Review*, *67*(August), 49–60. doi:10.1016/j.eiar.2017.08.005

Cavicchi, C., Oppi, C., & Vagnoni, E. (2022). Back and Forth on Sustainable Development: A Focus on Healthcare Organisations. *Sustainability (Basel)*, *14*(9), 1–15. doi:10.3390u14094958

Chen, C., Shih, H.-S., Shyur, H.-J., & Wu, K.-S. (2012). A business strategy selection of green supply chain management via an. *Computers & Mathematics with Applications (Oxford, England)*, *64*(1), 2544–2546. doi:10.1016/j.camwa.2012.06.013

Cullen, K. (2005). Workplace-Based Return-to-Work Interventions. *Journal of Occupational Rehabilitation*, *15*(4), 607–631. doi:10.100710926-005-8038-8 PMID:16254759

Daniels, T., Williams, I., Robinson, S., & Spence, K. (2013). Tackling disinvestment in health care services The views of resource allocators. *Journal of Health Organization and Management*, *27*(6), 762–780. doi:10.1108/JHOM-11-2012-0225 PMID:24422258

Desmond, S. (2018). Health service planning and sustainable development : Considering what, where and how care is delivered through a pro-environmental lens. *Australian Health Review*, *42*(2), 140–145. doi:10.1071/AH16217 PMID:28248634

Esayas Lelisho, M., Akessa, G. M., Kifle Demissie, D., Fikadu Yermosa, S., Andargie, S. A., Tareke, S. A., & Pandey, D. (2023). Application of Parametric Shared Frailty Models to Analyze Time-to-Death of Gastric Cancer Patients. *Journal of Gastrointestinal Cancer*, *54*(1), 104–116. doi:10.100712029-021-00775-y PMID:35064523

Falagas, M. E., Pitsouni, E. I., Malietzis, G. A., & Pappas, G. (2008). Comparison of PubMed, Scopus, Web of Science, and Google Scholar : Strengths and weaknesses. *The FASEB Journal*, *22*(338–342), 338–342. Advance online publication. doi:10.1096/fj.07-9492LSF PMID:17884971

Godbole, N. S., & Lamb, J. (2018) 'Research into Making Healthcare Green with Cloud, Green IT, and Data Science to Reduce Healthcare Costs and Combat Climate Change', *2018 9th IEEE Annual Ubiquitous Computing, Electronics and Mobile Communication Conference, UEMCON 2018*. IEEE, pp. 189–195. 10.1109/UEMCON.2018.8796529

Goh, C. Y., & Marimuthu, M. (2016) 'The Path towards Healthcare Sustainability: The Role of Organisational Commitment', Procedia - Social and Behavioral Sciences. The Author(s), 224(August 2015), pp. 587–592. doi:10.1016/j.sbspro.2016.05.445

Graham, A. L., Burke, M. V., Jacobs, M. A., Cha, S., Croghan, I. T., Schroeder, D. R., Moriarty, J. P., Borah, B. J., Rasmussen, D. F., Brookover, M. J., Suesse, D. B., Midthun, D. E., & Hays, J. T. (2017). 'An integrated digital / clinical approach to smoking cessation in lung cancer screening : Study protocol for a randomized controlled trial', *BioMed Central. Trials*, *18*(568), 1–14. doi:10.118613063-017-2312-x PMID:29179734

Greenhalgh, T., MacFarlane, F., Barton-Sweeney, C., & Woodard, F. (2012). "If we build it, will it stay?" A case study of the sustainability of whole-system change in London. *The Milbank Quarterly*, *90*(3), 516–547. doi:10.1111/j.1468-0009.2012.00673.x PMID:22985280

Greenhalgh, T., Robert, G., MacFarlane, F., Bate, P., & Kyriakidou, O. (2004). Diffusion of innovations in service organizations: Systematic review and recommendations. *The Milbank Quarterly*, *82*(4), 581–629. doi:10.1111/j.0887-378X.2004.00325.x PMID:15595944

Gruen, R. L. et al. (2008) 'Sustainability science: an integrated approach for health-programme planning', *The Lancet*. Elsevier Ltd, 372(9649), pp. 1579–1589. . doi:10.1016/S0140-6736(08)61659-1

Gupta, S., Nawal, M., Janu, N., & Goyal, D. (2022). IoT, Enabling Technologies, and Sensor Node Deployment Pattern in WSN. *ECS Transactions*, *107*(1), 7441–7455. doi:10.1149/10701.7441ecst

Gupta, S., Nawal, M., Janu, N., & Goyal, D. (2022). IoT, Enabling Technologies, and Sensor Node Deployment Pattern in WSN. *ECS Transactions*, *107*(1), 7441–7455. doi:10.1149/10701.7441ecst

Harris, C., Allen, K., Waller, C., & Brooke, V. (2017). 'Sustainability in health care by allocating resources effectively (SHARE) 3 : Examining how resource allocation decisions are made, implemented and evaluated in a local healthcare setting', *BMC Health Services Research. BMC Health Services Research*, *17*(340), 1–21. doi:10.118612913-017-2207-2 PMID:28486953

Harris, C., Green, S., Ramsey, W., Allen, K., & King, R. (2017). 'Sustainability in Health care by allocating resources effectively (SHARE) 1 : Introducing a series of papers reporting an investigation of disinvestment in a local healthcare setting', *BMC Health Services Research. BMC Health Services Research*, *17*(323), 1–13. doi:10.118612913-017-2210-7 PMID:28472962

Hovlid, E., Bukve, O., Haug, K., Aslaksen, A. B., & von Plessen, C. (2012). Sustainability of healthcare improvement : What can we learn from learning theory? *BMC Health Services Research*, *12*(235), 2–13. doi:10.1186/1472-6963-12-235 PMID:22863199

Hussain, M. et al. (2018) 'Exploration of Social Sustainability in Healthcare Supply Chain', *Journal of Cleaner Production*. Elsevier B.V. . doi:10.1016/j.jclepro.2018.08.157

Iyyanar, P., Anand, R., Shanthi, T., Nassa, V. K., Pandey, B. K., George, A. S., & Pandey, D. (2023). A Real-Time Smart Sewage Cleaning UAV Assistance System Using IoT. In *Handbook of Research on Data-Driven Mathematical Modeling in Smart Cities* (pp. 24–39). IGI Global.

Janu, N., & .... (2022). *Cross-lingual Word Sense Disambiguation Using Multilingual Co-occurrence Graphs*. Graph Learning and Network Science for Natural Language Processing. doi:10.1201/9781003272649-9

Janu, N., ... "Query-Based Image Retrieval Using SVM." Emerging Technologies in Computer Engineering: Cognitive Computing and Intelligent IoT: 5th International Conference, ICETCE 2022, Jaipur, India, February 4–5, 2022, Revised Selected Papers. Cham: Springer International Publishing, 2022.

Johnston, R. B. (2016) *Transforming Our World: The 2030 Agenda For Sustainable Development, Global Sustainability - Proceedings of the 6th International Congress on Arsenic in the Environment, AS 2016*. 10.1201/b20466-7

Kantabutra, S., & Punnakitikashem, P. (2020). Exploring the Process Toward Corporate Sustainability at a Thai SME. *Sustainability (Basel)*, *12*(1), 2–19. doi:10.3390u12219204

Kashyap, A. (2012) 'Building Sustainable Healthcare Knowledge Systems by Harnessing Efficiencies from Biomedical Linked Open Data', in IEEE, pp. 16–18. doi:10.1109/INDCON.2011.6139343

Kenny, C., & Priyadarshini, A. (2021). Review of Current Healthcare Waste Management Methods and Their Effect on Global Health. *Health Care*, *9*(284), 2–17. PMID:33807606

Kumar, A., Dadheech, P., Goyal, D., Patidar, P. K., Dogiwal, S. R., & Janu, N. (2021). A novel scheme for prevention and detection of black hole & gray hole attack in VANET network. *Recent Patents on Engineering*, *15*(2), 263–274. doi:10.2174/1872212114999200512120211

Kumar, S., & Preetha, G. S. (2012). Health promotion: An effective tool for global health. *Indian Journal of Community Medicine*, *37*(1), 5–12. doi:10.4103/0970-0218.94009 PMID:22529532

Leape, L. (2009). *Transforming healthcare : a safety imperative*. Harvard School of Public Health., doi:10.1136/qshc.2009.036954

Lee, M. Y., Albright, S., O'Leary, L., Terkla, D. G., & Wilson, N. (2008). Expanding the reach of health sciences education and empowering others : The OpenCourseWare initiative at Tufts University. *Medical Teacher*, *30*(1), 159–163. doi:10.1080/01421590701881665 PMID:18464140

Lee, Y., & Hu, J. (2018). Integrated Approaches for Business Sustainability : The Perspective of Corporate Social Responsibility. *Sustainability (Basel)*, *10*(2), 2–20. doi:10.3390u10072318

Lelisho, M. E., Pandey, D., Alemu, B. D., Pandey, B. K., & Tareke, S. A. (2023). The negative impact of social media during COVID-19 pandemic. *Trends in Psychology*, *31*(1), 123–142. doi:10.100743076-022-00192-5

Liu, C., Liu, Y., Liu, J., Yang, G., Zhao, X., & Quan, W. (2016). Electro-hydraulic servo plate-inclined plunger hydraulic transformer. *IEEE Access : Practical Innovations, Open Solutions*, *4*(1), 8608–8616. doi:10.1109/ACCESS.2016.2628355

Lo, H. et al. (2018) 'An integrated model for solving problems in green', *Journal of Cleaner Production*. Elsevier B.V. . doi:10.1016/j.jclepro.2018.04.105

Ma, G. A., Diaby, V., & Xiao, H. (2014) 'Application of multicriteria decision analysis in health care : a systematic review and bibliometric analysis', health Epectations, 18, pp. 1894–1905. doi:10.1111/hex.12287

Machado, C. M. L. (2015) 'Sustainability at the healthcare organizations: An analysis of the impact on the environment, society, and economy', Chemical Engineering Transactions, 45(June 2017), pp. 727–732. doi:10.3303/CET1545122

Moher, D. (2009). Academia and Clinic Annals of Internal Medicine Preferred Reporting Items for Systematic Reviews and Meta-Analyses. *Annals of Internal Medicine*, *151*(4), 264–269. doi:10.7326/0003-4819-151-4-200908180-00135 PMID:19622511

Mongeon, P., & Paul-Hus, A. (2016). The journal coverage of Web of Science and Scopus: A comparative analysis. *Scientometrics*, *106*(1), 213–228. doi:10.100711192-015-1765-5

Montour, A., Baumann, A., Blythe, J., & Hunsberger, M. (2009). The changing nature of nursing work in rural and small community hospitals. *Rural and Remote Health*, *9*(1), 1089–1092. http://usblogs.pwc.com/industrialinsights/2017/05/09/megatrends-to-watch-in-the-paper-and-packaging-industry/. doi:10.22605/RRH1089 PMID:19199373

Ngulube, P., Pandey, B. K., Pandey, D., Nassa, V. K., George, A. S., Gupta, M. K., & Dadheech, P. (2022). Evaluation of the SARS-COV-2 Omicron variant's features, prospective consequences, and upcoming prospects. An Interdisciplinary Approach in the Post-COVID-19 Pandemic Era, 63-76.

Pandey, D., Hasan, A., Pandey, B. K., Lelisho, M. E., George, A. H., & Shahul, A. (2023). COVID-19 epidemic anxiety, mental stress, and sleep disorders in developing country university students. CSI Transactions on ICT, 1-9.

Pandey, D., Lelisho, M. E., Hasan, A., Rajput, N., Kumar, B., George, A. H., & Shahul, A. (2023). Factors Associated With Sleep Disorders Among Indian and Ethiopian Students in the Digital World in the Midst of the COVID-19 Pandemic. In Applications of Neuromarketing in the Metaverse (pp. 137-154). IGI Global.

Pinzone, M., Lettieri, E. and Masella, C. (2012) 'Sustainability in Healthcare : Combining Organizational and Architectural Levers Regular Paper', *Intech open science*, 4(4), pp. 1–9. . doi:10.5772/54841

Proctor, E., Luke, D., Calhoun, A., McMillen, C., Brownson, R., McCrary, S., & Padek, M. (2015). 'Sustainability of evidence-based healthcare : Research agenda, methodological advances, and infrastructure support', *Implementation Science. Implementation Science : IS*, *10*(88), 1–13. doi:10.118613012-015-0274-5 PMID:26062907

Punnakitikashem, P., & Hallinger, P. (2020). Bibliometric review of the knowledge base on healthcare management for sustainability, 1994-2018. *Sustainability (Basel)*, *12*(1), 1–17. doi:10.3390u12010205

Qureshi, M. I., Qayyum, S., Nassani, A. A., Aldakhil, A. M., Qazi Abro, M. M., & Zaman, K. (2019). Management of various socio-economic factors under the United Nations sustainable development agenda. *Resources Policy*, *64*(October), 101515. doi:10.1016/j.resourpol.2019.101515

Rabanimotlagh, A., & … . (2014) 'Healthcare facilities: Planning for increasing extreme heat events', *IIE Annual Conference and Expo 2014*, pp. 1810–1816.

Ramirez, B., Oetjen, R. M., & Malvey, D. (2011). Sustainability and the health care manager: Part i. *The Health Care Manager*, *30*(2), 133–138. doi:10.1097/HCM.0b013e318216f4e5 PMID:21537134

Ramirez, B., West, D. J., & Costell, M. M. (2013). Development of a culture of sustainability in health care organizations. *Journal of Health Organization and Management*, *27*(5), 665–672. doi:10.1108/JHOM-11-2012-0226 PMID:24341182

Ray, S. and Ray, B. (2018) 'Business Group Affiliation and Corporate Sustainability Strategies of Firms : An Investigation of Firms in India', *Journal of Business Ethics*. Springer Netherlands, 0(0), p. 0. . doi:10.1007/s10551-018-3917-z

Riboni, D., Bettini, C., Civitarese, G., Janjua, Z. H., & Helaoui, R. (2016). 'Artificial Intelligence in Medicine SmartFABER : Recognizing fine-grained abnormal behaviors for early detection of mild cognitive impairment', *Artificial Intelligence In Medicine. Artificial Intelligence in Medicine*, *67*, 57–74. doi:10.1016/j.artmed.2015.12.001 PMID:26809483

Robinson, S., Williams, I., Dickinson, H., Freeman, T., & Rumbold, B. (2012). Social Science & Medicine Priority-setting and rationing in healthcare : Evidence from the English experience. *Social Science & Medicine*, *75*(12), 2386–2393. doi:10.1016/j.socscimed.2012.09.014 PMID:23083894

Rodriguez, R., Svensson, G., & Eriksson, D. (2021). Priorities determining future directions of sustainable development in business models of the healthcare industry—Findings and framework. *Sustainability (Basel)*, *13*(11), 1–11. doi:10.3390u13116507

Ruggeri, K., & … . (2020). 'Well-being is more than happiness and life satisfaction: A multidimensional analysis of 21 countries. Health and Quality of Life OutcomesHealth and Quality of Life Outcomes [revista en Internet] 2020 [acceso 4 de julio de 2021]; 18(1): 1-16.', *Health and Quality of Life Outcomes. Health and Quality of Life Outcomes*, *1*(2), 1–16. https://hqlo.biomedcentral.com/track/pdf/10.1186/s12955-020-01423-y.pdf

Sahoo, S. et al. (2022) *Blockchain for sustainable supply chain management: trends and ways forward, Electronic Commerce Research*. Springer US. . doi:10.1007/s10660-022-09569-1

Salem, M. A., Shawtari, F. A., Shamsudin, M. F., & Hussain, H. I. (2016). 'The relation between stakeholders ' integration and environmental competitiveness'. *Social Responsibility Journal*, *12*(4), 755–768. doi:10.1108/SRJ-12-2015-0189

Scheirer, M. A. (2005). Is sustainability possible? A review and commentary on empirical studies of program sustainability. *The American Journal of Evaluation*, *26*(3), 320–347. doi:10.1177/1098214005278752

Shediac-rizkallah, M. C., & Bone, L. R. (1998). Planning for the sustainability of community-based health programs : Conceptual frameworks and future directions for research, practice and policy. *Health Education Research*, *13*(1), 87–108. doi:10.1093/her/13.1.87 PMID:10178339

Sheth, A., ... (2008) 'Reviewing the sustainability of existing healthcare facilities', Loughborough university, pp. 1193–1202.

Stirman, S. W., & ... . (2012). *The sustainability of new programs and innovations : a review of the empirical literature and recommendations for future research.* Implmentation Science.

Torkayesh, A. E. et al. (2021) 'An integrated BWM-LBWA-CoCoSo framework for evaluation of health-care sectors in Eastern Europe', *Socio-Economic Planning Sciences*. Elsevier Ltd, 78(September 2020), p. 101052. . doi:10.1016/j.seps.2021.101052

van Eck, N. J., & Waltman, L. (2017). Citation-based clustering of publications using CitNetExplorer and VOSviewer. *Scientometrics*, *111*(2), 1053–1070. doi:10.100711192-017-2300-7 PMID:28490825

Vest, J. R., & Gamm, L. D. (2009). 'A critical review of the research literature on Six Sigma, Lean and StuderGroup ' s Hardwiring Excellence in the United States : The need to demonstrate and communicate the effectiveness of transformation strategies in healthcare'. *Implementation Science : IS*, *4*(35), 1–9. doi:10.1186/1748-5908-4-35 PMID:19570218

Wan. (2007). *International Journal of Health Care Quality Assurance*, *20*(3), 253–263.

White, H. D., & McCain, K. W. (1998). Visualizing a discipline: An author co-citation analysis of information science, 1972–1995. *Journal of the American Society for Information Science*, *49*(4), 327–355.

Zupic, I., & Čater, T. (2015). Bibliometric Methods in Management and Organization. *Organizational Research Methods*, *18*(3), 429–472. doi:10.1177/1094428114562629

Chapter 12
Biogas Plants for Sustainable Municipal Waste Management: A Brief Review

Shalom Akhai
https://orcid.org/0000-0002-7533-457X
Department of Mechanical Engineering, Chandigarh Engineering College, India

ABSTRACT

Municipal waste management is a critical issue for sustainable development, and traditional methods are becoming unsustainable. Biogas plants offer a promising solution using organic waste to produce biogas for electricity or fuel. This review examines the benefits and challenges of biogas plants for municipal waste management, including the complex biochemical process, factors affecting efficiency, and challenges like high costs, etc. Biogas production diverts waste from landfills, reduces greenhouse gas emissions, and creates economic opportunities. Successful case studies from Germany, India, South Africa, etc. are also highlighted in this chapter. Future research directions are also discussed to optimize the process, assess environmental impacts, and evaluate economic viability. Overall, biogas plants represent a promising approach to sustainable municipal waste management that can contribute to the transition toward a more circular and sustainable economy.

INTRODUCTION: MUNICIPAL WASTE MANAGEMENT AND THE NEED FOR SUSTAINABLE SOLUTIONS

Municipal solid waste management (MSWM) is an essential service provided by municipalities and other public entities to manage solid waste generated by households, businesses, and institutions within their jurisdiction. Effective MSWM is crucial for protecting public health and the environment, conserving natural resources, and promoting sustainable development. This section provides an overview of the background, history, and current practices of MSWM (Brunner & Rechberger, 2015; Kulkarni, 2020; Shah et al., 2021; Vyas et al., 2022).

DOI: 10.4018/978-1-6684-8117-2.ch012

Background

MSWM has been a concern for humans since the beginning of civilization. Historically, people disposed of waste by throwing it into rivers, burying it in pits, or burning it. In ancient Rome, waste was thrown onto the streets, while in medieval Europe, waste was burned in large pits. However, with the rise of industrialization and urbanization in the late 19th and early 20th centuries, MSWM became more complex and challenging (Brunner & Rechberger, 2015; De Feo et al., 2014; Karak et al., 2012; Sandhu, 2014; Wilson, 1976).

In the United States, the first municipal dump was established in New York City in 1895. Other cities followed suit by creating their own landfills. However, landfills began to reach capacity, and the public became increasingly aware of the environmental and health hazards associated with improper waste disposal. In the 1960s and 1970s, the U.S. government passed a series of laws and regulations to address these issues, including the Solid Waste Disposal Act of 1965 and the Resource Conservation and Recovery Act of 1976 (Kollikkathara, Feng, & Stern, 2009; Kollikkathara, Feng, & Stern, 2009; Lee & Huffman, 1989; Oakes & Shank, 1979; Wilson, 1976).

MSWM Practices

MSWM practices typically include several steps: collection, transportation, processing, and disposal. The following are the general steps followed for the same.

- **Collection:** Waste is collected from households, industries, and institutions by waste haulers using specialized trucks. In some cases, residents may need to separate recyclable materials from non-recyclable waste, while in other cases, the waste is collected mixed together.
- **Transportation:** Waste is transported to a transfer station, where it is sorted and loaded onto larger trucks for transport to a landfill or recycling facility. The transfer station is also where the waste is weighed, and the haulers are charged a fee based on the weight of the waste.
- **Processing:** Waste is processed at a landfill or recycling facility, depending on its composition. Landfills are designed to safely contain waste and prevent contamination of the surrounding environment. Landfills typically involve compacting and burying the waste under several feet of soil. Some landfills also generate electricity by burning methane gas produced by decomposing waste. Recycling facilities, on the other hand, separate recyclable materials from non-recyclable waste. Recyclable materials such as paper, plastic, glass, and metal are then turned into new products.
- **Disposal:** Non-recyclable waste is disposed of at a landfill, where it is buried and covered with soil to prevent odors and pests. Landfills must be carefully engineered to prevent groundwater contamination and protect air quality. Modern landfills are required to have liners, leachate collection systems, and methane gas management systems.
- **Recycling:** Recyclable materials such as paper, plastic, glass, and metal are separated and sent to recycling facilities, where they are processed and turned into new products. Recycling helps conserve natural resources, reduce the amount of waste sent to landfills, and reduce greenhouse gas emissions.

As depicted in Figure 1, there are different kinds of biomass available. However, to ensure the safe and effective handling of waste, it is crucial to implement proper municipal solid waste management

Figure 1. Types of biomass

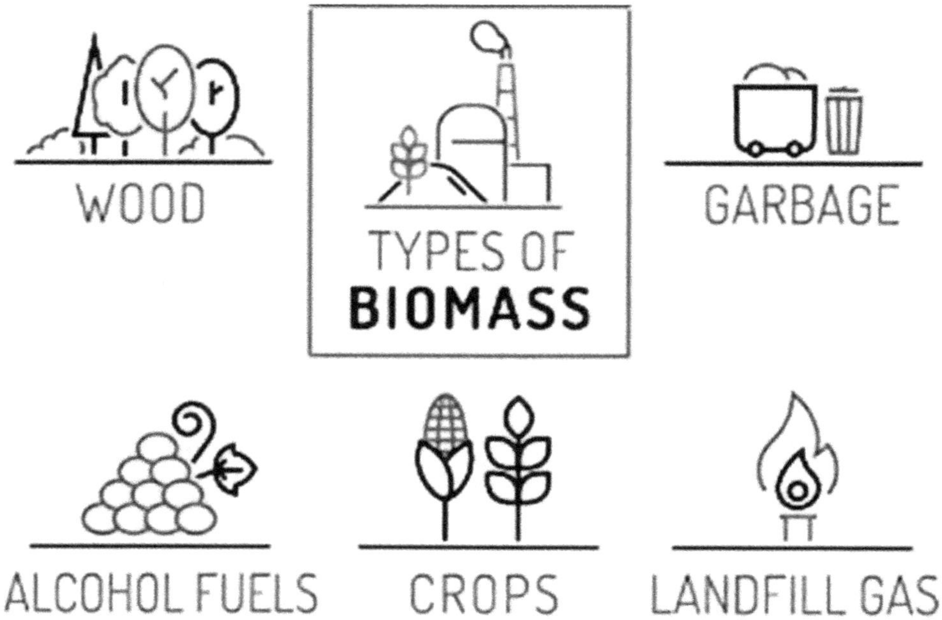

(MSWM) practices. These practices involve a series of steps, including collection, transportation, processing, and disposal, all aimed at reducing the negative impact of waste on the environment and public health. In addition, recycling is a vital component in minimizing waste and preserving our natural resources. It is imperative that individuals, industries, and governments collaborate to adopt sustainable waste management practices and reduce our collective waste footprint (Hirpe & Yeom, 2021; Joshi & Ahmed, 2016; Kaur & Deswal, 2019; Sharholy et al., 2008; Venkiteela, 2020).

Overview of Biogas Plants and Their Applications

Biogas plants are facilities that use anaerobic digestion to produce biogas from organic waste, such as agricultural waste, food waste, and sewage sludge. The biogas produced can be used for various applications, including:

1. **Electricity Generation:** Biogas can be used to generate electricity through the use of gas turbines or internal combustion engines. This electricity can be used on-site or exported to the grid.
2. **Heat Generation:** Biogas can be used to generate heat through the use of boilers or cogeneration systems. The heat can be used for space heating or industrial processes.
3. **Vehicle Fuel:** Biogas can be purified and compressed to produce bio-gasoline or bio-methane, which can be used as a vehicle fuel in natural gas vehicles.
4. **Cooking Fuel:** Biogas can be used as a cooking fuel in households, particularly in rural areas where traditional biomass stoves are still used.
5. **Organic Fertilizer Production:** The digestate, which is a by-product of biogas production, can be used as an organic fertilizer in agriculture.

Figure 2. Biogas applications

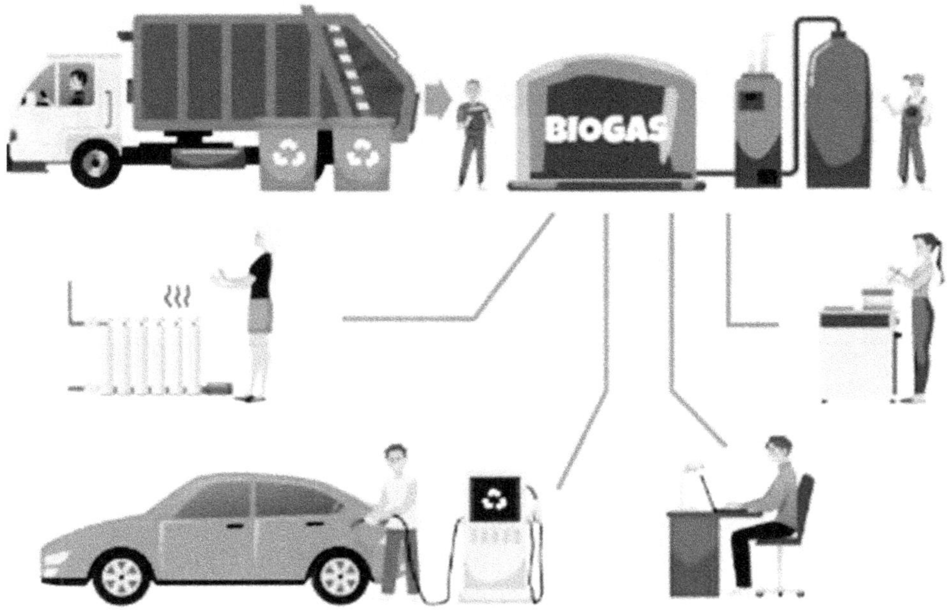

6. **Waste Management:** Biogas plants can help to reduce the amount of organic waste that is sent to landfills, reducing the environmental impact of waste disposal.

As illustrated in Figure 2, biogas plants are highly versatile facilities that offer a range of benefits, including the production of renewable energy, the reduction of greenhouse gas emissions, and sustainable waste management. With their diverse range of applications, biogas has the potential to promote a more circular and sustainable economy (Abanades et al., 2022; Ajay et al., 2021; Kabeyi & Olanrewaju, 2022; Kaur & Deswal, 2019; Pandyaswargo, Jagath Dickella Gamaralalage, Liu, Knaus, Onoda, Mahichi, & Guo, 2019; Tabatabaei et al., 2020; Thiruselvi et al., 2021; Tian et al., 2021).

BIOGAS PLANTS: PROMISING TECHNOLOGY FOR SUSTAINABLE MUNICIPAL WASTE MANAGEMENT

Communities worldwide are facing a growing challenge when it comes to municipal solid waste management. With the volume of waste produced on the rise and landfill sites becoming increasingly scarce, finding sustainable waste management solutions is more important than ever. Fortunately, biogas plants offer a promising option for tackling this issue. By using organic waste to generate biogas, a renewable energy source that can be used for electricity generation or as fuel, biogas plants provide a range of benefits as illustrated in Figure 2. This review offers a comprehensive overview of the potential of biogas plants as a sustainable solution for managing municipal waste (Bong et al., 2017; Chaianong & Pharino, 2022; Khalil et al., 2019; Roubík et al., 2018; Thomas & Soren, 2020; Tsui & Wong, 2019).

Figure 3. Biogas advantages

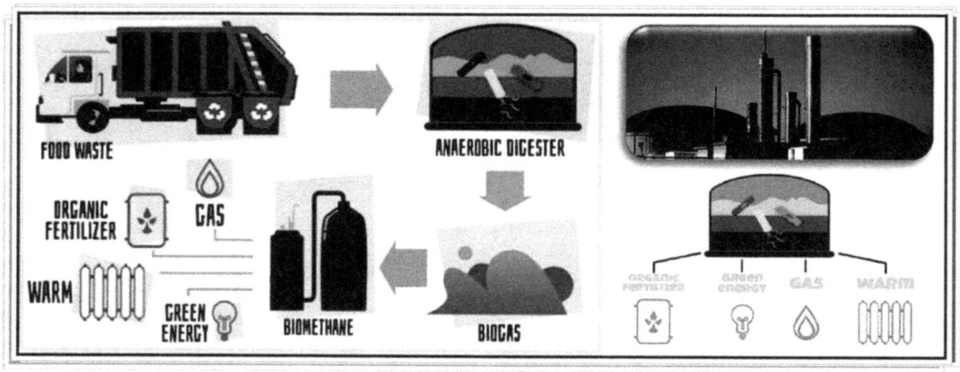

Types of Biogas Plants

There are several types of biogas plants, and the choice of plant depends on various factors, such as the type and quantity of feedstock, the desired energy output, and the availability of resources. Here are some common types of biogas plants (Bharathiraja et al., 2016; Caruso et al., 2019; Kalaiselvan et al., 2022; Kumar et al., 2021; Mahmudul et al., 2021; Nkuna et al., 2022; Tauseef et al., 2013; Van et al., 2020):

1. **Continuous Stirred Tank Reactor (CSTR):** The CSTR is the most common type of biogas plant and consists of a single tank that is continuously stirred. The feedstock is added to the tank, and biogas is produced as the feedstock is decomposed. The digested slurry is then removed from the tank, and fresh feedstock is added.
2. **Plug Flow Reactor (PFR):** The PFR is similar to the CSTR but has a different design that allows for better mixing and more efficient biogas production. The feedstock is added at one end of the reactor and flows through the tank as biogas is produced. The digested slurry is then removed from the other end of the tank.
3. **Batch Reactor:** The batch reactor is a simple type of biogas plant that operates in batches. The feedstock is added to the reactor, and biogas is produced over a fixed period of time. Once the biogas production is complete, the digested slurry is removed, and fresh feedstock is added.
4. **Two-Stage Reactor:** The two-stage reactor consists of two tanks that operate in series. The first tank is used for hydrolysis and acidogenesis, while the second tank is used for methanogenesis. This type of biogas plant is useful for feedstocks that are difficult to digest, such as lignocellulosic biomass.
5. **Upflow Anaerobic Sludge Blanket (UASB):** The UASB reactor is a high-rate anaerobic digestion process that operates without any mixing. The feedstock is added at the bottom of the reactor, and biogas is produced as the feedstock rises through the tank. The digested slurry is then removed from the top of the tank.
6. **Fixed Dome Biogas Plant:** The fixed dome biogas plant is a low-cost biogas plant that consists of a dome-shaped digester made of brick or concrete. The feedstock is added to the digester, and biogas is produced as the feedstock is decomposed. The digested slurry is then removed from the bottom of the digester.

Each type of biogas plant has its advantages and disadvantages, and the choice of plant depends on various factors, such as the type and quantity of feedstock, the desired energy output, and the availability of resources.

FACTORS AFFECTING BIOGAS PLANT PERFORMANCE AND EFFICIENCY

There are several factors that can affect the performance and efficiency of a biogas plant, including:

1. **Feedstock Quality:** The quality and composition of the feedstock being used in the biogas plant can have a significant impact on its performance and efficiency. Ideally, the feedstock should have a high organic content and be free of contaminants such as plastics, metals, and glass.
2. **Temperature:** The temperature of the biogas plant is important for the growth and activity of the microorganisms responsible for the anaerobic digestion process. The optimal temperature range for most anaerobic microorganisms is between 30°C and 40°C.
3. **Retention Time:** The retention time is the amount of time that the feedstock remains in the biogas plant, and it is an important factor for the efficiency of the biogas production. The optimal retention time depends on the type of feedstock, the operating temperature, and the design of the biogas plant.
4. **Mixing:** Proper mixing of the feedstock is important to ensure that all the material is digested evenly and efficiently. Good mixing can improve the production of biogas and reduce the retention time required.
5. **pH Control:** The pH of the biogas plant is another important factor that affects the performance and efficiency of the plant. The optimal pH range for most anaerobic microorganisms is between 6.5 and 7.5.
6. **Nutrient Balance:** Adequate nutrient levels (e.g., nitrogen, phosphorus, and potassium) are important for the growth and activity of the microorganisms responsible for the anaerobic digestion process. A lack of nutrients can limit biogas production, while excess nutrients can cause problems with the digestate.
7. **Biogas Storage and Utilization:** Proper storage and utilization of the biogas is important to maximize the efficiency of the biogas plant. The biogas can be used to generate electricity or heat, or it can be upgraded and injected into the natural gas grid.

Overall, the performance and efficiency of a biogas plant are influenced by several factors, and optimizing these factors is essential for achieving maximum biogas production and utilization (Abuabdou et al., 2020; Chiu & Lo, 2016; Hagos et al., 2017; Klassen et al., 2016; Liu et al., 2021; Mao et al., 2015; Nsair et al., 2020; Pöschl et al., 2010; Shen et al., 2015).

CHALLENGES ASSOCIATED WITH BIOGAS PRODUCTION FOR MUNICIPAL WASTE MANAGEMENT

Biogas production can be a promising technology for municipal waste management. However, there are several challenges associated with biogas production for municipal waste management. Some of these challenges include:

1. **Feedstock Availability and Quality:** One of the primary challenges associated with biogas production for municipal waste management is the availability and quality of the feedstock. Municipal waste streams can be highly variable, and it can be difficult to ensure a consistent supply of high-quality feedstock.
2. **Contaminants in the Feedstock:** Municipal waste streams can contain a wide range of contaminants, including heavy metals, plastics, and other non-organic materials. These contaminants can interfere with the biogas production process and reduce the quality of the biogas produced.
3. **Technical Challenges:** Biogas production for municipal waste management can be a technically complex process that requires specialized equipment and expertise. The construction and maintenance of biogas plants can be expensive and require ongoing technical support.
4. **Regulatory Challenges:** There can be regulatory challenges associated with biogas production for municipal waste management. For example, there may be permitting requirements and other regulatory hurdles that need to be addressed before a biogas plant can be built and operated.
5. **Public Perception:** Biogas production for municipal waste management can sometimes face public resistance due to concerns about odors and other environmental impacts associated with the process.
6. **Economic Viability:** The economic viability of biogas production for municipal waste management can be a challenge, particularly in areas where there is a low demand for biogas. Without a viable market for the biogas produced, the costs of building and operating a biogas plant may outweigh the benefits.

Overall, biogas production for municipal waste management can be a promising technology for addressing the environmental and economic challenges associated with waste disposal. However, addressing the challenges associated with the technology will be critical for its successful adoption and implementation (Bachmann et al., 2015; Bhatt & Tao, 2020; Bundhoo, 2018; Ghosh et al., 2020; Igliński et al., 2015; Moreda, 2016; Pandyaswargo, Jagath Dickella Gamaralalage, Liu, Knaus, Onoda, Mahichi, & Guo, 2019; Patinvoh & Taherzadeh, 2019; Tabatabaei & Ghanavati, 2018).

By utilizing biogas as a power source, it may be possible to revolutionize the way we approach air conditioning. Traditionally, these energy-intensive appliances can consume up to 80% of the electricity used in indoor dwellings during the summer months (Aggarwal et al., 2022; Akhai, Mala, & Jerin, 2020; Akhai, Mala, & Jerin, 2021; Akhai, Singh, & John, 2016; Akhai, Singh, & John, 2016; Akhai, Srivastava, & Sharma, 2020; Akhai, Srivastava, Sharma et al, 2021; Akhai et al., 2017; Kumar & Akhai, 2022; Tanwar & Akhai, 2017). However, by coupling biogas plants with other renewable technologies like solar or wind energy, we can generate even more electricity to support high-consumption industries and sectors like air conditioning. This is especially crucial since air conditioning alone can account for 62 to 80 percent of the electricity used in indoor dwellings during the summer season. The potential for biogas to make a significant impact on reducing our carbon footprint while simultaneously providing a reliable source of energy is truly exciting.

BENEFITS OF BIOGAS PLANTS FOR SUSTAINABLE MUNICIPAL WASTE MANAGEMENT

Biogas plants can play an important role in sustainable municipal waste management by providing a range of benefits, including:

1. **Reducing Waste Disposal in Landfills:** Biogas plants can effectively reduce the amount of waste that ends up in landfills. Organic waste is broken down in the biogas plant and converted into biogas, leaving behind a smaller amount of solid residue that can be used as fertilizer.
2. **Producing Renewable Energy:** Biogas is a renewable energy source that can be used for heating, electricity generation, and transportation fuel. The energy produced from biogas can reduce the reliance on fossil fuels and help to mitigate greenhouse gas emissions.
3. **Improving Air Quality:** Biogas plants can significantly reduce air pollution from organic waste, which emits harmful gases like methane and ammonia when left untreated. By capturing these gases and converting them into biogas, biogas plants can help improve air quality in the surrounding area.
4. **Creating Employment Opportunities:** Biogas plants require skilled labor and can create employment opportunities in the local community.
5. **Providing Organic Fertilizer:** The solid residue left behind after the production of biogas can be used as a high-quality organic fertilizer, reducing the need for synthetic fertilizers and improving soil quality.
6. **Job Creation and Economic Benefits:** Biogas plants can create jobs in the waste management and renewable energy sectors, contributing to local economic development and job creation.
7. **Improved Public Health:** Biogas plants can reduce the environmental impacts of waste management practices, such as air and water pollution, which can have negative impacts on public health. By reducing these impacts, biogas plants can contribute to improved public health.

Thus, biogas plants offer several benefits in municipal waste management, including renewable energy generation, reduction of greenhouse gas emissions, waste reduction and diversion, production of high-quality organic fertilizers, job creation and economic benefits, and improved public health. Overall, biogas plants can provide a sustainable and environmentally-friendly solution to municipal waste management, while also providing economic and social benefits to local communities (Ali et al., 2012; Ansar Ali et al., 2022; Arvola et al., 2012; Baeyens et al., 2016; Czekała, 2022; Guo et al., 2019; Hettiarachchi et al., 2018; Lamb, 2020; Nielsen & Oleskowicz-Popiel, 2008; Shankar & Khandelwal, 2017).

FEW CASE STUDIES OF BIOGAS PLANTS FOR MUNICIPAL WASTE MANAGEMENT

Biogas plants have become an increasingly popular solution for municipal waste management, as they allow for the generation of renewable energy while also reducing waste disposal costs. Here are some successful case studies of biogas plants for municipal waste management ():

1. **Copenhagen, Denmark:** Denmark is leading the charge in sustainable energy with nearly 20 centralised biogas plants currently in operation. The largest biogas facility in the Nordic region, Maabjerg BioEnergy, is capable of treating a whopping 800,000 tonnes of biomass each year and producing 21 million m^3 of biogas enough biogas to power 5,000 households. It's clear that Denmark is setting the standard for renewable energy and making great strides in creating a cleaner, greener future for all.
2. **Slovenia:** Croatia, Serbia, and Slovenia are exploring the potential of EBA biogas in their energy mix. These nations are studying a range of feedstock combinations, including energy crops, food waste, biowaste, and food and beverage waste, to develop innovative ways of producing clean and renewable energy. Given their predominantly agricultural regions and high agricultural productivity, it is unsurprising that these three nations already have 49 biogas plants with a total installed capacity of 8.38 MWel. Slovenia is leading the way with a remarkable 27.2 MW installed capacity, followed by Croatia and Serbia with 18 MWel and 8.33 MWel, respectively. Additionally, these countries are examining the feasibility of using biogas and biomethane for transportation and evaluating existing gas networks to optimize their use of this sustainable energy source. It is clear that Croatia, Serbia, and Slovenia are pioneering the biogas revolution, and the potential for growth and advancement in this industry is immense.
3. **Germany:** Germany is leading the charge in sustainable energy with its state-of-the-art biogas plants! These remarkable facilities process organic waste to generate a staggering 28.8 million cubic meters of raw biogas per year. This clean, renewable energy source is then fed directly into the gas grid, providing a reliable source of energy for countless households and businesses.
4. **Pune, India:** The city of Pune operates a biogas plant that processes organic waste from households and markets. The plant has a capacity of 200 tons of organic waste per day and produces enough biogas to support 3,000 households.
5. **Assam, India:** Sonpur, Assam is taking a big step towards sustainable energy! On February 25th, 2023, a foundation stone was laid for a biogas plant that will revolutionize the way organic waste is managed. With a capacity to process 5 tons of waste per day, this state-of-the-art plant will produce biogas that can be used as a renewable energy source. Residents can look forward to the plant becoming operational in November 2023, bringing a new era of sustainable energy to the city.
6. **United States:** The city of San Francisco opened a biogas plant that processes organic waste from households, restaurants, and grocery stores. The plant has capacities 600 tons of organic waste per day and produces enough biogas to support around 5000 households.

Overall, these successful case studies demonstrate the potential of biogas plants as a sustainable solution for municipal waste management, providing renewable energy while reducing waste disposal costs and environmental impact.

FUTURE RESEARCH DIRECTIONS FOR BIOGAS PLANTS FOR SUSTAINABLE MUNICIPAL WASTE MANAGEMENT

Biogas plants are a promising technology for sustainable municipal waste management, as they can help reduce greenhouse gas emissions and generate renewable energy. There is a need for more research on the performance evaluation of biogas plants. This includes the study of different performance indicators, such as biogas yield, biogas composition, and effluent quality, to evaluate the effectiveness of biogas plants and identify areas for improvement. Here are some future research directions that can further improve the efficiency and sustainability of biogas plants:

1. **Optimization of Feedstock Composition:** The composition of feedstock has a significant impact on biogas production and quality. Future research can focus on identifying optimal mixtures of different organic waste streams, such as food waste, yard waste, and biosolids, to improve biogas yield and quality.
2. **Advanced Pre-Treatment Methods:** Pre-treatment methods such as thermal hydrolysis and ultrasound can improve the breakdown of complex organic compounds and increase biogas production. Future research can focus on developing more efficient and cost-effective pre-treatment methods that can be integrated into existing biogas plants.
3. **Enhanced Digestate Management:** Digestate, the byproduct of the biogas process, is a valuable fertilizer that can be used to improve soil health and fertility. Future research can focus on developing new technologies to improve digestate quality and reduce pathogens and contaminants.
4. **Integration With Other Renewable Energy Sources:** Biogas plants can be integrated with other renewable energy sources, such as solar and wind power, to provide a more stable and reliable energy supply. Future research can focus on optimizing the integration of biogas plants with other renewable energy sources to maximize energy output and reduce costs.
5. **Life Cycle Assessment:** A comprehensive life cycle assessment can provide a holistic view of the environmental and economic impacts of biogas plants. Future research can focus on developing more accurate and reliable life cycle assessment methodologies to help policymakers and industry stakeholders make informed decisions about the deployment and operation of biogas plants.
6. **Process Control and Automation:** Process control and automation can improve the efficiency and reliability of biogas plants. Future research can focus on developing advanced process control and automation systems that can optimize biogas production and quality while minimizing energy and resource consumption.
7. **Biogas Plant Design and Operation:** More research is needed on the design and operation of biogas plants. This includes the study of factors that affect plant performance, such as mixing, temperature, pH, and hydraulic retention time, to optimize the design and operation of biogas plants.
8. **Economic and Environmental Sustainability of Biogas Plants:** There is a need for more research on the economic and environmental sustainability of biogas plants. This includes the study of the costs and benefits of biogas production, the environmental impacts of biogas production, and the potential for scaling up biogas production to meet increasing demand.
9. **Public Acceptance and Stakeholder Engagement:** More research is needed on public acceptance and stakeholder engagement in biogas plant development and operation. This includes the study of factors that affect public perception and acceptance of biogas plants, as well as strategies for engaging stakeholders in the planning and implementation of biogas projects.

These research directions can help to further advance the development and deployment of biogas plants for sustainable municipal waste management. As seen above, there are several research gaps and opportunities in the area of biogas plants for sustainable municipal waste management. Addressing these gaps can help to enhance the effectiveness of biogas plants and contribute to a more sustainable and circular economy.

CONCLUSION

In conclusion, biogas plants have the potential to play a significant role in sustainable municipal waste management. By using organic waste as feedstock, biogas plants can reduce the amount of waste sent to landfills, which not only reduces the environmental impact of waste disposal but also provides a source of renewable energy. Biogas plants can also produce organic fertilizer, which can be used in agriculture to improve soil quality and reduce the use of chemical fertilizers.

Furthermore, biogas plants offer a flexible and scalable solution that can be adapted to different feedstocks and energy demands. Biogas can be used for electricity generation, heat production, vehicle fuel, and cooking fuel, among other applications. This versatility makes biogas plants an attractive option for municipalities looking to diversify their energy sources and reduce their carbon footprint. However, to fully realize the potential of biogas plants for sustainable municipal waste management, there are some challenges that need to be addressed. These include the need for adequate feedstock supply, efficient and cost-effective technology, and favorable policy and regulatory frameworks. In addition, there is a need for more research and development to optimize the biogas production process, improve the quality of the biogas produced, and explore new applications for biogas. Biogas plants offer a promising solution for sustainable municipal waste management, and continued investment in research and development can help to further advance this technology and realize its full potential.

Overall, biogas plants are a promising solution for sustainable municipal waste management that can provide a range of environmental and economic benefits. With the right support and investment, biogas plants can play a crucial role in the transition to a more sustainable and circular economy.

REFERENCES

Abanades, S., Abbaspour, H., Ahmadi, A., Das, B., Ehyaei, M. A., Esmaeilion, F., El Haj Assad, M., Hajilounezhad, T., Hmida, A., Rosen, M. A., Safari, S., Shabi, M. A., & Silveira, J. L. (2022). A conceptual review of sustainable electrical power generation from biogas. *Energy Science & Engineering*, *10*(2), 630–655. doi:10.1002/ese3.1030

Abuabdou, S. M., Ahmad, W., Aun, N. C., & Bashir, M. J. (2020). A review of anaerobic membrane bioreactors (AnMBR) for the treatment of highly contaminated landfill leachate and biogas production: Effectiveness, limitations and future perspectives. *Journal of Cleaner Production*, *255*, 120215. doi:10.1016/j.jclepro.2020.120215

AffairsCloud. (2023, February 28). *Current Affairs*. AffairsCloud. https://affairscloud.com/current-affairs-28-february-2023/

Aggarwal, P., Rana, M., & Akhai, S. (2022). Briefings on e-waste hazard until COVID era in India. *Materials Today: Proceedings*, *71*(2), 389–393. doi:10.1016/j.matpr.2022.09.507

Ajay, C. M., Mohan, S., & Dinesha, P. (2021). Decentralized energy from portable biogas digesters using domestic kitchen waste: A review. *Waste Management (New York, N.Y.)*, *125*, 10–26. doi:10.1016/j.wasman.2021.02.031 PMID:33667979

Akhai, S., Mala, S., & Jerin, A. A. (2020). Apprehending Air Conditioning Systems in Context to COVID-19 and Human Health: A Brief Communication. *International Journal of Healthcare Education & Medical Informatics, 7*(1&2), 28-30.

Akhai, S., Mala, S., & Jerin, A. A. (2021). Understanding whether air filtration from air conditioners reduces the probability of virus transmission in the environment. *Journal of Advanced Research in Medical Science & Technology, 8*(1), 36-41.

Akhai, S., Singh, V. P., & John, S. (2016). Investigating Indoor Air Quality for the Split-Type Air Conditioners in an Office Environment and Its Effect on Human Performance. *Journal of Mechanical Civil Engineering, 13*(6), 113–118.

Akhai, S., Singh, V. P., & John, S. (2016). Human performance in industrial design centers with small unit air conditioning systems. *Journal of Advanced Research in Production Industrial Engineering, 3*(2), 5–11.

Akhai, S., Srivastava, P., & Sharma, S. (2020). Developments in Horizontal Axis Wind Turbines - A Brief Review. *Journal of Critical Reviews, 7*(19), 255–260.

Akhai, S., Srivastava, P., Sharma, V., & Bhatia, A. (2021). Investigating weld strength of AA8011-6062 alloys joined via friction-stir welding using the RSM approach. [). IOP Publishing.]. *Journal of Physics: Conference Series, 1950*(1), 012016. doi:10.1088/1742-6596/1950/1/012016

Akhai, S., Thareja, P., & Singh, V. P. (2017). Assessment of Indoor Environment Health Sustenance in Air Conditioned Class Rooms. *Journal of Advanced Research in Civil and Environmental Engineering, 4*(1&2), 1–9. doi:10.24321/2393.8307.201701

Al-Mansour, F., Sucic, B., & Pusnik, M. (2014). Challenges and prospects of electricity production from renewable energy sources in Slovenia. *Energy, 77*, 73–81. doi:10.1016/j.energy.2014.04.106

Ali, G., Nitivattananon, V., Abbas, S., & Sabir, M. (2012). Green waste to biogas: Renewable energy possibilities for Thailand's green markets. *Renewable & Sustainable Energy Reviews, 16*(7), 5423–5429. doi:10.1016/j.rser.2012.05.021

Ansar Ali, S. K., Kumar, P., Sharma, V. K., Akhai, S., Kumar, V., & Joshi, R. S. (2022). Determination of friction factor and heat transfer rate in laminar flow in an agitated vessel. *Proceedings of the Institution of Mechanical Engineers, Part E: Journal of Process Mechanical Engineering*.

Arvola, J., Belt, P., Harkonen, J., Kess, P., & Imppola, R. (2012). Biogas as an option for industrial applications. *International Journal of Sustainable Economy, 4*(1), 71–88. doi:10.1504/IJSE.2012.043996

Bachmann, N., la Cour Jansen, J., Bochmann, G., & Montpart, N. (2015). *Sustainable biogas production in municipal wastewater treatment plants*. IEA Bioenergy.

Baeyens, J., Appels, L., Peng, L. Y., & Dewil, R. (2016). Biogas as a renewable fuel. *Renewable Energy, 96*(Part B), 1055-1152.

Bharathiraja, B., Sudharsanaa, T., Bharghavi, A., Jayamuthunagai, J., & Praveenkumar, R. (2016). Biohydrogen and Biogas–An overview on feedstocks and enhancement process. *Fuel, 185*, 810–828. doi:10.1016/j.fuel.2016.08.030

Bhatt, A. H., & Tao, L. (2020). Economic perspectives of biogas production via anaerobic digestion. *Bioengineering (Basel, Switzerland)*, *7*(3), 74. doi:10.3390/bioengineering7030074 PMID:32674480

Biogas Forum India. (2015). *E–News Letter, 6*(2). Biogas Forum. http://www.biogasforum.in/newsletter/BFI_Newsletter_Jan-Jun_2015.pdf

Bong, C. P. C., Ho, W. S., Hashim, H., Lim, J. S., Ho, C. S., Tan, W. S. P., & Lee, C. T. (2017). Review on the renewable energy and solid waste management policies towards biogas development in Malaysia. *Renewable & Sustainable Energy Reviews*, *70*, 988–998. doi:10.1016/j.rser.2016.12.004

Brunner, P. H., & Rechberger, H. (2015). Waste to energy–key element for sustainable waste management. *Waste Management (New York, N.Y.)*, *37*, 3–12. doi:10.1016/j.wasman.2014.02.003 PMID:24630214

Bundgaard, S. S., Kofoed-Wiuff, A., Herrmann, I. T., & Karlsson, K. B. (2014). Experiences with biogas in Denmark. *DTU Management Engineering*.

Bundhoo, Z. M. (2018). Solid waste management in least developed countries: Current status and challenges faced. *Journal of Material Cycles and Waste Management*, *20*(3), 1867–1877. doi:10.100710163-018-0728-3

Caruso, M. C., Braghieri, A., Capece, A., Napolitano, F., Romano, P., Galgano, F., Altieri, G., & Genovese, F. (2019). Recent updates on the use of agro-food waste for biogas production. *Applied Sciences (Basel, Switzerland)*, *9*(6), 1217. doi:10.3390/app9061217

Central Public Health and Environmental Engineering Organisation (CPHEEO). (2016). *SWACHH BHARAT MISSION Municipal Solid Waste Management Manual Part II: The Manual*. Ministry of Urban Development. http://moud.gov.in/sites/default/files/SBM-II-2016.pdf

Chaianong, A., & Pharino, C. (2022). How to design an area-based prioritization of biogas production from organic municipal solid waste? Evidence from Thailand. *Waste Management (New York, N.Y.)*, *138*, 243–252. doi:10.1016/j.wasman.2021.11.042 PMID:34906780

Chiu, S. L., & Lo, I. M. (2016). Reviewing the anaerobic digestion and co-digestion process of food waste from the perspectives on biogas production performance and environmental impacts. *Environmental Science and Pollution Research International*, *23*(24), 24435–24450. doi:10.100711356-016-7159-2 PMID:27380183

Czekała, W. (2022). Biogas as a Sustainable and Renewable Energy Source. *Clean Fuels for Mobility*, 201-214.

De Feo, G., Antoniou, G., Fardin, H. F., El-Gohary, F., Zheng, X. Y., Reklaityte, I., Butler, D., Yannopoulos, S., & Angelakis, A. N. (2014). The historical development of sewers worldwide. *Sustainability (Basel)*, *6*(6), 3936–3974. doi:10.3390u6063936

Dimitrov, R., & Zlateva, P. (2019, September). Possible Uses of Biogas for Power Purposes. In *2019 11th Electrical Engineering Faculty Conference (BulEF)* (pp. 1-4). IEEE. 10.1109/BulEF48056.2019.9030756

Gardiner, B. (2014). Biogas, a low-tech fuel with a big payoff. *The International New York Times. Nova Iorque, 30*.

Ghatge, K. (2014). Waste to energy: A way forward for sustainable solid waste management in India. *iCrier.* https://icrier.org/Urbanisation/events/30-5-14/Waste_Energy_Ketaki%20Ghatge.pdf

Ghosh, P., Shah, G., Sahota, S., Singh, L., & Vijay, V. K. (2020). Biogas production from waste: Technical overview, progress, and challenges. *Bioreactors*, 89-104.

Guo, Z., Sun, Y., Pan, S. Y., & Chiang, P. C. (2019). Integration of green energy and advanced energy-efficient technologies for municipal wastewater treatment plants. *International Journal of Environmental Research and Public Health, 16*(7), 1282. doi:10.3390/ijerph16071282 PMID:30974807

Hagos, K., Zong, J., Li, D., Liu, C., & Lu, X. (2017). Anaerobic co-digestion process for biogas production: Progress, challenges and perspectives. *Renewable & Sustainable Energy Reviews, 76*, 1485–1496. doi:10.1016/j.rser.2016.11.184

Hettiarachchi, H., Meegoda, J. N., & Ryu, S. (2018). Organic waste buyback as a viable method to enhance sustainable municipal solid waste management in developing countries. *International Journal of Environmental Research and Public Health, 15*(11), 2483. doi:10.3390/ijerph15112483 PMID:30405058

Hirpe, L., & Yeom, C. (2021). Municipal solid waste management policies, practices, and challenges in Ethiopia: A systematic review. *Sustainability (Basel), 13*(20), 11241. doi:10.3390u132011241

Igliński, B., Buczkowski, R., & Cichosz, M. (2015). Biogas production in Poland—Current state, potential and perspectives. *Renewable & Sustainable Energy Reviews, 50*, 686–695. doi:10.1016/j.rser.2015.05.013

Jacobsen, B. H., Laugesen, F. M., & Dubgaard, A. (2014). The economics of biogas in Denmark: a farm and socioeconomic perspective. *International Journal of Agricultural Management, 3.*

Joshi, R., & Ahmed, S. (2016). Status and challenges of municipal solid waste management in India: A review. *Cogent Environmental Science, 2*(1), 1139434. doi:10.1080/23311843.2016.1139434

Kabeyi, M. J. B., & Olanrewaju, O. A. (2022). Biogas production and applications in the sustainable energy transition. *Journal of Energy*. doi:10.1155/2022/8750221

Kalaiselvan, N., Glivin, G., Bakthavatsalam, A. K., Mariappan, V., Premalatha, M., Raveendran, P. S., Jayaraj, S., & Sekhar, S. J. (2022). A waste to energy technology for Enrichment of biomethane generation: A review on operating parameters, types of biodigesters, solar assisted heating systems, socio economic benefits and challenges. *Chemosphere, 293*, 133486. doi:10.1016/j.chemosphere.2021.133486 PMID:35016951

Karak, T., Bhagat, R. M., & Bhattacharyya, P. (2012). Municipal solid waste generation, composition, and management: The world scenario. *Critical Reviews in Environmental Science and Technology, 42*(15), 1509–1630. doi:10.1080/10643389.2011.569871

Kaur, A., & Deswal, S. (2019). Sustainable solid waste management in Indian cities. In *Recycled Waste Materials: Proceedings of EGRWSE 2018* (pp. 239-251). Research Gate.

Khalil, M., Berawi, M. A., Heryanto, R., & Rizalie, A. (2019). Waste to energy technology: The potential of sustainable biogas production from animal waste in Indonesia. *Renewable & Sustainable Energy Reviews, 105*, 323–331. doi:10.1016/j.rser.2019.02.011

Klassen, V., Blifernez-Klassen, O., Wobbe, L., Schlüter, A., Kruse, O., & Mussgnug, J. H. (2016). Efficiency and biotechnological aspects of biogas production from microalgal substrates. *Journal of Biotechnology*, *234*, 7–26. doi:10.1016/j.jbiotec.2016.07.015 PMID:27449486

Kollikkathara, N., Feng, H., & Stern, E. (2009). A purview of waste management evolution: Special emphasis on USA. *Waste Management (New York, N.Y.)*, *29*(2), 974–985. doi:10.1016/j.wasman.2008.06.032 PMID:18796347

Kollikkathara, N., Feng, H., & Stern, E. (2009). A purview of waste management evolution: Special emphasis on USA. *Waste Management (New York, N.Y.)*, *29*(2), 974–985. doi:10.1016/j.wasman.2008.06.032 PMID:18796347

Kulkarni, B. N. (2020). Environmental sustainability assessment of land disposal of municipal solid waste generated in Indian cities–A review. *Environmental Development*, *33*, 100490. doi:10.1016/j.envdev.2019.100490

Kumar, P., & Akhai, S. (2022). Effective energy management in smart buildings using VRV/VRF systems. *Additive Manufacturing in Industry*, *4*(0), 27–35. doi:10.1201/9781003360001-2

Kumar, R., Neerudu, U. K., Gothalwal, R., Mohapatra, S., Deshpande, P. K., Vani, M. M., & Merugu, R. (2021). Bioprocess parameters for thermophilic and mesophilic biogas production: recent trends and challenges. *Bioenergy Research: Basic and Advanced Concepts*, 225-256.

Lamb, J. J. (2020). *Biogas and the Energy Sector*. Anaerob. Dig. From Biomass to Biogas. doi:10.48216/9788269203325ch1

Lazaro, K., Malempati, S., & Varela, S. (2015). *Analysis of the production and usage of biogas at the San Francisco agricultural school. An Interactive Qualifying Project*. Worcester Polytechnic Institute. https://web.wpi.edu/Pubs/E-project/Available/E-project-051515-102329/unrestricted/Analysis_of_the_Production_and_Usage_of_Biogas_at_the_San_Francisco_Agricultural_School.pdf

Lee, C. C., & Huffman, G. L. (1989). Incineration of solid waste. *Environmental Progress*, *8*(3), 143–151. doi:10.1002/ep.3300080312

Liu, M., Wei, Y., & Leng, X. (2021). Improving biogas production using additives in anaerobic digestion: A review. *Journal of Cleaner Production*, *297*, 126666. doi:10.1016/j.jclepro.2021.126666

Mæng, H., Lund, H., & Hvelplund, F. (1999). Biogas plants in Denmark: Technological and economic developments. *Applied Energy*, *64*(1-4), 195–206. doi:10.1016/S0306-2619(99)00067-7

Mahmudul, H. M., Rasul, M. G., Akbar, D., Narayanan, R., & Mofijur, M. (2021). A comprehensive review of the recent development and challenges of a solar-assisted biodigester system. *The Science of the Total Environment*, *753*, 141920. doi:10.1016/j.scitotenv.2020.141920 PMID:32889316

Mao, C., Feng, Y., Wang, X., & Ren, G. (2015). Review on research achievements of biogas from anaerobic digestion. *Renewable & Sustainable Energy Reviews*, *45*, 540–555. doi:10.1016/j.rser.2015.02.032

Ministry of Housing and Urban Affairs. (2015). Swachh Bharat Mission (Urban). In *Handbook for Swachh Bharat Mission (Urban)* (Vol. 2). MoHUA. https://mohua.gov.in/upload/uploadfiles/files/Part2.pdf

Moreda, I. L. (2016). The potential of biogas production in Uruguay. *Renewable & Sustainable Energy Reviews, 54*, 1580–1591. doi:10.1016/j.rser.2015.10.099

Mundhe, N., Jaybhaye, R., & Dorik, B. (2014). Assessment of municipal solid waste management of Pune city using geospatial tools. *International Journal of Computer Applications, 100*(10), 24–32. doi:10.5120/17562-8184

Nielsen, J. B. H., & Oleskowicz-Popiel, P. (2008, December). Biogas–a promising renewable energy source for Europe. In *AEBIOM Workshop-European Parliament, Brussels*. Scientific Research Publishing.

Nkuna, R., Roopnarain, A., Rashama, C., & Adeleke, R. (2022). Insights into organic loading rates of anaerobic digestion for biogas production: A review. *Critical Reviews in Biotechnology, 42*(4), 487–507. PMID:34315294

Nsair, A., Onen Cinar, S., Alassali, A., Abu Qdais, H., & Kuchta, K. (2020). Operational parameters of biogas plants: A review and evaluation study. *Energies, 13*(15), 3761. doi:10.3390/en13153761

Oakes, T. W., & Shank, K. E. (1979). *Brief review of the Resource Conservation and Recovery Act (RCRA) 1976*. OSTI.

Pandyaswargo, A. H., Jagath Dickella Gamaralalage, P., Liu, C., Knaus, M., Onoda, H., Mahichi, F., & Guo, Y. (2019). Challenges and an implementation framework for sustainable municipal organic waste management using biogas technology in emerging Asian Countries. *Sustainability (Basel), 11*(22), 6331. doi:10.3390u11226331

Patinvoh, R. J., & Taherzadeh, M. J. (2019). Challenges of biogas implementation in developing countries. *Current Opinion in Environmental Science & Health, 12*, 30–37. doi:10.1016/j.coesh.2019.09.006

Poje, T. (2010). Slovenian legislation-driving power and obstacle to the development of agricultural biogas plants. *Novi izzivi v poljedelstvu 2010. Zbornik simpozija, Rogaška. Slatina, 2-3*(December), 277–283.

Poje, T. (2016). Situation and development trends of agricultural biogas plants in Slovenia. *51. hrvatski i 11. međunarodni simpozij agronoma, 15.-18. veljače 2016. godine, Opatija, Hrvatska. Zbornik radova*, 502-506.

Pöschl, M., Ward, S., & Owende, P. (2010). Evaluation of energy efficiency of various biogas production and utilization pathways. *Applied Energy, 87*(11), 3305–3321. doi:10.1016/j.apenergy.2010.05.011

Rao, K. C., Parthan, B., & Doshi, K. (2018). *Power from municipal solid waste at Pune Municipal Corporation (Pune, Maharashtra, India)*. CGIAR.

Raven, R. P., & Gregersen, K. H. (2007). Biogas plants in Denmark: Successes and setbacks. *Renewable & Sustainable Energy Reviews, 11*(1), 116–132. doi:10.1016/j.rser.2004.12.002

Roubík, H., Mazancová, J., Le Dinh, P., Dinh Van, D., & Banout, J. (2018). Biogas quality across small-scale biogas plants: A case of central Vietnam. *Energies, 11*(7), 1794. doi:10.3390/en11071794

Sandhu, K. (2014). Historical trajectory of waste management; an analysis using the health belief model. *Management of Environmental Quality, 25*(5), 615–630. doi:10.1108/MEQ-09-2013-0098

Saur, G., & Milbrandt, A. (2014). Renewable hydrogen potential from biogas in the United States (No. NREL/TP-5400-60283). National Renewable Energy Lab (NREL).

Shah, A. V., Srivastava, V. K., Mohanty, S. S., & Varjani, S. (2021). Municipal solid waste as a sustainable resource for energy production: State-of-the-art review. *Journal of Environmental Chemical Engineering*, *9*(4), 105717. doi:10.1016/j.jece.2021.105717

Shankar, Y. S., & Khandelwal, R. (2017). Sustainable waste management strategy for a campus: A case study of JUET, Guna. *Management of Environmental Quality*, *28*(5), 6. doi:10.1108/MEQ-01-2016-0008

Sharholy, M., Ahmad, K., Mahmood, G., & Trivedi, R. C. (2008). Municipal solid waste management in Indian cities–A review. *Waste Management (New York, N.Y.)*, *28*(2), 459–467. doi:10.1016/j.wasman.2007.02.008 PMID:17433664

Shen, Y., Linville, J. L., Urgun-Demirtas, M., Mintz, M. M., & Snyder, S. W. (2015). An overview of biogas production and utilization at full-scale wastewater treatment plants (WWTPs) in the United States: Challenges and opportunities towards energy-neutral WWTPs. *Renewable & Sustainable Energy Reviews*, *50*, 346–362. doi:10.1016/j.rser.2015.04.129

Štangar, U. L., & Kranjčevič, E. (2008). *Renewable energy sources in slovenia: facts and plans*. Springer Netherlands.

Sunil, M., Narayan, A., Bhat, V., & Vinay, S. (2013). Smart biogas plant. *International Journal of Innovative Technology and Exploring Engineering*, *3*(3), 62–66.

Tabatabaei, M., Aghbashlo, M., Valijanian, E., Panahi, H. K. S., Nizami, A. S., Ghanavati, H., & Karimi, K. (2020). A comprehensive review on recent biological innovations to improve biogas production, part 2: Mainstream and downstream strategies. *Renewable Energy*, *146*, 1392–1407. doi:10.1016/j.renene.2019.07.047

Tabatabaei, M., & Ghanavati, H. (Eds.). (2018). *Biogas: fundamentals, process, and operation* (Vol. 6). Springer. doi:10.1007/978-3-319-77335-3_6

Tanwar, N., & Akhai, S. (2017). Survey Analysis for Quality Control Comfort Management in Air Conditioned Classroom. *Journal of Advanced Research in Civil and Environmental Engineering*, *4*(1&2), 20–23. doi:10.24321/2393.8307.201702

Tauseef, S. M., Premalatha, M., Abbasi, T., & Abbasi, S. A. (2013). Methane capture from livestock manure. *Journal of Environmental Management*, *117*, 187–207. doi:10.1016/j.jenvman.2012.12.022 PMID:23376302

Thakare, S., & Nandi, S. (2016). Study on potential of gasification technology for municipal solid waste (MSW) in Pune city. *Energy Procedia*, *90*, 509–517. doi:10.1016/j.egypro.2016.11.218

Thiruselvi, D., Kumar, P. S., Kumar, M. A., Lay, C. H., Aathika, S., Mani, Y., Jagadiswary, D., Dhanasekaran, A., Shanmugam, P., Sivanesan, S., & Show, P. L. (2021). A critical review on global trends in biogas scenario with its up-gradation techniques for fuel cell and future perspectives. *International Journal of Hydrogen Energy*, *46*(31), 16734–16750. doi:10.1016/j.ijhydene.2020.10.023

Thomas, P., & Soren, N. (2020). An overview of municipal solid waste-to-energy application in Indian scenario. *Environment, Development and Sustainability*, *22*(2), 575–592. doi:10.100710668-018-0235-7

Tian, H., Wang, X., Lim, E. Y., Lee, J. T., Ee, A. W., Zhang, J., & Tong, Y. W. (2021). Life cycle assessment of food waste to energy and resources: Centralized and decentralized anaerobic digestion with different downstream biogas utilization. *Renewable & Sustainable Energy Reviews*, *150*, 111489. doi:10.1016/j.rser.2021.111489

Tsui, T. H., & Wong, J. W. (2019). A critical review: Emerging bioeconomy and waste-to-energy technologies for sustainable municipal solid waste management. *Waste Disposal & Sustainable Energy*, *1*(3), 151–167. doi:10.100742768-019-00013-z

Van, D. P., Fujiwara, T., Tho, B. L., Toan, P. P. S., & Minh, G. H. (2020). A review of anaerobic digestion systems for biodegradable waste: Configurations, operating parameters, and current trends. *Environmental Engineering Research*, *25*(1), 1–17. doi:10.4491/eer.2018.334

Venkiteela, L. K. (2020). Status and challenges of solid waste management in Tirupati city. *Materials Today: Proceedings*, *33*, 470–474. doi:10.1016/j.matpr.2020.05.044

Vyas, S., Prajapati, P., Shah, A. V., & Varjani, S. (2022). Municipal solid waste management: Dynamics, risk assessment, ecological influence, advancements, constraints and perspectives. *The Science of the Total Environment*, *814*, 152802. doi:10.1016/j.scitotenv.2021.152802 PMID:34982993

Wilson, D. G. (1976). A brief history of solid-waste management. *The International Journal of Environmental Studies*, *9*(2), 123–129. doi:10.1080/00207237608737618

Chapter 13
Natural Gas:
A Solution to Sustainable Energy Issues – Safe Methods for Sustainable Environment

Meenakshi Sharma
https://orcid.org/0000-0002-6958-8741
University of Petroleum and Energy Studies, India

Mukul Kumar Rana
University of Petroleum and Energy Studies, Dehradun, India

ABSTRACT

According to current estimations, natural gas can be found in many parts of the world in sufficient amounts to provide energy for human activities for about 250 years. Natural gas accounted for around 21 percent of the total revenue generated by the global energy sector in 2011, trailing only coal and oil in terms of importance. That corresponds to hundreds of billions of dollars in financial terms each year. The purpose of this essay is to compare the utilization of various fuels to that of natural gas. The results would aid the government, business leaders, and decision-makers in deciding whether to switch from gasoline and diesel to natural gas. There is a need for more research to be done in this field by academics and scholars.

1. INTRODUCTION

A sustainable energy transition is defined as a shift toward a high-efficiency energy system that is well-managed to balance environmental and social costs, risks, and benefits such that the shift is deemed to be sustainable (Chen et al., 2019). The transition also includes the shift from utilizing fossil fuel to exploiting renewable energy sources in energy generation and the evolution from centralized to decentralized energy systems (Guidolin and Alpcan, 2019). A fossil fuel that developed deep within the earth's crust is natural gas. Numerous distinct chemicals can be found in natural gas. Petroleum resources are still regarded as a major asset for the provision of energy. Enhancing the extraction of oil from the underutilised reserves is required to meet the rising demand for energy. (Mariyate & Bera, 2023)Petroleum also falls under the

DOI: 10.4018/978-1-6684-8117-2.ch013

category of traditional energy source. Hence, once these energy sources have been used up and are no longer available, they cannot be reused. The significance of sustainability in the production of energy is now apparent. Sustainability is the concept of creating and utilising items in accordance with demand while keeping in mind the needs of future generations. Natural gas also contains trace amounts of non-hydrocarbon gases including carbon dioxide and water vapour, as well as natural gas liquids (NGL), which are also hydrocarbon gas liquids. Natural gas is mostly used for heating and producing power, but other consumer sectors also use it for other purposes. Natural gas is used in the electric power sector to create useful thermal output and generate electricity. In reality, according to BP's "Rapid" future scenario, natural gas will have a bright future and contribute 22% of primary energy by 2050, compared to renewables' 45%. Similar to what DNV predicts, gas prices will remain stable between 2020 and 2050. The spot market and the futures market are the two separate marketplaces for natural gas. Natural gas is regarded as pure because, in contrast to other gases, it releases very few byproducts into the environment that are pollutants. As a result, the air we breathe is kept pure. Natural gas is less expensive than other burning fuels, making it economical.

Economic challenges in sustainable utilization of natural gas include capital intensity, financing, and returns on investment of the natural gas pipelines and LNG projects within the region (Sovacool, 2009). It is important to seek cost-optimal pathways for sustainable natural gas utilization, which includes electricity generation, transmission, and storage to maintain robust economic growth within the region. There is a dearth of studies that analyse natural gas, oil, gold,

exchange rate, and stock market simultaneously. In the context of India, there is only one study (Kumar et al., 2019) which tends to explore the dynamic relationship among oil, natural gas, and stock market prices.

When oil enters the environment, one of the biggest detrimental effects is harm to lands and waterways. Being exposed to harmful petroleum products generally causes lower reproduction rates, organ damage, and mortality in wildlife caught in an oil spill. For the sake of the nation and society at large, it is therefore necessary to assess the situation and decide whether to switch from an oil-based to a gas-based economy. (Salisu et al., 2023)Researchers working to offer investors and decision-makers technical support to help them make judgements have continued to pay attention to the growing concern over climate change and the risk it entails. The behaviour of the energy market, which is distinguished by massive emissions with more severe implications on climate change than most of the other non-energy sectors, further emphasises the necessity for this study.

2. LITERATURE REVIEW

Since energy is unavoidable source in production process, the rapid increase in oil price leads to several economic problems such as security of energy supply, rising foreign-dependency, increase in energy bills, current account deficit and a decrease in economic growth (Kesikoğlu & Yıldırım, 2014). A better quality of life is dependent on access to energy from any energy resource, and almost 1.6 billion people, 20% of the world's population, do not have access to modern energy services (Safari et al., 2019). On empirical front, studies have evolved to analyze the possible nexus between climate change and the volatility of resources prices. Examining the out-of-sample predictive value of climate-risk factors for the realized volatility of returns of some precious metals, (Gupta and Pierdzioch 2022) employ OLS and shrinkage estimators and find that climate-risk factors forecast the realized variance of returns of metal prices under examination. Similar to this, albeit focusing on oil price volatility is the work of (Bouri et al. 2021). According to current estimates, there is enough natural gas present in many parts of the planet to provide energy security for human activities for around 250 years. According to a report by IRENA in 2018, natural gas contributed the largest share (41%) to the power generation mix for ASEAN in 2015, followed by coal (33%) and hydropower (16%) (IRENA, 2018). Exploiting localized natural gas supplies from unconventional sources and looking for new ones is a scientific and economic challenge, but it also poses major environmental risks and shifts geopolitical power dynamics (Eia, 2012). When compared to other fossil fuels, natural gas burning is the cleanest, especially in terms of CO2 emissions, yet even these minor emissions continue to contribute to the greenhouse effect (Fenguin Li & Xingmai Li, 2016). Natural gas is a desirable energy source due to the variety of uses for which it is used, the relatively low cost of its production, the ongoing expansion of pipeline networks, and the transportation of liquefied natural gas. These factors, along with projections for rising global energy needs, form a cycle of mutually reinforcing actions. (Long, David, 2003). Natural gas has a significant role in determining the energy policies of numerous nations, particularly the producers, as well as of international organizations. (2014) Ralph Heibutzki. Energy technology advancements have the potential to bring about a revolution in the near future by providing cheap, abundant, and clean energy sources that would displace more traditional sources. (Darien Alexander Williams & Amy K. Glasmeier, 2005). This will undoubtedly lead to confrontation between strong lobbying groups, some of which want to introduce innovations while others want to maintain the status quo. 2020 (Keyu Luo).

By 2025, the Indian economy is forecast to be the third largest consumer economy in the world, driven by rising consumption, which is projected to increase thrice to US$ 4 trillion (IBEF, 2019). To maintain economic growth, the Indian government is undertaking a number of initiatives, including Made in India, Digital India, Skill India, and others. Demand for resources is rising as a result of intensified economic growth (Luthra et al., 2015). A crucial resource for economic growth is energy (Luthra et al., 2015). And the primary energy sources that contribute most to global warming are fossil fuels (Mustapha et al., 2017). (Olson and Lenzmann, 2016). In the past, India relied on coal and liquid fuels to meet its energy needs, but as environmental concerns have grown, it is now important for India to concentrate on cleaner fuels, and natural gas meets this criterion well (Ministry of Petroleum and Natural Gas, 2019). Gas market in India is still regulated (Kumar et al., 2020). As a result, between 2018 and 2030, the Indian government expects to significantly shift from crude oil to natural gas and intends to increase the amount of natural gas in the country's energy mix from the present level of 6.5%–15%.

3. MATERIALS AND METHODS

With the aim to compare the utilization of various fuels to that of natural gas, extant literature review is done from various sources like Scopus and Web of Science database, research gate and also different news information sites related to energy resources. The collected information is discussed and validated with academicians, industry experts and fellow colleagues. Further the study has been structured to answer the following research question.

Research Question

Is it Beneficial to change our energy resource from Oil to Gas?

To answer the above question the study has been conducted to achieve the following objectives:
Objectives

1. To study the natural gas market
2. To analyse the gas-based economy over oil products

4. DISCUSSION

Effects of COVID – 19

During the first half of 2020, the COVID-19 pandemic caused significant price swings for all refined petroleum products. However, variations in the timing and intensity of the pandemic's effects on petroleum products, such as jet fuel, diesel fuel, and gasoline, were caused by shifts in demand and refiners' responses trying to limit output and shift toward more profitable fuels. In April 2020, the PPIs for gasoline and jet fuel experienced record declines because of lower fuel demand brought on by consumer volition, government-mandated stay-at-home orders, and restrictions on international travel. These factors were combined with the oversupply of crude oil that resulted from the OPEC-Russia price war. After then, prices began to rise because of three events: OPEC+ deciding to restrict oil production (as previously mentioned); refiners cutting back on capacity utilization and closing some plants; and an increase in demand as a result of the lifting of travel and business restrictions. Due to these circumstances, the PPI for gasoline in May and the PPI for jet fuel in June both experienced records monthly rises.

Natural Gas Markets

Most of the world's natural gas markets are in the process of changing from long-term contracts with natural monopolies to considerably shorter-term competitive energy commodity markets. Government policy, rather than the gas business itself, is the driving force behind these significant structural changes, which began in the US in the late 1970s and in the UK during the 1990s. Since the EU Electricity and Gas Directives went into effect, they have been expanding into Continental Europe. As a result, formulas connected to the price of oil or other competitive fuels are no longer used to set natural gas prices in the short term. However, the link to oil still exists in many long-term contracts, upstream joint production, and downstream inter-fuel competition, especially in power generation. The emergence of gas trading, along with physical and technical limitations, causes natural gas prices to fluctuate greatly in the near term. The demand for gas varies greatly during the day and the year, and the weather has a significant impact. Even though these variations are often predictable, such as at seasonally typical temperatures, the inherent uncertainty of the weather (both for heating and cooling needs) makes exact forecasting difficult, even over as short a time horizon as one day. Over the period 2019 to 2024 an average growth rate of 2.2% is expected. However, after the drop during the Covid-19 pandemic, a higher growth rate of about 3.6% in the natural gas consumption is expected. This sharp short-time growth will be balanced out by 2024 with an expected increase of 1.7% per year. In general, a reason for the market growth is the economic growth in particular the activities in the industrial sector. But for the current development the climate crisis is a major driver for natural gas, as natural gas has lower CO2 emissions than other primary energy sources like oil or coal. However, natural gas still counts to the fossil energy sources and contributes to the CO2 emission and therefore, can be seen as bridging energy source to renewable energies.

Price Comparison

When it comes to fuel prices, gas is less expensive than oil. Natural gas has a steadier supply and demand than oil, which makes it less volatile and sensitive to supply and demand fluctuations. When commodity prices rise, there is an impact on how people travel, how goods are shipped, and how people formulate their budgets. In comparing the rise of natural gas vs. oil prices, both have an impact on the consumer. For instance, when home heating prices climb, people have to decide whether or not they can afford to turn up their thermostats. Furthermore, when various goods have become more expensive because their components also cost more, people have to make difficult choices on what to buy.

Table 1. Oil Vs Gas Price comparisons

Energy Prices			
INDEX	**PRICE**	%	**TIME(GMT)**
WTI Crude	$80.12	**+2.64**	9:40 AM
Brent Crude	$86.69	**+2.59**	9:39 AM
Natural Gas	$2.47	**+1.56**	9:40 AM
10 February 2023 9:52 AM GMT			

Source: https://www.rigzone.com/news/commodity/

Is Natural Gas a Better Fuel?

Due to its cleaner combustion, natural gas is the most environmentally benign fossil fuel. Natural gas emits between 50 and 60 percent less carbon dioxide (CO2) in power plants than conventional oil or coal-fired power plants. Additionally, it releases into the environment greenhouse gases with a shorter life cycle. But combustion also results in methane emissions and worsens air quality. Natural gas is a cleaner fuel in the sense that burning it produces fewer conventional air pollutants, like sulfur dioxide and particulates, than does burning coal or oil. How much more depends on the characteristics of the fuel, the combustion technology, how well equipment is maintained and operated, and other factors. In general, burning natural gas also produces less carbon dioxide per unit of energy – about half compared to the best coal technology – and by this measure it's better from a climate perspective.

Is Natural Gas Bad for the Environment?

Every extraction of fossil fuels damages the ecosystem and expands our economic footprint. Therefore, the answer is yes to the question, "Is natural gas worse for the environment than solar power?" However, the response is also true if we question, "Is natural gas the most environmentally beneficial fossil fuel?"

The method of fracking poses the greatest threat to natural gas extraction since it uses a lot of water from nearby water reservoirs and pollutes waterways. The process also releases methane into the atmosphere. While the amount of carbon dioxide released from burning natural gas is not very great, methane, a potent greenhouse gas that escapes to the environment in significant quantities, is also released.

Carbon monoxide, nitrogen oxides (NOx), and Sulphur dioxide are other byproducts of burning natural gas (SO2). If it is not extracted or transported safely, it is also harmful. If natural gas is not transported properly, it could result in an explosion. Natural gas storage is problematic because of the bulk of the fuel, which necessitates larger, more expensive-to-maintain storage facilities.

The fact that it is not renewable is a major drawback. There are only 52 years' worth of natural gas reserves left to be extracted, according to World meters. You need to consider alternative energy sources besides natural gas if you never want to run out of energy.

However, according to the Intergovernmental Panel on Climate Change (IPCC), the world's natural gas resource base is large and geographically more scattered than oil –so it is a more reliable source of energy

6. ENVIRONMENTAL EFFECTS OF NATURAL GAS

The most environmental-friendly fossil fuel is natural gas. A gas-fired power plant's emissions have 50 percent less carbon dioxide (CO2) during combustion compared to a coal power plant

Key advantages

- Cheap
- Has less carbon dioxide emissions than other fossil fuels
- Reliable

Disadvantages

- Costly to store
- Dangerous: it could explode
- Has much more CO2 and methane emissions than solar or wind power
- Hydraulic fracturing can break natural habitats

Reserves Comparison

The depletion of fossil fuels is anticipated to occur this century. Coal may last for up to 114 years, whereas natural gas can last for up to 53 years for oil. However, the demand for renewable energy isn't high enough, therefore we may quickly deplete our supplies.

Advantages of Natural Gas

- **Natural gas is environmentally friendly:** Of all the fossil fuels that may be burned on Earth, natural gas is the most eco-friendly. Natural gas is regarded as pure because, in contrast to other gases, it releases very few by-products into the environment that are pollutants. As a result, the air we breathe is kept pure.
- **It is affordable:** Compared to other burning fuels, natural gas is less expensive.

Figure 1. Reserves comparisons

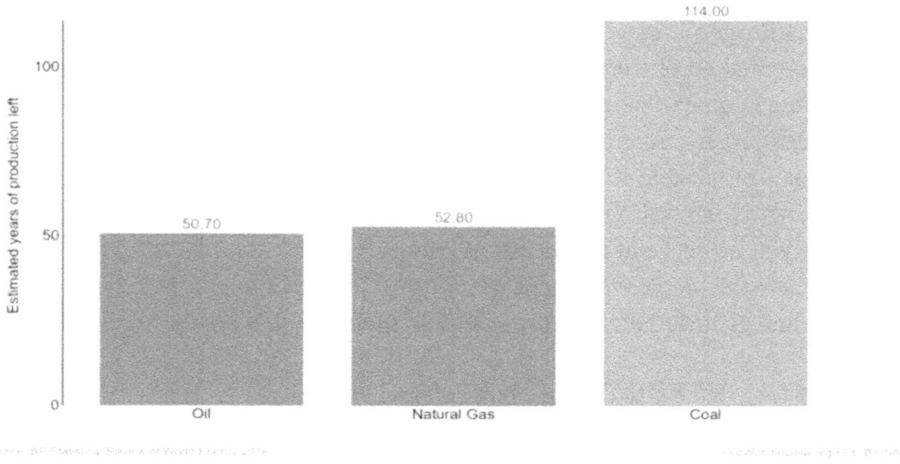

- **It is Convenient:** Natural gas can be directed connected to the consumer's home with the use of pipelines if they use it for cooking or electric purposes. By doing this, the time-consuming procedure of gathering, processing, packing, and delivering the finished product to customers is eliminated.
- **Natural gas may be used safely:** Leakage is one of our main concerns when utilising LPG cylinders or any other kind of fuel. Gases can be extremely dangerous since they can burn nearby objects and humans. Natural gas is lighter than air, though. In the event of a leak, it immediately evaporates into the air, preventing fire. That seems to be very safe.
- **It is Easily accessible:** According to experts, there is currently more natural gas accessible than crude oil or other similar products.
- **It is More Efficient:** When compared to substances like propane gas, natural gas burns more efficiently since it produces more energy.
- **Natural gas is simple to transport:** If we consider the natural gasas the preferred fuel for cooking, its delivery would be much easier since all that is needed to deliver natural gas is a network of pipes.

Table 2. New Explorations: - Shale oil and Gas

Parameters of Comparison	*Oil*	*Gas*
Found	Oil is found below the earth's surface.	Gases are formed from fossil fuels and they are mixture of gases.
State	Oil is usually in liquid state.	Gases are in gaseous state.
Soluble	Oil is not soluble in water	Gases are soluble in water
Harmful	Oil is hazardous to nature because they produce more smoke.	Gases are nature-friendly because when heated there is no smoke evolved.
Reliability	Oil pipelines are not that good in reliability.	Gas pipelines are very good in reliability and are safe for underground pipelines.

Global Shale Gas Resources Scenario

Supplies made from unconventional gas will be able to supply a growing portion of the world's natural gas consumption. Unconventional gas is predicted to increase from 10% of global production in 2010 to 30% by 2040. At present demand levels, the use of existing technology on these unconventional deposits has enabled the global gas supply to be extended by up to 250 years. According to the International Energy Agency (2013), there are an estimated 920 trillion cubic meters of unconventional gas in the world. Shale gas, tight gas, and coal bed methane are all expected to play a significant role in supplying the world's energy needs in the future. The Asia Pacific area, which continues to consume the largest portion of the world's liquefied petroleum gas, will experience the greatest increase in demand.

Indian Shale Gas Scenario

According to the US Energy Information Administration, India, the fourth-largest energy user in the world, may possess 96 trillion cubic feet (tcf) of exploitable shale gas reserves, which would satisfy nearly 26 years' worth of the nation's gas needs. For India, which imports a large portion of its energy, shale production is still a way off. According to information now available, shale gas potential exists in six basins: Cambay (in Gujarat), Assam-Arakan (in the North-East), Gondwana (in central India), KG onshore (in Andhra Pradesh), Cauvery onshore, and Indo-Gangetic basins. The Upper Paleocene to Middle Eocene Cambay Shale in the Cambay Basin, the Upper Jurassic to Cretaceous Raghavapuram Shale and its stratigraphic equivalents in the Krishna Godavari Basin, the Lower to Upper Cretaceous Andimadam and Sattapadi Shales in the Cauvery Basin, and the Barren Measures Shale in the Damodar Basin are all mentioned in the 2013 EIA report as having significant In addition to lowering India's reliance on pricey energy imports, the development of domestic shale gas deposits may be able to meet the country's rising energy needs. The growth of the domestic shale gas industry may also help the economy.

Recent Shale Gas Developments in India

To investigate and extract the natural gas trapped within the shale formations found in the Cambay basin, which is estimated to have a shale gas potential of 20TCF (trillion cubic feet), ONGC has dug the first well in Jambusar in the last week of October 2013. The US-based oil company ConocoPhilips, a pioneer in deep-water and shale gas exploration, is a partner in the shale gas exploration activities. One of the basins identified as having gas-bearing shale deposits is Cambay. In addition to the Cambay basin, ONGC will soon investigate the shale gas potential of the Krishna-Godavari, Cauvery, and Vindhyan sedimentary basins. ONGC has been investing Rs 2,000 rore per annum on the exploration of oil and gas in Gujarat and Rajasthan, as per official statement.

Shale Gas Exploration Challenges in India

Modern shale gas development differs significantly from conventional natural gas development in that it makes considerable use of horizontal drilling and high-volume hydraulic fracturing. Although shale gas has been produced in the United States for more than a century, the wells were frequently only moderately profitable. Shale gas wells are now more profitable due to rising natural gas prices, new developments in hydraulic fracturing, and horizontal completions. Due to the high cost of horizontal drilling and the extensive hydraulic fracturing necessary to create shale gas, shale gas typically costs more than gas from conventional wells. The minimal risk associated with shale gas wells, however, frequently offsets this.

5. CONCLUSION

The primary fuel for producing heat and power is natural gas. CNG, or compressed natural gas, is utilized as a fuel for automobiles. Across the world, it is utilized as fuel for air conditioners and boilers. Additionally, this is used to create fertilizers, primarily ammonia.

The study's conclusions point to the need for a switch from oil to gas. First off, there is more gas than oil in our country, and using natural gas is safe and good for the environment. Most importantly, it is far less expensive than oil. Additionally, it is more effective and simpler to dispense, among other benefits. Therefore, the government should consider switching to natural gas and develop regulations and establish strategies that would aid company owners and industrialists in analyzing and making preparations to switch from an economy based on oil to one based on gas.

6. SUGGESTIONS

Natural gas and even solar energy are the future of the energy markets. We must preserve our planet clean and employ resources that are not damaging. We ought to put our efforts into improving their usefulness to us. To increase the efficiency of this industry, new technologies like AI can be considered. It is projected that the increase in global carbon emission between 2005 and 2030 will be 30%. This projection was made with consideration of factors such as the expected growth in energy effectiveness and the usage of more sustainable sources of energy (Kumar et al., 2011). Therefore, there would be a need for more to be done for the projected increase in demand to meet sustainability.

REFERENCES

Chen, B., Xiong, R., Li, H., Sun, Q., & Yang, J. (2019). Pathways for Sustainable Energy Transition. *Journal of Cleaner Production*, *228*, 1564–1571. doi:10.1016/j.jclepro.2019.04.372

Guidolin, M., & Alpcan, T. (2019). Transition to Sustainable Energy Generation in Australia: Interplay between Coal, Gas and Renewables. *Renewable Energy*, *139*, 359–367. doi:10.1016/j.renene.2019.02.045

IBEF. (2019). *Indian economy growth rate & statistics*. IBEF. https://www.ibef. org/economy/indian-economy-overview

IRENA. (2018). *Renewable Energy Market Analysis: Southeast Asia*. International Renewable Energy Agency.

Kesikoğlu, F., & Yıldırım, E. (2014). The Causal Effect of Shifting Oil to Natural Gas Consumption on Current Account Balance and Economic Growth in 11 OECD Countries: Evidence from Bootstrap-corrected Panel Causality Test. *Procedia: Social and Behavioral Sciences*, *143*, 1064–1069. doi:10.1016/j.sbspro.2014.07.556

Kevin, M. (2020). Camp, David Mead, Stephen B. Reed, Christopher Sitter, and Derek Wasilewski, "From the barrel to the pump: The impact of the COVID-19 pandemic on prices for petroleum products," Monthly Labor Review. *U.S. Bureau of Labor Statistics*, (October). doi:10.21916/mlr.2020

Kumar, S., Kwon, H. T., Choi, K. H., Lim, W., Cho, J. H., Tak, K., & Moon, I. (2011). I. Moon LNG: An eco-friendly cryogenic fuel for sustainable development. *Applied Energy*, *88*(12), 4264–4273. doi:10.1016/j.apenergy.2011.06.035

Kumar, S., Pradhan, A. K., Tiwari, A. K., & Kang, S. H. (2019). Correlations and volatility spillovers between oil, natural gas, and stock prices in India. *Resources Policy*, *62*, 282–291. doi:10.1016/j.resourpol.2019.04.004

Kumar, V. V., Shastri, Y., & Hoadley, A. (2020). A consequence analysis study of natural gas consumption in a developing country: Case of India. *Energy Policy*, *145*, 111675. doi:10.1016/j.enpol.2020.111675

Luthra, S., Mangla, S. K., & Kharb, R. K. (2015). Sustainable assessment in energy planning and management in Indian perspective. *Renewable & Sustainable Energy Reviews*, *47*, 58–73. doi:10.1016/j.rser.2015.03.007

Mariyate, J., & Bera, A. (2023). Paradigm shift towards the sustainability in upstream oil industry for enhanced recovery - A state-of-art review. In *Journal of Cleaner Production* (Vol. 386). Elsevier Ltd., doi:10.1016/j.jclepro.2022.135784

Ministry of petroleum and natural gas. (2019). *About Natural Gas*. MPNG. http://petroleum.nic.in/natural-gas/about-natural-gas.

Mustapha, M. A., Manan, Z. A., & Alwi, S. R. W. (2017). Sustainable Green Management System (SGMS) - an integrated approach towards organisational sustainability. *Journal of Cleaner Production*, *146*, 158–172. doi:10.1016/j.jclepro.2016.06.033

Norsyahida, M., Mohamad, I. W. W., Indati, M. S., & Victor, A. B. (2021). Natural Gas as a Key Alternative Energy Source in Sustainable Renewable Energy Transition: A Mini Review. *Frontiers in Energy Research*, *9*. https://www.frontiersin.org/articles/10.3389/fenrg.2021.625023

Olson, C., Lenzmann, F. (2016). The social and economic consequences of the fossil fuel supply chain. *MRS Energy & Sustainability, 3*.

Gupta, R., & Pierdzioch, C. (2022). Climate risks and forecastability of the realized volatility of gold and other metal prices. *Resour. Pol., 77*.

Safari, A., Das, N., Langhelle, O., Roy, J., & Assadi, M. (2019). Natural gas: A transition fuel for sustainable energy system transformation? *Energy Science & Engineering*, *7*(4), 1075–1094. doi:10.1002/ese3.380

Salisu, A. A., Ndako, U. B., & Vo, X. V. (2023). Transition risk, physical risk, and the realized volatility of oil and natural gas prices. *Resources Policy*, *81*, 103383. doi:10.1016/j.resourpol.2023.103383

Sovacool, B. K. (2009). Energy Policy and Cooperation in Southeast Asia: The History, Challenges, and Implications of the Trans-ASEAN Gas Pipeline (TAGP) Network. *Energy Policy*, *37*(6), 2356–2367. doi:10.1016/j.enpol.2009.02.014

The Federal Institute for Geosciences and Natural Resources (BGR). (2011). *Energy Study 2012*. BGR. http://www.bgr.bund.de/DE/Gemeinsames/Produkte/Downloads/DERA_Rohstoffinformationen/rohstoffinformatio nen-15e.pdf?__blob=publicationFile&v=3\

US IEA. (2013). *Key World Energy Statistics 2013*. US IEA. http://www.iea.org/publications/freepublications/publication/KeyWorld2 013.pdf https://www.eia.gov/energyexplained/oil-and-petroleum-products/ https://education.seattlepi.com/causes-negative-effects-oil-pollution-4661.html

Chapter 14
Smart Vehicle-Emissions Monitoring System Using Internet of Things (IoT)

Lakshmana Phaneendra Maguluri
Koneru Lakshmaiah Education Foundation, India

J. Ananth
AMET University (Deemed), India

S. Hariram
Rajalakshmi Engineering College, India

C. Geetha
R.M.K. Engineering College, India

Archana Bhaskar
REVA University, India

S. Boopathi
Mythayammal Engineering College (Autonomous), India

ABSTRACT

Eco-friendly environment concern contributes to the survival of life on earth. Monitoring chemicals produced by growth requires a multidisciplinary scientific approach. Climate change is a result of pollution and overuse, particularly in the air. Although government and industry have taken initiatives to limit air pollution, it is still necessary to check the quality of the air at ground level. The goal of this endeavour is to build intercommunication-based individual vehicle air pollution monitoring and identification by internet of things (IoT). Idling and accelerating conditions were recorded using tailpipe emission sensing. Authorities can act and assure upkeep if they are able to identify a car at a busy intersection; mobile connectivity enabled route and vehicle identification. In India, fewer people will die from air pollution because to data gathered in the cloud.

DOI: 10.4018/978-1-6684-8117-2.ch014

INTRODUCTION

The goal of this endeavour is to build intercommunication-based individual vehicle air pollution monitoring and identification systems. The input used to record the data for idle and accelerating conditions was the detection of tailpipe emissions. Finding the car near the main intersection will enable the responsible authorities to act right away and launch a campaign for the rest of the community on vehicle upkeep. The authority will be assisted in eliminating the gross fleet by the data gathered in the cloud, which will identify and track the vehicle with the largest pollutant. This might revolutionise the present risk of air pollution-related fatalities. Although research on wireless networking and vehicle density has been ongoing for a while, adoption of these technologies is difficult because of how mobile cars are on the road. To provide the greatest possible data transfer, researchers have given arbitrary motion patterns for vehicles with a limit on speed. Although there is no information on the pollution brought on by specific cars, the Pollution Control Board of State and Center indicated that transportation was the main contributor to air pollution. Even the government-maintained "Vahan" and "Sarathy" software provide information on registration, cases filed, and associated facts. There is no particular software available for the car to identify the emission levels(Vashisht & Rakshit, 2021).

The first stage of the job, which is to detect and remove the fleet of cars that create more pollution than the intended limit, was made possible by the construction of a microcontroller and a simplified version of the IoT module. With an emphasis on "fit for purpose" and an evaluation of network expansions and capacity increases, the NTDPC advised using a comprehensive approach to creating an integrated transport network. Although a logistical infrastructure must be constructed, there is no specialised software available to identify the fleet's cars. Researchers have been able to determine the optimal path/route for cars in a fleet using the Global Positioning System (GPS) thanks to disruptive technical advancements in the automobile industry. However, new developments with high-resolution street maps and heterogeneous interoperability with standard interfaces have made this a promising study area. Most researchers have investigated route distance, multiple vehicle routing, and operation research methodologies(Lee & Yoo, 2000).

Network models that depict the spatial characteristics of transportation infrastructure have been developed as a result of quantitative approaches to transportation phenomena. Concave cost network flow issues are used to characterise behaviour patterns or normative attempts to use the transportation infrastructure efficiently, whereas the network equilibrium problem may be used to estimate an origin/destination matrix. The shortest path, assignment, and transportation are all optimised using the Minimum Cost Network Flow Problem. The Max-flow Min cut Algorithm was used to create a road map for parallel and distributed systems. To use the cores on a single computing node or a distributed memory technique, parallelization entails transforming sequential code into multi-threaded or "vectorized" code. When it comes to hardware, parallel computers may accommodate more than one core and processor on a single system. By consuming less power, multi-core CPUs with many processors aid in improving the effectiveness of processing through multitasking. A method for handling complicated calculations and multitasking is parallel processing. It entails giving the processor smaller jobs to do while putting together solutions using the same tool. The driver is verified to reveal information about vehicles in closer proximity in order to identify the individual user.

The conventional Mobile Adhoc routing methods offered include Proactive, Reactive, or Location-based. While reactive protocols only permit the new node to enter if the old node becomes disconnected, proactive protocols construct new routes or update existing ones in a set amount of time. In order to arrange all destination modes according to their locations, location-based protocols rely on a location server. Solutions for communication between vehicles or between vehicles and infrastructure are used to replace mobile devices with ones that have more storage and processing capability, enhancing the user experience. To offer information about the closest end-user linked at the time of the emergency, the nearest car is connected to the emission analysis. Road safety, traffic regulation, mobile data services, and driving assistance all depend on vehicular communication, yet vehicle pollution represents at least 50% of all air pollution. The amount of data that must be gathered from automobiles is rising, enabling the development of big data technology. The objective is to determine the vehicle's emission analysis and to gather and analyse data to confirm the weekly and monthly predictions of emissions from transportation. The mobile unit is cross-verified with the sensing unit linked to the hardware module, and the application recognises and manages the mobile user within the vehicular limit. The collected data may be used to analyse various types of data.

Problem Statement

To lessen air pollution in India brought on by two-wheelers and public transit, research is required. The primary focus is on calculating the pollution the vehicle produces and communicating that information to the mobile device linked to the driver. Finding the vehicle that has to be removed from the fleet might be assisted by an analysis of the pollutant level. The Central Pollution Control Board (CPCB) processed the data, which is one of the most crucial facts, and the hardware component was developed to sense the emission level from the car, which is the main cause of pollution. To report the data and identify each individual car in the fleet, mobile application development was also made available. Researchers from all around India are attempting to offer a succinct, significant, and simple remedy to the escalating consequences of air pollution.

OBJECTIVES

Humans are really concerned about air pollution, so it's critical out how to make the air better. This may be achieved by looking at the pollution level of the car, locating the specific vehicle and the route it is taking, and looking at the overall pollution level. The hardware module that detects the main pollutant from the car, as well as the communication within the vehicle, must be cross-verified with the authentication method with which the hardware module is associated. The information that must be communicated from the hardware module to the mobile device can be updated in the cloud, where more information may be accessible and the history of vehicle pollution may be accessed. The routing mechanism faces a number of issues in order to guarantee that data is approved for transmission from the source to the intended destination with acceptable performance.

A REVIEW OF THE WORKS

This study intends to give an overview of how the Smart Transportation System (ITS) will be implemented. It entails detecting pollution from vehicular sources, sending the data to the cloud, coordinating vehicle interactions, and identifying the offending vehicle. This chapter examines the effects of technical progress on the expansion of ITS, from software development to hardware development.

Air Pollution and Embedded Networking

The dependability of early electronics equipment was limited, but recent advances in miniaturising and integrating hardware technology have reduced cost and size. It was noted in a study by many academics that internet-enabled gadgets were used to gather data to facilitate communication. Using the Zig-bee protocol and a grid structure, presented a prototype of a microclimate monitoring system, but they neglected to consider the communication made with the zig-bee and the second kind of grid structure. Zhang studied the use of IoT to track the environment for animals. The system used three different types of nodes and was based on a heterogeneous communication network(Zhang et al., 2011). To optimise the working and power management approaches, changes were implemented. Real-time data collection is well outlined, yet their approach has drawbacks when applied across a broad spectrum(Kasar et al., 2013).

A prototype for fine-grained real-time air quality assessment utilising a Bluetooth module and geo-tagging was presented. The social community sensing approach produced data based on participation, whereas the public infrastructure model was focused on the particular route. To calculate the pollution produced by each unit, personal involvement is required(Devarakonda et al., 2013).

Instead of focusing on healthcare systems, they paid attention to the Internet of Things (IoT). A real-time operating system is required to regulate low energy devices because Padova Smart City in Italy offers air pollution monitoring and management services. The author evaluated the various platforms and operating systems available for IoT(Malche & Maheshwary, 2015).

Researchers used an Arduino UNO board to measure the level of environmental pollution and to track changes in the CO level over time, both during the day and at night. The cloud platform's excel file was utilised to store and analyse data(Guthi et al., 2016).

It was used an EPISODE dispersion model and a low-cost electrochemical sensor to deploy an air quality mesh structure to monitor air quality in Norway. They determined the mean bias for contaminants and concluded that the model was based on fewer, more evenly distributed geo-statistical observations(Schneider et al., 2017). A four-level design was suggested to cut down air pollution in tourism by up to 55%. The paradigm has four main levels: application, service, virtualization, and real-world, plus two cross layers for security and reliability(Nitti et al., 2017). A situation involving a travelling salesperson was used to test the use case. 90 percent of occupational health depends on the interior environment, hence, they have concentrated on it. Real-time PM monitoring has shown encouraging outcomes, but hospitals, schools, and business organisations still require better gear and software(Ikumapayi et al., 2018). Individuals may get information regarding air quality using a cloud environment dubbed "Thing-speak,". Graphs are displayed in a think-speak setting.

Sami Kaivonen used a stationary data gathering technique to evaluate the deployment of a low-cost sensor for real-time air pollution monitoring in Sweden. The CO sensor functioned better than the NO2 sensor, which had a success rate of only 1%. Without resending, the data transmission rate across the 4G network was 70%(Sami & Edith, 2019). Route optimization, parking, and accident detection were the main topics study of how machine learning methods and IoT technologies affected the ITS. They concluded that the growth in IoT devices and volume scale-up needed the development of machine learning applications(Zantalis et al., 2019). Only two nations, according to Fiore, have air quality sensors in the top five smartest cities in the world, suggesting that the vehicular traffic control and monitoring system is insufficient because numerous players are required for interaction(Copat et al., 2020).

The evaluation gives information on inefficient vehicle maintenance, energy use, and capacity analysis of traffic flow. Additionally, it provides additional ways to reduce pollution levels, like using public transportation and evaluating the health effects of both indoor and outdoor pollution. There are no exceptions offered.

Vehicular Communication System

This paper examines the ways in which previous scholars have explained the necessity of vehicular interconnection for intelligent transportation systems. It was concentrated on determining a possible score based on the motion, energy, and density of cluster nodes at a location. Although efficiency may be raised by sending fewer packets and expanding the network, numerous situations and applications still call for improvement. Information was discussed on the ITS for Asia and the Pacific's sustainable development. They spoke about its uses and goods, regulatory and policy services, and regional collaboration. They also discussed how the implementation of ITS in Korea is anticipated to result in a 12 percent reduction in carbon emissions. In the United States, Canada, the European Union, and Japan, there was a completely integrated system with a certain level of intelligence. The present research has been divided into three categories: static sensor networks (SSN), community sensor networks (CSN), and vehicle sensor networks (VSN). It is challenging to create a low-cost, portable air pollution monitoring device due to the wide variation in pollutant concentrations within a given magnitude. VSN is advised for unmanned aerial vehicles, active monitoring sensor nodes, and the uncontrolled mobile node because it offers higher performance, sensor node cost, and data quality(Yi et al., 2015).

To deal with a shortage of communication infrastructure, developed a vehicle cloud architecture. In order to escape traffic, he offered detours and used the mathematical approach to offer an alternative course of action. Establishing a system between cars for managing and searching resources just between vehicles is more difficult. Allocation-related problems have not been addressed by the author(Meneguette, 2016). Air pollution surveillance has been covered; however, networking-related concerns, including geographic locations, sensing length, and data storage and availability, have not. In order to pinpoint the root of the congestion, concentrated on the source-to-destination spatiotemporal distribution. To better regulate urban traffic, a trip matrix based on the stay areas of the moving vehicle was created(Tan et al., 2018). The growth of the road transport organisation and the importance of intelligent transportation systems were highlighted. They emphasised technical tools to assist the system and concentrated on creating a system that uses technology to operate intelligently. Implementation may have an impact on daily life and significantly contribute to environmental issues like traffic and pollution(Umap & Deshmukh, n.d.). Thiago Sobral and Borges investigated how urban transportation is managed using visualisation tools and sensor data.Urban traffic flow and monitoring, crowd dynamics, traffic accidents, air pollution, travel patterns on public transportation systems, service levels, and trip patterns are all visualised(Sobral et al., 2019). A three-level architecture for ubiquitous Internet connectivity was put out and includes client, connection, and cloud levels. While fibre optic cable was used to maintain the connection level and the cloud level acted as a backbone, the client level was outfitted with a multi-interfaced global ID enabled vehicle. This solution would meet the VANET's requirements for widespread connectivity(Sherazi et al., 2019). The general assessment explains why categorical separation of wireless sensor networks is necessary and how to handle the transportation issue. The majority of the study concerned collecting data while the vehicle was stationary or in a place unrelated to the particular vehicle. The author mentioned energy optimization and network lifespan maintenance, but neglected to consider total network coverage, connection time enhancement, and resource allocation during the failure.

Role of Big Data in Intelligent Transport System

Big data is crucial for the Internet of Things because it necessitates collecting and backing up enormous volumes of data for later analysis. Using "space-air-ground" technology, Gang projected a search that concentrated on data availability, consumption, and necessity(Xiong et al., 2014). The functional flow of the processes involved in data processing is provided in detail by the transportation model. It was advocated for the use of numerical data with flexible limitations for data collection and processing in order to encourage novel procedures and alignment in data use. Environmental big data must be transformed from environmental sensors for air pollution monitoring(Gabrys, 2016). Interior reading is used to identify groupings based on the main data set for feature and outlay. Big data is a common instrument in company management, research, and government, and it is inextricably linked to local expertise in environmental sensing technology and supply conditions(Loukissas, 2016). To enhance the design of smart cities, Silva et al. created a three-level architecture. Heterogeneous data was gathered at the lowest level, and filtered data was maintained at the middle layer. Based on the choice made in that department, each department delivers a unique notice at the application layer. The authors concluded that an increase in data size improves scalability, which requires future attention. They reviewed the numerous applications and came to a few important conclusions in each of the smart settings(da Silva et al., 2017).

Zaree and Honarvar used k-means clustering as part of a large data mining method to improve the actual level forecast of air pollution. They discovered that the existing central-site monitoring technique was inferior to a data collection that was accessible during weekday daytime measurements and 10–20 driving days. Due to the development of quick response, inexpensive sensor technologies, the equipment's cost increased(Zaree & Honarvar, 2018). With the use of simulation and sensitivity analysis, investigated whether mobile nodes could gather data despite noise. They demonstrated the practical difficulties and viability of employing mobile device trajectory data for dynamic simulation(Sugania et al., 2018). The requirement for massive data analysis for vehicular networks was thoroughly explained by Cheng et al. (2018), who also focused on machine learning techniques that may be modified for poor communication settings. To determine accuracy based on the alternative opportunity data pipelines, they used the Naive Bayes and Support Vector Machine methods. According to this paper, big data technology will undergo a significant shift as a result of alternative means including increased mobile broadband, low latency communication, and enormous machine type communication. Existing communication architecture may suffer while sending high volumes of data. With an emphasis on cloud and fog computing, offered a thorough overview of how to delay sensitivity in real-time intelligent transport systems. They stressed the necessity of ITS's massive data processing and analysis in real-time(Benjamin et al., 2018). Machine learning prediction models were employed to identify the ozone level in Murcia, Spain, with 90% accuracy and RMSE values under 11 micros per metre cube. Sulfur dioxide and particulate matter (PM), which are major sources of air pollution, may be the focus of future attention. Open problems in big data analytics and ITS were described from data collection through storage and processing. Machine learning and big data tools, which may be used to alter data gathered from a different vehicle, are required for the usage of big data technology.

TRANSPORTATION, POLLUTION, AND ITS IMPACTS

One of the main causes of mortality in India is air pollution, so the government is taking steps to educate the populace about the quality of the air they breathe. The majority of Indian cities rely on motorised personal transportation, and diesel cars are the most polluting, with concentrations of particulate matter of 56% and nitrogen oxides of 70%. To decrease car hazardous gases, the government has developed measures. Improve public transit, increase road usage, increase fuel quality, decrease the use of private vehicles, and offer adequate parking. The Indian Supreme Court has ordered that an on-board diagnostic for vehicle emissions be put into place in tandem with the BS-IV injection programme. Because of uncontrolled urban growth, a lack of multi-sectoral policymaking, a lack of funding for research and development, and a lack of monitoring, air pollution is a bigger problem for low- and middle-income nations(Tan et al., 2018).

Health Risks and Air Pollution

The government of India committed $1 billion for the plan in 2015 to cut CO2 emissions by 1.2–1.5 percent in 2020. Although there are many other sources for vehicle fleet data, VAHAN and SARATHY software give the essentials. The largest fleet is located in Chennai, where 3.8 million automobiles are registered.

Air Pollution Modality

To control the air pollution brought on by moving vehicles, top-down or bottom-up approaches are applied. The top-down approach was based on qualitative investigation, whereas the bottom-up approach assumed that wind speed was constant with time and height. In a study, it was discovered that both strategies were consistent. Road transportation accounts for 80% of carbon dioxide emissions in India, which are produced by the transportation industry. A third of urban residents rely heavily on the road for transportation, and 60% of freight is transported by vehicles that can move quickly. The demand for transportation has increased at a rate that is higher than the growth of the GDP. The entire distance travelled by the passenger in all shared modes, along with the fuel intensity and carbon emission factor in litres of gasoline consumed, determine the low carbon transport emission factor. Due to the scarcity of resources, a bottom-up strategy is required to gather information on the number of cars, fuel efficiency per kilometre, and kilometres travelled per vehicle annually in India(Vashisht & Rakshit, 2021).

Transportation and Technology

Three methods are used to gather data for intelligent transportation: manual, application technologies, and dynamic low-cost cellular data. Additional wireless technologies Application Technologies on the Side of the Road: Infrared beams or CCTV cameras are used in this technique to gather data, which may then be analysed manually or using image processing software. However, the expense of installation and upkeep is substantial. dynamic, inexpensive mobile and cellular data. The three techniques used to identify a car while it is travelling are triangulation, vehicle re-identification, and GPS. all-weather compatibility, low cost, yet considerable delays and traffic congestion. Other wireless technology: There are costs connected with multi-sensor tracking, RFID, mobile telephony, and dedicated short-range communication.

This study focuses on the dynamic low-cost cellular/mobile data strategy, which calls for data collection, data integrity, and data veracity to guarantee pollution under the control test. Maharashtra developed a star ranking system for particle air pollution, while Tamil Nadu has 256 facilities for pollution testing.

Data Dissemination in Vehicle Sensor Network

Data on air monitoring is gathered from a single source, restricted to highways and their immediate surroundings, or applied to the entire city, state, or nation. It is employed to calculate the emission rate by source, fuel, technology, and usage and to comprehend how air pollution affects human health. Increasing research indicates a connection between air pollution and obesity, diabetes, and Alzheimer's disease, and strict guidelines include Older vehicles need to be phased out. Flyovers, underpasses, traffic management systems, production facilities, and strict standards all need to be put in place in order to minimise pollution in India.

Equip cars with mobile sensing devices to gather and analyse data, and use Google Street View and the ACLIMAS sensing platform to deliver real-time air quality. The measuring method is described in Standard Operating Procedures (SOP). SOPs list the measuring technique, guiding principles, level of accuracy and precision, and validity assumption. (Substances, tools, reagents, and suppliers; people in charge of conducting processes; a traceability trail; initial standards; and references supplied by the CPCB. The equipment, measurement principle, flow rate, sampling time, frequency, analytical technique, minimum detection limit, and minimum reportable value for each pollutant are the most crucial parameters (Figure 1).

Figure 1. Sensed data to cloud to monitoring purpose

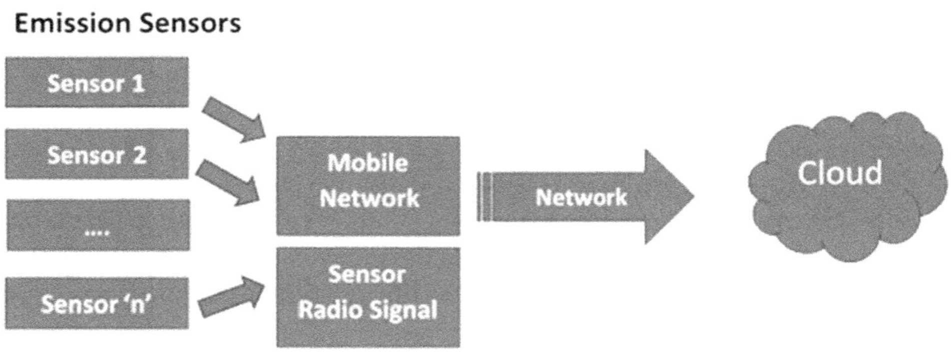

Types of Intelligent Transport System (ITS) Data Collection

Smart cards, GPS, video, roadside sensors, floating car sensors, wide-area sensors, connected and autonomous vehicles, passive collection, and other sources are some of the international data collection sources for ITS, with a focus on data like travel time, vehicle speed, density, speed, classification, coordinates, acceleration, and safety data.

- Automatic vehicle location: transit agencies connect with existing signpost structures via transmitters.
- Automatic Vehicle Identification: probe automobiles To identify and gather data on each vehicle's uniqueness, interact with roadside transceivers.
- Radio navigation: Similar to GPS, data is gathered by communication between probe vehicles and radio towers.
- Cellular Geo-location: This system tracks the transmissions of cellular phone calls to get statistics on trip times.
- GPS: In order to transfer signals from earth-orbiting satellites to a control centre for real-time position and trip time information, probe vehicles employ GPS receivers and two-way communication.

Air Fuel Consumption and Emission

The capacity of the engine to burn the mixture properly, which results in a poor combustion stage, influences air pollution. Theoretical air fuel ratios are 14.7:1, 13:1, and 15:1 for rich and lean air, respectively. The engine may perform well in a variety of operating circumstances, including starting, warming up, accelerating, cruising, under high load, and decelerating. As the fuel loss of automobiles may be caused by emissions or refilling, the idea of air-fuel consumption and air pollution is crucial. The work focuses on exhaust emissions, which can be either start-up emissions or running emissions, that are released via the exhaust pipe when the vehicle is started or running(Xiong et al., 2014).

The trip component, the highway network component, and the vehicle component are the three main elements that influence vehicle emission rates. The number of journeys, the distance travelled, and the type of driving are all aspects of travel. The geometry of the roadway, the size, horsepower, and weight of the vehicle, as well as the ambient temperature, the kind of engine, and urbanisation are all elements that affect highway networks. Predictions of emission rates are made using emission models such as MOBILE, the Motor Vehicle Emission Simulator (MOVES), the Emission Factor Model (EMFAC), the Computer Programme to Calculate Emissions from Road Transport (COPERT), and the Comprehensive Modal Emission Model (CMEM). Although they are more responsive to continuous cruising speed levels than to vehicle stops, vehicle fuel consumption and emission rates significantly increase as a vehicle stop is implemented.

Vehicle emission rates were significantly impacted by how quickly a vehicle stopped. While NOx emissions normally rose at acceleration rates in the range of 0.2 to 0.8 Amax, HC and CO emission rates were extremely sensitive to the amount of acceleration. With a 60% rise in HC emissions, an 80% increase in CO emissions, and a 40% increase in NOx emissions, raising the speed restriction at a plant might have severely detrimental environmental effects.

Nodal Communication Need

To increase the effectiveness of the nation's road transportation, the country needs a nodal agency for intelligent transportation systems (ITS). In addition to promoting mobility and safety, ITS technologies also help to lessen traffic congestion and improve environmental conditions. To boost the capacity, connectivity, efficiency, and safety of the highway network, the Tamil Nadu government has outlined ITS-related goals in its twelfth five-year plan. Road infrastructure is necessary for integrated multimodal urban transportation, the development of small and intermediate ports, and equitable socioeconomic growth. number of new vehicle registration certificates and registration certificates overall.

An integrated system of parking, alternative forms of transportation, and traffic management is offered as part of smart transportation solutions. Intelligent infrastructure, sensors, active traffic management, adaptive signal control technology, and digital technologies offer the best connections between people, cars, and infrastructure. With the use of a wireless network known as Vehicle to Everything (V2X), vehicles may communicate with one another and share information about their movements, including their speed, position, direction of travel, braking, and loss of stability. Vehicle-to-vehicle (V2V) technology, which employs specialised short-range communication and has several thousand operational prototypes or test automobiles with modifications, is still only a concept.

A smart system's cost and power consumption can be decreased by using a highly parallel heterogeneous embedded CPU, enabling real-time data processing. Traditional apps are used to gather sensor data, processed data, and the position of the vehicle. The data can be updated frequently. However, data clustering in the traffic zone is inadequate, and communication between sub-centres and main centres is frequently inaccessible. As a result, studies on the availability of vehicles in a certain location as well as information on the vehicle and its route are required.

Health Effects

In order to minimise pollution, the "avoid-shift-improve" approach suggests avoiding driving and switching to alternate forms of transportation. AAP exposure in the Indian population is probably responsible for both short- and long-term health impacts, including respiratory illnesses and premature mortality. The effects of short-term exposure to ambient pollution on chronic obstructive pulmonary disease, respiratory infections, and greater rates of hospital admission or visit have been studied in 59 studies in all. The main cause of respiratory health issues is particulate matter (PM2.5 and PM10), and most studies have found statistically significant links between AAP exposure and an increased risk of early mortality. To track the exposure to PM2.5 and PM10 concentrations in urban regions and rural areas across India, it is crucial to establish a consolidated data gateway. The exposure to and emission of dangerous air pollutants are not well addressed by the current air monitoring system, thus, the monitoring setup has to be improved to make it possible to determine how emissions affect population health. According to studies, both rural and urban people experience a substantial link between ambient air pollution sources and a wide spectrum of respiratory symptoms. The government's varied initiatives and financial investments in emission reduction are the main topics of this chapter. The most important information on the various forms of transportation and their technical implications is also highlighted. The need to go from data collection to vehicle identification with technical advancements is projected.

TAIL PIPE EMISSION ANALYSIS

To encourage a clean, convenient, and congestion-free India, the "Transforming India's Mobility" framework was formally established. It emphasises connectivity, travelling with a small carbon footprint, and adjusting to cutting-edge technology. Using current technology has the advantage of giving the mobile system improved connectivity to data centres. The amount of gasoline wasted in India was significant as a result of traffic jams and air pollution, and vehicle movement in Delhi and Bengaluru was slower than intended. Governments at the federal and state levels have taken actions to address this. Odd-Even Scheme, Traffic Signal Synchronization, the Eastern Peripheral Expressway, Ahmedabad Intelligent Traffic Management System, Automatic Vehicle ID Fining, 12 Automated Driver Testing Centres, and Fully Automated Smart Parking are just a few of the initiatives that have been implemented. Real-time data collection and traffic management have been proposed by the Singapore ITS and the Indian government. The study's main focus was on car tailpipe emissions, and the outlet of the tailpipe was considered when determining the emission criteria that the various vehicles must meet according to the emission regulating authorities. Many smartphone applications have been created in some places to aid in the development of the public transportation network. The regional transportation system will be able to govern the vehicle fleet with the use of individual vehicle analysis.

This study seeks to make vehicle owners aware of the need to perform timely vehicle maintenance checks. The data from the vehicle can be sensed and measured by external equipment, such as sensors, and input can be sent to the vehicle's user. The level of pollution may be significantly lowered if the car is properly maintained. A minor change in a physical phenomenon may be detected with a sensor, and market prices for sensors have decreased.

Sensors

Based on physical measurements, environmental characteristics are collected through sensors. Physical amount, active vs. passive, digital vs. analogue, null and deflection techniques, and input/output configuration may all be measured using analogue sensors. With input based on type, range, and resolution; a power supply based on voltage, current, or power dissipation; and output based on merit or signal type, sensors and microprocessors offer data acquisition and processing. Frequency, pulse width range, pulse width step size, symbol rate, serial format, voltage range, transfer slope, and settling time are among the output features(Boopathi, 2019; M et al., 2018).

Calibration of Sensor

The quality and amount of measurements made by the sensors, as well as how to adjust them to monitor certain characteristics, are the most crucial factors. Depending on the threshold, one point or two points of calibration are both options. To calibrate a sensor at one point, take a measurement using it, compare it to the standard reference measure, remove the measured value from the standard, and add the offset to each reading.

Sensors Used

The sensors utilised in this study to monitor carbon monoxide at low temperatures are MQ7 and MQ135. The conductivity of the sensor rises along with the gas concentration, and this change in conductivity is what determines the output signal for gas concentration. Regular sensor cleaning is necessary since prolonged exposure to high levels of contamination might have negative effects. For detecting NOx, smoke, and CO2, the MQ135 gas sensor is a reliable, long-lasting, quick-responding, high sensitivity simple driving circuit. The range of the sensing concentration is 10 ppm to 1000 ppm(Chakravarthi et al., 2022; Jeevanantham et al., 2022; Samikannu et al., 2022).

Micro Controller

It is necessary to connect processing media, such as a microprocessor or microcontroller, to sensor data collection, processing, and cloud transmission. Many boards employ micro-controllers because they can manage things using bits and internal storage. The interface and connection of microcontroller boards have been streamlined, making them compatible with the placement, accessibility of commonly used boards, and requirements. Examples include the Arduino UNO, Raspberry Pi, Intel boards, Node, MCU, and Arduino(Chakravarthi et al., 2022; Kumara et al., 2023).

Data Storage for Sensor

The correct balance of power, coverage, data rate, and cost must be struck while creating a sensor-based infrastructure. The cellular network, which necessitates a costly, specialised radio on the mobile phone to connect to the mobile tower, is the first generally accessible radio connection for connecting the sensor to the cloud. The user is responsible for paying the mobile network provider. Government, automakers, and automotive businesses all strive to build a smart vehicle and a pollution-free environment. In order to do this, sensors that transmit detected data to the final candidate through the internet must be used. This data must be communicated and processed beforehand, with the final processed data being sent across the internet for final analysis. Every vehicle emits some amount of gases; however, the issue arises when the emission exceeds the permitted limits. The primary goal of this study is to develop an automated design for an automotive pollution control system utilising an Arduino UNO R3 and sensors to monitor CO (Carbon Monoxide) and NO2 levels (Nitrogen Dioxide). This project's goal is to monitor and manage vehicle pollution using a pollution control circuit made up of a controller, LCDs, IOT modules, smoke sensors, and other sensors.

Circuits

This concept is an automated approach to regulating a vehicle's emission levels in accordance with BS4 requirements. The master micro-controller, an Arduino UNO R3, receives sensor data from two distinct gas sensors while being supplied by a +5V DC supply. The Arduino IDE is used to programme the board, and it offers a functional design perspective. The sensor is a potentiometer attached to an analogue input port.

The LCD unit attached to the other sensor uses the potentiometer as a bridge and a voltage divider that may be adjusted. The detected information is transformed into an analogue format before being locally shown on the LCD. The LCD in use is a 16x2 alphanumeric display called the JHD162A. The Arduino continuously monitors sensor data, converts it to PPM levels, and displays the levels on the LCD. A DC buzzer alerts the violator when the sensor value exceeds the threshold value. BS4 standards are used to programme the emission values. IoT modules submit data to IoT cloud servers and remotely monitor sensor values.

The standards mentioned to utilise g/km, mg/km, and g/km while projecting the detected data in ppm (parts per million). Data is expressed in ppm since the actual measured data must be added together and averaged over the course of the vehicle's whole run, as well as the kind of vehicle being used. Since the data are from an online business, they are presented in grams per kilometre in the second portion of the big data analytics. Embedded apps may be created thanks to the GSM modem's ability to connect to any GSM network and function as a SIM card. The SIM800 offers GSM/1800MHz performance for voice, SMS, data, and fax in a tiny form factor at an inexpensive price. The SIM900A is a dual-band GSM/GPRS solution in an SMT module with an industry-standard interface. The SIM800 is a small gadget with dimensions of 24 x 24 x 3 mm. It has been tested by BSNL and Airtel and offers improved connectivity and causes the mobile application to load more quickly.

Algorithm and Result

When the car is moving, the design starts the GSM Module, which picks up the network signal, gathers sensor data, displays it on the LCD, and sends it to the server. The emission table's data expands and contracts based on the data available, not on data from roadside vehicles. Through the use of existing network infrastructure, the IoT enables remote sensing and control of items. This more direct integration of the real world with computer-based systems leads to gains in efficiency, accuracy, and financial gain. Cyber-physical systems, which also include smart grids, smart homes, intelligent transportation systems, and smart cities, include IoT as an example. The cloud, big data, and other cutting-edge technologies are integrated into smart transportation systems to improve user service and coordination.

Application Development

The ability to access the internet, hail a car, stay in touch with friends, and view movies all depend on cellular networks. However, they are inappropriate for IoT applications that demand remote sensors and devices to persist for months or years since they have a trade-off between power consumption, range, and bandwidth. The base station's cell tower is in charge of transmitting and receiving data in the 746-787 MHz range. The creator of a mobile application must make it compatible with the memory and screen size of the device, and use PHP, CSS, and JavaScript to send data to the cloud. PHP was used to construct a pop-up window for registering the login and password, and CSS was called to display the latest value of CO2, NOx, log date, and log time in a tabular format. The programme used the associated JSON file that included the cloud data to show it in the table and chart. The real setting, which required data to be presented once every second, might be changed depending on the degree of application.

Big Data

Big data may be utilised to simplify conventional data management systems, boost business processes, and improve income. It is used to query a sizable dispersed aggregate of loosely structured data from sources including servers, mobile devices, databases, and apps. Big Data presents three key challenges: data, process, and management. The difficulties encountered during the processing of the data include data aggregation and integration, analysis and modelling, data interpretation, acquisition and warehousing, mining, and cleaning. The key to overcoming management issues is to have a strong set of analytical and comprehension abilities. The Spark platform was used to find the top big data solutions for environmental problems. Programming in Python and Scala are contrasted in terms of output, execution time, safe refactoring, spark integration, statistics packages, and community. The six execution levels of big data technology—the data layer, the data processing layer, the data intake layer, the data presentation layer, the scheduling layer, security, and governance—are utilised for analytics and visualisation.

Big Data Analytics

Big and open data are anticipated to be essential to the creation of cutting-edge ITS services like navigation and urban services. The government must establish regulations for the collection, processing, and use of data in order to protect privacy. Big data analysis assists in finding trends and gathering useful data that may be utilised to lessen traffic and pollution. Big data analytics needs a machine learning model to handle complex applications, and product development methodology and design need further study. Implementation of machine learning, a recommendation system, a dashboard, and ad hoc analysis A standard of SO2, NO2, and PM standard values were used to normalise the online data set, which was presumedly taken from the OGD platform. In Tamilnadu, the datasets include 12352 rows, while in AP, 15353 rows. Cleaning, resolving missing values, identifying attribute dependencies, and data organisation are all included in normalisation. Big and open data are anticipated to be essential to the creation of cutting-edge ITS services like navigation and urban services. The government must establish regulations for the collection, processing, and use of data in order to protect privacy. Big data analysis assists in finding trends and gathering useful data that may be utilised to lessen traffic and pollution. Big data analytics needs a machine learning model to handle complex applications, and product development methodology and design need further study. Implementation of machine learning, a recommendation system, a dashboard, and ad hoc analysis A standard of SO2, NO2, and PM standard values were used to normalise the online data set, which was presumedly taken from the OGD platform. In Tamilnadu, the datasets include 12352 rows, while in AP, 15353 rows. Cleaning, resolving missing values, identifying attribute dependencies, and data organisation are all included in normalisation(Loukissas, 2016; Xiong et al., 2014).

Apache Spark

Scala was the most used API, while Apache Spark was the most widely used framework for big data analysis. There are existential types, macros, and implicates in Scala, but there are no local data transformations or visualisations. Python includes libraries for natural language processing and machine learning. A separate classifier was used to assess the data set, and the accuracy and processing time were reported. Although the data is altered to a smaller view and the time difference between Python and Scala is calculated, the disparity increases if the data set exceeds 2 MB. Big data technology is making the world a better place by addressing environmental problems.

The data obtained via outsourcing isn't always clean, so missing value handling is handled as follows:

- Recover: Missing values can be filled in if data isn't there.
- Educated guessing: is the process of inferring missing values from the information at hand.
- Imputation: Average filling is a technique for filling in the gaps left by missing values.
- Common Point: The most crucial notion is that the midpoint is the most crucial point.
- Removing: If unsuccessful, remove the whole field value.

A categorical variable's execution speed and correctness have been assessed using machine learning. Initially, the parameters were discussed, and a select handful of them were considered before checking the outcomes. A particular idea in machine learning was used to guide the model selection, and regression was the method of choice. For severely skewed data, tree structure was used, while for imbalanced data, a random forest technique was used.

Machine Learning Models

Classification and regression issues are solved using several machine learning models, with each method performing differently on various data sets. Logistic regression, Naive Bayes, Decision Tree, and Random Forest are some categorization algorithms.

Logistic Regression

Early in the 20th century, biological scientists employed this statistical technique to confirm the data set's contamination threshold. It considers the logical connection between the explanatory factors and the outcomes. With a maximum iteration of 10, this study employs multinomial logistic regression with the assumption of 80% training data and 20% test data. The regularisation parameter, with a value of 0.3, offers the trade-off between minimising the loss and minimization of model complexity. The evaluator function employed is multiclass classification. The regularisation parameter value will often be higher than zero(Zantalis et al., 2019).

Naive Bayes

A straightforward classification technique built on the Bayes theorem is called Naive Bayes. When the existence of one characteristic in a class has no relationship to the presence of any other feature and the data set is sufficiently vast, this technique is employed. It may be broken down into Bernoulli, multinomial, and Gaussian distributions. In this study, the frequency at which particular events are produced by a single-nominal distribution is represented by feature vectors created using multi-nominal naive Bayes. The ratio of the training and testing data sets was 80:20. As it predicts the posterior probability using the joint distribution of the input and target features, this model is generative.

Decision Tree

Both classification and regression tree techniques classify a tree using a decision tree. It is used to forecast the class or value of target variables based on historical data and is a component of a supervised learning system. Due to its portrayal as a tree, it is simple to grasp. In order to build a decision tree, it is necessary to assume that all training data is a root node, prefer categorical feature values, distribute records recursively based on attribute values, and statistically identify the root or internal node.

There are two stages to decision tree implementation:

- Pre-process the dataset, partition the dataset, and train the classifier during the construction phase.
- Operational Phase: The most crucial principle is that training and testing data should be distributed in a 70/30 ratio.

Proposed Work

By using learning trees based on a randomly selected sample of input variables and data instances, ensemble techniques de-correlate the base learners. Random forests are effective classifiers for huge databases because they are constructed using bagging and random attribute selection. Choosing f features from a data collection, creating a decision tree for each feature, repeating the process n times, and determining the final prediction based on the most popular projected target are all steps in the random forest ensemble pattern. By using learning trees based on a randomly selected sample of input variables and data instances, ensemble techniques de-correlate the base learners. Random forests are effective classifiers for huge databases because they are constructed using bagging and random attribute selection. Choosing f features from a data collection, creating a decision tree for each feature, repeating the process n times, and determining the final prediction based on the most popular projected target are all steps in the random forest ensemble pattern.

In the future study, the various parameters to control the air pollution using various machine learning techniques(Babu et al., 2022; Boopathi et al., 2022; Boopathi, Khare, et al., 2023; Boopathi, Siva Kumar, et al., 2023; Domakonda et al., 2022; Jeevanantham et al., 2022; Kumara et al., 2023; Palaniappan et al., 2023; S. et al., 2022; Samikannu et al., 2022; Sampath et al., 2022; Senthil et al., 2023; Vanitha et al., 2023). The TOPSIS, VIKOR, NSGA-II and MOEA have been used to predict the result to minimize the air emission from the system(Boopathi, 2022c, 2022a; Kannan et al., 2022; Myilsamy & Sampath, 2021; Sampath et al., 2022; Yupapin et al., 2022). The experiments have been performed using statistical approaches such as Taguchi, Response surface methodology and mixture methods(Boopathi, 2021, 2022b; Kalidas et al., 2012; Sampath et al., 2021; Saravanan et al., 2022; Trojovský et al., 2023).

The numbers are produced for several inputs, such as residential, industrial, and sensitive zones, to establish the pollution level in a certain location and on a specific day of the week. The government official may assure a quicker fleet purge with the use of these numbers. In order to assist the end-user in keeping the vehicle in good working order, this chapter examines the necessity of measuring the pollution levels from specific automobiles. The information was gathered from the Indian government's general repository for additional analysis in order to give the administration the information it needs to control the flow of automobiles. To lower the quantity of pollutants, a mobile server application can be set to remind car owners to maintain their vehicles at certain intervals.

SUMMARY

Eco-friendly environmental concerns help ensure that life on earth continues. A comprehensive scientific approach is required to monitor the compounds that growth produces. Even if government and business have taken steps to reduce air pollution, it is still important to monitor the quality of the air below ground. The objective of this project is to develop intercommunication-based Internet of Things individual vehicle air pollution monitoring and identification systems (IoT). Tailpipe emission sensing was used to capture the circumstances during idling and acceleration. If authorities are able to recognise an automobile at a busy crossroads, they may act and ensure maintenance. Route and vehicle identification were made possible through mobile connectivity. Due to data acquired in the cloud, fewer individuals in India will pass away from air pollution.

REFERENCES

Babu, B. S., Kamalakannan, J., Meenatchi, N., Karthik, S., & Boopathi, S. (2022). Economic impacts and reliability evaluation of battery by adopting Electric Vehicle. *IEEE Explore*, 1–6.

Benjamin, E. J., Virani, S. S., Callaway, C. W., Chamberlain, A. M., Chang, A. R., Cheng, S., Chiuve, S. E., Cushman, M., Delling, F. N., Deo, R., de Ferranti, S. D., Ferguson, J. F., Fornage, M., Gillespie, C., Isasi, C. R., Jiménez, M. C., Jordan, L. C., Judd, S. E., Lackland, D., & Muntner, P. (2018). Heart disease and stroke statistics—2018 update: A report from the American Heart Association. *Circulation*, *137*(12), e67–e492. doi:10.1161/CIR.0000000000000558 PMID:29386200

Boopathi, S. (2019). Experimental investigation and parameter analysis of LPG refrigeration system using Taguchi method. *SN Applied Sciences*, *1*(8), 892. doi:10.100742452-019-0925-2

Boopathi, S. (2021). Improving of Green Sand-Mould Quality using Taguchi Technique. *Journal of Engineering Research*, 1–12. doi:10.36909/jer.14079

Boopathi, S. (2022a). An Extensive Review on Sustainable Developments of Dry and Near-Dry Electrical Discharge Machining Processes. *Journal of Manufacturing Science and Engineering*, *144*(5), 1–37. doi:10.1115/1.4052527

Boopathi, S. (2022b). (in press). Cryogenically treated and untreated stainless steel grade 317 in sustainable wire electrical discharge machining process: A comparative study. *Environmental Science and Pollution Research International*, 1–10. doi:10.100711356-022-22843-x PMID:36057706

Boopathi, S. (2022c). Experimental investigation and multi-objective optimization of cryogenic Friction-stir-welding of AA2014 and AZ31B alloys using MOORA technique. *Materials Today. Communications*, *33*, 104937. doi:10.1016/j.mtcomm.2022.104937

Boopathi, S., Arigela, S. H., Raman, R., Indhumathi, C., Kavitha, V., & Bhatt, B. C. (2022). Prominent Rule Control-based Internet of Things: Poultry Farm Management System. *IEEE Explore*, 1–6.

Boopathi, S., & Khare, R. KG, J. C., Muni, T. V., & Khare, S. (2023). Additive Manufacturing Developments in the Medical Engineering Field. In Development, Properties, and Industrial Applications of 3D Printed Polymer Composites (pp. 86–106). IGI Global.

Boopathi, S., Siva Kumar, P. K., & Meena, R. S. J., S. I., P., S. K., & Sudhakar, M. (2023). Sustainable Developments of Modern Soil-Less Agro-Cultivation Systems. In Human Agro-Energy Optimization for Business and Industry (pp. 69–87). IGI Global. doi:10.4018/978-1-6684-4118-3.ch004

Chakravarthi, P. K., Yuvaraj, D., & Venkataramanan, V. (2022). IoT-based smart energy meter for smart grids. *ICDCS 2022 - 2022 6th International Conference on Devices, Circuits and Systems*, (pp. 360–363). IEEE. 10.1109/ICDCS54290.2022.9780714

Copat, C., Cristaldi, A., Fiore, M., Grasso, A., Zuccarello, P., Santo Signorelli, S., Conti, G. O., & Ferrante, M. (2020). The role of air pollution (PM and NO2) in COVID-19 spread and lethality: A systematic review. *Environmental Research*, *191*, 110129. doi:10.1016/j.envres.2020.110129 PMID:32853663 da Silva, R. F. B., Batistella, M., & Moran, E. F. (2017). Socioeconomic changes and environmental policies as dimensions of regional land transitions in the Atlantic Forest, Brazil. *Environmental Science \& Policy, 74*, 14–22.

Devarakonda, S., Sevusu, P., Liu, H., Liu, R., Iftode, L., & Nath, B. (2013). Real-time air quality monitoring through mobile sensing in metropolitan areas. *Proceedings of the 2nd ACM SIGKDD International Workshop on Urban Computing*, (pp. 1–8). ACM. 10.1145/2505821.2505834

Domakonda, V. K., Farooq, S., Chinthamreddy, S., Puviarasi, R., Sudhakar, M., & Boopathi, S. (2022). Sustainable Developments of Hybrid Floating Solar Power Plants: Photovoltaic System. In Human Agro-Energy Optimization for Business and Industry (pp. 148–167). IGI Global.

Gabrys, J. (2016). *Program earth: Environmental sensing technology and the making of a computational planet* (Vol. 49). U of Minnesota Press.

Guthi, V. R., Arepalli, S., & Ganapa, P. (2016). Study of socio demographic factors among persons affected by leprosy in Kurnool division of Kurnool district, Andhra Pradesh, India. [Internet]. *International Journal of Community Medicine and Public Health, 3*(12), 3548–3555. doi:10.18203/2394-6040.ijcmph20164290

Ikumapayi, O. M., Okokpujie, I. P., Afolalu, S. A., Ajayi, O. O., Akilabi, E. T., & Bodunde, O. P. (2018). Effects of quenchants on impact strength of single-vee butt welded joint of mild steel. *IOP Conference Series. Materials Science and Engineering, 391*(1), 12007. doi:10.1088/1757-899X/391/1/012007

Jeevanantham, Y. A., Saravanan, A., Vanitha, V., Boopathi, S., & Kumar, D. P. (2022). Implementation of Internet-of Things (IoT) in Soil Irrigation System. *IEEE Explore*, 1–5.

Kalidas, R., Boopathi, S., Sivakumar, K., & Mohankumar, P. (2012). Optimization of Machining Parameters of WEDM Process Based On the Taguchi Method. *IJEST, 6*(1\&2), 91.

Kannan, E., Trabelsi, Y., Boopathi, S., & Alagesan, S. (2022). Influences of cryogenically treated work material on near-dry wire-cut electrical discharge machining process. *Surface Topography : Metrology and Properties, 10*(1), 015027. doi:10.1088/2051-672X/ac53e1

Kasar, T., Barlas, P., Adam, S., Chatelain, C., & Paquet, T. (2013). Learning to detect tables in scanned document images using line information. *2013 12th International Conference on Document Analysis and Recognition*, (pp. 1185–1189). ACM.

Kumara, V., Mohanaprakash, T. A., Fairooz, S., Jamal, K., Babu, T., & B., S. (2023). Experimental Study on a Reliable Smart Hydroponics System. In *Human Agro-Energy Optimization for Business and Industry* (pp. 27–45). IGI Global. doi:10.4018/978-1-6684-4118-3.ch002

Lee, G. H., & Yoo, J. Y. (2000). Performance analysis and simulation of automobile air conditioning system. *International Journal of Refrigeration, 23*(3), 243–254. doi:10.1016/S0140-7007(99)00047-X

Loukissas, Y. A. (2016). A place for Big Data: Close and distant readings of accessions data from the Arnold Arboretum. *Big Data \& Society, 3*(2), 2053951716661365.

M., G., B., R., S., S., & Boopathi, S. (2018). Fabrication of Solar Irrigation System. *International Research Journal of Engineering and Technology (IRJET), 5*(2), 2106–2108.

Malche, T., & Maheshwary, P. (2015). Harnessing the Internet of things (IoT): A review. *International Journal (Toronto, Ont.)*, *5*(8).

Meneguette, R. I. (2016). A vehicular cloud-based framework for the intelligent transport management of big cities. *International Journal of Distributed Sensor Networks*, *12*(5), 8198597. doi:10.1155/2016/8198597

Myilsamy, S., & Sampath, B. (2021). Experimental comparison of near-dry and cryogenically cooled near-dry machining in wire-cut electrical discharge machining processes. *Surface Topography : Metrology and Properties*, *9*(3), 035015. doi:10.1088/2051-672X/ac15e0

Nitti, M., Pilloni, V., Giusto, D., & Popescu, V. (2017). IoT Architecture for a sustainable tourism application in a smart city environment. *Mobile Information Systems*, *2017*, 2017. doi:10.1155/2017/9201640

Palaniappan, M., Tirlangi, S., Mohamed, M. J. S., Moorthy, R. M. S., Valeti, S. V., & Boopathi, S. (2023). Fused Deposition Modelling of Polylactic Acid (PLA)-Based Polymer Composites: A Case Study. In Development, Properties, and Industrial Applications of 3D Printed Polymer Composites (pp. 66–85). IGI Global.

S., P. K., Sampath, B., R., S. K., Babu, B. H., & N., A. (2022). Hydroponics, Aeroponics, and Aquaponics Technologies in Modern Agricultural Cultivation. In *Trends, Paradigms, and Advances in Mechatronics Engineering* (pp. 223–241). IGI Global. doi:10.4018/978-1-6684-5887-7.ch012

Sami, K., & Edith, N. (2019). Real-time air pollution monitoring with sensors on city bus. *Digital Communications and Networks.*

Samikannu, R., Koshariya, A. K., Poornima, E., Ramesh, S., Kumar, A., & Boopathi, S. (2022). Sustainable Development in Modern Aquaponics Cultivation Systems Using IoT Technologies. In *Human Agro-Energy Optimization for Business and Industry* (pp. 105–127). IGI Global.

Sampath, B., Naveenkumar, N., Sampathkumar, P., Silambarasan, P., Venkadesh, A., & Sakthivel, M. (2021). Experimental comparative study of banana fiber composite with glass fiber composite material using Taguchi method. *Materials Today: Proceedings*, *49*(5), 1475–1480. doi:10.1016/j.matpr.2021.07.232

Sampath, B. C. S., & Myilsamy, S. (2022). Application of TOPSIS Optimization Technique in the Micro-Machining Process. In Trends, Paradigms, and Advances in Mechatronics Engineering (pp. 162–187). IGI Global. doi:10.4018/978-1-6684-5887-7.ch009

Saravanan, M., Vasanth, M., Boopathi, S., Sureshkumar, M., & Haribalaji, V. (2022). Optimization of Quench Polish Quench (QPQ) Coating Process Using Taguchi Method. *Key Engineering Materials*, *935*, 83–91. doi:10.4028/p-z569vy

Schneider, P., Castell, N., Vogt, M., Dauge, F. R., Lahoz, W. A., & Bartonova, A. (2017). Mapping urban air quality in near real-time using observations from low-cost sensors and model information. *Environment International*, *106*, 234–247. doi:10.1016/j.envint.2017.05.005 PMID:28668173

Senthil, T. S., Puviyarasan, M., Babu, S. R., Surakasi, R., Sampath, B., & Associates. (2023). Industrial Robot-Integrated Fused Deposition Modelling for the 3D Printing Process. In Development, Properties, and Industrial Applications of 3D Printed Polymer Composites (pp. 188–210). IGI Global.

Sherazi, H. H. R., Khan, Z. A., Iqbal, R., Rizwan, S., Imran, M. A., & Awan, K. (2019). A heterogeneous IoV architecture for data forwarding in vehicle to infrastructure communication. *Mobile Information Systems*, *2019*, 2019. doi:10.1155/2019/3101276

Sobral, T., Galvão, T., & Borges, J. (2019). Visualization of urban mobility data from intelligent transportation systems. *Sensors (Basel)*, *19*(2), 332. doi:10.339019020332 PMID:30650641

Sugania, S. S., Vishalishwaran, D. S., & Kumar, J. V. (2018). Automated System for Road Safety Enhancement using Big. *Data Report.*

Tan, J., Zhang, M., Li, H., Guo, W., & Wang, L. (2018). A Bidirectional Green Wave Band Method under Asymmetric Phase Sequence Mode based on Mobile Navigation data. *2018 IEEE 7th Data Driven Control and Learning Systems Conference (DDCLS)*, (pp. 1191–1196). IEEE.

Trojovský, P., Dhasarathan, V., & Boopathi, S. (2023). Experimental investigations on cryogenic friction-stir welding of similar ZE42 magnesium alloys. *Alexandria Engineering Journal*, *66*(1), 1–14. doi:10.1016/j.aej.2022.12.007

Umap, V. S., & Deshmukh, P. H. (n.d.). Intelligent Transportation System—Recent Trends in Transportation System. *RE:view*.

Vanitha, S. K. R., & Boopathi, S. (2023). Artificial Intelligence Techniques in Water Purification and Utilization. In *Human Agro-Energy Optimization for Business and Industry* (pp. 202–218). IGI Global. doi:10.4018/978-1-6684-4118-3.ch010

Vashisht, S., & Rakshit, D. (2021). Recent advances and sustainable solutions in automobile air conditioning systems. *Journal of Cleaner Production*, *329*, 129754. doi:10.1016/j.jclepro.2021.129754

Xiong, G., Zhu, F., Fan, H., Dong, X., Kang, W., & Teng, T. (2014). Novel ITS based on space-air-ground collected big-data. *17th International IEEE Conference on Intelligent Transportation Systems (ITSC)*, (pp. 1509–1514). IEEE. 10.1109/ITSC.2014.6957647

Yi, S., Li, C., & Li, Q. (2015). A survey of fog computing: concepts, applications and issues. *Proceedings of the 2015 Workshop on Mobile Big Data*, (pp. 37–42). ACM. 10.1145/2757384.2757397

Yupapin, P., Trabelsi, Y., Nattappan, A., & Boopathi, S. (2022). Performance Improvement of Wire-Cut Electrical Discharge Machining Process Using Cryogenically Treated Super-Conductive State of Monel-K500 Alloy. *Iranian Journal of Science and Technology - Transactions of Mechanical Engineering*, 1–17. doi:10.1007/s40997-022-00513-0

Zantalis, F., Koulouras, G., Karabetsos, S., & Kandris, D. (2019). A review of machine learning and IoT in smart transportation. *Future Internet*, *11*(4), 94. doi:10.3390/fi11040094

Zaree, T., & Honarvar, A. R. (2018). Improvement of air pollution prediction in a smart city and its correlation with weather conditions using metrological big data. *Turkish Journal of Electrical Engineering and Computer Sciences*, *26*(3), 1302–1313.

Zhang, P., Dong, G., Sun, B., Zhang, L., Chen, X., Ma, N., Yu, F., Guo, H., Huang, H., Lee, Y. L., Tang, N., & Chen, J. (2011). Long-term exposure to ambient air pollution and mortality due to cardiovascular disease and cerebrovascular disease in Shenyang, China. *PLoS One*, *6*(6), e20827. doi:10.1371/journal.pone.0020827 PMID:21695220

Chapter 15
Green Environmental Sustainability:
Initiatives by Government and Corporate Sectors

Jyoti Verma
https://orcid.org/0000-0002-7559-4312
Chitkara Business School, Chitkara University, India

Gagandeep
Chitkara Business School, Chitkara University, India

Monica Gupta
Chitkara Business School, Chitkara University, India

ABSTRACT

Green environmental sustainability is a responsible human interaction with the ecosystem which allows for long-term environmental quality and avoids depletion of our natural resources. It focuses on environmentally friendly products that satisfy the needs of ultimate customers and its vision is to protect the environment. Nowadays, customers must be self-socially responsible and concerned about the environment. A massive amount of plastic waste has been generated due to increasing dependency and consumption of its wide range, which is increasingly being used in the packaging of different food items, cosmetic products, pharmaceuticals, and other production sectors. If not processed and managed appropriately, the after-use of generated plastic waste has many poor impacts on the environment. Data had been collected from multiple sources of evidence. This chapter aims to discuss some initiatives taken by the Indian Corporate Sector and the Indian Government in this context. It also discusses various methods adopted by the authorities to manage plastic waste.

DOI: 10.4018/978-1-6684-8117-2.ch015

1. INTRODUCTION

Green environmental sustainability is the ability to conserve the most significant qualities in the physical environment (Cronin et al, 2011; Cherian & Jacob, 2012). Threats to these natural resources imply that there is a risk that these items will not be kept stored (Eltayeb & Zailani, 2014). For instance, the excavation of large non-renewable resources can create threats of severe destruction or even its extinction. In the past time, when environmental problems arise, environmental authorities deal with some innovative ways to reduce damage (Eneizan, 2019; Tirabeni et al.,2019). But it is not always easy to know when and where the threats will have consequences and often the effects are hard to undo. So, environmental managers are increasingly using techniques that are intended to prevent damage done in the first place (Abbaszadeh et al., 2006; Dabija et al.,2018). Sustainability issues arise whenever there is a risk of severe or irreversible loss of property or natural resources that people value, and whenever such dangers exist, there is a measure of urgency to take action. Now a days, customers are required to be self socially responsible and concerned about environment. The companies started marketing green products to this segment customer (Singhal et al., 2018) and try to showcase their attempt in preventing disposal of hazardous products into the ecosystem (Peattie, 2001).

1.1 Role of Plastic In Everyone's Life and its Impact on Ecosystem

Starting from the beginning, in the year 1907, plastic has been playing a critical role in our routine lives and it continues to propound its presence till date. This amazing product of the 20th century is gradually obstructing our environment and destroying our natural resources like rivers, seas, oceans etc. Plastic is inevitable and it seems to be used by everyone. The word-plastic, is being derived from the word "pliable" which means "easily shaped". It can be easily converted various shapes depending upon the desired functions, physical and chemical properties. Plastic products range from items such as containers, packaging, durable items, and fragile goods. There is a confusion that all plastics products can be recycled, so they are recycled. But there are many problems with this view. Although the use of plastic is increasingly high, doubts about effective recycling options, recycling and disposal are also growing. Problems such as the growing number of additives that change strength, texture, flexibility, color, germ resistance, and other plastic properties, make plastics unusable. In addition, there is very little market value for some plastics, leading municipalities to dispose of or burn plastics as waste. Many researchers and environmentalists are paying close attention to environmental issues with indications of a sustainable development perspective. The international community is being challenged by many problems associated with pollution and the continued use of energy resources. The economic progress in the world by promoting the ideas of civilization and industry is based on the use of too much energy (AlKhidir & Zailani, 2009). Traditional methods of obtaining the required amount of energy to connect the needs of developed quotations and industries cannot be effectively discussed according to the threat of pollution resulting from the developed energy industry. In addition, traditional methods of energy systems do not consider into account the concept of sustainable development. There is a strong need to focus on alternative energy sources. Green energy can be discussed as an effective way to promote the required amount of energy in industries, urban areas, and businesses without harming the environment (Pradeep & Akhilesh, 2016; Schoolman et al.,2016). From this point on, the concept of green energy is closely linked to renewable energy that is often discussed as a source of energy with minimal side effects (Yasmin & Shamshuddin, 2014; Zhang et al.,2020). Sustainable development is the most important requirement in modern society in relation to the general trend of global progress towards addressing environmental and environmental problems (Abdulsahib et al, 2019; Song & Yu, 2018). The ultimate goal of sustainable development is based on addressing the needs of modern people by focusing on the potential needs of future generations (Polonsky, 2011; Verma & Gagandeep, 2020).

1.2 Innovative Strategies to Tackle This Issue

Plastic wastes are rapidly increasing because of the continuous growth of the development of the industrial sector and the population growth rate in the world (McCullough et al., 2016). Bio-degradable/decomposable and non-degradable/decomposable wastes are generated from the natural processes of the living creatures on the earth. Local authorities, Social communities, and Government municipalities have introduced various measures and rules for environmental safety that can assist people to dispose properly of all plastic waste after utilization. There are many strategies for waste management like Recycling, Landfills, Incineration, Pyrolysis, and Bioremediation. These methods are adopted to have a clean, hygienic, and safe environment.

1.2.1 Recycling

Recycling, also known as reusing/renewing, is one of methods of management of waste that actually collects waste materials, converts them into raw materials, and further reused it to form another valuable product. It helps to protect us from the harmful effect of waste products on the environment. The plastics products are non-biodegradable and contains materials which can be easily melted, converted into another plastic products like furniture items etc. The following steps are being used for its completion. It concludes collection of waste plastics materials, its sorting into different categories, being washed properly to omit impurities, shredding, resizing, separating usable plastics, and then compounding. When plastic is reused properly, it provides number of benefits rather than disposing it to non-desirable places. The combustion of waste plastic material causes the emission of harmful gases in the atmosphere. The main advantage of recycling is to reduce the content of carbon dioxide (CO_2) and other harmful gases and protect the human life on earth. This method reduces the level of pollution across the environment and maintain ecological balance. It requires less amount of energy and also helps in conserving natural resources. It also protects steadily depleting landfill areas and enables the requirements of fossil fuels. Thus, it helps to promote a sustainable life-style culture and better contributes towards the development of economy.

1.2.2 Incineration

In the incineration method, the burning of wastes took place in the presence of O2, which is also termed complete combustion. It releases H2O and CO2 into the atmosphere. After incineration, the waste produced is composed of various chemicals (which are volatile in nature), a less quantity of Hydrochloric acid (HCl), and ash. Normally, all kind of plastic waste is not suitable for the combustion process. To select appropriate plastics from household waste plastic, one must be alert to avoid any kind of accidents. This is not mandatory that burning could be used to tackle every kind of waste plastic. The fuels, in the form of energy, can be produced by the combustion of organic molecules. This fuel is found in three states such as solid, liquid, and gas, which can be used by all kinds of vehicles. The incineration method has several positive impacts on the ecology. It can also be used as a waste reduction method as it produces electricity from the waste items, which has great demand in the industrial sector. It has a significant role in generating renewable energy sources from bio-mass resources. This method has many advantages for waste management of plastic products. These include reduction in waste material quantity in the ecology, generating heat and electricity, reducing pollution, elimination of harmful germs and chemicals, and saving transportation costs of the waste movement.

1.2.3 Bioremediation

It is the process by which microorganisms decompose waste. It has main principles towards decontamination, which biodegrade all-natural compounds with the help of microorganisms. In order to facilitate microorganism's growth, it requires specified conditions for medium such as enzymes, temperature, pressure, nutrients, etc which get finally settled at an optimal level. Polyvinyl chloride (PVC) plastic under natural decomposition method results in the chain of monomers of phthalates such as dioxins, chlorofluorocarbons (CFC), vinyl monomers. The solid waste management depends on the methods of incineration (burning) and landfills. This method is not suitable for the environment. Hence, a huge funds are required to carry on the degradation of synthetic material by using enzymes.

1.2.4 Landfills

It refers to those areas that can be used for disposable of plastics waste before going deep buried under the earth's surface, proper landfill mechanism has been adopted. Many preventive steps should be considered to circumvent any kind of another side effects like soil degradation and groundwater contamination. The prime objective of this method is to introduce a safe method of plastic waste disposal and protect environment. This method requires huge amount of work to be done in the specified area, like digging a deep hole, dumping waste in high depths and letting it decompose. It is more time consuming and can take more than one-year time. In the processing stage, each organic molecule ensures the movement through decomposition or biodegradation. Different type of plastic wastes took long time of degradation due to their chemical properties and environment conditions. This method is an excellent source of energy as it releases methane gas (CH_4) during process of biodegradation. This method keeps the surroundings clean and segregates harmful waste from other wastes. Therefore, the first option for disposing plastic products should be either reuse or recycle.

1.2.5 Pyrolysis

It refers to those areas that can be used for disposable plastics waste before going deep buried under the earth's surface, proper landfill mechanism has been adopted. Many preventive steps should be considered to circumvent any kind of other side effects like soil degradation and groundwater contamination. The prime objective of this method is to introduce a safe method of plastic waste disposal and to protect the environment. This method requires a huge amount of work to be done in the specified area, like digging a deep hole, dumping waste in high depths, and letting it decompose. It is more time-consuming and can take more than one-year time. In the processing stage, each organic molecule ensures movement through decomposition or biodegradation. Different types of plastic wastes took a long time to degrade due to their chemical properties and environmental conditions. This method is an excellent source of energy as it releases methane gas (CH4) during the process of biodegradation. This method keeps the surroundings clean and segregates harmful waste from other wastes. Therefore, the first option for disposing of plastic products should be either reused or recycle.

1.3 Initiatives Taken by Individuals

Individuals can do wonders to bring such dramatic changes in ecosystem. They have a significant role to perform in combating the impacts of plastic pollution and thus create more sustainable environment. Following points are few ways that an individual can do-

- pick up littered items and keep surroundings clean
- use cloth/jute bags for shopping
- use of reusable cups to get your tea/coffee
- support 'plastic free' businesses only
- choose products which contain less amount of plastic content or that can be easily recycled
- avoid single use decorative products or crockeries for your events/parties
- reduce Your Use of Single-Use Plastics
- recycle properly
- save and grow trees and do participate in such programmes
- participate in clean water campaign and create awareness to others
- support bans imposed by government on any product and create awareness
- avoid products containing microbeads
- support organizations who are addressing the issue like Plastic Pollution, Water Pollution, Air Pollution etc.

1.4 Initiatives Taken by Government

There is a strong need to take some initiatives to maintain ecological balance. For this purpose, our Indian Prime Minister has started a campaign to eliminate single use plastic by 2022, keeping into consideration the poor effects of waste plastic on aquatic and terrestrial system. Single use plastic products have become a cause of pollution for all countries. India is now ready to take serious action for mitigating pollution. In 2019, fourth UN Environment Assembly, India had brought a resolution on addressing the issue of single-use plastic and also highlighting the urgency for entire population to seriously focus on this issue and take remedial action. Various plastic products even in the processing phase of manufacturing, importing of plastic products, stocking or storing, sale, final use shall be prohibited[1] w. e. f 1st July, 2022. Hence, it is everyone's duty to bring a positive change in the environment. Lots of other plastic related products like earbuds, candy sticks, ice-cream sticks, plastic sticks for balloons, cigarette packets, sweet box packing, invitation cards, plastic banners (of less than 100 micron), stirrers etc need to be prohibited. The thickness of plastic bags has also been increased from 50 to 75 microns and to 120 microns w.e.f. December 31, 2022. Due to this increase in the thickness of plastic products, reuse of plastic can be possible. 'Swachh Bharat Mission' introduced by government has strengthened the waste management system in all States and union territories. Along with this, initiatives have also been taken to strengthen the implementation of Plastic Waste Management Rules, 2016. Government is taking various measures for the elimination of single use plastics and also the effective implementation of the rules (Yacob et al., 2018). On the theme for spreading awareness, ministry has conducted Pan India essay writing competition amongst school students in the country.

Under Pradhan Mantri Ujjwala Yojana scheme, gas connections were provided to women of BPL families. It has resulted in a drastic change of respiratory disease as it contributed to the upgraded air quality as per recent medical reports[2]. India stands 3rd in the renewable power, 4th in wind power, and 5th in solar power across the world. The climate strategies call for efficient and clean energy systems and planned eco restoration.

India has reduced carbon dioxide gas emissions upto 38 million tons annually by using energy efficient appliances, provide clean fuel for cooking purpose and provide electricity to 100% of its villages, will restore 26 million hectares of land by 2030 and target to install 450 GW of renewable energy[3]. India has launched global partnerships for climate action.

1.5 Initiatives by Corporates

Companies play an important role in revamping the way we use plastics and are coming up to handle this prime responsibility. In early 2000, when production of plastic is at its peak and it actually mismanaged the amount of plastic waste. More than 300 million tons of plastic waste every year is being generated across world. It is estimated that 8 million tons of plastic waste flow into the ocean annually.

Many world's leading companies are pledging to resolve this issue for the betterment of life. Companies like PepsiCo and Unilever have also pledged to convert to 100 percent reusable, and compostable packing by 2025. The Coca-Cola Company has increased the recycled content in their bottles from 25% to 50% and ensures the successful implementation of this new sustainable strategy. The Ban of plastic is not only the solution as it will affect the lives of many people, especially who are falling under the category of lower income group. The consumers have been shifted to conventional plastic to recycled plastic and thus resulting the escalating growth of plastic waste management market.

Retailers have started the use of paper bags/jute bags/cloth bags and are replacing plastic bags to large extent. Along with this, they are also using bioplastics that might play a major role in daily lives. The other alternatives are cardboard packaging, bamboo cutlery, etc., which ought to be well practiced and implemented further. For instance, Marriott International, largest hotel chain in the world, has announced that it would stop doing business of plastic toiletries to its customers by 2020. Amazon also committed to avoid the use of packaging of single use plastic across its operations in India. In order to tackle the issue of plastic waste and pollution, immense innovative and entrepreneurial ideas came into picture that can be positively considered as an opportunity for market players.

Many private and public sector banks have taken initiatives to support environment and ensure the sustainability of ecology. For instance, State Bank of India (SBI) is entered into "Green Channel Counter" in 2010. It is a green service and ensures paperless banking. SBI launched carbon disclosure project in collaboration with Suzlon Energy Ltd. and used wind-energy to reduce carbon emissions. Bank of Baroda had added new delivery channel like e-banking, Mobile banking, Internet banking, ATM to allow paperless banking. IndusInd Bank had opened the first solar-powered ATM. Canara Bank has collaboration with UNEP for the solar project in the year 2003. The bank is also providing solar power biometric ATM in a rural area. Punjab National Bank (PNB) has initiated energy efficient appliances and also launched green audit sheet to access the impact of the various green initiatives taken by banks. The bank also provides a loan to a business unit producing renewable energy. HDFC Bank has also contributed in reducing paper by issuing and encouraging e-transaction services to its corporate and retail customers.

Kotak Mahindra bank start "Think Green" initiatives to encourage the consumers to sign for e-statement and by doing so Trees.com plants a tree on behalf of Kotak Mahindra bank.

Many corporates who are dealing with electronic products have also forward to contribute in sustainable development. For instance, Philips Electronics launched 'Philips Eco Vision' program. The company has 91 new Green Products in the market. Panasonic provides appliances by using sensors and has a series of products like a refrigerator, washing machine, cooker that are made from recycled material. IBM establishes green data centers that consumed less energy. Apple has taken steps to take back old apple products, by giving a discount on next purchase. Videocon Group has upgraded their Research & Development centers to green technology emit less harmful substance and save energy. Idea cellular has also initiated a campaign 'Use Mobile Save Paper'. It organized Green pledge campaign and decorated buses with plants to convey the message of greenery so that everybody can contribute to saving paper and trees. Nokia's has taken the initiative to manage the e-waste material. It starts to take back scheme and establish a collection center to recover useful materials and dispose of waste. Samsung was the first company to offer a mobile phone that is made of renewable material. These mobile handsets –W510 and F268 are environment-friendly products. In India, Wipro Infotech was the first company to introduce eco-friendly computer. It has also launched a new version of related products. These products are helping in reducing e-waste.

Many corporates who are dealing with Auto Sector has also initiated and contributed towards Green Initiative. For instance, Mahindra Group has launched an innovative project 'Mahindra Hariyali' that leads to the theme of afforestation, which means millions of trees will be planted by all stakeholders in India. Hero Honda Motor has a continuous focus on innovation in green products and makes a balance between business and nature. Hyundai has offered a Santro car based on CNG. Maruti has launched a 'WagonR' based upon LPG. It frames policy based upon 3R Reduce, Reuse, Recycle. Tata's new mantra is Going Green. It has built up a showroom using the natural material in their design. It shows natural atmosphere. Tata group also launched a purifier that is made from natural material. Honda introduced its civic Hybrid car in India.

Many corporates who are dealing with FMCG Sector has also initiated and contributed towards Green Initiative. For instance, Cocacola, instead of plastic, used syrup directly from the tank. New Surf Excel is very effective because it reduces water consumption. Proctor and Gamble (P& G) has initiated refills for its Handwash and Detergent packs that can throw away after using. ITC company has introduced Papercraft, which is eco-friendly paper. It also takes farm-forestry initiative in an arid land. Kirloskar Copeland Limited has introduced the eco-friendly R404A gas compressor recently. Kansai Nerolac has worked on the project which is extracting hazardous metals from paints. The hazardous heavy metals like mercury (Hg), chromium (Cr), lead (Pb), etc have bad effects on human's health.

Almost every sector is working on this noble theme. For Instance, In McDonald's restaurant, the use of napkins, bags which are made up of recycled paper, have been encouraged to support environment. KFC has taken new initiative to build a green restaurant by using elements that represents the value of environmental design and leadership in energy. In order to fetch benefits from carbon trading, many companies like Essar Power, Gujarat Fluorochemicals Ltd., Tata Steel, etc have obtained clearance to continue with specifically designed projects. Suzlon Energy manufactures provides an innovative source of energy, which is based on wind power source of energy. The purpose of this green initiative is to reduce the carbon footprint. Oil and Natural Gas Corporation (ONGC) has introduced Mokshada Green Crematorium, that saves 60-70 percent of wood. Taj hotel Group runs the Taj chain has created Eco rooms and they have been using energy efficient resources for delivering services. Indian Railways has taken an initiative of E-tickets and allowed passengers to carry PNR no. of e-tickets on their devices. Agartala is first green city in India. Tripura govt announced that all vehicles run on compressed natural gas in Agartala. Bhardwara power station of NTPC in Delhi used coal ash in a different way to reduce pollution. Reva India is a Bangalore based company that generate an electronic car and it is first in the world. A electronic rickshaw/ e-rick/battery operated rickshaw, which is under the sponsorship of cellular service providers, promote environment friendly mode of transport in Delhi.

CONCLUSION

Sustainable growth of a business is quite a challenging task nowadays. Making strategies to introduce green products to the customers is the prime objective of every business as the environmental problems are massively growing day by day. The environmental problem is growing at a rapid rate in India. Every stakeholder wants to play some important role to conserves our environment. Green marketing is a recent phenomenon and helps to create awareness among consumers and business firms to reduce the detrimental impact on the environment. It is very important to implement green practices in business. Every business is entering into the green era at high speed to maintain strong relations with customers and maintain demands via reducing their environmental impacts. Green marketing makes a drastic change in the world. Now consumer prefers eco-friendly products. This chapter helps us to study various initiatives taken by the government as well as green practices adopted by different firms like Electronic firms, the Banking sector, FMCG firms, and other initiatives simultaneously in India.

REFERENCES

Abbaszadeh, S., Zagreus, L., Lehrer, D., & Huizenga, C. (2006). Occupant satisfaction with indoor environmental quality in green buildings. Proceedings of Healthy Buildings, Lisbon, Vol. III, 365-370.

Abdulsahib, J. S., Eneizan, B., & Alabboodi, A. S. (2019). Environmental concern, health consciousness and purchase intention of green products: An application of extended theory of planned behavior. *The Journal of Social Sciences Research*, *5*(4), 868–880. doi:10.32861/jssr.54.1203.1215

AlKhidir, T., & Zailani, S. (2009). Going green in supply chain towards environmental sustainability. *Global Journal of Environmental Research*, *3*(3), 246–251.

Cherian, J., & Jacob, J. (2012). Green marketing: A study of consumers' attitude towards environment friendly products. *Asian Social Science*, *8*(12), 117–126. doi:10.5539/ass.v8n12p117

Cronin, J. J. Jr, Smith, J. S., Gleim, M. R., Ramirez, E., & Martinez, J. D. (2011). Green marketing strategies: An examination of stakeholders and the opportunities they present. *Journal of the Academy of Marketing Science*, *39*(1), 158–174. doi:10.100711747-010-0227-0

Dabija, D. C., Bejan, B. M., & Grant, D. B. (2018). The impact of consumer green behaviour on green loyalty among retail formats. *Moravian geographical reports,*26(3),173-185.

Eltayeb, T., & Zailani, S. (2014). Going green through green supply chain initiatives toward environmental sustainability. *Operations and Supply Chain Management: an International Journal*, *2*(2), 93–110. doi:10.31387/oscm040019

Eneizan, B. M., Abd-Wahab, K., & Obaid, T. F. (2016). Effects of green marketing strategy on the financial and non-financial performance of firms: A conceptual paper. *Oman Chapter of Arabian Journal of Business and Management Review*, *5*(12), 14–27. doi:10.12816/0031457

McCullough, B. P., Pfahl, M. E., & Nguyen, S. N. (2016). The green waves of environmental sustainability in sport. *Sport in Society*, *19*(7), 1040–1065. doi:10.1080/17430437.2015.1096251

Peattie, K. (2001). Golden goose or wild goose? The hunt for the green consumer. *Business Strategy and the Environment*, *10*(4), 187–199. doi:10.1002/bse.292

Polonsky, M. J. (2011). Transformative green marketing: Impediments and opportunities. *Journal of Business Research*, *64*(12), 1311–1319. doi:10.1016/j.jbusres.2011.01.016

Pradeep, M.D. & Akhilesh Suresh A Kuckian (2016). Green Marketing to meet consumer demands and sustainable development-Challenges and Opportunities. International journal of Advanced Trends in Engineering and Technology (IJATET), 1(1), 34-41.

Schoolman, E. D., Shriberg, M., Schwimmer, S., & Tysman, M. (2016). Green cities and ivory towers: How do higher education sustainability initiatives shape millennials' consumption practices? *Journal of Environmental Studies and Sciences*, *6*(3), 490–502. doi:10.100713412-014-0190-z

Singhal, S., Deepak, A., & Marwaha, V. (2018). Green Initiatives Practices in Indian Hotels. *IOSR Journal of Business and Management*, *20*(8), 10–13.

Song, W., & Yu, H. (2018). Green innovation strategy and green innovation: The roles of green creativity and green organizational identity. *Corporate Social Responsibility and Environmental Management*, *25*(2), 135–150. doi:10.1002/csr.1445

Tirabeni, L., De Bernardi, P., Forliano, C., & Franco, M. (2019). How can organisations and business models lead to a more sustainable society? A framework from a systematic review of the industry 4.0. *Sustainability (Basel)*, *11*(22), 6363. doi:10.3390u11226363

Verma, J. (2020). A holistic approach of quality in higher education system vis-a-vis sustainable development. [IJM]. *International Journal of Management*, *11*(10), 433–440.

Yacob, P., Wong, L. S., & Khor, S. C. (2018). An empirical investigation of green initiatives and environmental sustainability for manufacturing SMEs. *Journal of Manufacturing Technology Management*, *30*(1), 2–25. doi:10.1108/JMTM-08-2017-0153

Yasmin Begum, R. N., & Shamshuddin, R. N. (2014). Green Marketing: Challenges and Strategies for Indian Companies in 21st Century. *International Journal of Research in Business Management*, *2*(5), 91–104.

Zhang, H., He, J., Shi, X., Hong, Q., Bao, J., & Xue, S. (2020). Technology characteristics, stakeholder pressure, social influence, and green innovation: Empirical evidence from Chinese express companies. *Sustainability (Basel)*, *12*(7), 2891. doi:10.3390u12072891

ENDNOTES

[1] https://pib.gov.in/PressReleaseIframePage.aspx?PRID=1745433

[2&3] https://economictimes.indiatimes.com/industry/renewables/green-investment-sustainability-a-priority-in-public-policy-mosenvironment/articleshow/87316071.cms?from=mdr

Chapter 16
E-Waste Recovery and Utilization Processes for Mobile Phone Waste

S. Selvakumar
https://orcid.org/0000-0001-8056-097X
Aarupadi Veedu Institute of Technology, Vinayaka Mission Research Foundation, India

R. Shankar
https://orcid.org/0000-0003-1863-3685
Teegala Krishna Reddy Engineering College (Autonomous), India

P. S. Ranjit
Aditya Engineering College, India

Sumanta Bhattacharya
https://orcid.org/0000-0003-2563-2787
Maulana Abul Kalam Azad University of Technology, India & NALSAR University of Law, India

Amara S. A. L. G. Gopala Gupta
Koneru Lakshmaiah Education Foundation, India

Sampath Boopathi
https://orcid.org/0000-0002-2065-6539
Muthayammal Engineering College, India

ABSTRACT

Mobile phones are among the most widely used devices worldwide, but they also present a serious environmental contamination issue. Mobile phones often contain dangerous materials that, if not properly handled and disposed of, will leak into the environment. To ascertain the level of contamination, risk assessment techniques must be developed. In this research, national and international scenarios and regulating measures have been discussed for the harmfulness and management of mobile waste materials. The various characteristics of mobile wastes and materials flow for the processing of mobile related wastes have been elaborated. The different types of frameworks for the indicating systems, selection criteria, and risk assessment indication systems were also illustrated. The various stages of mobile waste handling and processing to recover polymers have also been represented.

DOI: 10.4018/978-1-6684-8117-2.ch016

INTRODUCTION

Since the 1990s, there has been a considerable increase in electronic garbage ("e-waste"), making it one of the most rapidly expanding pollution issues globally. Since mobile phones are compact and portable electronic devices, they greatly contribute to this stream of e-waste. People use their mobile phones as essential communication tools in their daily lives, and they switch phones regularly. Waste mobile phones can seriously affect the environment if they are not disposed of properly since they contain a number of dangerous and harmful materials, including heavy metals and brominated flame retardants. The frequency of replacement is currently less than two years in industrialised countries and less than three years in underdeveloped nations. Electronic goods that are no longer functional or have simply been thrown away are referred to as "e-waste." Wasted mobile phones contain a variety of valuable materials, including gold, silver, and copper. The recycling process, however, can also have a negative impact on the environment and people's health if there are insufficient effective controls on pollutants. Mobile phone waste is referred to as "e-waste," and due to the quantity created, many laws and programmes have been implemented to manage this type of waste(Cherukuri et al. 2018). The WEEE Directive (on waste electrical and electronic equipment) and the RoHS Directive are significant pieces of legislation in the European Union.

Certain hazardous compounds cannot be used in electrical and electronic equipment, according to the RoHS Directive. To find practical answers, a global initiative called Solving the E-waste Problem (StEP) has been created. It also offers advice on the best ways to control the usage and disposal of mobile devices. The process of evaluating the risks associated with used mobile devices is seen as challenging and ambiguous. According to the World Health Organization, risk assessments can evaluate the extent of environmental pollution and enable risk managers to develop strategies for risk reduction that are grounded in science. Establishing a risk assessment indicator system to measure the danger level in the context of mobile phones looks preferable because it can streamline the risk assessment process. A good indicator can successfully transform complex data or events into clear, understandable information. An indicator is a single measure of a characteristic. Hazardous Laboratory used TCLP, WET, and TTLC to conduct leaching experiments on a variety of electronic devices, including LCD monitors, cell phones, and plasma TV panels. Pb was the metal that surpassed the threshold limit the most frequently, the study showed. In one study, 34 mobile phones were used to compare TCLP, WET, and TTLC in order to assess the consistency of hazardous waste classification. It was discovered that one sample exceeded the Lead regulatory limit. The improved technique leached less Pb than the TC limit but more Pb than the acceptable limit (5 mg/L). Disc drives, expansion cards, motherboards, and power supply units in outdated PCs were examined by Li et al. using TCLP extraction fluid. The following 18 elements were examined: As, Au, Ag, Al, Ba, Cr, Cu, Fe, Be, Cd, Cr, Cu, Fe, Ga, Pb, Sb, Se, Ni, Pd, Sn, and Zn. According to the study, old computers continuously produce Pb when they are dumped or exposed to rain. Pb extraction from motherboards was found to be 30–100 times greater than the permitted limit set by the EU (5 mg/L). 50 to 80 percent of the Pb that leached out of PWBs was caused by the motherboards. Heavy metal toxicity in the TCLP leachates generated by electronic gadgets was examined(Elleuch et al. 2018). Pb content was less than 5 mg/L in solidified PWBs. The dangerous PWBs were changed into non-hazardous PWBs via high compaction and cement solidification. The study's findings indicated that using modified TCLP increased Pb extraction by 50.9 percent.

According to the study, even after extensive transit, Pb would continue to leak. The metal content of PWBs and spent batteries (Lithium and nickel-metal hydride) from ten different manufacturers are discussed using EPA 3050B. Co and Ni both exceeded the cut-off point by a factor of and 38, respectively. In 90% of the samples, the concentration was less than 100 mg/kg. According to the findings of this study, the plastic used in cell phone manufacturing poses no immediate threat. Elements like Cd, Pb, Ag, and Ni are available in a mobile phone's plastic. The highest concentrations of PBDEs, PCBs, and HBCDs were found in samples from Bui Dau city. In comparison to landfill leachates, TCLP tests produced more lead, according to a study conducted in the Philippines. The threshold limit was surpassed by factors of 100 and 20, respectively, by Cu and Pb. The disposal of mobile phones was regarded as harmful. Chromobacterium violaceum was capable of leaching Cu, Au, Zn, Fe, and Ag at rates more than 79, 69, 46, 9, and 7%, respectively. Pb and Cr concentrations discharged from used printed circuit boards (PWBs) complied with the requirement. For all of the samples, the Cd and Hg metal concentrations were below the EU-permitted limit. According to the EPA 3050B approach, the amounts of Cu, Pb, Ni, and Zn in PCBs of 10 abandoned laptops have been examined. Laptop PCBs that were thrown were classified as hazardous trash. The investigation discovered a drop in the levels of Cu, Pb, and Zn(Priya Jeyaraj 2021). On the other hand, for TCLP and WET, respectively, the concentration was found to be 13 and 11 times higher than the EPA limit.

The compositional changes, redox chemistry, and character of the leaching fluid all have an impact on how metals leach. In comparison to WET, TCLP was determined to be a more aggressive test. PWBs failed the Pb and Se tests, while batteries failed the Co and Cu tests. Using Milli Q water, Yadav et al. studied the metal toxicity of the plastics, PWBs, and batteries of the used mobile phones. In discarded motherboards, CRT monitors, LCD monitors, and LCD touch-screen displays, the study found Cd, Cr, Hg, and Pb. According to a study, cell phones are a reliable supply of precious metals like ag and Au. With the exception of Pb, the study found that motherboards had higher metal concentrations than plastic housing and glass samples. Metal concentrations mentioned in the RoHS Directive were discovered at lower levels than allowed by law. Computers, TVs, cell phones, and other electrical and electronic devices have made our lives easier. Environmentalists, waste management experts, and governments all around the world are facing a significant difficulty in managing the trash produced toward the end of life. E-waste, often known as electronic waste, makes up 5% of municipal garbage and is rapidly increasing in wealthy nations. Due to their short lifespan, mobile phones are one of the main sources of electronic waste. In industrialised countries, the average lifespan of a mobile phone is less than two years, but it is three years in underdeveloped nations(Cherukuri et al. 2018; Needhidasan et al. 2014). The second-highest number of mobile phone subscribers is in India. Without the battery, the mobile phone is made up of between 15 and 55 percent plastic. The e-waste (management rules were redrafted by the Ministry of Environment, Forestry, and Climate Change (MoEFCC).

Particular attention should be paid to rechargeable batteries made of Cd, Li, and Pb. Various regulatory authorities have put out and adopted a variety of strategies. The Environmental Protection Agency (USEPA) in the United States of America has suggested numerous techniques for determining the discharge of dangerous compounds from e-waste. Utilizing the TCLP, WET, and SPLP tests, the concentrations of several environmentally relevant metals have been evaluated. The e-waste domain represented by mobile phones is the one that is expanding the fastest. PWBs, plastics, and batteries from mobile phones have been chosen as the three main components. The WET and TCLP tests were chosen and used because they illustrate how the waste behaves after it is dumped. SPLP was chosen because it simulates how rain would affect e-waste leaching if it were dumped on the ground. Waste is proven to have failed the test and classified as hazardous if the amounts of metals present in it are higher than the threshold levels. In this study, we used TCLP, WET, and SPLP to analyse the metal concentration in mobile phone components for 9 metals: Ba, Cd, Cr, Cu, Fe, Ni, Pb, Sn, and Zn. Additionally, we investigated how various tests affected metal leaching as well as how metal content varied by manufacturer, model, and year of production(Julander et al. 2014).

This study intends to quantify the environmental danger brought on by e-waste by employing a collection of acceptable indicators to build an indicator system for risk assessment. How to evaluate the environmental concerns of mobile phones is covered in these questions and answers. How can a generic indicator system for evaluating the intricate environmental dangers of used mobile phones be created? What are the next requirements and difficulties for creating a strong, user-friendly, scientific indicator system to assess the environmental danger of trash mobiles? It is believed that the creation of a thorough and tested indication system would take too much time for a 10-week-long thesis project. Even if an indicator is straightforward and simple to grasp, it might still have certain drawbacks. Since a set of indicators includes a variety of perspectives on a problem, it can compensate for some shortcomings. It is challenging to include all pertinent indicators in an indicator system, though, as doing so will make the system's functioning and interpretation more complex.

SCENARIO OF MOBILE WASTES

Global Scenario of E-Waste

E-waste production worldwide was 93.5 Mt in 2016 and is expected to increase to 130 Mt in 2018 at a CAGR of 17.6 percent. The five biggest manufacturers of electronic waste are the US, China, Japan, Germany, and India, according to a 2014 UN report. As technology develops, more people will buy electrical and electronic devices, which will lead to an increase in e-waste creation each year. According to estimates, Europe and the U.S. manufacture 3 metric tonnes of electronic garbage annually, which includes 100 million phones and 30 million PCs. African nations use EEE at a lower rate than those in other parts of the world. India and China import e-waste from developed nations in addition to producing it. For African nations dealing with Asian nations like China and India, the main route is through West Africa(Blinová and Godovčin 2021).

Scenario of E-Waste in India

In India, there are few scientifically sound and environmentally responsible methods for treating e-waste. According to research, India's commercial and residential sectors throw out 1.38 million personal computers annually. In 2005, it was estimated that India's daily e-waste emissions were around 1.47 lakh t (0.573 Mt). According to a 2007 Greenpeace report, India produced 380,000 metric tonnes of electronic garbage. According to research provided by the Electronics Industry Association of India, by the end of 2009, India's total e-waste creation will soar to a staggering 4.34 lakh metric tonnes every day (ELCINA). The survey was carried out during Componex Nepcon 2009, a trade show for the electronics sector. Approximately 30,000 metric tonnes of e-waste were created in Delhi each year, and more than 1.5 lakh people were employed in both formal and unofficial recycling facilities. India currently produces 2 MT of electronic garbage, but treats barely 0.036 of it. India has very inadequate e-waste management, which poses significant environmental problems. Recycling techniques such as acid digestion and open burning are currently used to extract valuable metals and other materials. Additionally, compared to other types of trash, the health implications are significantly worse; as a result, it requires more attention and action. India is rapidly becoming the e-waste industry's main disposal hub. In 2009, Tamil Nadu produced over 22 thousand metric tonnes of electronic garbage, of which personal computers made up roughly 60 percent. India's top ten e-waste-producing metro areas are Mumbai, Delhi, Bangalore, Chennai, Hyderabad, Ahmedabad, Kolkata, Surat, Pune, and Jaipur(Cherukuri et al. 2018; Needhidasan et al. 2014).

HARMFUL IMPACTS OF MOBILE PHONES

Many dangerous compounds found in e-waste have an impact on human health as well as other environmental factors like soil, air, and water. One of the largest electronic garbage dumps in the world is at Guiyu Town, China. When compared to the reference locations, the concentration of heavy metals is frequently considerably high. Even the legal standards established by the US Environmental Protection Agency have been exceeded by the concentration of Hg, Sb, Cu, and Pb in trash from e-waste recycling facilities (EPA). Vietnamese mothers who lived near two e-waste disposal sites had breast milk concentrations of polybrominated diphenyl ethers (PBDEs) that were three times higher than those at a reference location. Polychlorinated dibenzodioxins and furans (PCDDs and F) were 18 times more abundant than PBDEs(Priya Jeyaraj 2021).

Compared to reference sites, those living close to an e-waste recycling facility were more exposed to organic contaminants. The concentrations of PBDEs (median: 27 ng/g in human hair) and hexabromocyclododecane (HBCDs) are higher than non-recycling locations. In comparison to a reference location, residents of an e-waste recycling facility in Vietnam were exposed to higher concentrations of PCBs and BFRs in interior dust and air. In contrast to PCBs, which were shown to be the most common in the air, PBDEs were found to be the most plentiful in dust, followed by deca-bromo-di-phenylethane (DBDPE). The type, quantity, processing, and dismantling procedures of the e-waste all affect how much contamination is present in the environment at any given time. A higher concentration of polyaromatic hydrocarbons was found. The correlations between the Pb and Cd levels in children's blood and the locations where recycling is carried out using informal, non-technical methods There were 115 ng/g of PBDEs identified in mixed muscles of tilapia (Oreochromis spp.) from the Lianjjang and Nanyang rivers. The investigation was conducted at Guiyu, the largest recycling facility in China.

REGULATIONS FOR E-WASTE (MOBILE WASTE) MANAGEMENT

The number of dump sites has decreased as the amount of e-waste has increased. The majority of nations in the world have strong laws in place governing how electronic trash is to be disposed of and managed.

International Regulations

Basel Convention

The Basel Convention, which deals with e-waste definition, general as well as specific obligations of the country for the disposal of hazardous waste, and trans-boundary movement, is the first legally binding international instrument on hazardous wastes. The Basel Convention Treaty was adopted on May 5,

1992, following its signing on March 22, 1989. The majority of industrialised nations joined the Basel Convention on Hazardous and Toxic Materials (BCT) in 1992, although the US, Afghanistan, Haiti, and other US territories did not. The UN has not regulated used goods, including certain e-waste, but it has contributed to the shipment of out-of-date goods being prohibited(Chen et al. 2018).

Restriction of Hazardous Substances (RoHS)

Directive on the Prohibition of Certain Hazardous Substances from Use in Electrical and Electronic Equipment (RoHS). The European Commission and the European Union Directive on Waste from Electrical and Electronic Equipment jointly adopted this directive.

Bamako Convention

In reaction to Basel Convention Article 11, the Bamako Convention was created on January 30, 1991, and it went into effect in 1998. It now has 25 parties and 29 signatories. The treaty has an impact on 12 African Union member countries.

EU Landfill Directive

The member states of the European Community were supposed to implement the Landfill Directive (1999/31/EC). The directive's primary goal is to stop contamination of the air, groundwater, soil, and surface waters. Additionally, it tries to stop the risk that landfilling trash poses to people's health(Ilankoon et al. 2018).

Waste from Electrical and Electronic Equipment (WEEE)

The European Union's Trash from Electrical and Electronic Equipment Directive (Directive 2002/96/EC) addresses how electrical and electronic waste is gathered, handled, recovered, and recycled in Europe. The directive's primary goals are to minimise e-waste production and encourage its reuse and recycling. The directive mandates that manufacturers recycle equipment rather than simply disposing of it in landfills at the end of its useful life. Following the adoption of the directive, manufacturers of electronic devices in the EU are required to specify a location for equipment collection and recycling(Qu et al. 2019).

Indian Regulations

Ministry of Environment, Forest, and Climate Change (MoEF)

The MoEF authorised the procedures for managing e-waste disposal without harming the environment. The letter's main goal was to list and categorise different sources of electronic waste. Waste management regulations in 2008 addressed both hazardous and non-hazardous waste. Given that the producer is in charge from the point of manufacture till the item is disposed of, this responsibility is known as Extended Producer Responsibility (EPR). The "Batteries (Management and Handling Rules), 2001" established under the Environment (Protection) Act, 1986, apply to batteries. In 2016, changes were made to the 2010 e-waste Management and Handling Rules. These regulations for the first time included protection for CFL and mercury-containing lights. On October 1, 2016, these regulations went into effect. Producers must return the goods under Extended Producer Responsibility, which is a change from the former regulations(Arya and Kumar 2020).

CHARACTERIZATION OF WASTE MOBILE PHONES

Material flow processes for mobile waste management are illustrated in Figure 1.

Composition

The substance composition of mobile phones can differ slightly due to varied brands, models, producers, and functionalities. The majority of mobile phones are made of plastic and metal, with more than 20 metallic components making up 35–40% of the overall weight. A typical cell phone is made up of a battery, display enclosure, printed circuit board, charger, and accessories(N Singh et al. 2018).

Potential Value

A tonne of cell phones contains 130 g of copper, 140 g of palladium, 340 g of gold, and 3.5 g of silver. Additionally, it is possible to reuse the plastic from outdated cell phones. They can be employed as heat sources, reducing agents, or fuel substitutes for hydrocarbons. It can be recovered and reused since the material's mechanical performance can be compared to that of virgin materials like glass and metal. The classification of plastics is difficult; thus, the process is still in its infancy(Hira et al. 2018).

Material Flow of Mobile Wastes

Since they can be recycled, smartphones can last longer. Customers can give their phones away or gift them to friends, family, and other people. The majority of consumers keep their old, useless smartphones at home. Occasionally, customers just throw their phones away or lose them. Mobile phones can either be repaired and refurbished for sale and reuse, or they can be recycled. Customers can also sell their

Figure 1. Material flow processes for mobile waste management

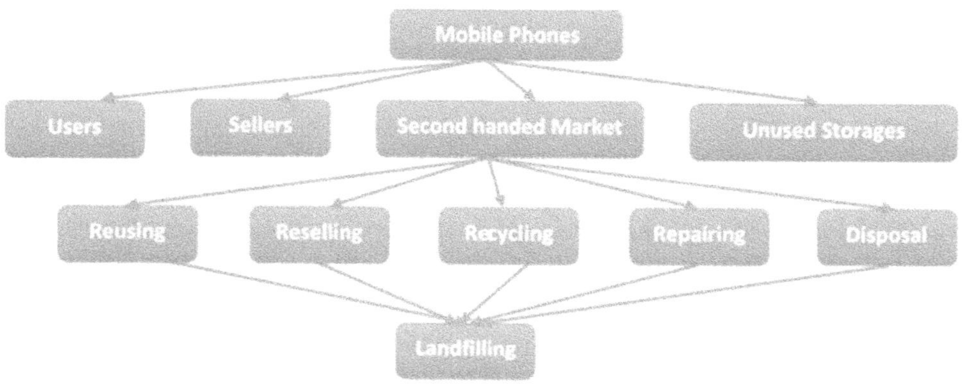

mobile phones to street vendors and at flea markets. When mobile phones are shipped for landfilling or burning, the environment is put in danger.

Environmental Impacts of Mobile Waste

As they retain electrical charge to some extent, batteries have the potential to produce short circuits and fires. High temperatures could cause the gadgets to burst and catch fire, releasing poisonous compounds like dioxins and furans in the process. The environmental risk in the transportation process is connected to weather conditions and the manner of transportation. Equipment used for waste disposal can damage the environment by igniting on exposure to sunshine or high temperatures, and polluting soil and underground water on exposure to rain. In the event of adverse weather, the devices can disperse and pollute the water(Nnorom and Osibanjo 2009b).

Recycling

Prior to disposal, used mobile phones should be destroyed since recycling them without first removing the batteries might result in short circuits and fires. Noise and dust particles from this process could contain any materials found in mobile phones. Batteries must be removed prior to shredding in order to prevent the release of caustic chemicals. Three phases are intended to separate various materials and segregate them based on their physical characteristics. Mobile phone plastics can be burned in this place to generate energy and serve as a reducing agent. Individual metals must go through a number of processes after being smelted in order to be upgraded and refined. In order to recover the desired metal, these methods involve heating and melting materials at extremely high temperatures in furnaces. Additionally, they entail dissolving the metal in leaching solutions like sulfuric and nitric acid before recovering it through precipitation or electrolysis(Nnorom and Osibanjo 2009a).

Safety Disposals

Mobile phones are one of the final possibilities for garbage disposal after incineration and landfill. During burning, plastics and other hydrocarbon components included in used mobile phones may totally oxidise and mix with halogens. Lead and cadmium are two metals that can melt inside of mobile phones and produce microscopic metal oxide particles or odours. Hazardous compounds will result in pollution if appropriate steps to stop leakage are not taken. Information on individual contaminants and their transmission to soil is quite scarce. Mobile phone garbage produces pollutants that can be emitted into the atmosphere and water directly.

FRAMEWORK OF AN INDICATOR SYSTEM

An evaluation indicator system with one goal, six elements, and additional criteria for each factor for brownfield rehabilitation projects. to determine the environmental risk factors contributing to water eutrophication in rural areas. An indication system can be represented by a table or an image. Instead of using a table to depict an indication system, a picture is frequently used, and the hierarchy is built at three levels. It comprises a goal/target/object level, a factor/criteria level, and an indicator/sub-sub-criteria level. The indicator system may be more transparent and understandable thanks to the hierarchy structure(Forti et al. 2018; Widmer et al. 2005).

Methodology

The construction of an indicator starts with agreement on a statement, which is a description of the issue, and is based on reading literary works. The construction of indicators can be guided by a DPSIR framework, as shown in Fig. 1. It is required to (4) interpret the indications after they have been picked, stating the rationale behind the choice. You can get more specific information by reading the pertinent literature, which will be provided here.

DPSIR Framework

Economic and social progress are the driving forces (D) behind the changes that put pressure (P) on the environment. Identification of indicators is done using the DPSIR framework, which was created by the European Environmental Agency for reporting on environmental issues. These are chosen based on how these actions influence and advance the other four components (Figure 2).

Figure 2. DPSIR framework

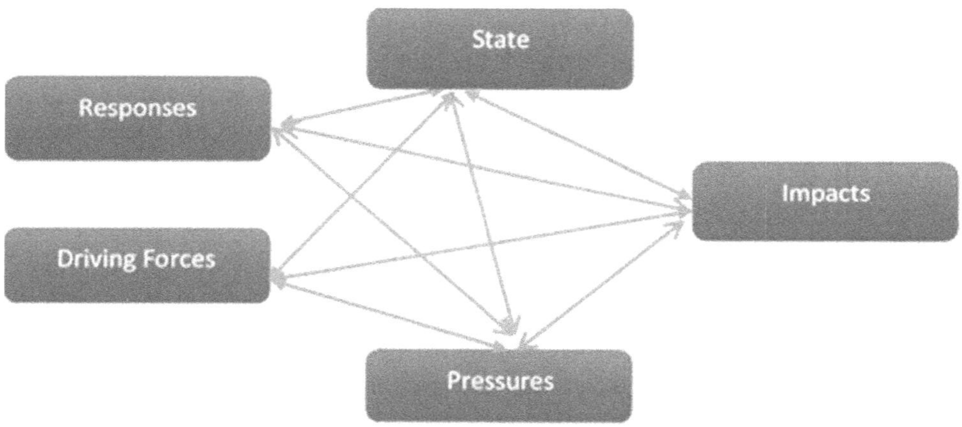

FORMATION OF RISK ASSESSMENT INDICATOR SYSTEM

Selection Criteria Indicator

The various stages for indicator system for mobile wastes is depicted in Figure 3. The following guidelines should generally guide the selection of indicators:

(1) Based on science and accurate: The indications should be realistic and theoretically sound.

(2) Comprehensiveness: Because the risk assessment process incorporates many different variables, it is challenging to evaluate the risk using only one indicator. As a result, a variety of influencing factors need to be carefully considered.
(3) Operability and general applicability: The indicators chosen should be simple to comprehend, interpret, and use. Additionally, they must be receptive to human activity and environmental changes.
(4) Measurability: To make the indicators more understandable and obvious, they should ideally be expressed quantitatively. Furthermore, the information used as indicators should be easily accessible and able to be compared to a standard value.
(5) Validity: The indicators chosen should be representative, as it is challenging to choose all of the relevant indicators and too many indicators would make the indicator system more complex. Furthermore, there should be no overlap between any two indicators.

Figure 3. Stages for indicator system for mobile wastes

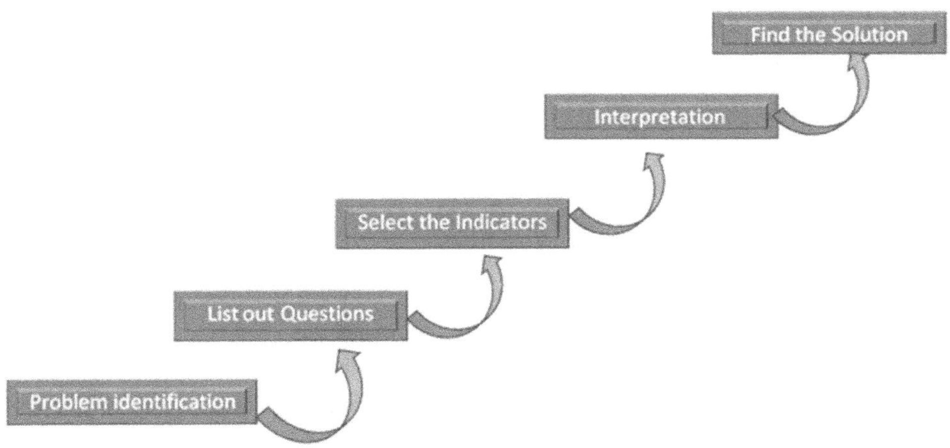

Mobile Phone Wastes

Poor handling of this waste results in a variety of contaminants that put a lot of pressure (P) on the ecosystem. The production and upgrading of mobile phones are driven (D) by rapid economic development, cutting-edge technologies, and significant market needs. Every step of the life cycle of used mobile phones poses a risk to the environment. Heavy metals in particular will contaminate the air, water, and soil when they are released, among other hazardous and destructive elements. As a result, the ecosystem's ability to function will be negatively impacted, which may also have an influence on human health. Because they contain elements that may be recycled, used mobile phones have become highly popular in many nations. The recycling process can be hazardous to the environment and people if it is not done correctly, and this issue may be more common in developing nations. To encourage the management of e-waste, legislation, regulations, and programmes have been devised; nevertheless, many regions have not complied with the demands, such as collection and recycling targets. In order to improve the situation, more targeted and efficient actions should be taken, considering the local conditions of various locations(A Singh 2022).

Indicator Selection and Interpretation

The process of transportation and storage will be impacted by weather conditions, particularly high temperatures and precipitation. Leachate may also move into groundwater if a landfill's liner system and garbage containers are continuously damaged by heavy rain. Consumers are quickly influenced by vogue trends and drawn to new mobile phone features, which shortens the lifespan of phones. Each region has its own notion of what constitutes a high temperature. Mobile phone garbage generation and disposal are largely under the control of consumers. Regarding this topic, two markers of consumer behaviour are the frequency of mobile phone replacements and the manner in which used mobile devices are disposed of. The degree of environmental danger will vary as a result of these various remediation options. System indicator and risk assessment for mobile wastes are displayed in Figure 4.

The whole life cycle of used mobile phones emits a large number of hazardous substances. This study chose CO2, SO2, NOx, heavy metals, dioxins, and particle matter as its markers. Consumer surveys can be used to get the information needed for these metrics. The recycling of used mobile phones creates a variety of air pollutants, including SO2, NOx, and particulate matter. Dioxins are extremely dangerous chemicals that are generated during incomplete combustion or smelting processes when halogens like fluorine and bromine are present. Given that mobile phones include more than 20 metallic elements, heavy metals are particularly representative. When they are released into the air, water, or soil and accumulate in the tissues of plants, they can be absorbed by them. Additionally, heavy metals can be ingested through contaminated meat and vegetables, skin contact, and inhalation. In the leachability tests, which imitate landfill conditions for metal leaching, Pb, Cu, and Ni from mobile phone parts exceeded the standards of various leaching test techniques. Pb is present in modest amounts in mobile phones and is frequently used in tin-lead solder, but it is nevertheless a possible carcinogen and neurological toxin, raising safety concerns(Boopathi Arigela et al. 2022; Boopathi Balasubramani et al. 2023; Boopathi Siva Kumar et al. 2023; Gunasekaran et al. 2022; Jeevanantham et al. 2022; Kumara et al. 2023; Vanitha et al. 2023).

Pb, the element with the highest concentration in a cell phone, can harm the liver and kidneys at extremely high levels. As is a material that is considered harmful and poses serious risks to human health, and the United States Environmental Protection Agency has classified it as a carcinogen. Although the heavy metal Hg is exceedingly poisonous, it is not considered to be a carcinogen. When it enters the environment, it will be left behind in the soil and water. Methyl mercury is an extremely poisonous chemical that can bioaccumulate in living things, particularly fish. Ni is a necessary nutrient at low amounts, but at excessive concentrations, it can be harmful to human health and lead to birth abnormalities and dizziness. Consuming vegetables can expose humans and animals to nickel since plants can accumulate the metal. Additionally, it has harmful effects on the respiratory tract, causes lung and nose malignancies, and is probably a human carcinogen(Babu et al. 2022; Boopathi Thillaivanan et al. 2022; Chakravarthi et al. 2022; Sampath et al. 2022). Even with careful planning and construction, landfill leachate generation can contaminate groundwater. The environmental effects of a landfill depend on a number of variables, including the trash type, size, service time, and leakage protection system. A risk value will be assigned to each circumstance. Two indicators are used for management measures: producer responsibility and legal regulation. Data for the latter two variables can be easily acquired from basic landfill information. Two quantitative key indicators that can be measured in terms of operational and financial flexibility can be used to break down legal regulation. Each condition is assigned a value so that it can be expressed both quantitatively and qualitatively depending on whether the regulation satisfies these requirements.

Figure 4. System indicator and risk assessment for mobile wastes

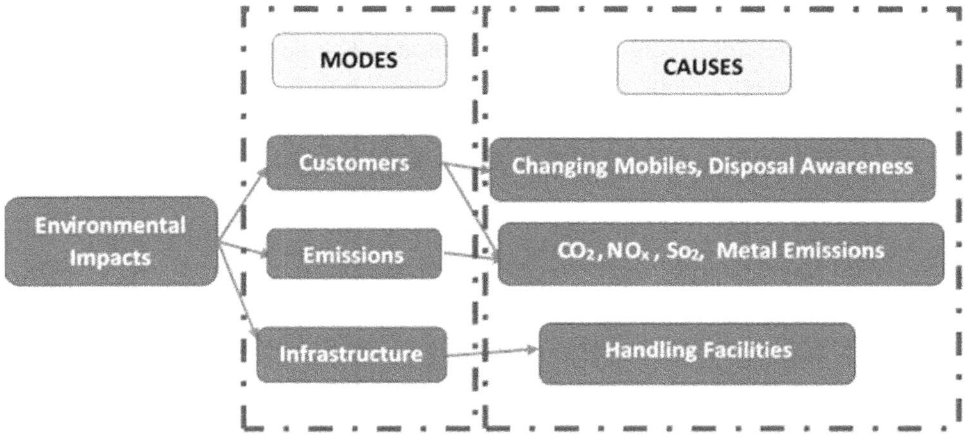

Indicator System Structure

An indicator system for evaluating the risk of wasting mobile phones has been developed through research. The object layer, factor layer, indicator layer, and sub-indicator layer are the four layers of this system. The only elements in the object layer are the assessment goal and five elements that influence the amount of environmental risk.

Results Interpretation

The DPSIR framework, which may characterise an environmental issue from a comprehensive perspective through five components, is an effective method for selecting indicators. The only sources of information used in the literature were reputable official websites or scientific journals. Combining indicators from many elements has led to the establishment of an indicator system for the risk assessment of discarded mobile phones. Due to time constraints, a preliminary indicator system was created, which calls for the improvement of indicators through expert consultation and consideration of real-world circumstances. When it is thought to be used, there might be certain issues. Despite the fact that the system was created using science and easily accessible data, it is not yet known if or when it will be used in practise. The environmental danger posed by used mobile phones in a given area may be evaluated using an upgraded indicator system, which could then be used to direct and enhance the management of that area. Despite the fact that this is only a preliminary study, it should be a useful effort for future research on how mobile phones are disposed(Boopathi Khare et al. 2023; Chakravarthi et al. 2022; Jeevanantham et al. 2022; Palaniappan et al. 2023; Samikannu et al. 2022; Senthil et al. 2023). The study aims to contribute to the development of a reliable indicator system for measuring the environmental risks associated with mobile phone use, but additional data collection is required before it can be used. In order to quantify the risk, it is crucial to consider the significance of each signal and how they connect to one another. In order to make sure that there is no "leakage" in the analysis, the data used for the indicators should also be compared with a reference value.

POLYMERS RECOVERED FROM MOBILE WASTES

Compared to most other electronic equipment, mobile phones include a larger proportion of polymers—between 30 and 50 percent—in their construction. Among the materials recovered from used mobile phones are engineering polymers like polycarbonate and Acrylonitrile Butadiene Styrene. Plastics recovered from WEEE must first be identified, sorted, and characterised before being recycled. Better qualities are provided by a homogeneous blend of the same plastic type than by a heterogeneous blend of different plastic types. The goal of the current study was to identify and describe plastic parts that were recovered from used mobile phones. Depending on the type of base polymer used, it can be difficult to classify mobile phone components into different groups. It is also crucial to recognise any dangerous substances, such as brominated flame retardants, that may be present in such plastics. Regulations like the EU Directive place restrictions on the use, recovery, and recycling of plastics that include certain constituents(Forti et al. 2018; A Singh 2022).

Process Flow

Plastics From Mobile Wastes

Each day, it processes around six tonnes of electronic garbage and sends the finished product to further recycling facilities. The materials that are accepted by E-Parisaraa are divided into many categories based on the source, such as small WEEE, major home appliances, etc. Following grouping, every piece of equipment goes through a de-manufacturing stage in which the parts are taken apart, destroyed, and then separated into metals and non-metals. Polymers are present in both battery and PCB assemblies; however, due to their complexity, it is difficult to remove these polymers. More recently, a significant number of layers with specialised functionalities have been built into such systems as a result of developments in display technology. As a result, their recycling requires specialised tools and plans, which must also be affordable. Its main goal is to recover and add value to the polymers found in the shells and keypads of used mobile phones. Nearly 35 percent of a mobile phone is made up of the keypad and case, which may be used to create a 54-part recycling process that is commercially viable. More than 80 percent of the total polymers used in the production of a mobile phone are made of polycarbonate (PC) and PC/ABS blends. In comparison to these plastics, the shares of other polymers are insignificant(Forti et al. 2018; Sarath et al. 2017). The various stages for recovering Polymers from mobile wastes are illustrated in Figure 5.

Preliminary Stages

In this section, each level is described in detail. The initial step involved classifying the plastic garbage from collected cell phone parts into three categories: front casings, back casings, and keypads. Plastic keypads and elastomeric keypads were the next categories of keypads to be classified. It is simple to identify and sort a huge number of plastic components, enabling easy recycling of these waste parts. Components with markings and parts without markings were added to the list of subcategories. The markings that must be provided on a plastic item are specified by several standards, including RIC and ASTM D1600. In stage two, melt mixing was used to create a single, consistent composition out of all the front casing components that lacked generic markings. With this type of melt-mixing, plastics are blended uniformly and homogeneously(Jha et al. 2020; Kasper et al. 2011). In situations where a single

Figure 5. Stages: recovery of polymer from mobile wastes

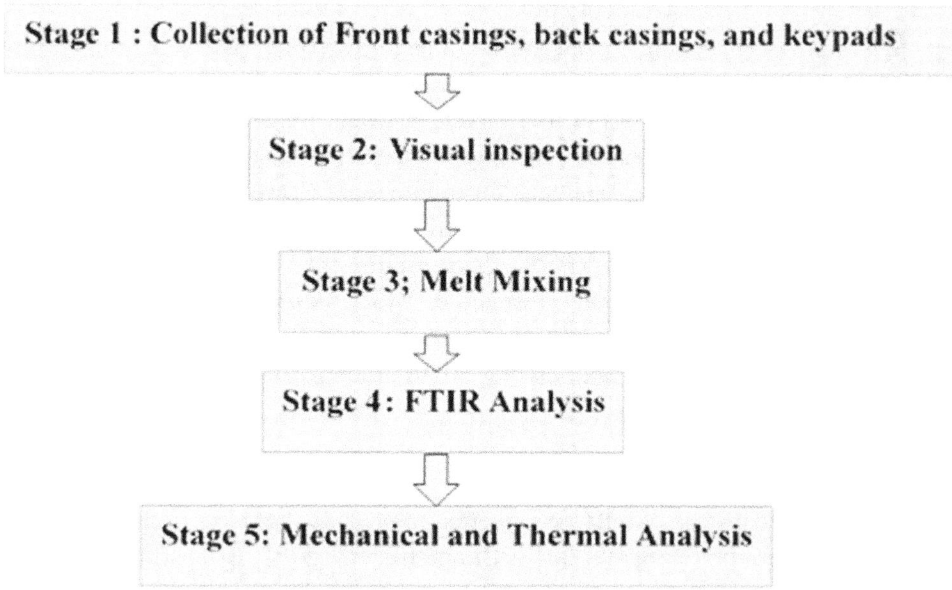

sample would serve as a good representation of the entire class, it is also useful for further identification and characterization. The plastic and elastomeric keypads were coded as FC-N, BC-N, and K-P, respectively, after melt mixing. Before being sent for additional analysis, samples were scanned and GC/MS-analyzed to make sure they were devoid of brominated flame retardants. To ensure adequate drying, mobile phone parts were dried in an air oven for two hours at 110°C and then for 12 hours at 80°C. After washing and cleaning, the components were divided into front casings, back casings, and keypads.

Visual Method

The market's shift toward touch displays as the dominant technology is responsible for this trend. Back casings account for over 45% of the total components collected and have a large amount of 1197 parts. The back casing, display brackets, and screen components are the main plastic parts of touch screen phones. With 1635 components, keypads looked to be the component with the second largest volume. The study's plastic garbage was retrieved from old phones that were discarded in the latter half of the previous decade. With 788 pieces, front casings were discovered to be on the sparse side. Most items are made with PC, followed by PC/ABS blends. Melt-mixing and FTIR analyses were used to identify a significant portion of the recovered pieces that lacked generic markings(Gu et al. 2019; Sarath et al. 2017).

FTIR Analysis

With the exception of a few black pieces, the majority of the 4400 parts were light-coloured, such as white, silver, and cream. They discovered that neither the intensity nor colour of these photons varied noticeably, proving that the melt mixing procedure worked as intended.

Thermal Analysis

A single glass transition (Tg) is visible in the thermograms for FC-N and BC-N at temperatures between 140°C and 145°C. Given that the Tg of this material is likewise within the same temperature range, this supports the FTIR results indicating both of these materials were made primarily using polycarbonate. The DSC thermogram of K-P reveals a strong melting point of 225 °C and a glass transition temperature of 103 °C, both of which testify to the existence of polystyrene. However, the material also had large polycarbonate peaks that were visible in the FTIR spectra. According to a study, recovered plastics don't contain any brominated flame retardants (BFR), so recycling them poses no risk for environmental toxicity. An earlier study that looked at the presence of BFR in different electronic gadgets from the WEEE stream came up with results that were similar(Jha et al. 2020; Kasper et al. 2011). Future work is extended using Taguchi method(Boopathi 2019a, 2019b; Fatouh and Kafafy 2006).

Future Work Extensions

The optimization of e-waste process would be extended to provide the best settings to improve the process and helpful to the operators(Boopathi 2013; Boopathi et al. 2021; Boopathi 2022c; Boopathi and Sivakumar 2013). The multi-objective optimization and different DOE methods have been applied for these tasks(Boopathi 2021, 2022a, 2022b; Boopathi Thillaivanan et al. 2022; Boopathi and Sivakumar 2016; Saravanan et al. 2022; Trojovský et al. 2023; Yupapin et al. 2022).

SUMMARY

This chapter has discussed national and international issues as well as legislative activities with regard to the harm and management of mobile waste materials. It has been developed on the various characteristics of mobile waste and the materials flow for treating mobile wastes. The various framework types for the risk assessment indication systems, selection criteria, and indicating systems were also demonstrated. In order to recover polymers, the various steps of mobile waste processing and handling have also been illustrated.

REFERENCES

Arya, S., & Kumar, S. (2020). E-waste in India at a glance: Current trends, regulations, challenges and management strategies. *Journal of Cleaner Production*, *271*, 122707. doi:10.1016/j.jclepro.2020.122707

Babu, B.S., Kamalakannan, J., Meenatchi, N., Karthik, S., Boopathi, S. (2022). Economic impacts and reliability evaluation of battery by adopting Electric Vehicle. *IEEE Explor.* 1–6.

Blinová, L., & Godovčin, P. (2021). Importance of Recycling the Waste-Cables Containing Copper and PVC. *Res. Pap. Fac. Mater. Sci. Technol. Slovak Univ. Technol.*, *29*(48), 1–21. doi:10.2478/rput-2021-0001

Boopathi, S. (2013). Experimental Study and Multi- Objective Optimization of Near-Dry Wire-Cut Electrical Discharge Machining Process Faculty of Mechanical Engineering. In *Shodhganga@INFLIBNET* (Issue August)http://hdl.handle.net/10603/16933

Boopathi, S. (2019a). Experimental investigation and parameter analysis of LPG refrigeration system using Taguchi method. *SN Applied Sciences*, *1*(8), 892. doi:10.100742452-019-0925-2

Boopathi, S. (2019b). Experimental investigation and parameter analysis of LPG refrigeration system using Taguchi method. *SN Applied Sciences*, *1*(8), 892. doi:10.100742452-019-0925-2

Boopathi, S. (2021). Improving of Green Sand-Mould Quality using Taguchi Technique. *J. Eng. Res.* doi:10.36909/jer.14079

Boopathi, S. (2022a). An experimental investigation of Quench Polish Quench (QPQ) coating on AISI 4150 steel. *Eng. Res. Express*, *4*(4), 045009. doi:10.1088/2631-8695/ac9ddd

Boopathi, S. (2022b). An Extensive Review on Sustainable Developments of Dry and Near-Dry Electrical Discharge Machining Processes. *Journal of Manufacturing Science and Engineering*, *144*(5), 1–37. doi:10.1115/1.4052527

Boopathi, S. (2022c). Experimental investigation and multi-objective optimization of cryogenic Friction-stir-welding of AA2014 and AZ31B alloys using MOORA technique. *Materials Today. Communications*, *33*, 104937. doi:10.1016/j.mtcomm.2022.104937

Boopathi, S., Arigela, S.H., Raman, R., Indhumathi, C., Kavitha, V., & Bhatt, B.C. (2022). Prominent Rule Control-based Internet of Things: Poultry Farm Management System. *IEEE Explor.* 1–6.

Boopathi, S., Balasubramani, V., & Sanjeev Kumar, R. (2023). Influences of various natural fibers on the mechanical and drilling characteristics of coir-fiber-based hybrid epoxy composites. *Eng. Res. Express*, *5*(1), 15002–15020. doi:10.1088/2631-8695/acb132

Boopathi, S., & Khare, R. K.G., J.C., Muni, T.V., & Khare, S. (2023). Additive Manufacturing Developments in the Medical Engineering Field. In Development, Properties, and Industrial Applications of 3D Printed Polymer Composites (pp. 86–106). IGI Global

BoopathiS.MyilsamyS.SukkasamyS. (2021) Experimental Investigation and Multi-Objective Optimization of Cryogenically Cooled Near-Dry Wire-Cut EDM Using TOPSIS Technique. *Ijamt Prepr.* doi:10.21203/rs.3.rs-254117/v1

Boopathi, S., Siva Kumar, P. K., & Meena, R. S. J. SI, P., S.K., & Sudhakar, M. (2023) Sustainable Developments of Modern Soil-Less Agro-Cultivation Systems. In Human Agro-Energy Optimization for Business and Industry (pp. 69–87). IGI Global doi:10.4018/978-1-6684-4118-3.ch004

Boopathi, S., & Sivakumar, K. (2013). Experimental investigation and parameter optimization of near-dry wire-cut electrical discharge machining using multi-objective evolutionary algorithm. *International Journal of Advanced Manufacturing Technology*, *67*(9–12), 2639–2655. doi:10.100700170-012-4680-4

Boopathi, S., & Sivakumar, K. (2016). Optimal parameter prediction of oxygen-mist near-dry Wire-cut EDM. *Int. J. Manuf. Technol. Manag.*, *30*(3–4), 164–178. doi:10.1504/IJMTM.2016.077812

Boopathi, S., Thillaivanan, A., Azeem, M. A., Shanmugam, P., & Pramod, V. R. (2022). Experimental investigation on abrasive water jet machining of neem wood plastic composite. *Funct. Compos. Struct.*, *4*(2), 025001. doi:10.1088/2631-6331/ac6152

Chakravarthi, P. K., Yuvaraj, D., & Venkataramanan, V. (2022) IoT-based smart energy meter for smart grids. *ICDCS 2022 - 2022 6th Int. Conf. Devices, Circuits Syst.* (pp. 360–363). IEEE. 10.1109/ICDCS54290.2022.9780714

Chen, Y., Chen, M., Li, Y., Wang, B., Chen, S., & Xu, Z. (2018). Impact of technological innovation and regulation development on e-waste toxicity: A case study of waste mobile phones. *Scientific Reports*, *8*(1), 1–9. doi:10.103841598-018-25400-0 PMID:29740013

Cherukuri, I., Sultana, N., & Podila, S. P. (2018). Status Of E-Waste In India- A Review. *IOSR Journal of Environmental Science, Toxicology and Food Technology*, *12*(11), 8–16. doi:10.9790/2402-1211010816

Elleuch, B., Bouhamed, F., Elloussaief, M., & Jaghbir, M. (2018). Environmental sustainability and pollution prevention. *Environmental Science and Pollution Research International*, *25*(19), 18223–18225. doi:10.100711356-017-0619-5 PMID:29971753

Fatouh, M., & El Kafafy, M. (2006) Experimental evaluation of a domestic refrigerator working with LPG. In Applied Thermal Engineering, 26(14–15), 1593–1603. doi:10.1016/j.applthermaleng.2005.11.026

Forti, V., Balde, C., & Kuehr, R. (2018) Guidelines on Classification, Reporting and Indicators. *E-Waste Stat. 2*. https://ewastemonitor.info/

Gu, F., Summers, P. A., & Hall, P. (2019). Recovering materials from waste mobile phones: Recent technological developments. *Journal of Cleaner Production*, *237*, 117657. doi:10.1016/j.jclepro.2019.117657

Gunasekaran, K., Boopathi, S., & Sureshkumar, M. (2022). Analysis of a Cryogenically Cooled Near-Dry Wedm Process Using Different Dielectrics. *Materiali in Tehnologije*, *56*(2), 179–186. doi:10.17222/mit.2022.397

Hira, M., Yadav, S., Morthekai, P., Linda, A., Kumar, S., & Sharma, A. (2018). Mobile Phones—An asset or a liability: A study based on characterization and assessment of metals in waste mobile phone components using leaching tests. *Journal of Hazardous Materials*, *342*, 29–40. doi:10.1016/j.jhazmat.2017.08.008 PMID:28822247

Ilankoon, I. M. S. K., Ghorbani, Y., Chong, M. N., Herath, G., Moyo, T., & Petersen, J. (2018). E-waste in the international context – A review of trade flows, regulations, hazards, waste management strategies and technologies for value recovery. *Waste Management (New York, N.Y.)*, *82*, 258–275. doi:10.1016/j.wasman.2018.10.018 PMID:30509588

Jeevanantham, Y.A., Saravanan, A., Vanitha, V., Boopathi, S., & Kumar, D.P. (2022). Implementation of Internet-of Things (IoT) in Soil Irrigation System. *IEEE Explor.* 1–5.

Jeyaraj, P. (2021). Management of E-waste in India – Challenges and recommendations. *World J. Adv. Res. Rev.*, *11*(2), 193–218. doi:10.30574/wjarr.2021.11.2.0329

Jha, R., Rao, M. D., Meshram, A., Verma, H. R., & Singh, K. K. (2020). Potential of polymer inclusion membrane process for selective recovery of metal values from waste printed circuit boards: A review. *Journal of Cleaner Production*, *265*, 121621. doi:10.1016/j.jclepro.2020.121621

Julander, A., Lundgren, L., Skare, L., Grandér, M., Palm, B., Vahter, M., & Lidén, C. (2014). Formal recycling of e-waste leads to increased exposure to toxic metals: AN occupational exposure study from Sweden. *Environment International*, *73*, 243–251. doi:10.1016/j.envint.2014.07.006 PMID:25300751

Kasper, A. C., Bernardes, A. M., & Veit, H. M. (2011). Characterization and recovery of polymers from mobile phone scrap. *Waste Management & Research*, *29*(7), 714–726. doi:10.1177/0734242X10391528 PMID:21382879

Kumara, V., Mohanaprakash, T. A., Fairooz, S., Jamal, K., & Babu, T. B. S. (2023). Experimental Study on a Reliable Smart Hydroponics System. In *Human Agro-Energy Optimization for Business and Industry* (pp. 27–45). IGI Global. doi:10.4018/978-1-6684-4118-3.ch002

Needhidasan, S., Samuel, M., & Chidambaram, R. (2014) Electronic waste – an emerging threat to the environment of urban India. *J. Environ. Heal. Sci. Eng.* 1–9.

Nnorom, I. C., & Osibanjo, O. (2009a). Heavy metal characterization of waste portable rechargeable batteries used in mobile phones. *International Journal of Environmental Science and Technology*, *6*(4), 641–650. doi:10.1007/BF03326105

Nnorom, I. C., & Osibanjo, O. (2009b). Toxicity characterization of waste mobile phone plastics. *Journal of Hazardous Materials*, *161*(1), 183–188. doi:10.1016/j.jhazmat.2008.03.067 PMID:18448243

Palaniappan, M., Tirlangi, S., Mohamed, M. J. S., Moorthy, R. M. S., Valeti, S. V., & Boopathi, S. (2023). Fused Deposition Modelling of Polylactic Acid (PLA)-Based Polymer Composites: A Case Study. In Development, Properties, and Industrial Applications of 3D Printed Polymer Composites (pp. 66–85). IGI Global

Qu, Y., Wang, W., Liu, Y., & Zhu, Q. (2019). Understanding residents' preferences for e-waste collection in China - A case study of waste mobile phones. *Journal of Cleaner Production*, *228*, 52–62. doi:10.1016/j.jclepro.2019.04.216

Samikannu, R., Koshariya, A. K., Poornima, E., Ramesh, S., Kumar, A., & Boopathi, S. (2022). Sustainable Development in Modern Aquaponics Cultivation Systems Using IoT Technologies. In *Human Agro-Energy Optimization for Business and Industry* (pp. 105–127). IGI Global.

Sampath, B., Pandian, M., Deepa, D., & Subbiah, R. (2022). Operating parameters prediction of liquefied petroleum gas refrigerator using simulated annealing algorithm. *AIP Conference Proceedings*, *2460*(1), 070003. doi:10.1063/5.0095601

Sarath, P., Bonda, S., Mohanty, S., & Nayak, S. K. (2017). Identification and thermomechanical characterization of polymers recovered from mobile phone waste. *Journal of Material Cycles and Waste Management*, *19*(4), 1391–1399. doi:10.100710163-016-0525-9

Saravanan, M., Vasanth, M., Boopathi, S., Sureshkumar, M., & Haribalaji, V. (2022). Optimization of Quench Polish Quench (QPQ) Coating Process Using Taguchi Method. *Key Engineering Materials*, *935*, 83–91. doi:10.4028/p-z569vy

Senthil, T. S., Puviyarasan, M., Babu, S. R., Surakasi, R., & Sampath, B. (2023). Industrial Robot-Integrated Fused Deposition Modelling for the 3D Printing Process. In Development, Properties, and Industrial Applications of 3D Printed Polymer Composites (pp. 188–210). IGI Global

Singh, A. (2022). Indicators and ICTs application for municipal waste management. *Waste Management & Research*, *40*(1), 24–33. doi:10.1177/0734242X211010367 PMID:33836633

Singh, N., Duan, H., Yin, F., Song, Q., & Li, J. (2018). Characterizing the Materials Composition and Recovery Potential from Waste Mobile Phones: A Comparative Evaluation of Cellular and Smart Phones. *ACS Sustainable Chemistry & Engineering*, *6*(10), 13016–13024. doi:10.1021/acssuschemeng.8b02516

Trojovský, P., Dhasarathan, V., & Boopathi, S. (2023). Experimental investigations on cryogenic friction-stir welding of similar ZE42 magnesium alloys. *Alexandria Engineering Journal*, *66*(1), 1–14. doi:10.1016/j.aej.2022.12.007

Vanitha, S. K. R., & Boopathi, S. (2023). Artificial Intelligence Techniques in Water Purification and Utilization. In *Human Agro-Energy Optimization for Business and Industry* (pp. 202–218). IGI Global. doi:10.4018/978-1-6684-4118-3.ch010

Widmer, R., Oswald-Krapf, H., Sinha-Khetriwal, D., Schnellmann, M., & Böni, H. (2005). Global perspectives on e-waste. *Environmental Impact Assessment Review*, *25*(5), 436–458. doi:10.1016/j.eiar.2005.04.001

Yupapin, P., Trabelsi, Y., Nattappan, A., & Boopathi, S. (2022). Performance Improvement of Wire-Cut Electrical Discharge Machining Process Using Cryogenically Treated Super-Conductive State of Monel-K500 Alloy. *Iranian Journal of Science and Technology. Transaction of Mechanical Engineering*, 1–17. doi:10.100740997-022-00513-0

Chapter 17
Survey on the Artificial Intelligence and Machine Learning Techniques on the Applications of Wastewater Treatment for Sustainable Environment

Yahya Asmar Zakur
University of Mazandaran, Iran

Biswadip Basu Mallik
Institute of Engineering and Management, Kolkata, India

Yousif Asmar Zakoor
Department of Civil Engineering, University of Wasit, Iraq

Digvijay Pandey
https://orcid.org/0000-0003-0353-174X
Department of Technical Education, IET, Dr. A.P.J. Abdul Kalam Technical University, Lucknow, India

ABSTRACT

Critical water and wastewater treatment applications have been optimized, modelled, and automated using artificial intelligence (AI) techniques and machine-learning models. Also, it describes the cases in which machine learning algorithms have been applied to evaluate the water quality in different water environments, such as surface water, groundwater, drinking water, sewage, and seawater. Wastewater characteristics prediction in wastewater treatment plants (WWTPs) is valuable and can reduce the number of samplings, energy, and cost. The study reviews machine learning, deep learning, and smart technologies used in wastewater treatment for generation, prediction enhancement, and classification

DOI: 10.4018/978-1-6684-8117-2.ch017

tasks, providing a guide for future water resources challenges. These models can be used to make decisions in water resources management and governance, but ethics and future directions need to be addressed and focused.

INTRODUCTION

It is crucial to consider all possible measures to lessen the overexploitation of the finite freshwater resources because the globe is currently experiencing a water scarcity problem (Salehi, 2022). One of the crucial sources of water as the population and activity rise[5] is wastewater (Saleh et al., 2022; Müller et al., 2023; Smetana & Grosser, 2023; Salgot & Folch, 2018). It is desirable and technically possible to take a selective approach to multiple wastewater recovery, supplying water that meets precise quality specifications for each reuse goal (Ungureanu et al., 2020). By avoiding needless treatment and long-distance conveyance, this method might conserve water and lower manufacturing costs and energy demand (Giammar et al., 2021). Because of this, it is essential to use the best ways for treating wastewater utilizing artificial intelligence and machine learning techniques (Lowe et al., 2022; Maamoun et al., 2023; Zhu et al., 2022; Bagherzadeh et al., 2021; Sit et al., 2020)

Artificial intelligence (AI) has shown its ability to address the problems associated with the treatment of drinking water (DWT). It helps manage and run DWT processes mechanically, rather than depending on human actions. AI-based data analysis and evolutionary learning methods enable water quality diagnoses, autonomous decision making, and process optimisation (Li et al., 2021).Wastewater treatment is given an example of how to create ANN models in clause (Jawad et al., 2021). Membrane and thermal processes, which account for 65% and 35%, respectively, of the need for freshwater, are used to fill the gap (Kress, 2019). Several membrane technologies, including reverse osmosis, forward osmosis, membrane distillation, microfiltration, ultrafiltration, and nanofiltration, are used in desalination. Wastewater and sewage are treated using the activated sludge technique in a membrane bioreactor with MF and UF (Silva et al., 2023). Mathematical models are necessary for the process' modelling and optimisation. Permeate flow, salt rejection, and membrane fouling are a few of the processing variables for membrane filtration that have received the most attention (Niu et al., 2022). Because of assumption (Badrnezhad & Mirza, 2014), theoretical or transport-based models cannot reliably anticipate these properties. Artificial intelligence (AI) methods, including ANN, fuzzy logic, ANFIS, genetic programming, and support vector machines, offer an alternate method to accurately describe membrane processes(Garud et al., 2020).

A machine learning technique called an artificial neural network (ANN) is based on biological neurones and uses a learning mechanism like that of the human brain to address a variety of issue (Lagaros, 2023). Fuzzy logic is a type of many-valued logic that permits partial truths, enabling us to take errors and uncertainties into account while making decisions. Uses for this technique range from AI and control systems to NLP and medical diagnostics through image processing and computer vision (Esmaeili et al., 2023). To integrate knowledge and learning, ANFIS (fuzzy inference system) combines the best aspects of fuzzy systems with neural networks (Yosif et al., 2022). Genetic programming (GP) is a field of artificial intelligence that uses evolutionary algorithms to find answers to issues that people are unable to solve without the help of human biases or prejudices (Santoso et al., 2018). Non-linear classifier techniques like Support Vector Machine (SVM) are frequently cited as having superior classification performance over other techniques. It maximises the margin of difference between positive and negative examples, leading to improved classification accuracy, by using a hyper plane as a decision surface. SVMs have both a strong theoretical foundation and contemporary success in practical applications (Waqas et al., 2022).

LITERATURE REVIEW

For membrane bioreactor (MBR) technologies to be commercialised, they must be highly efficient, sustainable, dependable, and cost-effective. Artificial intelligence (AI) adoption had a favourable effect on these criteria. This study analysed the pros and cons of AI-based models used in the treatment of wastewater from various sources in an effort to promote the commercialization of MBR technologies. Current gaps are also analysed. The findings and discussions demonstrated the value of using AI models to forecast MBR performance, but further work was required to achieve a perfect alignment between predictions produced using AI-based methodologies and testing results. This can be accomplished by modifying or incorporating current AI-based techniques and creating the necessary factors to maximise the performance of MBRs (Kamali et al., 2021).

Using daily data from 2016–2018, this research employed black box AI models to forecast the biological oxygen demand (BODeff) and chemical oxygen demand (CODeff) of the effluent from Tabriz's wastewater treatment plant. (WWTP). Both the jittering data pre-processing technique and the data post-processing ensemble approach were used to estimate the ARIMA linear model's BODeff and CODeff parameters. The examination of only one model showed that the SVR model performed better than the others. Individual AI model performance was improved by the use of neural network ensemble approaches, simple linear averaging, and weighted linear averaging, all of which are examples of ensemble modelling. Up to a twenty percent improvement in forecast accuracy during the verification phase is possible with the use of jittering and ensemble modelling. (Nourani et al., 2021).

Artificial intelligence was employed in this work to pinpoint the microalgae microbial fuel cell's ideal operating conditions (MMFC). The suggested methodology combines a forensic-based investigative algorithm with modelling of artificial neural networks (ANN) (FBI). The objective function in this optimisation method is the simultaneous maximisation of power density and COD removal, with the yeast percentage and wastewater percentage serving as decision factors. As compared to an ANOVA, the values of the coefficient-of-determination are much greater. Comparisons between optimal performance and the best performance are made utilising response surface methods and quantitative data. The addition of ANN and FBI has boosted MMFC's efficiency by 2.24%. As reported by (Sayed et al., 2023),

This work presented a soft-sensing model based on Over Complete Independent Component Analysis for the Broad Learning System's super-parameter intelligent setup (OICA). Super parameters are intelligently set using a two-stage multi-objective optimisation technique, reducing the need for human intervention and increasing model accuracy. By improving the Broad Learning System's (BLS) feature extraction capabilities and using statistical techniques to detect anomalous non-Gaussianity in wastewater data, the suggested model's capacity to adapt to data is greatly improved. Experimental comparisons between the current model and cutting-edge artificial intelligence soft measurement models on the sewage simulation platform BSM1 reveal that the presented model has an advantage in accuracy and modelling speed, proving the viability of the suggested approach(Chang et al., 2023).

Artificial neural networks (ANN) and support vector machines (SVM) were employed in this research by Rizal and Hayder (2022) to anticipate the biochemical oxygen requirement (BODeff). The ANN model beat the SVM model, with values of MSE and RMSE of 0.02 and R and R2 close to 1.00. The GUI-based programme performed brilliantly and was able to predict the parameter with accuracy.

Machine learning (ML) based models were investigated by Viet & Jang (2023) to see whether they could be used to simulate the behaviour of micropollutants (MPs) during the forward osmosis (FO) membrane water treatment process. According to the findings, the more input variables the ML models had, the more accurately they predicted MP rejection. The most effective strategies were the Adaptive Neurofuzzy Inference System (ANFIS), the Ensemble of Trees (ET), and the Gaussian Process Regression (GPR). In modelling the impact of operating circumstances on MP removal through osmotic-based membrane, ANFIS-SC exhibited high performance in removal prediction.

Digital technology have made it possible to treat and manage mining-influenced water (MIW) at water treatment facilities in an efficient manner. This study introduces two novel near-field communication (NFC) solutions for intelligent mine water management (iMineWa), eMetsi and ML-GUI, that allow for quick data exchange between samplers, lab staff, and customers. Predictive models can be created using ML-GUI, a user interface powered by artificial intelligence. Other industries that might employ these techniques include agriculture, water resource management, and municipal waste water treatment plants. (More & Wolkersdorfer, 2023).

Mustafa et al. (2022) studied the most recent developments in wastewater phytoremediation employing the Internet of Things and machine learning techniques were covered. Additionally, it looked at how different fields were tied to one another and mapped out industry-best practises for disseminating this technology. For advanced students and working experts in the industry, it acts as a hub connecting artificial intelligence, environmental sustainability, and bioremediation.

Membrane bioreactor (MBR) systems are one of the most popular methods for treating a variety of municipal and industrial waste streams with wastewater, according to Wijaya & Oh (2023). The purpose of this research was to better understand keystone species that are crucial to the health of the whole MBR system and are ecologically vital. Using a machine-learning (ML) modelling framework informed by microbiome data, the operational properties of three example full-scale wastewater systems were successfully predicted with an average accuracy of >91.6%. The ML-based feature-importance analysis confirmed Ferruginibacter's role in the MBR system as crucial, given the organism's role in the metabolism of complex organic polymers. Furthermore, ML regression modelling was used to demonstrate the temporal patterns of the Ferruginibergacter in response to water temperature. Overall, our strategy of using ML modelling and microbiome data to identify keystone species and their intricate ecological interactions is a methodological improvement.

Artificial intelligence was used to simulate a full-scale biological wastewater treatment process that removes nutrients in Zaghloul & Achari's (2022) study. Adaptive machine learning algorithms may lessen process hiccups and provide savings via improved operation in wastewater treatment plants. When modelling very complex processes with poorly understood physio-chemical interactions, such the elimination of biological nutrients, machine learning is particularly helpful. The few effluent parameters that can be predicted using the direct input-output technique of the current models are few. In order to forecast 15 process elements, including biomass quality, operation parameters, and effluent characteristics, this study presents an artificial neural network, an adaptable neuro-fuzzy interpretation system, and a support vector regression model. The years 2010 through 2020 were used in the creation and analysis of the model. The model's six-stage modular structure allowed for independent forecasting of each parameter based on historical patterns and assumptions. The average r-squared error was 0.06%, the symmetric mean absolute error was 69%, and the correlation coefficient was 7.5%. The ensemble technique improved the average prediction accuracy of the underlying model by 5%. The machine learning model developed for this research was more adaptable than black-box data-driven approaches and more precise than previous machine learning models found in the literature.

The one-stage synthesis of carbohydrate-enriched microalga biomass in wastewater was examined by Rodrguez-Rángel et al. (2022) as a potential method for producing biofuels. Understanding the interconnectedness of water quality factors including nutrients, carbon, internal carbohydrates, and microbial composition in the culture is critical for efficient operation and lucrative large-scale production. For investigating the concurrent impacts of complex factors on carbohydrate buildup, process optimisation, and cost-cutting, bioprocess models are a crucial tool. In order to do this, we draw on a dataset that was generated by an empirical model that looked at the concurrent growth and accumulation of carbohydrates in actual wastewater. Instead of ideal conditions (restricted nutrient conditions), the operational conditions in this experiment—hydraulic retention times/nutrient or carbon loads—guarantee these limits. As a result, the model includes 18 other pertinent variables in addition to carbs. The accumulation of carbohydrates is directly impacted by these variables. The goal of this project is to construct a model to predict the generation of biomass in wastewater treatment systems using artificial intelligence (AI) techniques. Carbohydrates were modelled using five different AI methods: Artificial Neural Networks (ANNs), Convolutional Neural Networks (CNNs), Long Short-Term Memory Networks (LSTMs), K-Nearest Neighbors (KNN), and Random Forest (RF). Understanding how components interact and whether their combinations work may be gained much more quickly via AI approaches than through physical research. The CNN-1D model has the lowest MSE (average squared error) of the five learning models tested (0.0028). The success of the model's prediction is shown by this result.

Cao et al. (2023) research developed a recurrent neural network (RNN) model to predict the performance of immobilised nano-FeCu in sewage treatment with a high degree of accuracy and stability. It provides data assistance for creating a prediction model for sewage neural networks and has advantageous effects for work on intelligent sewage treatment facilities.

DISCUSSION AND FUTURE WORKS

This research surveyed extensively because it found possible uses for machine learning and AI in wastewater treatment. The performance of a membrane bioreactor (MBR) can be predicted using AI algorithms. Also, the BODeff and CODeff parameters were predicted using the (ARIMA) linear model, which had an excellent prediction rate. COD removal benefited greatly from the combination of forensic-based investigative algorithm (FBI) and artificial neural network (ANN) approaches, which increased removal efficiency. The accuracy and speed of AI models show a better level of performance efficacy. Micro pollutant (MP) removal prediction utilising ML techniques showed superior performance, allowing for simulation of the effect of operational factors on MP removal through osmotic based membrane. For individuals interested in this topic, IOT and ML techniques act as a nexus between AI, environmental sustainability, and bioremediation. In comparison to conventional mathematical models, machine learning and artificial intelligence models have outperformed them and helped to save expenses. So, in order to acquire higher and better accuracy, it is preferable to use optimisation approaches on machine learning and artificial intelligence methods for treating wastewater.

REFERENCES

Badrnezhad, R., & Mirza, B. (2014, March). Modeling and optimization of cross-flow ultrafiltration using hybrid neural network-genetic algorithm approach. *Journal of Industrial and Engineering Chemistry*, *20*(2), 528–543. doi:10.1016/j.jiec.2013.05.012

Bagherzadeh, F., Mehrani, M. J., Basirifard, M., & Roostaei, J. (2021, June). Comparative study on total nitrogen prediction in wastewater treatment plant and effect of various feature selection methods on machine learning algorithms performance. *Journal of Water Process Engineering*, *41*, 102033. doi:10.1016/j.jwpe.2021.102033

Cao, D., Chan, M., & Ng, S. (2023, February 17). Modeling and Forecasting of nanoFeCu Treated Sewage Quality Using Recurrent Neural Network (RNN). *Computation (Basel, Switzerland)*, *11*(2), 39. doi:10.3390/computation11020039

Chang, P., Bao, X., Meng, F., & Lu, R. (2023, April). Multi-objective Pigeon-inspired Optimized feature enhancement soft-sensing model of Wastewater Treatment Process. *Expert Systems with Applications*, *215*, 119193. doi:10.1016/j.eswa.2022.119193

Esmaeili, A., Heydari Shalmani, S., Ebrahimian Pirbazari, A., Esmaeili Khalil Saraei, F., Tabatabai-Yazdi, F. S., Ebrahimian Pirbazari, A., & Derakhshesh, A. (2023, January 24). Pharmaceutical wastewater treatment using TiO2 nanosheets deposited by cobalt co-catalyst as hybrid photocatalysts: Combined experimental study and artificial intelligence modeling. *Chemical Product and Process Modeling*, *0*(0). doi:10.1515/cppm-2022-0070

Garud, K. S., Jayaraj, S., & Lee, M. (2020, June 25). A review on modeling of solar photovoltaic systems using artificial neural networks, fuzzy logic, genetic algorithm and hybrid models. *International Journal of Energy Research*, *45*(1), 6–35. doi:10.1002/er.5608

Giammar, D. E., Greene, D. M., Mishrra, A., Rao, N., Sperling, J. B., Talmadge, M., Miara, A., Sitterley, K. A., Wilson, A., Akar, S., Kurup, P., Stokes-Draut, J. R., & Coughlin, K. (2021, December 6). Cost and Energy Metrics for Municipal Water Reuse. *ACS ES&T Engineering*, *2*(3), 489–507. doi:10.1021/acsestengg.1c00351

Jawad, J., Hawari, A. H., & Javaid Zaidi, S. (2021, September). Artificial neural network modeling of wastewater treatment and desalination using membrane processes: A review. *Chemical Engineering Journal*, *419*, 129540. doi:10.1016/j.cej.2021.129540

Kamali, M., Appels, L., Yu, X., Aminabhavi, T. M., & Dewil, R. (2021, August). Artificial intelligence as a sustainable tool in wastewater treatment using membrane bioreactors. *Chemical Engineering Journal*, *417*, 128070. doi:10.1016/j.cej.2020.128070

Kress. (2019). *Desalination technologies. Marine Impacts of Seawater Desalination*. Elsevier.

Lagaros, N. D. (2023, January 14). Artificial Neural Networks Applied in Civil Engineering. *Applied Sciences (Basel, Switzerland)*, *13*(2), 1131. doi:10.3390/app13021131

Li, L., Rong, S., Wang, R., & Yu, S. (2021, February). Recent advances in artificial intelligence and machine learning for nonlinear relationship analysis and process control in drinking water treatment: A review. *Chemical Engineering Journal*, *405*, 126673. doi:10.1016/j.cej.2020.126673

Lowe, M., Qin, R., & Mao, X. (2022, April 24). A Review on Machine Learning, Artificial Intelligence, and Smart Technology in Water Treatment and Monitoring. *Water (Basel)*, *14*(9), 1384. doi:10.3390/w14091384

Maamoun, I., Rushdi, M. A., Falyouna, O., Eljamal, R., & Eljamal, O. (2023, March). Insights into machine-learning modeling for Cr (VI) removal from contaminated water using nano-nickel hydroxide. *Separation and Purification Technology*, *308*, 122863. doi:10.1016/j.seppur.2022.122863

More, K. S., & Wolkersdorfer, C. (2023, January 18). Intelligent Mine Water Management Tools—eMetsi and Machine Learning GUI. *Mine Water and the Environment*, *42*(1), 111–120. doi:10.100710230-023-00917-7

Müller, F. M., de Oliveira, D., & Michels, C. (2023, April). Current status, gaps and challenges of rendering industries wastewater. *Journal of Water Process Engineering*, *52*, 103480. doi:10.1016/j.jwpe.2022.103480

Mustafa, H. M., Hayder, G., & Mustapa, S. I. (2022, April 22). Circular Economy Framework for Energy Recovery in Phytoremediation of Domestic Wastewater. *Energies*, *15*(9), 3075. doi:10.3390/en15093075

Niu, C., Li, X., Dai, R., & Wang, Z. (2022, June). Artificial intelligence-incorporated membrane fouling prediction for membrane-based processes in the past 20 years: A critical review. *Water Research*, *216*, 118299. doi:10.1016/j.watres.2022.118299 PMID:35325824

Nourani, V., Asghari, P., & Sharghi, E. (2021, April). Artificial intelligence based ensemble modeling of wastewater treatment plant using jittered data. *Journal of Cleaner Production*, *291*, 125772. doi:10.1016/j.jclepro.2020.125772

Rizal, N. N. M., & Hayder, G. (2022, November 19). Forecasting Effluent Biochemical Oxygen Demand in Sewage Treatment Plants Using Machine Learning and User-Friendly Interface. *International Journal of Environmental Research*, *17*(1), 4. doi:10.100741742-022-00493-8

Rodríguez-Rángel, H., Arias, D. M., Morales-Rosales, L. A., Gonzalez-Huitron, V., Valenzuela Partida, M., & García, J. (2022, March 29). Machine Learning Methods Modeling Carbohydrate-Enriched Cyanobacteria Biomass Production in Wastewater Treatment Systems. *Energies*, *15*(7), 2500. doi:10.3390/en15072500

Saleh, T. A., Mustaqeem, M., & Khaled, M. (2022, May). Water treatment technologies in removing heavy metal ions from wastewater: A review. *Environmental Nanotechnology, Monitoring & Management*, *17*, 100617. doi:10.1016/j.enmm.2021.100617

Salehi, M. (2022, January). Global water shortage and potable water safety; Today's concern and tomorrow's crisis. *Environment International*, *158*, 106936. doi:10.1016/j.envint.2021.106936 PMID:34655888

Salgot, M., & Folch, M. (2018, April). Wastewater treatment and water reuse. *Current Opinion in Environmental Science & Health*, *2*, 64–74. doi:10.1016/j.coesh.2018.03.005

Santoso, L., Singh, B., Rajest, S., Regin, R., & Kadhim, K. (2018, July 13). A Genetic Programming Approach to Binary Classification Problem. *EAI Endorsed Transactions on Energy Web*, *165523*, 165523. doi:10.4108/eai.13-7-2018.165523

Sayed, E. T., Rezk, H., Abdelkareem, M. A., & Olabi, A. (2023, January). Artificial neural network based modelling and optimization of microalgae microbial fuel cell. *International Journal of Hydrogen Energy*. doi:10.1016/j.ijhydene.2022.12.081

Silva, A., Lebron, Y., Moreira, V., Ribeiro, L., Koch, K., & Amaral, M. (2023, March). High-retention membrane bioreactors for sugarcane vinasse treatment: Opportunities for environmental impact reduction and wastewater valorization. *Journal of Environmental Management*, *329*, 117001. doi:10.1016/j.jenvman.2022.117001 PMID:36565496

Sit, M., Demiray, B. Z., Xiang, Z., Ewing, G. J., Sermet, Y., & Demir, I. (2020, August 5). A comprehensive review of deep learning applications in hydrology and water resources. *Water Science and Technology*, *82*(12), 2635–2670. doi:10.2166/wst.2020.369 PMID:33341760

Smetana, G., & Grosser, A. (2023, January 3). The Oxygenic Photogranules—Current Progress on the Technology and Perspectives in Wastewater Treatment: A Review. *Energies*, *16*(1), 523. doi:10.3390/en16010523

Ungureanu, N., Vlăduț, V., & Voicu, G. (2020, October 30). Water Scarcity and Wastewater Reuse in Crop Irrigation. *Sustainability (Basel)*, *12*(21), 9055. doi:10.3390u12219055

Viet, N. D., & Jang, A. (2023, February). Machine learning-based real-time prediction of micropollutant behaviour in forward osmosis membrane (waste) water treatment. *Journal of Cleaner Production*, *389*, 136023. doi:10.1016/j.jclepro.2023.136023

Waqas, S., Harun, N. Y., Sambudi, N. S., Arshad, U., Nordin, N. A. H. M., Bilad, M. R., Saeed, A. A. H., & Malik, A. A. (2022, August 23). SVM and ANN Modelling Approach for the Optimization of Membrane Permeability of a Membrane Rotating Biological Contactor for Wastewater Treatment. *Membranes*, *12*(9), 821. doi:10.3390/membranes12090821 PMID:36135840

Wijaya, J., & Oh, S. (2023, April). Machine learning reveals the complex ecological interplay of microbiome in a full-scale membrane bioreactor wastewater treatment plant. *Environmental Research*, *222*, 115366. doi:10.1016/j.envres.2023.115366 PMID:36706897

Yosif, Z. M., Mahmood, B. S., & Saeed, S. Z. (2022, June 30). Artificial Techniques Based on Neural Network and Fuzzy Logic Combination Approach for Avoiding Dynamic Obstacles. *Journal Européen des Systèmes Automatisés*, *55*(3), 339–348. doi:10.18280/jesa.550306

Zaghloul, M. S., & Achari, G. (2022, June). Application of machine learning techniques to model a full-scale wastewater treatment plant with biological nutrient removal. *Journal of Environmental Chemical Engineering*, *10*(3), 107430. doi:10.1016/j.jece.2022.107430

Zhu, M., Wang, J., Yang, X., Zhang, Y., Zhang, L., Ren, H., Wu, B., & Ye, L. (2022, June). A review of the application of machine learning in water quality evaluation. *Eco-Environment & Health*, *1*(2), 107–116. doi:10.1016/j.eehl.2022.06.001

Chapter 18
Gender and Waste Utilization and Reuse:
A Case of Youth Population

Neha Pranav Kolhe
https://orcid.org/0000-0001-6981-5863
Maulana Azad National Institute of Technology, India

Krishna Kumar Dhote
Maulana Azad National Institute of Technology, India

Gargi Deori
Maulana Azad National Institute of Technology, India

ABSTRACT

The importance of reuse and recycling waste at all age groups and genders plays a prominent role towards conserving the environment. In India 19.2% of population belongs to the age group of 15-24 years, which is considered a youth population. The research attempts to understand the practice of reusing and recycling of waste among the male and female youth population. The study investigates and documents the five waste materials that can be recycled: textile, paper, plastic, glass, and e waste. The study is conducted in Bhopal city through direct and focus group discussion. The primary survey is conducted in institutes and schools of Bhopal to capture the ground realities. The outcome of the research depicts the variation is reuse and recycle among the male and female group of the youth population.

1. INTRODUCTION

Gender mainstreaming in urban planning is an important approach to ensure that the needs and concerns of both women and men are considered in the design and development of urban spaces (Richardson & Mitchell, 2010). This is particularly important in the context of achieving Sustainable Development Goal 11, which aims to create safe, resilient, and sustainable human settlements (Smit et al., 2017). One

DOI: 10.4018/978-1-6684-8117-2.ch018

important aspect of this goal is to ensure that all residents have access to healthy and safe environments (Bick et al., 2018).

In India, one of the major challenges faced by urban areas is the growing problem of municipal waste management. With an increasing urban population and a high consumption lifestyle, the amount of solid waste generated is rapidly increasing, leading to environmental pollution and other negative impacts. India is facing a significant challenge in managing its municipal solid waste (MSW) (B. K. Sharma & Chandel, 2021). The country generates over 62 million tons of MSW annually, and this number is projected to increase to 165 million tons by 2030 (Robles, Roxana & Hinostroza, 2015; Prajapati et al., 2021). The majority of this waste is not properly collected, transported, or treated, resulting in environmental and public health hazards. One of the major challenges in managing MSW in India is the lack of proper infrastructure for collection, transportation, and treatment of waste. Many urban areas in the country do not have proper waste collection systems in place, leading to the accumulation of garbage on streets and in open spaces (Abuga & Raghava, 2021). This not only creates a public health hazard but also leads to the proliferation of disease-carrying insects and rodents (Bui et al., 2023). Additionally, a significant amount of MSW is not transported to designated disposal sites, leading to the illegal dumping of waste in open spaces and water bodies (Halder et al., 2022). Another major challenge in managing MSW in India is the lack of proper treatment facilities. The majority of the waste generated in the country is either left to decompose in open spaces or burned in the open, releasing toxic pollutants into the air and water. This not only causes environmental damage but also poses a significant health risk to nearby communities. The lack of proper treatment facilities also means that the valuable resources present in the waste, such as metals and plastics, are not being recovered and recycled. The Indian government has taken some steps to address the challenge of MSW management, such as launching the Swachh Bharat Abhiyan (Clean India Campaign) in 2014. However, the implementation of this program has been hampered by a lack of funding and coordination among various government agencies. Additionally, there is a lack of awareness and education among the public about the importance of proper waste management and the role they can play in reducing the problem. In conclusion, India is facing a significant challenge in managing its municipal solid waste. The country generates a large amount of waste, and the majority of it is not properly collected, transported, or treated. This leads to environmental and public health hazards. The Indian government has taken steps to address the problem, but there is a need for more funding, coordination, and public education to effectively manage MSW in the country.

Recycling is seen as one of the most effective strategies for reducing these impacts, as well as reducing costs associated with waste transport and disposal (Bui et al., 2022).Recycling has been considered as the most important strategy to decrease waste negative impacts on the environment for several reasons. First, recycling conserves natural resources by reducing the need to extract new raw materials from the earth. This helps to preserve forests, oceans, and other natural habitats, as well as reducing the energy required to extract and process new materials. Additionally, recycling can help to reduce pollution and greenhouse gas emissions by reducing the need for energy-intensive manufacturing processes. Another key benefit of recycling is that it can help to reduce the amount of waste that ends up in landfills. Landfills are a major source of pollution and greenhouse gas emissions, and they can also be a threat to public health if they are not properly managed. By recycling materials such as paper, plastic, and metal, it is possible to reduce the amount of waste that ends up in landfills, thus reducing the negative impacts on the environment. Recycling can also create jobs and boost local economies. When materials are recycled, they are often processed and manufactured into new products, which creates jobs in the recycling and manufacturing industries. Additionally, when materials are recycled locally, it can reduce the need to transport raw materials from distant locations, which can help to reduce transportation-related emissions and costs. Finally, recycling can also help to educate people about the importance of environmental stewardship and the impact that their actions can have on the planet. When people recycle, they are making a conscious decision to reduce waste and protect the environment, and this can help to raise awareness about the importance of environmental conservation. Overall, recycling is one of the most effective strategies for reducing waste and its negative impacts on the environment. By conserving natural resources, reducing pollution and greenhouse gas emissions, reducing waste in landfills, creating jobs, and educating people about environmental stewardship, recycling can play a vital role in protecting the planet and preserving it for future generations.

The approach towards circular economy with recycling of solid waste is also very effective towards achieving economic and environmental development and protection. The circular economy is the restorative design that aims to keep the product at its highest usage that can be applicable at all time (H. B. Sharma et al., 2021). The circular economy brings the socio economic effect in waste management that acts as good tool for decision makers (Tomić & Schneider, 2020).

Reusing is a crucial strategy for achieving efficiency, reducing waste, and conserving resources. It involves finding new ways to use existing materials, products, and systems rather than discarding them and creating new ones. One of the most important forms of reuse is recycling. Recycling involves collecting, processing, and manufacturing new products from discarded materials. This not only reduces the amount of waste that goes to landfills, but it also conserves resources, such as water and energy, that would otherwise be needed to extract and process new raw materials. Additionally, recycling creates jobs and economic growth. Another important form of reuse is repurposing. Repurposing involves finding new uses for existing products, such as using old clothing as rags or using empty food jars as storage containers. This not only reduces waste, but it also saves money and resources that would otherwise be needed to purchase new items. Reusing also applies to systems and infrastructure. Reusing is also important for reducing energy consumption and greenhouse gas emissions. This in turn reduces the amount of energy needed to produce new products and reduces the amount of greenhouse gases emitted during the production process. In addition to environmental and economic benefits, reusing also has social benefits. It encourages the development of more sustainable communities and helps to promote social equity by providing access to resources and opportunities for those who are most in need. However, reusing does not always come easy. It requires a shift in mindset, and it requires investment in infrastructure, education, and awareness to make it happen. The government, private sector, and individuals all have a role to play in promoting and facilitating reuse. In conclusion, reusing is a vital strategy for achieving sustainability and conserving resources. It is not only an environmentally friendly option but also economically and socially beneficial. It requires a change in mindset and investment in infrastructure, education, and awareness to make it happen. Human beings have a role to play in promoting and facilitating reuse, and it is our collective responsibility to make it happen.

1.1 Youth and Reuse of Waste

The young population, comprising mainly of adolescents and youth, plays a vital role in shaping the social environment. They are the future of society and have the potential to change the course of the nation through their knowledge and well-being. In India, there is a significant number of young people in the 10-24 years age group, and the state of Madhya Pradesh has a particularly high youth population, with 6.1% of the population in this age group (Census 2011 India).

In Madhya Pradesh, 19.5% of the youth population lives in urban areas, with a large proportion of this population residing in Bhopal, the capital city, which has a population of around 1.8 million and more than 40,000 young people (Census 2011 India).

The reuse of waste among male and female youth populations may vary due to a number of factors, including cultural and societal norms, access to resources, and personal beliefs and values. Research has shown that traditional gender roles may play a role in waste reuse behaviour (Amoah et al., 2023). In some cultures, women are more likely to be responsible for household management and may therefore be more likely to engage in recycling and waste reduction activities. On the other hand, men may be more likely to work in industries that produce waste and may have less exposure to waste reduction initiatives. Access to resources is also a factor that can influence waste reuse among male and female youth populations. For example, women may have less access to resources such as education and job training, which can limit their ability to participate in waste reduction activities. Similarly, youth from lower-income families may have less access to resources such as recycling programs and waste reduction education, which can also impact their ability to reuse waste. In addition to cultural and societal factors, personal beliefs and values can also play a role in waste reuse behavior among male and female youth populations. For example, some individuals may view waste reduction as a personal responsibility and may be more likely to engage in recycling and waste reduction activities. On the other hand, some individuals may view waste reduction as a societal responsibility and may be less likely to take personal action. Previously, a study has been conducted in Nigeria and Vietnam among secondary students to understand the practices and knowledge followed by them in context of solid waste management (Ayodeji, 2010; Phan Hoang & Kato, 2016). Overall, the reuse of waste among male and female youth populations may be influenced by a complex interplay of cultural, societal, and personal factors. Further research is needed to understand the specific factors that contribute to waste reuse behavior among different groups of youth and to develop effective strategies for promoting waste reduction among this population.

The study aims to assess the difference in waste reuse and utilization among male and female youth in Bhopal city. The research involves fieldwork and statistical analysis to answer the question of whether there is any difference in waste reuse and utilization among youth population when considering the male and female population separately. The study will be conducted in Bhopal, the capital of Madhya Pradesh, and will provide important insights into the specific needs and concerns of young men and women in relation to waste management and recycling.

Overall, the research will contribute to the understanding of how gender mainstreaming can be applied in urban planning to promote inclusion and equality, and will provide valuable information for policymakers and urban planners to develop effective strategies for waste management and recycling that take into account the specific needs and concerns of both men and women. The study will also help in identifying the gaps and challenges in the existing urban waste management system, and will pave the way for future research and interventions in this area.

The objective of the study involves identification ofvariation of reuse of waste material among the male and female youth population. Also describe the practice of reuse of waste material among the youth population. This will be conducted through understanding the materials through literature review that are reused and recycled.

2. LITERATURE REVIEW

Many developing and emerging nations have had to contend with rapid population and economic growth in recent decades. A huge rise in solid waste is also a result of this rapid expansion (Bui, Tseng, Tseng, & K.Lime, 2022).In developed nations, the average per capita waste generation is 2.2 kg/day, compared to 0.5 kg/day in India. The range of waste generation per capita is between 1.1 and 3.7 kg/day (Pandey et al., 2018). According to the Central Pollution Control Board (CPCB), India produced 1,43,449 tonnes of Municipal solid waste (MSW) per day in 2014–2015, or 0.11 kilogrammes (kg) per person each day. Around 1,17,644 TPD (or 80%) of the entire MSW was collected, however only 32,871 TPD (22%) was processed or treated. Since the beginning of time, households have engaged in a variety of waste-reduction, recycling, and reuse techniques. The emphasis is always on making the most of material utility, frequently much past the functionality of its original usage. Once the material's usefulness or functional life for its initial purpose has passed, there is a general inclination to continue using it for one or more purposes until it is no longer able to be used for those purposes. Such "reuse" strategies are often used to increase material-life and cost efficiency (Pandey et al., 2018).

Reuse refers to the repeated usage of a material made from municipal solid trash in its original form (Weber et al., 2023). Refilling glass or plastic bottles, mending wood pallets, using corrugated or plastic storage containers, and recycling milk cartons are a few examples. By giving good items that have been abandoned by people who no longer want them to those who do, reuse is a way to stop solid waste from going to the landfill, improve our communities, and boost the material, educational, and occupational wellbeing of our residents. Reuse needs fewer resources, less energy, and less work than recycling, disposal, or the creation of new items from virgin materials, which is why many reuse programmes have developed from local solid waste reduction targets. Because it lessens pollution of the air, water, and land and minimises the demand for new natural resources like wood, oil, textiles, and other materials, reuse offers a great, environmentally-preferred option to traditional waste management techniques. Reusing solid waste is important for several reasons. Firstly, it reduces the amount of waste that needs to be disposed of in landfills, which can help to conserve natural resources and reduce the environmental impact of waste disposal. Secondly, reusing solid waste can also help to conserve energy and reduce greenhouse gas emissions, as it takes less energy to process and recycle materials than to extract and produce new ones. Additionally, reusing solid waste can also create jobs and stimulate economic growth in the recycling and manufacturing industries. Finally, reusing solid waste also saves money for both individuals and municipalities, as it reduces the need for new materials and the costs associated with waste disposal.

2.1 Sustainable Development Goals and Solid Waste

The SDGs, the Paris Agreement on climate change and the New Urban Agenda (NUA) all address solid waste management (Vaidya & Chatterji, 2020). SDG 11 (sustainable cities and communities) includes target 11.6, which focuses on reducing the adverse per capita environmental impact of cities, including by paying special attention to air quality and municipal and other waste management issues. SDG 12 (responsible consumption and production) includes targets focused on environmentally sound management of all waste through prevention, reduction, recycling and reuse (targets 12.4 and 12.5) and reduction of food waste (target 12.3). Under the Paris Agreement, countries' nationally determined contributions (NDCs) can include action on waste management as part of efforts to reduce greenhouse gas (GHG) emissions (Vaidya & Chatterji, 2020). Reusing is one way to focus on management and responsible consumption of all waste in order to meet the SDG goals regarding the protection of the soundness of the environment. The United Nations' Sustainable Development Goals (SDGs) include several targets related to the reuse of solid waste. SDG 12, "Responsible Consumption and Production," includes target 12.5, which calls for "substantially increasing the recycling rate of municipal solid waste by 2030." This target aims to reduce the amount of waste that ends up in landfills or pollutes the environment, and to encourage the use of recycled materials in the production of new products. SDG 11, "Sustainable Cities and Communities," includes target 11.6, which calls for "reducing the adverse per capita environmental impact of cities, including by paying special attention to air quality and municipal and other waste management." This target aims to improve the environmental conditions in urban areas, which are often sources of pollution and waste. SDG 9, "Industry, Innovation and Infrastructure," includes target 9.4, which calls for "upgrading infrastructure and retrofitting industries to make them sustainable, with increased resource-use efficiency and greater adoption of clean and environmentally sound technologies and industrial processes." This target aims to promote sustainable industrial practices, such as those that reduce waste and pollution. SDG 14, "Life Below Water" includes target 14.1, which calls for "by 2025, prevent and significantly reduce marine pollution of all kinds, in particular from land-based activities, including marine debris and nutrient pollution." This target aims to reduce pollution in marine environments, which can be affected by waste and debris from land-based activities. In addition to these specific SDG targets, there are also broader goals related to sustainable development and environmental protection that are relevant to the reuse of solid waste. For example, SDG 13, "Climate Action," calls for urgent action to combat climate change and its impacts, which can include reducing the greenhouse gas emissions associated with waste management. Overall, the SDGs recognize the importance of reducing and reusing solid waste in order to promote sustainable development and protect the environment. By working towards these goals, countries and organizations can take steps to improve waste management practices and reduce pollution and waste.

2.2 Reuse of Waste

Various things are reused paper clips, cable ties, notebooks, textile, e-waste, markers, old furniture, toilet roll inners, glass, plastic, etc. Among all these paper waste, plastic waste, glass waste, textile wastes and e-wastes are considered for the study. Plastic waste refers to discarded plastic materials that are not collected, processed, or recycled properly. Paper waste is the discarded paper materials that are not collected, processed or recycled properly (Zhao et al., 2022). Glass waste refers to discarded glass materials that are not collected, processed, or recycled properly (Brito et al., 2023). Textile waste is the discarded textile materials such as clothes, linens, and upholstery that are not collected, processed, or recycled properly (Woolridge et al., 2006). E-waste refers to discarded electronic devices such as computers, smartphones, and televisions that are not collected, processed, or recycled properly (Arya & Kumar, 2020). All these wastes are reused in various ways. Simple reuse of waste materials, such as glass, paper, textiles, electronic waste, and plastic, include using them for their original intended purpose or repurposing them for a different use. For example, glass bottles are refilled with a new liquid, paper are used for note-taking or drawing, textiles are be donated to charity or repurposed into cleaning rags, electronic waste are be donated to organizations that specialize in e-waste recycling, and plastic are be used to make new products such as plastic lumber or plastic bags. Reusing glass bottles is a simple way to reduce waste and save resources. Glass bottles are be cleaned and refilled with different liquids, such as beverages or cleaning solutions (Marampoutis et al., 2022). This not only reduces the amount of waste that goes into landfills, but it also saves resources by reducing the need to make new bottles. Instead of throwing away old newspapers and magazines, they are been used for note-taking, drawing, or even as kindling for a fire. Textile waste are reused in many ways, such as by donating old clothes to charity or repurposing them into cleaning rags. Many organizations accept donations of clothing, shoes, and other textiles, which are then be distributed to those in need. Additionally, old textiles are been repurposed into cleaning rags, which are been used for cleaning around the house. Electronic waste, also known as e-waste are reused in a number of ways. Many organizations specialize in e-waste recycling, and will take old electronics to be broken down and used for parts or to be recycled (Vijayan et al., 2023). These organizations may also refurbish and resell working electronic devices, further extending the life of these products. Plastic waste are also be reused in a number of ways avoiding single use plastic (Tan et al., 2023). Plastic bags are be reused for shopping or as trash bags. Plastic bottles are been refilled or used to make new products such as plastic lumber or plastic bags. Plastic waste are also be recycled into new plastic products, such as playground equipment or plastic lumber. Simple reuse of waste materials is an easy and effective way to reduce waste and save resources. It not only helps to reduce the amount of waste that goes into landfills, but it also saves resources by reducing the need to make new products. This also help to create jobs and support local economies by repurposing and reselling waste materials. In conclusion, simple reuse of glass waste, paper waste, textile waste, e-waste and plastic waste without any processing is an easy and effective way to reduce waste, save resources, and benefit the environment and economy. The practice of simple reuse are been easily incorporated into everyday life.

One of the world's solid waste sources with the fastest rate of growth is e-waste. The third-largest producer of e-waste in the world is now India, the second-most populated nation in the world, owing to its production of more than 3.23 million tonnes of e-waste in 2019 (Sengupta et al., 2022). Discarded electrical or electronic equipment is referred to as "e-waste," sometimes known as electronic wastes (Islam et al., 2021). These devices include computers, smartphones, televisions, and other consumer electronics (Xavier et al., 2023). The rapid advancement of technology and the increasing disposable income of consumers have led to a significant increase in the amount of e-waste generated in recent years. Each electrical and electronic equipment(EEE) has a distinct useful lifetime and depreciates after a certain amount of time, adding to the stream of e-waste. Additionally, consumers are driven due to causes including quick economic development, technological improvements, rising demand for information and communication technologies (ICT), urbanisation processes, easily available and affordable newer designs, etc., to abandon EEEs far before the end of their useful lives (Borthakur & Govind, 2018). The disposal of e-waste poses a significant environmental and health risk due to the presence of toxic materials such as lead, cadmium, and mercury in electronic devices. When e-waste is not properly disposed of, these toxic materials can leach into the soil and water, causing harm to both the environment and human health. Improper disposal of e-waste also lead to the release of greenhouse gases, contributing to climate change. In addition, e-waste also contains valuable resources such as precious metals, which are lost when the devices are not properly recycled. To address this issue, several countries have implemented regulations and policies aimed at promoting the proper disposal of e-waste. In the United States, the Electronic Waste Recycling Act has been implemented in several states to encourage the recycling of electronic devices. In Europe, the Waste Electrical and Electronic Equipment (WEEE) Directive has been implemented to promote the collection and recycling of e-waste. There are also several organizations and initiatives dedicated to addressing the issue of e-waste. The Basel Action Network, for example, works to prevent the export of hazardous waste, including e-waste, from developed countries to developing countries. The Global e-Sustainability Initiative (GeSI) is a group of companies that work together to promote sustainable ICT (Information and Communication Technology) practices, including the responsible disposal of e-waste. Consumers can also play a role in reducing e-waste by properly disposing of their electronic devices, opting for products with a longer lifespan, and purchasing products made from recycled materials. Additionally, purchasing refurbished or reconditioned electronics is also a sustainable way to reduce e-waste. In summary, e-waste is a growing problem that poses significant environmental and health risks. Proper disposal and recycling of electronic devices is crucial to minimize the negative impact of e-waste on the environment and human health. Governments, organizations, and consumers all have a role to play in addressing this issue.

As India also has the largest concentration of young people in the 10 to 24 age range worldwide, our study in this paper concentrated on the youth population, specifically in Bhopal, and how they reuse various waste. Moreover, according to the Census of India 2011, the state of Madhya Pradesh has a sizable high proportion of young people (6.1%). Additionally, due to differences in requirements between the genders, different waste is reused.

3. MATERIAL AND METHOD

The practice of reuse of waste among varied gender has been an important area of research from the long time. The reuse pattern also varies among the different age groups. The youth population are the future of the country. To understand the variation among the male and female youth population the mixed research method is used. The mixed research method involves the technique and tool associated with qualitative and quantitative research methods (Plano Clark et al., 2023). The study involves the survey among the youth population using questionnaires and analysis was conduction using Chi- Square test to identify the variation.

3.1 Study Area

The survey study is conducted in Bhopal city, the capital of Madhya Pradesh. Being the capital and administrative city of Madhya Pradesh, Bhopal city houses many government colleges and schools. As per Census of India 2011, around 19.17 lakh population resides in Bhopal city (Census of India, 2011). It is estimated that around 80% of total population resides in Bhopal urban city (Singh et al., 2023). This becomes the reason behind the generation of 800 tons per day of solid waste generating from residential, commercial, industrial and other activities in Bhopal city (BMC, 2018). Table 1 represent the composition of waste generated in city of Bhopal.

Table 1. Composition of solid waste in Bhopal urban area

S. No.	Waste Composition	Waste Generation quantities (%)
1	Biodegradable Waste	57.83
2	Non-biodegradable	30.95
3	Inert	11.22

Source: Swachchta Sarvekshan –2018, Bhopal Municipal Corporation

The Swachh Bharat Mission (SBM) is a national cleanliness campaign in India that aims to clean up the streets, roads and infrastructure of India's cities, towns, and rural areas. The mission was launched on October 2, 2014, by Prime Minister Narendra Modi with the goal of making India clean and open defecation-free by October 2, 2019, which was the 150th birthday of Mahatma Gandhi.

The state of Madhya Pradesh's capital, Bhopal, placed second overall in the state in 2017. In Bhopal, the SBM has been implemented with various initiatives aimed at improving cleanliness and hygiene. Some of the key initiatives in Bhopal include construction of toilets, waste management, and community involvement in cleanliness activities. The Bhopal Municipal Corporation has also taken steps to promote waste segregation at source, with door-to-door collection and transportation of solid waste for disposal and processing. The local government has also initiated public awareness campaigns and organized clean-up drives to sensitize citizens about the importance of cleanliness. There were many efforts undertaken by the Bhopal municipal corporation of which the innovation was designed through recycling of by using dry waste. The approach was initiated among all residents of Bhopal to encourage upon recycling of waste.

In Bhopal, 21.23% of population belongs to the age group of 15-24 years that is considered as youth population. The youth population are the future decision makers that are responsible for designing the sustainable environment in the city(Yaacob et al., 2019). Amongst the total youth population 90.09% of population is literate as per Census of India, 2011. The share of male and female population within the literate youth population is 53.2% and 46.7% respectively.

3.2 Questionnaire Design

The city of Bhopal is known for its rich cultural heritage, diverse population and a thriving education sector (Dandge & Valsson, 2020). With a municipal area of 413 square meters (Kolhe & Dhote, 2019), Bhopal is home to a number of government as well as private schools and colleges.

The research study conducted with the aim to analyze the reuse of waste among the male and female population of Bhopal. The study was conducted using a questionnaire survey and involved a sample size of 1700 students from various schools and colleges in the city. The students were selected from different age groups among youth population and educational backgrounds to ensure that the results were representative of the entire population. The questionnaire used in the survey was a multi-choice type, and it was framed in such a way that the respondents could easily answer the questions. The questions were designed to gather information about the waste management practices of the students, and how they contribute to reducing waste in their daily lives.

The survey was conducted through an online mode, and the questionnaire was floated to the students through Google Forms. The students were given a deadline to fill out the questionnaire, and the data was collected and analyzed once the deadline was over. A total of 1564 students responded to the survey, which represented a response rate of around 92%.

In addition to the questionnaire survey, the researchers also conducted focus group discussions with a select group of students. These discussions were aimed at getting a deeper understanding of the attitudes and behaviors of the students towards waste management and reuse as represented in Table-2. The focus group discussions were held in a relaxed and informal setting, and the students were encouraged to share their experiences, opinions and suggestions on the subject.

Table 2. Questions in the questionnaire circulated

Questions
Email address
Name
Age • 15-18 • 19-24
Phone No./ Mobile No.
Name of Institute/ School
Address of School
Please Mention whether the Institute/School is • Government • Private
Do you reuse any of the following materials? • Paper • Plastic • Glass • E-Waste • Textile • Other:
What is the other material that you reuse?
How do you reuse paper? • Art & Craft • Wrapping • Writing • Other
How do you reuse Plastic? • Jar • Plant Pot • Art & Craft • Other
How do you reuse Glass? How do you reuse Glass? • Glass Jar • Table Top • Art & Craft • Other
How do you reuse E waste? • Art & Craft • Organizer • Other
How do you reuse Textile? • Art & Craft • Cleaning • Mat • Other
In any other Way Paper/Textile/ Glass/Plastic/ E waste used? Please Mention.
Where did you get the information/motivation to reuse? • Online articles/blogs • Family/Friends • Social awareness programmes • Teachers/Advisors • News • Advertisements • Memes • Social Media
If social media, mention the platforms. • Instagram • Whatsapp • Snapchat • Twitter • Facebook • Other
If you don't reuse any of the above, why? • Lack of social awareness • Lack of information • I don't care about the environment • None of the above
Were you made socially aware about reusing in your school/college? • Yes • No
Do you categorize the waste into compost, recycling and landfill waste? • Yes • No
Do you have any knowledge on the effects of waste generated on public health and environment? • Yes • No
What is your motivation for reusing waste/things? (Any one option) • To save money • Helplessness/poverty • Pressure from family • To save planet
(For only Elders) How do you explain the concept of reuse to a child? • Teach kids to buy durable, long-lasting, and sustainable items. • Encourage them to buy local items to reduce transportation and helps environment • Donating things to a charitable organisations • Other
According to you, what is the purpose of reuse? • Resource conservation • Reduce negative environmental impacts Helps save landfill space • Mitigates pollution • Other

3.3 Chi-Square Test

To analyse the difference among the male and female youth population chi- square test is applied. The Chi-square test is a non-parametric tool that is designed to analysethe association between the various student responses. The approach has been used by varied research scholar to identify the differences (Tong & Kang, 2021; Choi et al., 2020; Daryabeigi et al., 2022; Daryabeigi et al., 2022). The chi-square test is a statistical test commonly used in the analysis of categorical data. It is used to determine if there is a significant association between two categorical variables or if there is a significant difference between observed and expected frequencies in a single categorical variable. The test is based on the chi-square distribution, which is a distribution of the sum of the squares of independent standard normal random variables. To perform a chi-square test, the data is first organized into a contingency table, where the rows represent one categorical variable and the columns represent the other. The entries in each cell of the table represent the number of observations that fall into that particular combination of categories. The observed frequencies are compared to the expected frequencies, which are calculated based on the marginal sums of the contingency table and the assumption of independence between the two variables. The test statistic used in the chi-square test is the sum of the squared differences between the observed and expected frequencies, divided by the expected frequencies. This test statistic follows a chi-square distribution with degrees of freedom equal to the number of cells in the contingency table minus the number of parameters estimated in the expected frequencies. The p-value of the chi-square test is the probability of obtaining a test statistic as extreme or more extreme than the one calculated from the data, assuming that the null hypothesis of independence or homogeneity is true. If the p-value is less than a predetermined significance level, typically 0.05, the null hypothesis is rejected and it is concluded that there is a significant association or difference between the variables. It is important to note that the chi-square test has certain assumptions, such as large expected frequencies, independence of observations, and the absence of systematic patterns in the residuals. In addition, the chi-square test is sensitive to sample size, and therefore the conclusions drawn from the test may be affected by small sample sizes. In conclusion, the chi-square test is a useful tool for analyzing categorical data and determining the presence of significant associations or differences between variables. However, it is important to understand its assumptions and limitations, and to carefully interpret the results in the context of the data and research question. Since the gender and reuse of waste material are categorical variable hence, chi-square test is applied. The chi-square independence test is a method that tests whether the two categorical variables are related in a given population (x^2). The gender is the independent variable and reuse of waste material is dependent variable in the present research. The analysis was conducted using Microsoft excel 2013 software.

The formula used to calculate the difference is

$$X_c^2 = \Sigma(O_i - E_i)^2 / E_i$$

Where,

C= Degree of freedom
O =Observed value
E= Expressed Value

4. ANALYSIS AND RESULT

4.1 Characteristics of Studied Students

The study under consideration encompasses a diverse array of students from various institutions, schools, and organizations, including both public and private entities. The participants are drawn from a broad cross-section of the student population and represent a range of genders. The majority of the students involved in the study fall within two key age brackets, namely 15 to 18 and 18 to 24 years old. This diversity in terms of institutional affiliation, gender, and age provides a comprehensive overview of the student population and allows for a well-rounded and in-depth examination of the subject matter. The profile of the students is shown in the Table 3, and the fourth and fifth numbers reflect the percentage of male and female students who reuse all five of the mentioned items—paper, plastic, glass, e-waste, and textiles—or any 4, any 3, any 2, or none at all.

Table 3. Profile of the students

S. No.	Categories	Sub-Categories	Percentage (%)
1	Gender	Male	52.1
		Female	47.9
2	Organization	Government	58
		Private	42
3	Age Group	15-18	24.4
		18-24	75.6
4	% of male	Reuse of articles	
		All	18.1
		Any 4	17.7
		Any 3	8.8
		Any 2	14.1
		Any 1	2.2
		0	1.1
5	% of Female	Reuse of articles	
		All	22.8
		Any 4	22.3
		Any 3	11.1
		Any 2	17.8
		Any 1	2.8
		0	1.1

4.2 Reuse of Material

The students majorly reuses papers for craft and writing purpose. The paper are the most demanded waste material among the youth that they prefer for reuse as the survey conducted (99.8%). The Table 4 states that apart from Paper, Plastic, textile and glass are also mostly reused among the youth population i.e, 85.7% 84.8% and 73.1% simultaneously. Only E waste is very less reused among the youth population. As per the literature also the most commonly resused material is paper and the same is observe among the youth population.

Among the male 54.35% of reuses paper and 45.65% of female reuses paper as per the survey (Table -5). The study depicts very less reuse of e-waste of which female youth population reuses e- waste more than male i.e, 57.54% and 42.86% simultaneously. Among these females, 55% of females reuses e- waste for art craft and organizers whereas 45% of male reuses e- waste for art and craft. The reuse pattern for plastic waste among female youth is higher in comparison to the male youth population. Also the similar reuse pattern is observed for glass waste, the females youth population reuses more of glass waste as compared to the male i.e 54.41% and 45.59% simultaneously.

Majorly students have reused plastic and glass as plastic jars for flowers and glass jar for art and craft. Majorly students have reused plastic and glass as plastic jars for flowers and glass jar for art and craft. Figure 1 states the way of reusing the material by the students. The textile are reused majorly among the

Table 4. The Reuse profile of five waste material among youth

Reuse Material	% of Reuse
Paper	99.8%
Plastic	85.7%
Glass	84.8%
Textile	73.1%
E-Waste	49.5%

Table 5. Waste reuse among male and female youth population

PAPER	Male	54.35%
	Female	45.65%
PLASTIC	Male	44.62%
	Female	55.38%
GLASS	Male	45.59%
	Female	54.41%
E WASTE	Male	42.86%
	Female	57.14%
TEXTILE	Male	40.00%
	Female	60.00%

Figure 1. Reuse of paper, plastic, glass by the youth

a. Reusing One sided Paper as Rough Book	b. Reuse of paper as Bag	c. Reuse of Drinking bottle as flower Vase	d. Reuse of Glass Jar as Art and Craft

youth as mat (38%). The 55% of male youth population reuses textile as mat. The 21% of youth reuses textile for cleaning purpose.

According to the survey conducted, 45% of the youth population learns about waste reuse through online articles and blogs. The survey highlights the importance of digital media in creating awareness and educating the youth about sustainable practices. With the availability of online resources, it has become easier for the youth to learn about the reuse of waste.

The internet has become an essential source of information for the younger generation. A vast amount of information is available online, and it is accessible to everyone. This means that the youth can learn about waste management practices from the comfort of their homes. There are several online resources available, such as blogs, videos, and podcasts that provide information about waste reuse. These resources are not only informative but also engaging and interactive, making it easier for the youth to learn about sustainable practices. The internet has also made it easier to spread awareness about waste management. Social media platforms like Facebook, Instagram, Twitter, and YouTube are popular among the youth. These platforms are used to create awareness about waste management practices. The youth can share information about waste reuse with their friends and followers, creating a ripple effect. By using social media, the youth can spread awareness about sustainable practices and encourage others to adopt them.

The family plays a crucial role in motivating the youth to adopt sustainable practices. The survey conducted revealed that 65.4% of the respondents stated that their family motivated them to reuse waste. This indicates the need to create awareness among families about the importance of waste management practices. The family plays a critical role in shaping the youth's behavior towards waste management. Parents can teach their children about the importance of waste reduction and reuse at an early age. Children are more receptive to new ideas and can quickly grasp the concept of waste management. Parents can also set an example by practicing waste reduction and reuse at home.

Schools are also playing a crucial role in educating the youth about sustainable practices. The survey conducted revealed that 39% of the youth learned about various methods of waste reuse from school. Schools can create a significant impact on the youth's behavior towards waste management. The incorporation of waste management in the school curriculum can make the younger generation more aware of the impact of waste on the environment. Eco-clubs in schools promotes sustainable practices among students. These clubs organizes events and activities that promote waste reduction and reuse. They also conduct awareness campaigns and educate students about the importance of waste management. By involving students in such activities, schools create a sense of responsibility towards the environment.

Apart from schools, various social awareness programs are conducted in developing countries like India. The Swachh Bharat Abhiyan is one such initiative that has influenced the youth majorly. The survey conducted revealed that 53% of the youth learned about the reuse of waste from various social awareness programs conducted at the city level. The government's role in creating awareness among the youth cannot be undermined. The government creates policies that promote sustainable practices and create a framework for waste management. The government also partners with NGOs and other organizations to create awareness about waste management. These organizations conduct awareness campaigns, organize events and activities, and educate the youth about the importance of waste management. By working together, the government and NGOs creates a significant impact on the youth's behavior towards waste management.

According to a recent survey, Instagram is the most popular social media platform used by the youth to learn about the reuse of waste. Around 86.3% of the youth use Instagram handles to gain knowledge about sustainable practices. This indicates that social media has a tremendous reach and influence on the younger generation. Social media provides a platform for individuals, organizations, and businesses to share information and educate the youth about sustainable practices. For instance, environmental organizations uses social media to share information about the impact of climate change and ways to mitigate it. They also use social media to promote eco-friendly products and services.

Social media influencers and celebrities also play a significant role in promoting sustainable practices among the youth. These individuals have a massive following on social media, and their influence can inspire the younger generation to adopt sustainable practices. Moreover, social media provides an opportunity for individuals to share their experiences and ideas about sustainable practices. People use social media to share their success stories and inspire others to adopt sustainable practices. Social media can also be used to create online communities where people can connect and exchange ideas about sustainable living.

Social media has made it easier to disseminate information and create awareness among the youth. Social media platforms are easy to use and provide a wide reach, making it easier to educate the younger generation about sustainable practices. Social media also provides a platform for individuals to engage with each other, share ideas, and create a movement towards sustainable living.

Indeed, the youth population's role in waste management is essential. The adoption of sustainable practices at an early age can create a significant impact on the environment. The survey highlights the various sources from which the youth learn about the reuse of waste. It is essential to create a framework that involves families, schools, and the government to create awareness among the youth. Digital media and social media can be used effectively to create awareness and educate the youth about sustainable practices.

4.3 Result of Chi Square Test

To identify the association between the gender and reuse of waste material chi square is applied as data collected is categorical in nature. The hypothesis considered for the analysis the result is:

$\mathbf{H_0}$ - There is no association between the male and female students and reuse of waste material.
$\mathbf{H_1}$ - There is association between the male and female students and reuse of waste material.

Table 6. Expected value and Chi square test value

	5	4	3	2	1	0
Male	193.35	170.47	129.04	139.83	32.80	9.49
Female	254.64	224.52	169.95	184.16	43.19	12.50
p value	0.05					
df	5					
Chi square value	1					

The significance value (p) assumed for analysis is 0.05%. Table 6 shows the expected value to perform Chi square test. As per the analysis the p value is lesser than 0.05, hence the null hypothesis (H_0) is rejected. This means there is a difference in between the reuse of waste material among the male and female youth population of Bhopal.

5. DISCUSSION

The study highlights the importance of understanding the waste utilization and reuse practices among the youth population, particularly in the context of India where 19.2% of the population belongs to the age group of 15-24 years. The study conducted in Bhopal city revealed that there are variations in the reuse and recycling practices among male and female youth population. The primary survey conducted in institutes and schools captured the ground realities of waste utilization and reuse practices in the youth population.

The study documented the five waste materials that can be recycled, including textile, paper, plastic, glass, and e-waste. The results also show high reuse of waste among female as compared to male youth population. Only for reuse of paper waste the male youth population reuses the waste more than female youth population. The study also reveals that for the motivation on reuse of waste, family play a very important role among the youth population. Also, the digital modes are playing very important role in order to learn the methods of reuse of which majorly students are using instagram in their daily life.

The findings showed that although there is a need for improvement in the waste utilization and reuse practices among the youth population, the efforts are commendable and provide a positive outlook for the future. However, there is a need for concerted efforts from various stakeholders, including the government, schools, and colleges, to raise awareness and educate the youth population on the importance of waste utilization and reuse. This will go a long way in conserving the environment and mitigating the negative impacts of waste.

The results of the study showed that a majority of the students were aware of the importance of waste management and reuse, and they had a positive attitude towards these practices. However, the results also revealed that there was a significant gap between their knowledge and actual practices, and the students needed to be more proactive in their efforts to reduce waste. The study also highlighted the need for more awareness campaigns and educational programs on waste management and reuse, so that the students can be equipped with the necessary knowledge and skills to make a positive impact on the environment.

The study highlights the importance of educating and raising awareness among the youth population on the importance of waste utilization and reuse. This education should not only be limited to the youth population but should also be extended to all age groups and genders. To achieve this, concerted efforts from various stakeholders, including the government, schools, and colleges, are necessary.

6. CONCLUSION

The purpose of the study was to first determine whether today's youngsters reuse their waste and whether they embrace such practices. The study's goal was to determine how wastereusage differs across male and female juvenile population groups. Waste management and reuse is a critical issue for the world's population, and it is particularly important for youth populations. Gender can play a significant role in how individuals approach and participate in waste management and reuse efforts. This research paper examined the relationship between gender and waste utilization and reuse, focusing on the perspectives and needs of youth populations.

The research highlights the importance of gender sensitivity in the implementation of waste utilization and reuse programs. There seemed to be variations in the way each gender reuse their waste. The findings suggest that the approach should be tailored to cater to the specific needs of male and female youth population, as the results showed a variation in the practices of waste utilization and reuse between the two groups. Hence, it is important to understand the gender dynamics of waste utilization and reuse and design programs that take into account these differences.

The role of the government in waste utilization and reuse cannot be overstated. The government has a critical role in implementing policies and regulations that promote waste reduction and encourage the reuse and recycling of waste materials. These policies and regulations can include tax incentives for businesses that adopt environmentally friendly practices, as well as penalties for those that engage in environmentally harmful activities. The government can also provide funding and support for research and development in the area of waste utilization and reuse, as well as for educational programs aimed at raising awareness among the youth population and the general public.

Schools and colleges play a vital role in educating and raising awareness among the youth population. These institutions can incorporate environmental education into their curricula, encouraging students to adopt environmentally friendly practices, such as reducing, reusing, and recycling waste. They can also organize workshops and events to raise awareness among students and the general public on the importance of waste utilization and reuse. In addition to the role of the government and educational institutions, the business sector also has a critical role to play in reducing waste and promoting the reuse and recycling of waste materials. Businesses can adopt environmentally friendly practices, such as reducing packaging and using environmentally friendly materials, as well as investing in research and development in the area of waste utilization and reuse. They can also partner with schools and colleges to organize events and workshops aimed at raising awareness among the youth population and the general public on the importance of waste utilization and reuse.

In conclusion, the study on "Gender and Waste Utilization and Reuse: A case of youth population" provides valuable insights into the waste utilization and reuse practices among the youth population in India, and highlights the importance of incorporating gender sensitivity in the implementation of waste utilization and reuse programs. The findings can be used as a reference for policy makers, educators, and other stakeholders in the development and implementation of waste utilization and reuse programs aimed at conserving the environment and mitigating the negative impacts of waste. The role of all stakeholders, including the government, educational institutions, and the business sector, is crucial in promoting waste utilization and reuse and reducing the negative impacts of waste on the environment. By working together, a sustainable future can be created for generations to come.

REFERENCES

Abuga, D., & Raghava, N. S. (2021). Real-time smart garbage bin mechanism for solid waste management in smart cities. *Sustainable Cities and Society*, *75*, 103347. doi:10.1016/j.scs.2021.103347

Amoah, J. O., Britwum, A. O., Essaw, D. W., & Mensah, J. (2023). Solid waste management and gender dynamics: Evidence from rural Ghana. *Research in Globalization*, *6*, 100111. doi:10.1016/j.resglo.2023.100111

Arya, S., & Kumar, S. (2020). E-waste in India at a glance: Current trends, regulations, challenges and management strategies. *Journal of Cleaner Production*, *271*, 122707. doi:10.1016/j.jclepro.2020.122707

Ayodeji, I. (2010). Exploring secondary school students' understanding and practices of waste management in Ogun State, Nigeria. *International Journal of Environmental and Science Education*, *5*(2), 201–215. https://files.eric.ed.gov/fulltext/EJ884419.pdf

Bick, I. A., Bardhan, R., & Beaubois, T. (2018). Applying fuzzy logic to open data for sustainable development decision-making: A case study of the planned city Amaravati. *Natural Hazards*, *91*(3), 1317–1339. doi:10.100711069-018-3186-2

BMC. (2018). *Swachchta Sarvekshan –2018*. Bhopal Municipal Corporation. https://cdn.cseindia.org/docs/photogallery/slideshows/02_20171212_BHOPAL_SBM_PPT.pdf

Borthakur, A., & Govind, M. (2018). Public understandings of E-waste and its disposal in urban India: From a review towards a conceptual framework. *Journal of Cleaner Production*, *172*, 1053–1066. doi:10.1016/j.jclepro.2017.10.218

Brito, L. B. Q., Agrawal, P., de Mélo, T. J. A., Brito, G. de F., & Morais, C. R. da S. (2023). Recycling of waste glass for the production of hollow blocks using the kiln-casting process. *Cleaner Waste Systems*, *4*, 100079. doi:10.1016/j.clwas.2023.100079

Bui, T. D., Tseng, J. W., Tseng, M. L., & Lim, M. K. (2022). Opportunities and challenges for solid waste reuse and recycling in emerging economies: A hybrid analysis. *Resources, Conservation and Recycling*, *177*, 105968. doi:10.1016/j.resconrec.2021.105968

Bui, T. D., Tseng, J. W., Tseng, M. L., Wu, K. J., & Lim, M. K. (2023). Municipal solid waste management technological barriers: A hierarchical structure approach in Taiwan. *Resources, Conservation and Recycling*, *190*, 106842. Advance online publication. doi:10.1016/j.resconrec.2022.106842

Census of India. (2011). *PCA_AY_2011_Revised (1)*. Government of India.

Choi, E., Bhandari, T. R., & Shrestha, N. (2020). Social inequality, noise pollution, and quality of life of slum dwellers in Pokhara, Nepal. *Archives of Environmental & Occupational Health*, *0*(0), 1–12. doi:10.1080/19338244.2020.1860880 PMID:33355030

Dandge, A., & Valsson, S. (2020). Conservation of Material, Technology and Practice in Heritage Structure and its Relevance in Today's Context. Encyclopedia of Renewable and Sustainable Materials, 39–49. doi:10.1016/B978-0-12-803581-8.10723-4

Halder, M., Rahman, T., Mahmud, A., Jim, S. A., Akbor, M. A., Siddique, M. A. B., & Joardar, J. C. (2022). Are the vegetables grown in the soil of municipal solid waste dumping sites safe for human health? An assessment from trace elements contamination and associated health risks. *Environmental Nanotechnology, Monitoring & Management*, *18*, 100731. doi:10.1016/j.enmm.2022.100731

Islam, M. T., Huda, N., Baumber, A., Shumon, R., Zaman, A., Ali, F., Hossain, R., & Sahajwalla, V. (2021). A global review of consumer behavior towards e-waste and implications for the circular economy. *Journal of Cleaner Production*, *316*, 128297. doi:10.1016/j.jclepro.2021.128297

Kolhe, N. P., & Dhote, K. K. (2019). Spatial assessment of rurban slums: A new perspective of urban villages. *International Journal of Recent Technology and Engineering*, *8*(3), 8688–8693. doi:10.35940/ijrte.A9215.098319

Marampoutis, I., Vinot, M., & Trilling, L. (2022). Multi-objective vehicle routing problem with flexible scheduling for the collection of refillable glass bottles: A case study. *EURO Journal on Decision Processes*, *10*, 100011. doi:10.1016/j.ejdp.2021.100011

Pandey, R. U., Surjan, A., & Kapshe, M. (2018). Exploring linkages between sustainable consumption and prevailing green practices in reuse and recycling of household waste: Case of Bhopal city in India. *Journal of Cleaner Production*, *173*, 49–59. doi:10.1016/j.jclepro.2017.03.227

Phan Hoang, T. T., & Kato, T. (2016). Measuring the effect of environmental education for sustainable development at elementary schools: A case study in Da Nang city, Vietnam. *Sustainable Environment Research*, *26*(6), 274–286. doi:10.1016/j.serj.2016.08.005

Plano Clark, V. L., Ivankova, N. V., & Yang, N. (2023). *Frameworks for conceptualizing mixed methods research* (4th ed.). International Encyclopedia of Education. doi:10.1016/B978-0-12-818630-5.11038-3

Prajapati, K. K., Yadav, M., Singh, R. M., Parikh, P., Pareek, N., & Vivekanand, V. (2021). An overview of municipal solid waste management in Jaipur city, India - Current status, challenges and recommendations. *Renewable & Sustainable Energy Reviews*, *152*(August), 111703. doi:10.1016/j.rser.2021.111703

Richardson, E. A., & Mitchell, R. (2010). Gender differences in relationships between urban green space and health in the United Kingdom. *Social Science & Medicine*, *71*(3), 568–575. doi:10.1016/j.socscimed.2010.04.015 PMID:20621750

Robles, A., Roxana, S., & Hinostroza, M. L. (2015). *Guidebook for the development of Nationally Appropriate Mitigation Actions (NAMAs)*. CORE.

Sengupta, D., Ilankoon, I. M. S. K., Dean Kang, K., & Nan Chong, M. (2022). Circular economy and household e-waste management in India: Integration of formal and informal sectors. *Minerals Engineering*, *184*(May), 107661. doi:10.1016/j.mineng.2022.107661

Sharma, B. K., & Chandel, M. K. (2021). Life cycle cost analysis of municipal solid waste management scenarios for Mumbai, India. *Waste Management (New York, N.Y.)*, *124*, 293–302. doi:10.1016/j.wasman.2021.02.002 PMID:33640669

Sharma, H. B., Vanapalli, K. R., Samal, B., Cheela, V. R. S., Dubey, B. K., & Bhattacharya, J. (2021). Circular economy approach in solid waste management system to achieve UN-SDGs: Solutions for post-COVID recovery. *The Science of the Total Environment*, *800*, 149605. doi:10.1016/j.scitotenv.2021.149605 PMID:34426367

Singh, S., Chakma, S., Alawa, B., Kalyanasundaram, M., & Diwan, V. (2023). Hazardous Materials Advances Identification, characterization, and implications of microplastics in soil – A case study of Bhopal, central India. *Journal of Hazardous Materials Advances*, *9*, 100225. doi:10.1016/j.hazadv.2022.100225

Smit, S., Musango, J. K., Kovacic, Z., & Brent, A. C. (2017). Conceptualising slum in an urban African context. *Cities (London, England)*, *62*, 107–119. doi:10.1016/j.cities.2016.12.018

Tan, Q., Yang, L., Wei, F., Chen, Y., & Li, J. (2023). Is reusable packaging an environmentally friendly alternative to the single-use plastic bag? A case study of express delivery packaging in China. *Resources, Conservation and Recycling*, *190*, 106863. doi:10.1016/j.resconrec.2022.106863

Tomić, T., & Schneider, D. R. (2020). Circular economy in waste management – Socio-economic effect of changes in waste management system structure. *Journal of Environmental Management*, *267*(April), 110564. doi:10.1016/j.jenvman.2020.110564 PMID:32421664

Tong, H., & Kang, J. (2021). Characteristics of noise complaints and the associations with urban morphology: A comparison across densities. *Environmental Research*, *197*(March), 111045. doi:10.1016/j.envres.2021.111045 PMID:33745930

Vaidya, H., & Chatterji, T. (2020). *SDG 11 Sustainable Cities and Communities*, 173–185. Springer. doi:10.1007/978-981-32-9927-6_12

Vijayan, R. V., Krishnan, M. M., Parayitam, S., Anantharaman Duraisami, S. P., & Saravanaselvan, N. R. (2023). Exploring e-waste recycling behaviour intention among the households: Evidence from India. *Cleaner Materials*, *7*, 100174. doi:10.1016/j.clema.2023.100174

Weber, S., Weber, O., Habib, K., & Dias, G. M. (2023). Textile waste in Ontario, Canada: Opportunities for reuse and recycling. *Resources, Conservation and Recycling*, *190*, 106835. doi:10.1016/j.resconrec.2022.106835

Woolridge, A. C., Ward, G. D., Phillips, P. S., Collins, M., & Gandy, S. (2006). Life cycle assessment for reuse/recycling of donated waste textiles compared to use of virgin material: An UK energy saving perspective. *Resources, Conservation and Recycling*, *46*(1), 94–103. doi:10.1016/j.resconrec.2005.06.006

Xavier, L. H., Ottoni, M., & Abreu, L. P. P. (2023). A comprehensive review of urban mining and the value recovery from e-waste materials. *Resources, Conservation and Recycling*, *190*, 106840. doi:10.1016/j.resconrec.2022.106840

Yaacob, M., Ibrahim, M. H., & Nasir, Z. M. (2019). Solid Waste Management in Malaysia: The Perspectives of Non-governmental Organizations (NGOs) and Youths on Consumers'. *Commitment*, 27–41. doi:10.1007/978-981-13-9173-6_3

Zhao, Y., Jia, G., Shang, Y., Zhao, P., Cui, X., & Guo, Q. (2022). Chlorine migration during hydrothermal carbonization of recycled paper wastes and fuel performance of hydrochar. *Process Safety and Environmental Protection*, *158*, 495–502. doi:10.1016/j.psep.2021.12.041

Chapter 19
E-Waste Management: A Transformation Towards a Circular Economy

H. R. Swapna
https://orcid.org/0000-0003-0362-409X
Jain University (Deemed), India

Geetika Madaan
https://orcid.org/0000-0001-8141-9935
Chandigarh University, India

Sanjeet Singh
Chandigarh University, India

Manoj B. Mandake
https://orcid.org/0000-0002-3002-0470
Department of Chemical Engineering, Bharati Vidyapeeth College of Engineering, India

Santosh Walke
https://orcid.org/0000-0001-8633-5757
College of Engineering, National University of Science and Technology, Muscat, Oman

Digvijay Pandey
https://orcid.org/0000-0003-0353-174X
Department of Technical Education, IET, Dr. A.P.J. Abdul Kalam Technical University, Lucknow, India

ABSTRACT

The global accumulation of electronic waste is a growing concern in terms of the health and safety of our planet. Electronic waste has a greater potential to create environmental contamination and human health problems due to the way it is often disposed of, making it a worldwide problem. 'Sustainable consumption and production' may help nations solve the problem of electronic waste management by forcing them to make fundamental changes in the way their communities manufacture and use commodities. Governments, NGOs, businesses, and citizens all work together to change unsustainable consumption and production habits. The effects of e-waste on human health and the natural environment will be discussed, as will various situations involving this kind of garbage. Developed and developing nations' (namely India's) approaches to managing discarded electrical and electronic equipment will be examined, as will appropriate pathways concerning recyclable components using extended producer responsibility that led to the creation of a circular economy.

DOI: 10.4018/978-1-6684-8117-2.ch019

INTRODUCTION

By 2045, it is projected that "Waste electrical and electronic equipment (WEEE)" or e-waste would have doubled as the fastest-growing waste source worldwide(Mor, Sangwan, Singh, Singh, & Kharub, 2021). Due to the ever-increasing demand for electronic goods, which has imposed significant costs on both developed and emerging countries throughout the globe, this issue is rapidly expanding into a worldwide crisis(Innocent C. Nnorom & Odeyingbo, 2019).Knowing the exponential growth rate of electrical and electronic equipment, which is three times quicker than that of other solid wastes, is a given(Das, 2013). E-waste consists of precious resources and dangerous components;thus, it must be handled with care as e-products have become more sophisticated and integrated into daily life (Iyyanar et al., 2023). But the garbage collection and management systems have not kept up yet, so there is a danger of resource loss and adverse effects on the environment and human health from improperly collecting and managing e-waste(Ahirwar & Tripathi, 2021; Halim & Suharyanti, 2019).

Many, often contentious, definitions of "e-waste" have been proposed. As stated in a UNEP study, "E-waste is a wide word that comprises a spectrum of electrical and electronic equipments which may be rendered as end-of-life electronic gadgets and do not entail value to their owners(Krishna & Saha, 2022)." Waste electrical and electronic equipment are sometimes referred to as WEEE. According to European Directive 2002/96/EC, "all trash electrical and electronic equipment, including with all its components, subassemblies, and consumables which constitute part of the product at the time of discarding." Any electric equipment that has reached the end of its useful life is considered "e-waste" by the OECD(M. V. Kaushik, 2018).

When it comes to its contents, WEEE is anything from uniform or simple; it includes both hazardous and non-hazardous materials and spans a wide range of classifications(Bhutta, Omar, & Yang, 2011). Electrical and electronic equipment include a variety of valuable materials that may be extracted via the process of e-waste recycling. As a result, WEEE recycling is a hot topic in the waste management and resource recovery communities (Pandey et al. (2023). To preserve natural resources and build a long-term solution for e-waste management, it is necessary to first identify and quantify valuable elements and dangerous compounds in order to understand the material features of this waste(Wath, Dutt, & Chakrabarti, 2011).

Electrical and electronic goods manufacturers often release new models on a regular basis. In addition, these items are made in a manner that limits their usefulness over time. Printed circuit boards are an integral part of almost all modern electrical and electronic devices (PCBs)(Bhatia & Baruah, 2020).

The concentrations of copper and gold in PCBs used in standard desktop computers are far higher than those found in either copper or gold ore, at around 20% and 250 g/t, respectively(Saravanamurthy, E, Sarangan, & Bharathkumar, 2018). Solutions based on physical, pyrometallurgical, bio-metallurgical, and hydrometallurgical techniques are available for recovering these metals from electrical and electronic wastes(Saravanamurthy et al., 2018).

Results have been usually dismal, despite a very well-established waste management infrastructure and significant collaborative efforts. This may be connected to the fact that not enough research has been put into developing evidence-based design tools, strategies, and technologies for recovering lost items(Constantinescu et al., 2022). Circular economy is seen as an innovative way to deal with the e-waste problem, which may help fix the aforementioned problems(Patel, 2021). With the help of circular economy incentives, manufacturers may restructure their operations and redesign their technologies to save money; however no anticipated incentives have been sought for the deployment of an EPR-based e-waste management system (Pandey et al. (2021), thus both individual operators and resource recovery have been stymied to a bare minimum by inadequate collection and subsequent material recycling operations. In addition, there is little hope that any of the materials that are thrown away may be recycled for use in later stages of manufacturing(Gaidajis, Angelakoglou, & Aktsoglou, 2010).

The goal of the circular economy is to reduce the rate at which electrical and electronic equipment (EEE) is consumed by the system, to keep waste materials within the system for as long as possible, and to reduce or eliminate the need for re-generation of these materials through more efficient product design and business practises. It's a part of the "7 R's" methodology, which entails"reduce, reuse, refuse, recycle, recovery, rethinking, and redesigning(Monika & Kishore, 2010). The COVID period saw the production of a lot of medical waste. (Pandey et al. (2021,2023) .

E-Waste Management: A Transformation Towards a Circular Economy

A Transition Towards a Circular Economy for E-Waste Management emphasises these three facets of the problem (in the case of India, Indian E-waste management rule, 2016)(Andeobu, Wibowo, & Grandhi, 2021). However, there are a number of issues that might slow down the process, including a lack of funding, a lack of trained workers in the informal sector, insufficient technology, and an inefficient collecting method. By optimising goods and cycling resources, the circular economy keeps them at their optimum usefulness and worthwhile working toward cleaner and renewable technology, revolutionary models, and developmental policies to "design out" waste(P. R. Kaushik & Herat, 2020). The improved goods that result from this optimization have a longer useful life, and there is a chance that the product

Figure 1. E-waste management

may be reused with resourceful material recovery because of the innovative waste management models that are being developed today. In Fig. 1, we see the overarching notion of a circular economy applied to the management of e-waste.

Researchers are still trying to learn more about the social and behavioural factors associated with the use of electronic products and how they might be adapted to help regular people support this circular economy(Kumar & Karishma, 2016). It is important to prioritise the sustainable manufacture of these items above the technicalities and recycling rates involved in the e-waste circular economic process. Circular economy is seen as crucial for the development and administration of e-products; however, it's going to need a mandatory effort from all parties involved (governments, enterprises, and consumers). Therefore, it is crucial to specify both the technical and cum economic side of the consumer's behaviour if the circular economy is to be sustained over the long run. Therefore, it is necessary to seek out promising avenues for developmental interventions to enhance e-waste management and bring about an enticing circular economy(Kahhat et al., 2008).

E-Waste Scenario at Global Level

The global E-waste Monitor 2020 reports that annual growth in e-waste output is between 3% and 4%. Production of electronic trash worldwide was around 44.7 million tonnes in 2016, and is projected to rise to 52.2 million tonnes by 2021(Kiddee, Naidu, & Wong, 2013). Only 17.4 percent of the 53.6 Mt of electronic garbage produced globally in 2019 was actually collected and recycled. Most of the 53.6 Mt of electronic trash comes from devices that use batteries or plugs (such as laptops and mobile phones). About 74 metric tonnes (Mt) of used batteries and discarded plugs are expected to accumulate worldwide every year from now until the year 2030. As a result, it seems that e-waste is the fastest-growing waste stream worldwide, mostly due to increased use and decreased durability. "Asia reportedly generated the largest volume of e-waste (2019) – around 24.9 Mt followed by the United States (13.1 Mt), Europe (12 Mt), Africa (2.9 Mt), and Oceania (0.7 Mt). Asia being the greatest producer of WEEE has a collection rate of 4.2 kg/inh where only 6% is properly collected and recycled(Lu et al., 2015; Mundada, Kumar, & Shekdar, 2004; I. C. Nnorom & Osibanjo, 2008)."

Due to inadequate facilities and insufficient rules and institutions, recyclable e-waste in India only accounts for 5% of the country's total waste. Because of this, those working in the recycling business face a paucity of resources and an acceleration of environmental damage(Lu et al., 2015). Sixty percent of India's e-waste comes from only 65 cities, with the majority coming from just three states: Maharashtra, Tamil Nadu, and Andhra Pradesh. Toxic additives and compounds like mercury (which is harmful to the human brain) and chlorofluorocarbons (CFCs) found in discarded electronics make them a health and environmental risk that must be addressed responsibly. Improving the rate at which garbage is recovered and treated is a primary goal of any waste management strategy. In an effort to legislate or regulate the recycling of electronic trash, several nations have adopted a national e-waste policy. However, progress and implementation of regulations are sluggish in many locations, and inadequate collection and management strategies persist in many of these areas(Mundada et al., 2004). An improved quality of life is observable in emerging countries as a result of stronger economic development, which in turn aids in reducing the poverty rate. Nonetheless, when these countries develop policies to deal with e-waste management difficulties as a critical environmental-health concern, it is not farsighted to focus on creating abundant possibilities for the locals via product recovery(I. C. Nnorom & Osibanjo, 2008).

As evaluated by the OECD, it is evident that impoverished nations produce much more electrical and electronic waste (WEEE) (Organization for Economic Co-operation and Development)(Bimir, 2020). It is estimated that by 2030, 550 million metric tonnes of WEEE would have been dumped worldwide from both developing and affluent countries. In order to get rid of their electronic waste, rich countries send it to poorer countries(Shumon, Ahmed, & Islam, 2014). The majority of this e-waste ends up in developing countries like China, India, and Africa, where it presents serious threats to the local ecosystems. An increase in the worldwide circulation of e-waste may be traced back to the lack of import taxes and the presence of unlicensed recyclers who use a broad array of inefficient ways for getting rid of the material. The Basel Convention on the Control of Trans-boundary Movements of Hazardous Wastes and their Disposal was ratified after public outrage over hazardous waste dumps carried to Africa and other developing areas(Nagi & Kurien, 2019). Its purpose is to reduce the potential harm that hazardous waste poses to people and the environment. WEEE in non-waste forms has been transferred despite the Basel Convention's goal of preventing such transfers. As an additional measure, European nations drafted the WEEE (Waste Electrical and Electronic Equipment) Directive and the REACH (Restriction of Hazardous Substances) Directive(Patil & Sharma, 2013).

E-Waste Control and Execution by Developed and Developing Nations

Collection, handling, processing via recycling or recovery, and disposal are the most important acceptable components in the worldwide framework for an efficient e-waste management system. In contrast, informal recycling and weak or non-existent e-waste regulation are cited as important challenges that pose significant dangers to the correct management of vast volumes of this e-waste in developing nations(Sivaramanan, 2013). The importance of the informal sector in low- and middle-income nations is likewise not to be underestimated(Wath, Vaidya, Dutt, & Chakrabarti, 2010).

Despite their lower rates of technical improvement, developing nations use WEEE at a quicker pace than industrialised countries. This results in twice as much e-waste being produced in poor countries. By 2030, it is predicted that 400–700 million dead computers would have been thrown away in underdeveloped nations and 200–300 million in rich ones(Mor et al., 2021). When compared to industrialised nations, the difficulty of e-waste disposal is further amplified by the fact that poor nations have less advanced technology and less effective norms and regulations(Innocent C. Nnorom & Odeyingbo, 2019). It has been argued by a number of scholars that wealthy countries give greater thought to the most important aspects of e-waste management, while developing countries have significant challenges in this area, particularly in the areas of policy development and institutional framework(Kahhat et al., 2008).

In any discussion of the relative merits of the legislative policies of industrialised and developing nations, the former are always seen as superior, since they are more strictly enforced. Nonetheless, several poor nations have taken the initiative to establish and implement unique rules for the management of electronic waste. Developed nations have established national register systems that include both robust logistical and data gathering infrastructures(Andeobu et al., 2021). For example, in Germany, the government has established programmes like the Packaging Trash Program and the Extended Producer Responsibility (EPR) programme, which impose financial requirements on manufacturers to collect and decrease packaging waste. The latter option was taken up and expanded to include nations like Sweden, Switzerland, Norway, and Taiwan. In 2003, the European Union passed a law encouraging the collection and recycling of e-wastes and limiting the use of hazardous materials in such waste(Kumar & Karishma, 2016). To keep track of the manufactured EEE and to provide a future manufacturer take-back mechanism, countries like China and India (without a national registry) have evidently adopted the EPR legislative system. In addition, the widespread availability of used EEE via underground marketplaces has contributed significantly to the current threat in underdeveloped nations. While China, India, and Pakistan have relatively cheap labour costs overall, managing their enormous amounts of e-waste (both import and local) and the dominance of the informal parties has proven to be a significant challenge(Wath et al., 2011).

The recycling of electronic garbage, the success of which depends heavily on consumer behaviour and the regional market, is already regulated on a global scale. However, a small number of studies undertaken by academics suggest that economically developing nations are the primary sources of e-waste(Bhatia & Baruah, 2020). The recycling and processing sector in Canada is mature and governed by strict regulations. Each of Europe's 27 member states has its own rules in place that are designed to encourage people to recycle WEEE. While the United States as a whole lacks a comprehensive plan for recycling WEEE, at least half of the country's states have enacted a wide variety of regulations governing the responsible disposal of electronic waste. Around 85 percent of excess electronic goods are reclaimed rather than wasted in Latin America, where Colombia, Mexico, Costa Rica, Peru, and Brazil all have e-waste legislation, while Africa lacks authorised mechanisms for e-waste recycling(Saravanamurthy et al., 2018). Although it is one of the leading producers of electronic garbage in Africa, Egypt does not have an effective system for managing this kind of waste(Constantinescu et al., 2022). Australia's National Television and Computer Recycling Scheme combined government mandates with industry participation to facilitate the collection and reuse of electronic trash, whereas only South Korea, Japan, China, Taiwan, and India in Asia had formal e-waste regulations. Taiwan recycles the most electronic trash, with a rate of around 82%, followed by Japan and South Korea, each of which recycles about 75%. Despite having a law in place for handling WEEE, China is still expanding its recycling infrastructure. Additionally to being the most common place for electronic trash to be dumped, it also ranks first in both electronic production and consumption. Many other nations' standards for handling e-waste are still in t(Shumon et al., 2014)he early stages of development(Wath et al., 2010).

Methods for the correct collection, preprocessing, and recycling of electronic trash are described by Balde et al. (2015). Although his research shows that established countries are more inventive in this respect than developing ones, there is still plenty of room for both to grow(Nagi & Kurien, 2019). Even now, for a variety of reasons, industrialised nations ship their discarded electronics to poorer ones. To name a few: high cost of labour and strict regulations on the disposal of hazardous waste in industrialised nations. In order to meet immediate needs, developing nations purchase e-waste from wealthy nations despite a lack of infrastructure, expertise, and resources necessary for effective recycling(Bimir, 2020). These recyclers use primitive techniques to salvage valuable components from discarded electronics. However, several nations have recently been working towards developing a more effective strategy for the management and recycling of electronic trash, and one such nation is Tunisia(I. C. Nnorom & Osibanjo, 2008). Large amounts of e-waste are sent outside of Europe, mostly to Asia and Africa, under the guise of being sent for reuse and rehabilitation. However, in many regions, even broken equipment are illegally branded as "used goods," causing severe damage to the local environment and population(Mundada et al., 2004).

Electronic Waste Management Approach Adopted by India

India's yearly increase in e-waste production to 2 million tonnes places it sixth worldwide. However, because the OECD nations are responsible for between 50 and 60 percent of this overall e-waste amount, measuring it precisely is a daunting challenge(Lu et al., 2015). It was predicted that by 2020, India's e-waste volume will rise to 5.2 million tonnes(Kiddee et al., 2013). In industrialised, middle-income metropolitan societies, rapid economic expansion and increased buying power are linked to a meteoric increase in the amount of electronic garbage produced. Poor asset recovery from obsolete items, poor integrated supply chain planning and design, and an unsuitable legal and policy climate are at the root of this problem. Thus, environmental and economic considerations must be included in throughout the planning and execution of e-waste treatment strategies(Kahhat et al., 2008). E-waste collection, green product design, multi-agency cooperation, proper disposal/recycling, and citizen involvement are just few of the procedures that make up a state-of-the-art system for managing e-waste. Formulating an integrated waste management system with long-term sustainability and environmental protection in mind is urgently required(Kumar & Karishma, 2016).

Mathematical modelling was used by Pathak et al. (2017) to make this forecast, and they determined that computers and mobile phones will be the two most marketable electronic goods in the Indian market(P. R. Kaushik & Herat, 2020). It is expected that computer trash will continue to make a significant contribution to e-waste until 2022, when it will begin to level off. However, it seems unlikely that the amount of electronic trash produced by mobile phones will ever reach a plateau. These problems with e-waste production and its purported antipathy to sustainability in India were addressed by developing a regulatory framework that took into account historical shifts(Andeobu et al., 2021). E-waste management and handling rules (schedule-I) were added to the Environment (Protection) Act 1986 in 2010, clarifying the responsibilities of producers (in the production, distribution, retail sale, and processing of EEE), as well as collection facilities, recyclers, re-furbishers, traders, dismantlers, auctioneers, and large-scale consumers. In spite of several setbacks (including widespread opposition and numerous suggestions for improvement), the regulation was eventually proposed in 2011(Monika & Kishore, 2010).

The legislation in India mandates the use of a system of collection centres, the distributor take-back system (DTBS), and authorised and licenced dismantlers or recyclers (PRO). The letter continues by stating that the potentially harmful substances may only be used in electronic devices that are powered by electricity or batteries. However, this was regarded insufficient to address the existing e-waste management issues, prompting calls for a revision and improvement of the legislation regulating e-waste management to include strategies for effective EPR management that mitigate adverse environmental consequences(Constantinescu et al., 2022; Gaidajis et al., 2010; Patel, 2021). After almost three decades of evolution into specialised e-waste laws, the law was revised in 2016 as E-waste Management Rule 2016, which includes the requirement of both makers and re-furbishers. Several local social organisations, recycling centres, and collection points have emerged as a direct consequence of the maturation of e-waste laws. The quantity of e-waste has dropped, however somewhat, and more valuable materials are being extracted from it. Somewhat slowly but hopefully, the informal sector is processing 90% of e-waste, with these recyclers implementing the benefits of recycling in an organised and scientific method(Monika & Kishore, 2010).

Ways to Manage E-Waste With a Strategic Mindset

Strategically, the circular economy (CE) is appropriate for environmental stress reduction and economic equilibrium because of the persistent focus on recycling. In light of the exponential growth of electronic goods and the shrinking profit margins associated with recycling their end-of-life (EoL) versions, the idea of CE was developed and adopted(Andeobu et al., 2021). In the context of a growing economic powerhouse like India, the recycling of EoL goods would help promote long-term economic growth. This entails making ongoing efforts and starting projects to deal with problems including pollution, trash, e-waste management, environmental sustainability, and protection. Environmental management systems (EMSs) are often regarded as the single most important instrument for shaping the landscape of other enablers already in place. It places a premium on discovering eco-friendly goods, drafting tougher laws, creating a green image, and assisting manufacturers in implementing CE procedures correctly. As a result, stakeholders and policymakers may lessen their impact on the environment and work to create an efficient e-waste management system based on the previously identified key enablers(Monika & Kishore, 2010).

Negative Effects of Electronic Waste on Humans and the Environment

Toxic and dangerous elements (lead, mercury, calcium, polybrominated biphenyls, and chromium) are continuously being added to our environment because to the increased rate of urban expansion, making e-waste the most pressing problem facing the contemporary world today. Concerns about e-waste centre on the fact that it may include a wide variety of toxic chemicals that pose serious risks to human health and the environment. If these wastes are not dealt with properly, harmful compounds may seep out and cause harm. Minerals like iron, copper, etc., are said to come mostly from this area, but it is also the source of many environmental dangers(Patel, 2021).

Electronic waste, also known as e-waste, is a composite material made up of several materials that underwent conformational changes during production, including metals (heavy and dangerous) including Cu, Co, Fe, Ni, Cd, Pb, Cr, Au, Ag, and Pd; plastics (polymers and additives); and ceramics. The absence of necessary services, legislations, and approaches towards e-waste management (particularly developing countries) raises serious problems, which in turn leads to mishandling and malpractices such open burning and dumping of garbage. The resulting environmental pollution has negative effects on human health and the balance of the ecosystem's plant and animal life. There is little doubt that the recycling or landfilling of these dangerous and poisonous e-waste contributes to air, soil, and water pollution. "The major contributors are the hazardous organic compounds like PCBs (polychlorinated biphenyls), PBDEs (polybrominated diphenyl ethers), and PCDD/Fs (polychlorinated dibenzo-p-dioxins and dibenzofurans) found to increase air and water contamination and affecting human health by disrupting food chain(Constantinescu et al., 2022)."

Countries like India, Ghana, Nigeria, China, Thailand, Indonesia, Vietnam, Pakistan, and Bangladesh are already struggling to cope with the effects of informal recycling. The customary activities, including the uncontrolled and unlawful melting of circuit boards, the open burning of wire heaps, and the discarding of metals containing acidic solutions, have been plain to see. Another ineffective method for getting rid of electronic garbage is dismantling, which may cause serious environmental and health problems if not done properly(Bhatia & Baruah, 2020). Open burning and dumping are the worst options for getting rid of garbage, yet they are still used by certain nations that don't have effective waste management laws and infrastructure.

Most often, this entails taking apart the electronic debris by hand using simple instruments like hammers, chisels, and screwdrivers. Gold and other precious metals are often mined through strip mining, in which the ores are dissolved in acid in open pits. Toxins such dioxins and furans are released during the process of recovering copper from burned electric wires in open pits, and during the process of shredding and liquefying plastics at high temperatures without enough ventilation(Constantinescu et al., 2022).

Undermining worker safety and the quality of the environment, these methods are very primitive. Even in affluent nations, unlawful dumping occurs because authorities fail to properly enforce environmental laws and WEEE management requirements.

Printed circuit boards, included in all electronic devices, are a major source of exposure to hazardous substances including lead, brominated flame retardants, and antimony oxide(P. R. Kaushik & Herat, 2020). For the purpose of removing components off printed circuit boards (PCBs), they are often heated. "Cathode ray tubes (CRTs), capacitors, mercury switches, batteries, printed circuit boards, liquid crystal displays (LCDs), cartridges, and electrolytes" are all examples of what may be found in electronic trash, and they all include dangerous substances including lead, mercury, and hexavalent chromium. Polychlorinated biphenyls (PCBs) may be found in transformers and capacitors, while lead and cadmium can be found in computer batteries and circuit boards(Mundada et al., 2004). Mercury can be found in switches and flat-screen displays. Growth, reproductive, thyroid, respiratory, and cellular dysfunction problems are only some of the ailments caused by these noxious chemicals. Most vulnerable are pregnant women, newborns, and young children who are exposed to these contaminants firsthand.

The most prevalent toxins identified in discarded electronics are "nickel (Ni), barium (Ba), beryllium (Be), cadmium (Cd), cobalt (Co), chromium (Cr), lead (Pb), lithium (Li), lanthanum (La), mercury (Hg) manganese (Mn), molybdenum (Mo), hexavalent chromium (Cr (VI)), and persistent organic pollutants (POPs) such as brominated flame retardants (BFRs)."Nickel (Ni) and arsenic (As) are both toxic metals that may lead to cancer and other health issues. Besides causing hair loss and infertility, antimony (Sb) harms the lungs and the heart(Constantinescu et al., 2022). Polybrominated diphenyl ethers (PBDE) are toxic chemicals that harm the thyroid, the liver, and the blood. Metals like cadmium and mercury are lung-damaging carcinogens. Tetra bromo bisphenol A (TBBPA) is a chemical that has been linked to cancer, mutations, and endocrine system dysfunction. In addition to having an effect on the thyroid, polybrominated biphenyls (PBB) are also known to have an effect on the kidneys. Paralysis is the result of barium (Ba) toxicity, which weakens muscles and upsets the stomach. A number of respiratory illnesses, pneumonia, and even lung cancer have been linked to exposure to beryllium (Be). Children who grow up in e-waste recycling zones are routinely exposed to very high concentrations of dangerous compounds, raising serious concerns about the potential for developmental delays and disruptions. These youngsters often have smaller frames than adults. Their physical weight may be little, but the amount of hazardous substances they inhale is enormous. There are negative effects on organ systems from these toxic chemicals as well(I. C. Nnorom & Osibanjo, 2008).

Resolving Difficulties Associated With E-Waste Recycling

Both the criteria and the nations used to determine which countries are developed and which are developing are broad categories with significant variation. Each country is either included in Annex VII of the Basel Convention (developed nations) or excluded from it (developing nations). When it comes to limiting the international transport of hazardous waste, it ranks among the most important international trade covenants (Basel Convention, 2016). However, WTO does not give a definition for its categorization of nations as developed or developing. The OECD, the European Union (EU), and Liechtenstein are classified as developed nations under the Basel Convention (Basel Convention, 2017), whereas other territories are classified as developing countries(Nagi & Kurien, 2019). The WTO places "the EU, North America (excluding Mexico), the European Free Trade Association (Iceland, Liechtenstein, Norway, and Switzerland), Japan, New Zealand, and Australia under developed countries while Africa,

Mexico, South and Central America, Caribbean, Europe (barring the EU and EFTA), the Middle East, and Asia (apart from Australia, Japan, and New Zealand) under developing countries."When nations are categorised in so many different ways, it becomes very difficult to sell electronic garbage across borders(Patil & Sharma, 2013).

Most industrialised economies act as e-waste exporters, whereas poor nations import these materials. The Basel agreement states that wealthy nations trade the vast majority of their hazardous e-waste with one another, whereas emerging economies trade just a small percentage. WTO, on the other hand, asserts that most affluent nations ship their e-waste to poorer nations. The developing countries of India, China, and Africa are major destinations for international WEEE dumping. The rapid pace of industrialisation, the increase in disposable income among the middle class, and the proliferation of the black-market recycling industry have all contributed to this issue in India. It's no secret that China is a major destination for international e-waste shipments. However, China has strictly prohibited the import of e-waste ever since the year 2000, and the country has also confirmed that it has not imported any waste printed boards (WPBs) since that time. However, several nations have reported exporting WPBs to China around the same period.

There is still resistance to the suggested recycling and disposal methods in emerging countries because to technical, infrastructural, and economical limitations. Bo2W theory, which considers the best of both worlds, seems like a reasonable choice to provide a workable answer to this dilemma. It is important to tailor the location of collection and screening sites to the specific needs of the community. Eco-friendly production designs, extended producer responsibility, product stewardship, recycling, and remanufacturing are all examples of sustainable solutions that should be implemented(Sivaramanan, 2013). The danger of exposure to hazardous compounds discharged owing to improper treatment of e-waste has risen as both the worldwide demand for WEEE manufacturing and its diverse composition have grown. Therefore, sustainable management practises must be implemented. In order to lessen the damage that waste electrical and electronic equipment (WEEE) has on people's health and the environment, several countries have adopted a policy known as extended producer responsibility (EPR). However, this approach is still lacking in terms of growing it into a circular economy, therefore there is room for development. The obstacles must be recognised and overcome if this goal is to be attained(Wath et al., 2010).

Features of Waste Electrical and Electronic Equipment

Since the content of WEEE is always changing as a result of technological development, careful research is necessary before effective management techniques can be formulated. Data on the concentration of materials in certain pieces of machinery is scarce. There is a dearth of data about the individual parts. Furthermore, there is a deficiency of generally accepted methodologies for determining the elemental content of WEEE. This necessitates the use of several procedures, which affects the final outcomes of the analytical conclusion(Mor et al., 2021). To get over this problem, a centralised database including information on WEEE components and uniform analysis techniques should be developed. We need strict rules and the right infrastructure to build a circular economy and shut the loop. Adaptability to various geographical settings is a key feature(Innocent C. Nnorom & Odeyingbo, 2019). Despite having specific law and using EPR theory in reality, high-income nations like the United Governments, Canada, India, and European states may benefit from higher collection rates. Introducing a system of competitive collection between the public and private sectors is one approach. As the amount of WEEE produced rises, the situation becomes even more dire in countries without enough regulations and infrastructure to deal with it. Recognizing effective policies that must be established taking into account the unique regional circumstances of each nation is a problem for those areas(Rajpurohit & Khan, 2018).

Due to the absence of adequate regulation, informal WEEE management industries have emerged, with their primary activity being the unlawful trafficking of both imported and local e-waste. The issue may be addressed by restricting the import of WEEE and increasing the pace of official collection. The foundation of circular economy is enhanced recycling techniques, thus it's important to take use of them(M. V. Kaushik, 2018). Current recycling procedures need pretreatment, mechanical separation, and refining to produce the desired materials, highlighting the need for innovation and improvisation in the business. However, not all processing plants have the expertise necessary to refine crude oil effectively. For lack of substantial returns, the recycling business remains undeveloped. Expected recycling procedures are ineffective for WEEE because of its complex composition, prompting the need for innovative recycling approaches. Focusing on maximising resource extraction from WEEE is central to the circular economy's viewpoint, which in turn helps to lessen environmental impact. There are still heavy metals and organic contaminants despite the fact that many dangerous compounds have been replaced. Restricting the usage of potentially harmful materials in electronics might improve the viability of recycling programmes(Bhutta et al., 2011).

The usage of bio-based electronic components is another solution recommended to remove the risks of WEEE recycling. Also, while making electronics, designers should think about how easily they can be broken down for recycling and reuse. As advocated by the circular economy method, this will not only make pre-treatment easier, but it will also guarantee that this obsolete equipment will be put to good use again. In addition to being good for the environment, recycling should be profitable. It has to be able to make use of materials that can successfully replace natural resources while also reducing environmental impact(Bhatia & Baruah, 2020). We need to switch from traditional refining methods like pyro- and hydrometallurgy to more sustainable practises like bio-metallurgy. Even though biosorption and bioleaching are tried and true methods, they are often overlooked because of their limited loading capacity and slow response time.

When WEEE is not properly disposed of, the toxic compounds it contains leak out into the atmosphere and pollute the surrounding area, posing a threat to human and animal health. When linked to the informal garbage recycling process, these results are very concerning. Workers engaging in such primitive and unsupervised activities are at risk for a wide range of occupational dangers. Air, water, and land are contaminated, and the local population is also at risk from the high levels of pollution in the area. Foresightful work by Cesaro et al. (2018) predicted a method for determining the relative consequences of different WEEE kinds in relation to their metal content, the pollutants of focus. One of the key difficulties with this technique is calculating the risks connected to the probable contamination conditions, since the strategy was based on priority criteria for WEEE management(Constantinescu et al., 2022). In order to determine the potential dangers to human health, it is possible to compile a comprehensive list of the many types of organic and inorganic contaminants generated during the unregulated disposal of WEEE. This will aid in the development of efficient recycling policies and procedures. WEEE categorization with regard to material composition is particularly difficult to address. Since new technologies allow for the use of a wide range of novel materials in the production of EEE, it is difficult to predict the long-term effects of these materials on human health and the natural environment.

Different socioeconomic factors mean that procedures used in established economies cannot be transplanted wholesale to underdeveloped or developing ones. A solution that may prove workable and productive is combining formal and informal systems with the backing of legal processes. By delegating tasks, the informal sector may be responsible for collecting and manually disassembling valuable WEEE components, while the formal sector can handle metal refinement and the disposal of the leftovers. By doing so, the quantity of WEEE produced on-site may be reduced, and the possibility of it being sold into the black market is reduced. However, the success of this strategy hinges on the introduction of stringent rules concerning the illicit import of WEEE. It is imperative that industrialised nations take on their fair share of responsibilities in this area(Ahirwar & Tripathi, 2021).

Due to a rise in the volume of e-waste being imported (legally and illegally) and weak and inappropriate e-waste policies, the end-of-life management of items has proved exceedingly challenging for developing countries. The fact that most individuals are unaware of the dangers posed by different forms of electronic trash is another issue. Recyclers of electronic garbage face scary and possibly hazardous conditions due to a lack of legal protections at recycling sites. They have to choose between being unemployed and working in dangerous circumstances. There is a severe lack of resources for recycling electronic trash, including experts, infrastructure, management, designed dump sites, machinery, and equipment. It's worth noting that there are collecting and disposal firms out there, but they only handle a tiny fraction of the entire amount of e-waste since there is a lack of standardised collection and pre-processing methods and relatively few certified wastes. Illegal dumping and cremation of hazardous e-waste together with other solid wastes pose a major threat to human health and the environment.

CONCLUSION

The sustainable development of nations that receive large quantities of e-waste is a major issue in the face of the worldwide problem of e-waste management. To properly address this issue, industrialised nations need a suitable legal framework and policies tailored to their circumstances. There have been many pieces of legislation passed in recent decades, and they all have a foundation in the circular economy model. Investments in training campaigns and capacity development should accompany WEEE legislations to improve recovery performance. Metals may be recovered from the complicated e-waste stream, but doing so will need novel advances in the processing systems already in use. It is essential to create a comprehensive database and map of e-waste recycling and processing facilities. To better manage global e-waste, rich nations should invest more in research and construction of new recycling facilities, while poorer nations should prioritise stricter regulation and more WEEE collection to increase their recycling capacity. In order to build shared treatment/recycling facilities, states with low e-waste production might band together. Due to the far greater concentration of valuable metals in e-waste compared to their native ores, there are significant opportunities for urban mining to recover these metals via e-waste recycling. Researchers, business leaders, and government officials are increasingly interested in the concept of a circular economy, and now the public has a moral duty to participate. Closing the loop and transforming the linear model into a circular one is crucial. For the circular economy to really succeed, it must be founded on a system that allows for the recycling and reuse of materials in order to achieve the SDGs (SDGs). Integrative recycling, which requires input from both the official and informal sectors, is key to achieving a considerable improvement in the e-waste management system in the context of the circular economy in the near future.

REFERENCES

Ahirwar, R., & Tripathi, A. K. (2021). E-waste management: A review of recycling process, environmental and occupational health hazards, and potential solutions. *Environmental Nanotechnology, Monitoring and Management, 15*(November 2020), 100409. doi:10.1016/j.enmm.2020.100409

Andeobu, L., Wibowo, S., & Grandhi, S. (2021). A systematic review of E-waste generation and environmental management of Asia Pacific countries. *International Journal of Environmental Research and Public Health, 18*(17), 9051. doi:10.3390/ijerph18179051 PMID:34501640

Bhatia, R., & Baruah, P. (2020). Exclusive talent management and its consequences: A review of literature. *Asian Journal of Business Ethics, 9*(2), 193–209. doi:10.100713520-020-00105-8

Bhutta, M. K. S., Omar, A., & Yang, X. (2011). Electronic Waste: A Growing Concern in Today's Environment. *Economic Research International, 2011*, 1–8. doi:10.1155/2011/474230

Bimir, M. N. (2020). Revisiting e-waste management practices in selected African countries. *Journal of the Air & Waste Management Association, 70*(7), 659–669. doi:10.1080/10962247.2020.1769769 PMID:32401631

Constantinescu, A., Platon, V., Surugiu, M., Frone, S., Antonescu, D., & Mazilescu, R. (2022). The Influence of Eco-Investment on E-Waste Recycling-Evidence From EU Countries. *Frontiers in Environmental Science, 10*(July), 1–15. doi:10.3389/fenvs.2022.928955

Das, A. K. (2013). E-waste management in India current scenario. *International Journal of Management, Technology and Engineering, 6*(6), 1–41. https://www.epa.gov/sites/default/files/2014-05/documents/india.pdf

Gaidajis, G., Angelakoglou, K., & Aktsoglou, D. (2010). E-waste: Environmental problems and current management. *Journal of Engineering Science and Technology Review, 3*(1), 193–199. doi:10.25103/jestr.031.32

Halim, L., & Suharyanti, Y. (2019). E-Waste: Current Research and Future Perspective on Developing Countries. *International Journal of Industrial Engineering and Engineering Management, 1*(2), 25–42. doi:10.24002/ijieem.v1i2.3214

Iyyanar, P., Anand, R., Shanthi, T., Nassa, V. K., Pandey, B. K., George, A. S., & Pandey, D. (2023). A Real-Time Smart Sewage Cleaning UAV Assistance System Using IoT. In *Handbook of Research on Data-Driven Mathematical Modeling in Smart Cities* (pp. 24–39). IGI Global.

Kahhat, R., Kim, J., Xu, M., Allenby, B., Williams, E., & Zhang, P. (2008). Exploring e-waste management systems in the United States. *Resources, Conservation and Recycling, 52*(7), 955–964. doi:10.1016/j.resconrec.2008.03.002

Kaushik, M. V. (2018). E- Waste and Its Management. *International Journal for Research in Applied Science and Engineering Technology, 6*(6), 170–178. doi:10.22214/ijraset.2018.6031

Kaushik, P. R., & Herat, S. (2020). Current state of e-waste management in India. *International Journal of Environment and Waste Management, 25*(3), 322. doi:10.1504/IJEWM.2020.106293

Kiddee, P., Naidu, R., & Wong, M. H. (2013). Electronic waste management approaches: An overview. *Waste Management (New York, N.Y.)*, *33*(5), 1237–1250. doi:10.1016/j.wasman.2013.01.006 PMID:23402807

Krishna, R., & Saha, S. (2022). *Study Paper On e-waste management.*

Kumar, R., & Karishma. (2016). Current Scenario of e-waste management in India:issues and strategies. *International Journal of Scientific and Research Publications, 6*(1), 424–430. www.ijsrp.org

Lu, C., Zhang, L., Zhong, Y., Ren, W., Tobias, M., Mu, Z., Ma, Z., Geng, Y., & Xue, B. (2015). An overview of e-waste management in China. *Journal of Material Cycles and Waste Management*, *17*(1), 1–12. doi:10.100710163-014-0256-8

Monika, & Kishore, J. (2010). E-waste management: As a challenge to public health in India. *Indian Journal of Community Medicine, 35*(3), 382–385. doi:10.4103/0970-0218.69251

Mor, R. S., Sangwan, K. S., Singh, S., Singh, A., & Kharub, M. (2021). E-waste Management for Environmental Sustainability: An Exploratory Study. *Procedia CIRP*, *98*, 193–198. doi:10.1016/j.procir.2021.01.029

Mundada, M. N., Kumar, S., & Shekdar, A. V. (2004). E-waste: A new challenge for waste management in India. *The International Journal of Environmental Studies*, *61*(3), 265–279. doi:10.1080/0020723042000176060

Nagi, D. S., & Kurien, D. J. V. (2019). International Journal of Advanced in Management, Technology and Engineering Sciences ISSN NO : 2249-7455. *International Journal of Management. Technology And Engineering*, *9*(IV), 2854.

Nnorom, Innocent C., & Odeyingbo, O. A. (2019). *Electronic waste management practices in Nigeria. Handbook of Electronic Waste Management: International Best Practices and Case Studies.* INC. doi:10.1016/B978-0-12-817030-4.00014-0

Nnorom, I. C., & Osibanjo, O. (2008). Overview of electronic waste (e-waste) management practices and legislations, and their poor applications in the developing countries. *Resources, Conservation and Recycling*, *52*(6), 843–858. doi:10.1016/j.resconrec.2008.01.004

Pandey, D., Babu, S. Z. D., Noibi, T. O., Pandey, B. K., Dey, P. T. D., Kassaw, C., & Canete, J. J. O. (2020). A study on novel covid-19 epidemic threat for universal physical situation in 2020. *Asian Journal of Advances in Research*, 27-33.

Pandey, D., Islam, T., & Malik, M. A. (2021). Novel coronavirus disease (SARS-COV-2): an overview. *Asian Journal of Advances in Medical Science*, 8-12.

Pandey, D., Aswari, A., Taufiqurakman, M., Khalim, A., & Azahrah, F. F. (2021). System Of Education Changes Due to Covid-19 Pandemic. *Asian Journal of Advances in Research*, 10-15.

Pandey, D., Gul, R., Canete, J. J. O., Rocha, I. C. N., Gowwrii, G., Pandey, B. K., & Peter, S. N. (2021). Mental Stress In Online Learning During The Pandemic: An Assessment Of Learners' perception. *Asian Journal of Advances in Research*, 37-49.

Patel, K. (2021). *Understanding the current E-waste management system in India: A study on strengthening the integration of E-waste management sectors in India under*. MDPI.

Patil, S. A., & Sharma, N. M. (2013). Electronic Waste - A Literature Review. *International Journal of Scientific Research*, *4*(4), 2319–7064. https://www.niir.org/information/content.phtml?content=

Rajpurohit, S. S., & Khan, T. (2018). A Review Paper on Blockchain and Cryptocurrency Based Transaction Systems. *International Journal on Latest Technology in Engineering, management, and Applied Sciences, 7*(3), 241–242.

Saravanamurthy, D., E, R. R. C. V, Sarangan, B., & Bharathkumar, R. (2018). *E-Waste Management, 6*(02).

Shumon, M. R. H., Ahmed, S., & Islam, M. T. (2014). Electronic waste: Present status and future perspectives of sustainable management practices in Malaysia. *Environmental Earth Sciences*, *72*(7), 2239–2249. doi:10.100712665-014-3129-5

Sivaramanan, S. (2013). E-Waste Management, Disposal and Its Impacts on the Environment. *Universal Journal of Environmental Research and Technology*, *3*(5), 531–537. Www.Environmentaljournal.Org. doi:10.13140/2.1.2978.0489

Wath, S. B., Dutt, P. S., & Chakrabarti, T. (2011). E-waste scenario in India, its management and implications. *Environmental Monitoring and Assessment*, *172*(1–4), 249–262. doi:10.100710661-010-1331-9 PMID:20151189

Wath, S. B., Vaidya, A. N., Dutt, P. S., & Chakrabarti, T. (2010). A roadmap for development of sustainable E-waste management system in India. *The Science of the Total Environment*, *409*(1), 19–32. doi:10.1016/j.scitotenv.2010.09.030 PMID:20951410

Chapter 20

Impact of Solid Waste Disposal on Inland Water Wetlands: Solid Waste Management

Rosy Bansal
GSSDGS Khalsa College, Patiala, India

Monika Hans
Govt P.G. College for Women, India

Esha Bansal
Red Roses Public School, Gurugram, India

ABSTRACT

The solid waste disposal affects the quality of water as it leaches down in deeper layers of water bodies having dissolved organic and inorganic constituents from solid waste. The resulting polluted liquid, popularly known as leachate, increases in its concentration levels when it seeps deep inside and appears brownish to black in colour with a rotten smell. It has more organic pollutants with ammoniacal nitrogen. The toxicity levels depend upon the kind of waste, like plastic or metals, being present in solid waste which has severe impacts on human health and aquatic animals. The unselective and mismanaged disposal of waste introduces various toxins comprising of heavy metals and degrades the environment and water resources. Solid waste not only affects the inland waters but also contributes to climatic changes like emission of greenhouse gases.

1. INTRODUCTION

The solid waste disposal affects the quality of water as it leaches down in deeper layers of water bodies having dissolved organic and inorganic constituents from solid waste. The resulted polluted liquid popularly known as leachate increases in its concentration levels when it seeps deep inside and appears

DOI: 10.4018/978-1-6684-8117-2.ch020

as brownish to black in colour having rotten smell. It has more of organic pollutants with ammoniacal nitrogen.

Since communities all over the world have started to realise the dangers that its management entails, solid waste issues and their management have received a great deal of environmental attention in the final quarter of the 20th century. For many years, landfilling has been the preferred form of garbage disposal due to a variety of factors, including the fact that it is likely the least expensive option and that there are plenty of open holes in the ground. Municipal solid waste landfilling is a widespread practise throughout the world.

Within subsurface strata or aquifers, groundwater is a component of the natural water cycle. Unfortunately, groundwater is all too frequently seen as invisible and unimportant. Over 98 percent of the freshwater that is readily available on the planet is groundwater that is kept in the pores and cracks of rock strata. Additionally, groundwater is a crucial resource for business, agriculture, and maintaining rivers with low flows. Groundwater is often crucial in supporting wetlands and their ecosystems, and it is not simply abstracted for supply or river regulation purposes. It also naturally feeds surface-waters through springs and tunnels into rivers. Within subsurface strata or aquifers, groundwater is a component of the natural water cycle. Unfortunately, groundwater is all too frequently seen as invisible and unimportant. Over 98 percent of the freshwater that is readily available on the planet is groundwater that is kept in the pores and cracks of rock strata. Additionally, groundwater is a crucial resource for business, agriculture, and maintaining rivers with low flows. Groundwater is often crucial in supporting wetlands and their ecosystems, and it is not simply abstracted for supply or river regulation purposes. It also naturally feeds surface-waters through springs and tunnels into rivers. Groundwater removal or diversion may alter total flow. Surface water quality and the achievement of water quality criteria can be greatly impacted by a decrease in the quantity or quality of groundwater that is discharged. As a result, the relationship between surface water and groundwater in the water cycle is quite close, with many shared problems. It is crucial to preserve the quality of groundwater. It is challenging, if not impossible, to clean up polluted groundwater. Any self-purification is restricted by the slow groundwater flow and minimal microbial activity.

The toxicity levels depend upon the kind of waste like plastic or metals being present in solid waste which has severe impacts on human health and aquatic animals. The unselective and mismanaged disposal of waste introduces various toxics comprising of heavy metals and degrades the environment and water resources (Meenakshi,2005). The solid waste generation continuously increases with increasing population and very less of it is being recycled. The waste generation also depends upon the economic conditions of particular country and normally it is considered to be kilogram of waste per day per person and this rate has increased to certain level in urban areas. Solid waste not only affects the inland waters but also contributes to climatic changes like emission of greenhouse gases. The climate is affected by emission of methane gas from landfill waste as it the waste decomposes in anaerobic conditions thus producing methane gas. The greenhouse gas emission will raise the average temperature which will further lead to melting of glaciers thus raising the sea level and coastal areas will be submerged in water. Solid waste is any kind of unutilized product generated from residential and commercial areas. Mainly solid waste includes rejected resources and also comprises of structure and building demolitions. The solid waste disposal in the form of landfills or incineration can lead to perilous effects on ecology and environment. Plastics generate various kinds of toxic substances like dioxins which are hazardous to human health. The release of gases from incineration of solid waste causes the release of poisonous gases polluting the environment and in turn results in acid rain thus polluting the rivers and streams. These toxins in inland water affects the growth of aquatic plants and also has a great impact on the reproductive performance of aquatic animals and fish. Solid waste mismanagement can barricade runoff of water along the streams thus forming stagnant water frames that develop as breeding sources of various flies and mosquitoes and many types of microorganisms causing diseases. This waste dumping can subsequently cause infectious diseases through ground water usage. The garbage can be transportable across the world through rivers and oceans and can be accumulated on beaches and whorls which can threaten the natural flora and fauna and also restricts the manly usage of rivers, stream, nautical and seaside environments.

The human life is dependent on water as it is an essential part of human existence. The municipal solid waste is being dumped unknowingly into river and streams transporting contaminants to water bodies (Kulikowska, and Klimiuk, 2008). The excess accumulation of heavy metals in surface water is a havoc to food chain (Damodharan, 2013). The waterways draining to agricultural areas hypothetically postulates hostile communal health and effects the environment if solid waste is not disposed of suitably. Developed nations have set some regulatory authorities and agencies and NGOs to look after the protection of environment and management of solid waste but developing nations are still throwing away the garbage and solid waste in waterways and are facing the water quality problems. In spite of rules and regulations for protecting waterways and inland waters, the waste is still dumped as garbage or litter. Refuse is also contributed by people who are involved in recreation activities related to water like fishing, water skiing etc.

The problem of solid waste management was acknowledged by UN conference on Environment and development in Rio de Janeiro in 1992 and UNEP in 2010 emphasized on the reuse and recycling of solid waste. According to World Health Organization (WHO), solid waste management should be prioritized over other environmental issues as all will be corrected if waste is dumped properly and inland wetlands will be safer and water will be of better quality.

The waste dumping without any differentiation can be hazardous as it can contain toxic materials from varied sources like Industry waste, household waste, agricultural wastes, medical waste, hotels and institutional waste. The pollutants from oil spills, household and poisonous substances can be passed to dumping sites and can threaten the human health (calvo et al., 2005). The masses of solid waste infested with insects and pests can be a major cause of spreading infections in urban areas. The ecological and environmental quality is degraded with indiscriminate solid waste disposal and is specifically affecting the population residing near the dumpsites. The Environmental complications connected to mismanaged waste disposal is related to health problems, aquatic contamination and soil degradation, bad odours and disgusting sight. Maximum wasteyards are situated around the neighbourhood of existing societies and marshlands. These wasteyards are the cause of severe environmental issues and thus threatening terrestrial, aquatic and aerial atmospheres.

These solid waste materials lead to increased toxicity levels of water. Water certainly contributes natural minerals required by human body and some of the impurities as well in form of salts and microorganisms but when these levels cross the verge limits then they lead to water pollution. The heavy metals like lead, manganese Arsenic Nickel, Mercury Cadmium are very much poisonous even at lower concentrations and their long-term exposure can lead to disrupt the body organ misfunctioning like mercury can cause impairment of brain and kidney and can interfere with nervous functions. The Zinc can make its way through non-natural lanes by manufacture of steel or power plants based on coals and waste material incineration (Hiscock and Grischek, 2002).

The pathogenic microorganisms present in water indicate the quality of water used for domestic purposes. The coliform bacteria present in water can pose serious health problems but on the other hand according to Environment Protection Agency coliforms need consistent examination and total count of coliforms determine the toxicity levels of potable water. The E.Coli indicates pathogenicity in water and they have Nitrogen carrying capacity by red cells of infants and can result in disorder identified as methemoglobinemia also commonly known as blue baby syndrome. Other water borne illnesses related to more nitrate's concentration are diarrhoea, Cholera, Hepatitis, typhoid, dermatitis and enteric fever etc. All these health risks can be reduced by controlling the solid waste disposal in water bodies thus improving the quality of potable water. The main objective of the chapter is to study the impact of solid waste dumping on in land water and in turn its consequences on human health as wetlands are the habitat of diverse species and protect the terrestrial and aquatic terrain. They also open a varied source for research studies and helps in decreasing the glacier melting caused by greenhouse gases which has become a major challenge in current situation. So, water quality can be improved by proper solid waste dumping and thus protecting our valuable water resources (Mor, et al., 2006).

2. SUPPLY OF GROUNDWATER

The ground water is available in the fissures and cracks of rocks and in the sediments underlying the surface which actually is originated in the form of rain or snow. This water permeates down the soil layers and is than procured by digging of wells or through oceans, streams or lakes for domestic and commercial usage. Groundwater is the main water supply system in the world. The ground water is less contaminated than surface water and 50 percent of world's population is dependent on ground water for potable purpose. The groundwater is vulnerable to contamination only when it is combined with geological and hydrological factors associated with pollution.

The groundwater Quality is dependent on the solid waste present in wetlands which can contaminate the water and this solid waste accumulation in uncontrolled manner can pose a great to the public residing near the dumpsites. These dumpsites are positioned in the locality of living societies and everglades mostly. The dumpsites are then responsible for discharging pollutant and toxic substances to the adjacent water bodies and are reserved as hazardous to the human health and other living beings. The dumping sites without any control are age old with unmanageable waste and no disposal techniques being adopted and have substantial impact on environment and poses deadly impact on three kinds of life that is Aerial, Aquatic and terrestrial.

This municipal solid waste, which is commonly referred to as "trash," is a natural by-product of human activity and is dumped. The most typical way of disposing of solid waste is land filling. The landfills closer to Pallavaram are open dumpsites since they have cheap operating expenses and no leachate collecting systems due to a lack of knowledge and equipment. Dumping in the open is unattractive, unhealthy, and generally unpleasant. They draw pigs, rodents, insects, scavenger animals, and other pests. Hazardous chemicals that are dissolved or leached out of dumpsites by surface or subsurface runoff can be moved away from the dumpsites by surface or subsurface water percolating through the rubbish. Heavy metals are among these substances that are particularly pernicious and can cause bioaccumulation and biomagnifications. If the leachate migrates into ground water, these heavy metals could pose a threat to the environment. Bore wells used to obtain ground water from garbage areas pose a concern of contaminating the water (Raman and Sathiya,2008).

Figure 1. Industrial process discharge

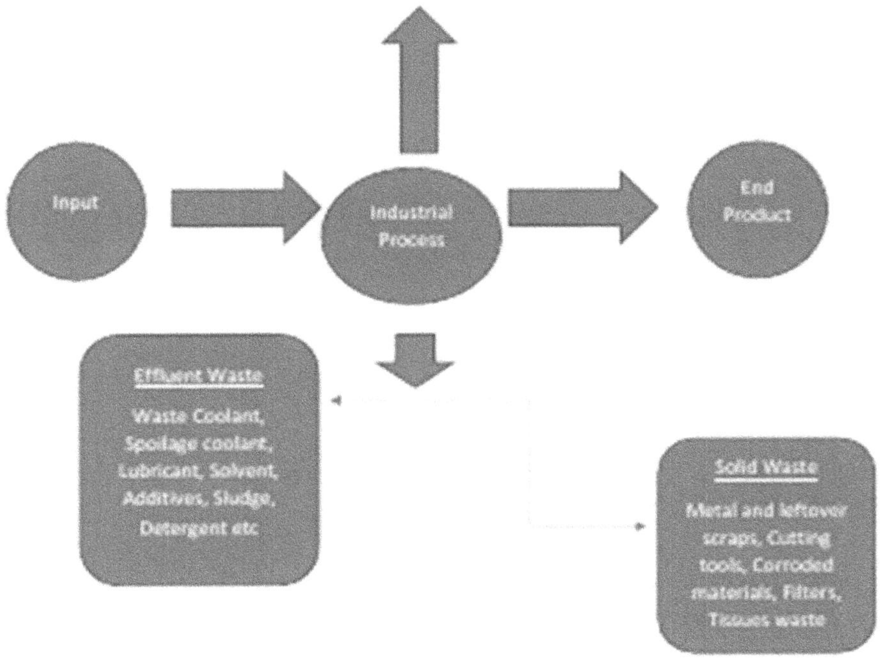

3. POLLUTANTS IN SOLID WASTE

The minerals like Zinc, Magnesium, Calcium and other materials like sand salt and microorganisms are naturally present in water and at certain level are safe in consumption but when they cross their verge limits result in water pollution. The complexes made by their ions leads to severe toxic problems to organisms which can be sublethal to lethal depending upon the exposure time and temperature conditions. Some metals like lead, Zinc, Manganese, Cadmium. Arsenic, Nickel and Mercury are poisonous even at low concentrations. The extensive contact of lead is the cause of anaemia by disrupting biosynthesis of haemoglobin, Kidney damage, Elevated Blood Pressure, Nervous disorders, Abortions and infertility in males. Mercury exposure is responsible for disruption of nervous system and can lead to brain and kidney damage. These pollutants when comprehensively dumped in landfills can find their way to groundwater and to surface water leading to polluted inland water wetlands (Khateeb,2013).

Impact on Water-Based Animals

The carelessness with disposal of waste and garbage affects animals in many ways, and they suffer from the effects of pollution caused by improper disposal of waste and rubbish.

The direct Consumption of Styrofoam and cigarette butts are known to be responsible for causing deaths in marine animals. Animals are also at hazard of intoxication while consuming grasses near polluted areas or landfills as the poisons dribble into the soil. The adverse Impact of solid waste to the wetland is shown in figure below:

Figure 2. Adverse Impact of solid waste to the wetland

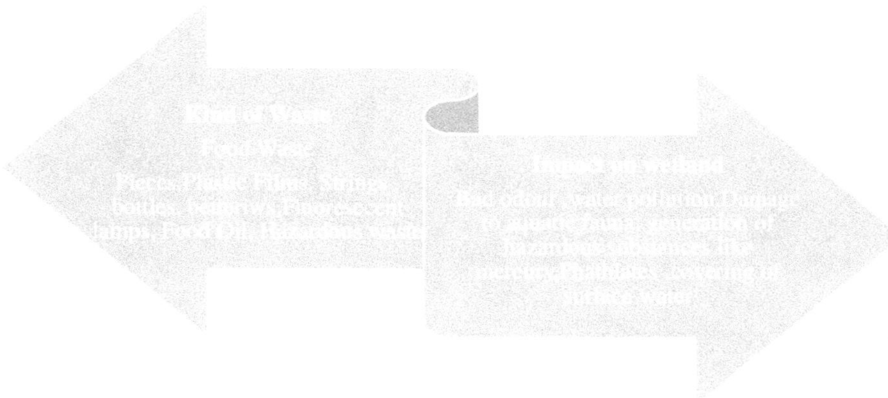

Impact of Solid Waste on Health

The population in places without suitable waste disposal methods, especially preschoolers, garbage workers, and employees of facilities that produce dangerous and infectious materials are among the groups at risk from the improper disposal of solid waste. Other high-risk populations include those who live close to trash dumps and those whose water supplies have been affected as a result of waste dumping or landfill site leaks.

Domestic organic waste in particular offers a severe risk because it ferments, fostering an environment that is conducive to microbial pathogen survival and expansion. Trash workers and rag pickers are particularly at risk for developing infectious and chronic diseases as a result of direct handling of solid waste. Human health can be impacted by exposure to hazardous waste, with children being more susceptible to these toxins. In actuality, since the dumping of chemical waste into the environment results in chemical poisoning, direct exposure can cause diseases through chemical exposure. To establish a link between health and hazardous waste, numerous research have been conducted around the globe. Human health can be impacted by exposure to hazardous waste, with children being more susceptible to these toxins. In actuality, since the dumping of chemical waste into the environment results in chemical poisoning, direct exposure can cause diseases through chemical exposure. To establish a link between health and hazardous waste, numerous research has been conducted around the globe. Additionally, hazardous health concerns can result from industrial and agricultural waste. In addition, mixing municipal garbage and industrial hazardous waste for disposal exposes humans to radioactive and chemical risks. In addition to preventing storm water discharge, uncollected solid waste can also cause stagnant water bodies, which serve as a breeding ground for disease. The water body or the ground water supply might get contaminated when waste is placed close to it. The accumulation of harmful compounds in the food chain through the plants and animals that feed on it happens as a result of the direct disposal of untreated garbage in rivers, seas, and lakes.

Hospital and other medical waste disposal call for extra care because it poses serious health risks. Syringe needles, bandages, cotton swabs, plasters, and other infectious waste produced by hospitals, healthcare facilities, medical laboratories, and research facilities are frequently disposed of alongside regular non-infected waste. Sites for waste treatment and disposal can also endanger the neighbourhood's health. Inefficiently run incineration facilities contribute to air pollution, while poorly run landfills draw pests and disease-carrying rodents of all kinds. These locations ought to be safe distances from any populated areas (Vasanthi et al., 2008). To prevent leaks into the neighbouring ground water sources, landfill sites should be properly lined and walled.

Recycling can also be harmful to your health if the right safeguards are not taken. Workers who handle garbage containing chemicals and metals risk exposure to toxins. Because abandoned syringe wounds can spread diseases like Hepatitis B and C, disposing of medical wastes needs to be done with extra care. Rag pickers and other individuals who rummage through trash dumps looking for recyclable items run the risk of getting hurt and coming into contact with these infected objects. Recycling can also be harmful to your health if the right safeguards are not taken. Workers who handle garbage containing chemicals and metals risk exposure to toxins. Because abandoned syringe wounds can spread diseases like Hepatitis B and C, disposing of medical wastes needs to be done with extra care. Rag pickers and other individuals who rummage through trash dumps looking for recyclable items run the risk of getting hurt and coming into contact with these infected objects.

Untreated releases of some chemicals, such as cyanides, mercury, and polychlorinated biphenyls, can cause illness or even death. Residents who were exposed to hazardous garbage had an increased risk of cancer, according to certain research (Saini et al., 2018). To establish a link between health and hazardous waste, numerous research have been conducted around the globe.

Classification of Waste

Waste can be of following types depending upon origin

1. Organic waste

It Includes waste from Kitchen, waste from food production, fruits and vegetables, flowers, leaves, fruits

2. Waste from Combustible

It includes combustible materials like Paper, wood, dried leaves, packaging for relief items etc. which have low moisture content.

3. Waste from Non-combustibles

It includes non-combustible materials like Metal, Tins, Cans, bottles, stones, etc.

4. Waste from Toxic Substances

It includes old medicines, paints, chemicals, bulbs, spray cans, fertilizer and pesticide containers, batteries, shoe polish.

5. Recyclable Waste

It includes the materials which can be recycled like Paper, glass, metals, plastics.

6. Waste from Ashes or Dust

It includes Residue from fires that are used for cooking.

7. Construction waste

It includes Rubble, roofing, broken concrete etc.

8. Hazardous waste

It includes Oil, battery acid, medical waste, industrial waste, hospital waste.

9. Dead animals

It includes waste from animal Carcasses of dead livestock or other animals.

10. Bulky waste

It includes branches of tree, tires etc.

11. Soiled waste

It includes hospital waste like cloth soiled with blood and other body fluids.

Classification of Wetland Impact Factors

The stakeholders should be educated enough regarding environmental issues and awareness should be created amongst them to participate maximally in resolving these issues. The main issues under consideration are:

Table 1. Classification of wetland impact factors

S.No.	Class	Subclass	Subdivision
1.	Human Factors	Land Area Impact	Agriculture and aquaculture, Natural System Modifications, Human settlements (non-agricultural), Transportation and service corridors
		Environment Impact	Pollution, Human Intrusions and disturbance, Energy production and mining
		Biodiversity Impact	Biological resource use, Invasive and other problematic species and genes
		Water Resources Impact	Water Regulation
2.	Natural Factors	-	Climate change and severe weather Geological events

1. Lack of understanding of environmental education
2. Lack of system of education related to environment
3. Lack of capability for delivering environment education
4. Lack of strategies for creating awareness and knowledge
5. Lack of participation in waste management programs
6. Lack of tools for effective outcomes
7. Lack of partnership between government departments to work in collaboration
8. Lack of decision makers
9. Lack of motivation and interest in stakeholders
10. Lack of imitativeness to assess the completion of work
11. Lack of finance and resources

4. RECOMMENDATIONS

It is necessary to utilise proper trash disposal techniques to make sure that it won't harm the local ecosystem or put the residents' health at risk. At the household level, garbage must be properly separated, and it must be made sure that all organic material is set aside for composting, which is unquestionably the best way to properly dispose of this category of waste. In actuality, the organic component of the waste produced decomposes more quickly, draws insects, and spreads illness. Composting organic waste allows it to be turned into fertiliser (Adesina, 2000).

Solid waste disposal, the creation of leachate, and its subsequent percolation can all contaminate groundwater. Water sources are under stress as a result of rapid industrialisation, increased population growth brought on by urbanisation, changing lifestyles, and growing standards of living. The availability and quality of water are becoming key issues. Managing supply and demand statistics is a highly challenging task.

5. CONCLUSION

The solution for solid waste management and its impact on inland wetlands should be observed analytically and rationally. The Integrated solid waste management is more significant to manage the waste which includes appropriate methods and programs with advance techniques pertaining to achieve the objectives of waste management. According to EPA that is Environment Protection Agency, the integrated solid waste management is a combination of different methods like waste sources reduction, recycling of waste, combustion of waste and landfills. All these methods can be adopted in combination or in hierarchy. It is concluded from the above studies that better solid waste management methods should be implemented by developing countries to protect the wetlands. Ironically only 50 percent of waste from cities and 25 percent from rural areas is collected and which is going to increase to 70 percent by 2050 (Smith, 2010). The efforts are required to improve the management of waste so that it can protect the environment as well as human health. It is our social and moral responsibility to keep the inland water wetland clean and should be concerned for the major disposal issues of waste by industrial processes in the name of development which leads to deterioration of environment. The authorities are planning many policies to reduce and recycle the waste and its stabilization so that final disposal in wetlands can be prevented.

The downstream of wetland can be at verge of danger with hazardous solid waste comprising of toxic substances including heavy metals and infectious waste so suitable measures are required to regulate the waste disposal activities. A wide range of actions in the form of reduction of waste, Information management system, Regulatory measures, licensing system should be implemented by government to control the disposal of solid waste in inland water wetlands in future.

REFERENCES

Adesina, A. O. G. (2000). *Integrated Waste Management System*. CORE.

Calvo, F., Moreno, B., Zamorano, M., & Szanto, M. (2005). Environmental Diagnosis Methodology for Municipal Waste Dumping sites. *Waste Management (New York, N.Y.)*, *25*(8), 768–779. doi:10.1016/j.wasman.2005.02.019 PMID:15905084

Damodharan, U. (2013). Bioaccumulation of heavy metals in contaminated river water-Uppanar, Cuddalore South East Coast of India. *Intech Open*. . doi:10.5772/5334

Hiscock, K. M., & Grischek, T. (2002). Attenuation of groundwater pollution by bank? ltration. *Journal of Hydrology (Amsterdam)*, *266*(139–144).

Khateeb, M.A. (2013). *Landfill leachate impact on groundwater quality report*. Institute of water and environmental science solid and hazardous waste management.

Kulikowska, D., & Klimiuk, E. (2008). The effect of dumping site age on municipal leachate composition. *Bioresource Technology*, *99*(13), 5981–5985. doi:10.1016/j.biortech.2007.10.015 PMID:18060769

Meenakshi, P. (2005). Elements of Environmental Science and Engineering. Prentice Hall of India Private limited, India.

Mor, S., Ravindra, K., Dahiya, R. P., & Chandra, A. (2006). Leachate characterization and assessment of groundwater pollution near municipal solid waste dumping site. *Environmental Monitoring and Assessment*, *118*(1-3), 435–456. doi:10.100710661-006-1505-7 PMID:16897556

Raman, N., & Sathiya,, D. (2008). Impact of solid waste effect on ground water and soil quality nearer to pallavaram solid waste landfill site in chennai. *Rasayan Journal of Chemistry*, *1*(4), 828–836.

Saini, Ritesh, & Siddiqui, Nihal & Bahukhandi, kanchan Deoli & J.Ahuja, Neelu. (2018). Study of effects of Municipal solid waste on water quality of Dehradun City. *Journal of Environmental Science and Health, Part A: Environmental Science and Engineering*, *7*, 409–416.

Smith. (2010). What is Solid Waste Management? *Greek.com* greek.com/ what-is-solid-waste-managment.htm)

UNEP. (2005). *Solid Waste Management*. United Nations Environment Programme - Division of Technology, Industry, and Economics.

Vasanthi, P., Kaliappan, S., & Srinivasaraghavan, R. (2008). Impact of poor solid waste management on ground water. *Environmental Monitoring and Assessment*, *143*(1-3), 227–238. doi:10.100710661-007-9971-0 PMID:17999155

WHO. (2006). *Protecting Groundwater for Health: Managing the Quality of Drinkingwater Sources* (O. Schmoll, G. Howard, J. Chilton, & I. Chorus, Eds.). Published by IWA Publishing.

WWAP (World Water Assessment Programme). (2012). *The United Nations World Water Development Report 4: Managing Water under Uncertainty and Risk*. UNESCO.

Chapter 21
A Holistic Study on E-Waste Management for Ensuring Sustainability of the Environment

Gurpreet Kaur
Chitkara Business School, Chitkara University, India

Sneha Sharma
Chitkara Business School, Chitkara University, India

ABSTRACT

Globalization and industrialization have led to drastic increase in the degree of solid waste generation; moreover, due to the migration of people, it is expected that in coming years demand of water and energy will increase manifold. This has led to the requirement of an efficient waste management mechanism. Though the ministry of environment and forest (MoEF) are continuously working for formulating conducive policies to curb the menace of waste disposal, still there is a need to have a central functionary to work at a wider level on this issue. There is a need to spread awareness among the masses so as to set accountability for each and every individual towards the environment. The present chapter is an attempt to provide a holistic approach towards e-waste management, which includes a systematic approach of collection, recycling, and minimization of e-waste resulting in bringing out sustainability in the environment. Finally, some suggestions and recommendations are provided which can be implemented for ensuring efficient e-waste management.

DOI: 10.4018/978-1-6684-8117-2.ch021

INTRODUCTION

When digitization and electronics were less prevalent in our lives before, there was scarcely any trash produced. The growing number of applications for IT in our daily lives produces enormous volumes of garbage, which has been a major contributor to recent environmental problems as the negative effects of E-waste is destroying the surroundings.

Each year, approx. 40 metric tons of e-waste are generated worldwide, of which 5% is global solid trash. UNEP predicted that by 2022, out of the 16 billion smartphones that are currently in use worldwide, 5.3 billion will be discarded. E-waste includes several hazardous elements in addition to priceless metals. The ecology becomes more toxic as a result of hazardous metals such (Cd, Hg, Pb, and Cr).

E-waste produced from household is often managed in one of three ways: formally collected by government agencies; dumped into bins; or collected by certain dealers or businesses. In order to safely and effectively remove valuable materials from the waste collected through municipal collection points, advanced technology and infrastructure are used (formal recycling). On the other hand, outside of the system, individual garbage firms or dealers recycle waste using subpar conditions and outdated methods without any safeguards to lessen the number of dangerous chemicals in the environment (informal recycling). When e-waste is placed in a landfill, it eventually contaminates the soil and groundwater posing problem for agricultural industry. Therefore, there is a dire requirement to implement an efficient waste management mechanism. Absence of an appropriate system of E-waste disposal will worsen the public health & will significantly deteriorate the environment.

A multi-faceted approach is needed to address the problem of e-waste management in order to maintain a balance among the detrimental effects of waste recycling on the human health, environment and financial stability. Presently, one of the most difficult problems in developing countries is managing e-waste Moreover, E-waste management is significantly required to satisfy international environmental standards. Therefore, this study will be effective in developing better understanding about e-waste disposal.

There have been many reviews on the generation of e-waste and its proper recycling, including ones on global e-waste management, e-waste recycling processes and methods. However, very few of these reviews are giving holistic view of E-Waste management.

To develop deep into the best practices & holistic approach for tackling the challenges of E-waste issue at global level, the present paper is an attempt to provide a comprehensive approach towards E-waste management which includes a systematic approach of collection, recycling, minimization of E-waste resulting to bring out sustainability in the environment. The paper takes into account various parameters influencing waste processing such as techniques employed, cost involved, modes of transportation and revenue generated by waste processing. The advanced techniques which can be applied for effectively handling and disposing of the waste are also discussed. The paper highlights the reason behind generation of E-waste & remedies taken to combat this issue in other countries. The limitations and challenges related to effective waste management are also over viewed.

This study gives detailed information on the current worldwide e-waste creation trend, the most recent e-waste recycling techniques, occupational and environmental health concerns, and suggestions for better trash recycling practices. The study is segregated into following sections. Section 2 deals with review of literature. Section3 describes research methodology. Section 4 &5 deals with the discussion, conclusion & future scope of the study.

REVIEW OF LITERATURE

Many researchers have undergone studies relating to E-waste in the past. Few studies are mentioned as follows: Kocak and Baglitas (2022) investigated how changes in economic, social, and technical developments impacted the amount of municipal solid waste (MSW) generated per person in OECD countries between 2003 and 2018. Zhou and Zhang (2022) evaluated the energy potential of trash using two varied waste disposal strategies, incineration and landfills, by 2030 and 2060, using the Greater Bay Area of China as an example.

An overview of the hazardous components of e-waste, their potential impact on the environment and people's health, and the present management strategies used in some countries are provided by Murthy and Ramakrishna (2022). Rasheed, et al., 2022 examined the financial benefits and environmental viability of such e-waste treatment in developing nations like Pakistan.Jangre et al.2022 prioritize the challenges in E-waste management consulting with the experts and systematic literature analysis.

Somani et al., 2021 examined the existing state of affairs, challenges, and approaches utilized to lessen waste, including collection/segregation, recycling, energy recovery (composting, biomethanation, and bioremediation), land filing of household and industrial waste, and its environmental effects.

According to Debrah, et. al.2021, suggested that in developing countries, environmental sustainability or waste sustainability can be achieved only when environmental attitudes, awareness, and knowledge are related to and transmitted from mentor to students through formal education. In the light of poor recycling methods for e-waste and their detrimental effects on the environment and human health, Rautela, et. al.2021 gives a complete examination of attempts made to address these challenges.

Ahirwar & Tripathi 2021 discussed the current global trend in e-waste generation, how it is recycled, and the harm that its poisons do to human health. Xavier, et al., 2021 give a general summary of the hazardous components found in e-waste, their possible effects on human health and the environment, and the current treatment techniques now in use in different nations. Shittu, et al., 2021 examines areas for reform while providing a critical and contemporary appraisal of the current state of the worldwide WEEE management system. The possible benefits and disadvantages of implementing an integrated water, waste, and energy management system in Curauma are discussed by Araya et al. in 2020.To treat electronic trash responsibly, Adanu et al. (2020) suggest that government assistance is required to subsidize the cost of sustainable solutions.A global review of the current condition of e-waste management as well as its historical growth was presented by Islam Et.al, 2020. Research has concentrated on creating a hydrometallurgical method replacement to recover metal from e-waste.

Many sustainable E-waste management strategies have been put forth, along with the potential challenges that may develop when these approaches are put into practise. (Masud et al,2018). By incorporating those technologies with Bangladesh's current waste management system, e-waste management-related difficulties can be rectified and a significant amount of renewable energy can be produced. Debnath et al. (2016) conducted a study wherein e-waste management was considered as a benchmark for environmental friendly computing. Perkins et al. (2014) in an research study stressed on E-waste as serious issue in the society& recommended on the recycle of e-waste in a safe and standardized way.

Sivaramanan 2013 examined the information acquired from several scholarly articles and websites, as well as a technical comment from Greenpeace International. The harmful components of e-waste, their potential impacts on human health and the environment, and the current management practices used in some countries were all examined by Kiddee et al. 2013's. According to Saoji, A. (2012) e-waste should be properly recycled, reused, or disposed of, otherwise it will pose difficult for both technologists and environmentalists to cope up with it.Extended producer responsibility and consumer-driven solutions are a few of the ideas that Gupta, S. (2011) looks at in order to reduce the harmful environmental effects of e-waste. According to Gaidajis et al. (2010), it is crucial to separate e-waste from other types of solid garbage and recycle it in order to recover valuable raw materials and vital metals. The E-waste management systems of industrialized and developing countries are analysed and evaluated by Wath et al. (2010), with a particular focus on Switzerland, the first country to design and run a complete E-waste management system. Focus is placed on the current state of affairs in Nigeria in relation to these issues as they relate to practices in developing countries in Osibanjo & Nnorom, 2007 study.

Thus the above reviews stresses the need to focus upon the E-Waste management so as to attain sustainable development. Many techniques and method to deal with E-waste are also enumerated. The detailed discussion of above review of literature is tabulated coming section.

RESEARCH METHODOLOGY

This section describes the studies conducted on E-waste and its management by different researchers at the global level during different time periods. The papers for the study were obtained from databases like scopus, web of sciences, prouest, science direct etc. the segregation of studies under different parameters are mentioned below:-

The table represents year-wise analysis of data, its source and name of the authors

The table shows the source of the data and a country-by-country analysis

The table represents the top 10 articles in terms of maximum citations

Table 1. Year-wise division of research.

Year of Publication	Name of the Journal	Author's Name
2022	Sustainable development 2022	Kocak & Baglitas
2022	Energy policy	Zhou & Zhang
2022	Sustainable and Advanced Remanufacturing Processes	Murthy & Ramakrishna
2022	Environmental science and pollution research	Rasheed et al.
2022	Environmental science and pollution research	Jangre et al.
2021	Journal of Physics: Conference Series	Somani et al.
2021	Recycling 2021	Debrah et al.
2021	Science of total environment	Rautela et al.
2021	Environmental Nanotechnology, Monitoring & Management	Ahirwar Tripathi
2021	Resources policy	Xavier et al.
2021	Waste management	Shittu et al.
2021	Environmental research	Ádám et al.
2020	Resources, conservation and recycling 156	Vergara et al.
2020	Journal of cleaner production	Islam et al.
2018	Environmental science and pollution research	Masud et al.
2016	Procedia environmental sciences	Debnath et al.
2014	Annals of global health	Perkins et al.
2013	Universal journal of environment	Sivaramanan
2013	Waste management 33	Kiddee & Wong
2012	National journal of medical research	Saoji
2011	Journal of education and practice	Gupta
2010	Journal of engineering science and technology review	Gaidajis et al.
2010	Science of the total environment	Wath et al.
2007	The journal of sustainable circular economy	Osibanjo& Nnorom

Authors Compilation.

Table 2. Sources of data collection and country - wise analysis

Country	Name of the Journal
China	Energy policy
Singapore	Sustainable and Advanced Remanufacturing Processes
Pakistan	Environmental science and pollution research
India	Journal of Physics: Conference Series
India	Environmental Nanotechnology, Monitoring & Management
Indonesia and Brazil	Resources policy
Africa	Waste management
China, India, Ghana and Nigeria.	Environmental research
chile	Resources, conservation and recycling 156
Bangladesh	Journal of cleaner production
Bangladesh	Environmental science and pollution research
Africa	Universal journal of environment
China, India and Ghana.	Waste management 33
India	National journal of medical research
USA and Canada	Journal of education and practice
China and India	Journal of engineering science and technology review
India	Science of the total environment
Africa	The journal of sustainable circular economy

Authors Compilation.

Table 3. Authors paper with maximum citations

Citation	Title of the Paper	Author's Name
798	Electronic waste management approaches: an overview	Kiddee & Wong
505	E-Waste: A Global Hazard	Perkins et al.
372	The challenge of electronic waste(E-waste)management in developing countries	Osibanjo& Nnorom
233	A roadmap for the development of sustainable E-waste management system in India	Wath et al.
225	E-waste environmental problems and current management	Gaidajis et al.
190	Advances in sustainable approaches to recover metals from e-waste-A review	Islam et al.
71	E-waste management and its effects on the environment and human health	Rautela
69	E-waste management: A review of recycling process, environmental and occupational health hazards, and potential solutions	Ahirwar& Tripathi
67	Raising Awareness on Solid Waste Management through Formal Education for Sustainability: A Developing Countries Evidence Review	Debrah et al.
65	Towards the effective E-waste management in Bangladesh: a review	Masud et al.
58	E-waste management, disposal and its impact on the environment	Sivaramanan

Authors' compilation

DISCUSSION

E-waste is a serious at the global level. Improper management of E-waste is critical for the growth of the country as it escalates pollution, causes health issues among the masses. To overcome the hazardous effects of E-waste on the environment, it is very crucial to implement & execute policies for E-waste management appropriately. The adoption of emerging technologies will act as a catalyst for developing new strategies which are capable of solving the problems of E-waste.

In the current study, an attempt is made to enumerate the effect of E-waste on the environment & measure to curb E-waste by managing it significantly. Figure 1 provides the overall consequences & remedies to combat which the problem of E-waste management.

Figure 1.
Authors' compilation

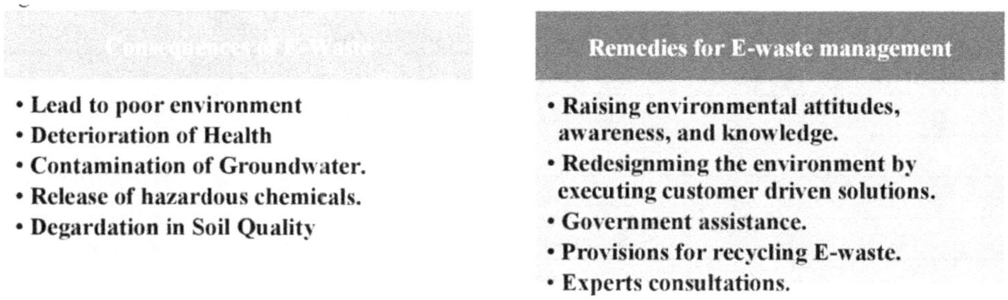

CONCLUSION

To recover the most precious metals and have the least amount of negative environmental impact, recycling is a vital trend. Pyrometallurgy is used to recover precious metals from e-waste. An economic engine for sustainable growth is the extraction of noble metals.

E-waste is of vital important to both human health and surroundings. Presently, one of the most difficult problems in developing countries is managing e-waste. E-waste management is significantly required to satisfy international environmental standards. Environmental sustainability or waste sustainability can be achieved in developing countries when environmental knowledge, attitudes and awareness are tied to and transmitted from mentor to students through formal education. A strategy needs to be developed to support the deployment of effective e-waste management in both industrialized and developing countries. This article gives a general overview of the potentially harmful elements included in e-waste, their possible effects on human health and the environment, and the current management techniques employed in different nations. It is essential to recycle e-waste, but it must be done so in a safe and standardized way. The overall objective is to provide leadership, guidance, and the application of international protocols and preventative measures that will aid in turning e-waste into safe and useful forms. The establishment of an informal sector is strongly encouraged by the fact that unregulated processing provides cash in underdeveloped countries, making it challenging to police regulations. The study concludes that effective handling of e-waste is essential for achieving environmental sustainability.

Scope for Future Research

The narrative literature studies describes the global issue of waste management, various sources of pollution affecting the surroundings and creating hindrances in sustainable development. The contents discussed will be helpful for the research scholars, policy makers and other stakeholders for assessing the overall impact. Besides this it will be helpful in planning for integrated solid waste & its treatment so as to improve sustainability worldwide.

The present paper is a compiled review comprising of references for which paper cited may be referred.

REFERENCES

Ádám, B., Göen, T., Scheepers, P. T., Adliene, D., Batinic, B., Budnik, L. T., & Au, W. W. (2021). From inequitable to sustainable e-waste processing for reduction of impact on human health and the environment. *Environmental Research*, *194*, 110728. doi:10.1016/j.envres.2021.110728 PMID:33444608

Adanu, S. K., Gbedemah, S. F., & Attah, M. K. (2020). Challenges of adopting sustainable technologies in e-waste management at Agbogbloshie, Ghana. *Heliyon*, *6*(8), e04548. doi:10.1016/j.heliyon.2020.e04548 PMID:32802976

Ahirwar, R., & Tripathi, A. K. (2021). E-waste management: A review of the recycling process, environmental and occupational health hazards, and potential solutions. *Environmental Nanotechnology, Monitoring & Management*, *15*, 100409. doi:10.1016/j.enmm.2020.100409

Debnath, B., Roychoudhuri, R., & Ghosh, S. K. (2016). E-waste management–a potential route to green computing. *Procedia Environmental Sciences*, *35*, 669–675. doi:10.1016/j.proenv.2016.07.063

Debrah, J. K., Vidal, D. G., & Dinis, M. A. P. (2021). Raising awareness on solid waste management through formal education for sustainability: A developing countries evidence review. *Recycling*, *6*(1), 6. doi:10.3390/recycling6010006

Gaidajis, G., Angelakoglou, K., & Aktsoglou, D. (2010). E-waste: Environmental problems and current management. *Journal of Engineering Science and Technology Review*, *3*(1), 193–199. doi:10.25103/jestr.031.32

Gupta, S. (2011). E-waste management: Teaching how to reduce, reuse and recycle for sustainable development-need of some educational strategies. *Journal of Education and Practice*, *2*(3), 2222–1735.

Islam, A., Ahmed, T., Awual, M. R., Rahman, A., Sultana, M., Abd Aziz, A., & Hasan, M. (2020). Advances in sustainable approaches to recover metals from e-waste-A review. *Journal of Cleaner Production*, *244*, 118815. doi:10.1016/j.jclepro.2019.118815

Jangre, J., Prasad, K., & Patel, D. (2022). Analysis of barriers in e-waste management in developing economies: An integrated multiple-criteria decision-making approach. *Environmental Science and Pollution Research International*, *29*(48), 72294–72308. doi:10.100711356-022-21363-y PMID:35696062

Kiddee, P., Naidu, R., & Wong, M. H. (2013). Electronic waste management approaches: An overview. *Waste management, 33*(5), 1237-1250.

Kocak, E., & Baglitas, H. H. (2022). The path to sustainable municipal solid waste management: Do human development, energy efficiency, and income inequality matter? *Sustainable Development (Bradford)*, *30*(6), 1947–1962. doi:10.1002d.2361

Masud, M. H., Akram, W., Ahmed, A., Ananno, A. A., Mourshed, M., Hasan, M., & Joardder, M. U. H. (2019). Towards the effective E-waste management in Bangladesh: A review. *Environmental Science and Pollution Research International*, *26*(2), 1250–1276. doi:10.100711356-018-3626-2 PMID:30456610

Murthy, V., & Ramakrishna, S. (2022). A review on global E-waste management: Urban mining towards a sustainable future and circular economy. *Sustainability (Basel)*, *14*(2), 647. doi:10.3390u14020647

Osibanjo, O., & Nnorom, I. C. (2007). The challenge of electronic waste (e-waste) management in developing countries. *Waste Management & Research*, *25*(6), 489–501. doi:10.1177/0734242X07082028 PMID:18229743

Perkins, D. N., Drisse, M. N. B., Nxele, T., & Sly, P. D. (2014). E-waste: A global hazard. *Annals of Global Health*, *80*(4), 286–295. doi:10.1016/j.aogh.2014.10.001 PMID:25459330

Rasheed, R., Rizwan, A., Javed, H., Sharif, F., Yasar, A., Tabinda, A. B., Mahfooz, Y., Ahmed, S. R., & Su, Y. (2022). Analysis of environmental sustainability of e-waste in developing countries—A case study from Pakistan. *Environmental Science and Pollution Research International*, *29*(24), 36721–36739. doi:10.100711356-022-18691-4 PMID:35064515

Rautela, R., Arya, S., Vishwakarma, S., Lee, J., Kim, K. H., & Kumar, S. (2021). E-waste management and its effects on the environment and human health. *The Science of the Total Environment*, *773*, 145623. doi:10.1016/j.scitotenv.2021.145623 PMID:33592459

Saoji, A. (2012). E-WASTE MANAGEMENT: AN EMERGING ENVIRONMENTAL AND HEALTH ISSUE. *National Journal of Medical Research*, *2*(01), 107–110.

Shittu, O. S., Williams, I. D., & Shaw, P. J. (2021). Global E-waste management: Can WEEE make a difference? A review of e-waste trends, legislation, contemporary issues and future challenges. *Waste Management (New York, N.Y.)*, *120*, 549–563. doi:10.1016/j.wasman.2020.10.016 PMID:33308953

Sivaramanan, S. (2013). E-Waste Management, Disposal and Its Impacts on the Environment. *Universal Journal of Environmental Research and Technology*, *3*(5).

Somani, P., Navaneethan, R. D., & Thangaiyan, S. (2021, May). Integrated solid waste management in urban India: A mini review. [). IOP Publishing.]. *Journal of Physics: Conference Series*, *1913*(1), 012084. doi:10.1088/1742-6596/1913/1/012084

Vergara-Araya, M., Lehn, H., & Poganietz, W. R. (2020). Integrated water, waste and energy management systems–A case study from Curauma, Chile. *Resources, Conservation and Recycling*, *156*, 104725. doi:10.1016/j.resconrec.2020.104725

Wath, S. B., Vaidya, A. N., Dutt, P. S., & Chakrabarti, T. (2010). A roadmap for development of a sustainable E-waste management system in India. *The Science of the Total Environment*, *409*(1), 19–32. doi:10.1016/j.scitotenv.2010.09.030 PMID:20951410

Xavier, L. H., Giese, E. C., Ribeiro-Duthie, A. C., & Lins, F. A. F. (2021). Sustainability and the circular economy: A theoretical approach focused on e-waste urban mining. *Resources Policy*, *74*, 101467. doi:10.1016/j.resourpol.2019.101467

Zhou, Z., & Zhang, L. (2022). Sustainable waste management and waste to energy: Valuation of energy potential of MSW in the Greater Bay Area of China. *Energy Policy*, *163*, 112857. doi:10.1016/j.enpol.2022.112857

Chapter 22
Biomedical Waste Handling Method Using Artificial Intelligence Techniques

D. Sengeni
C.K. College of Engineering and Technology, India

G. Padmapriya
School of Computing, SRM Institute of Science and Technology, India

S. Sagar Imambi
Koneru Lakshmaiah Education Foundation, India

D. Suganthi
Saveetha College of Liberal Arts and Sciences, SIMATS, India

Ashish Suri
Shri Mata Vaishno Devi University, India

S. Boopathi
Mythayammal Engineering College (Autonomous), India

ABSTRACT

The handling of genetic residues, biohazardous waste, and other non-shrinkable materials such as e-waste is essential for the preservation of Mother Nature and human wellbeing. This chapter will bring forth different methods of managing biomedical wastes, inspecting the roots of the spread of the hazardous virus, and protecting the front-line warriors. To properly react to this disaster, each administration must create a backup strategy based on regional circumstances and the seriousness of the coronavirus's spread. Technology is being used to reduce the need for in-person medical appointments and save on the cost of treatment, which is a boon for human civilization in the context of the pandemic. The Central Pollution Control Board's existing rules provide a framework for managing healthcare waste with respect for the safety of the public. Bio-waste management strategies using IoT and IoMT techniques are illustrated in this chapter.

DOI: 10.4018/978-1-6684-8117-2.ch022

INTRODUCTION

The bewilderment of the coronavirus in late 2019 created threats to individual health, affected our societies and economy with no uniform intensity, and transposed the general strata of perception of safety, health, and livelihood by the masses. In 2020, WHO declared COVID-19 an outbreak. What the planet did not foresee was the number of lives that would be lost. In respect to the above statement, the healthcare sector has then gained prominence over other services by saving lives, spreading awareness,, training their staff, and expecting a steadfast result from the battle of saving thousands of people against COVID 19. The Indian subcontinent has an unprecedented menace at hand: the rising pile of bio-medical waste. However, the jeopardy it has pushed upon the earth due to the mammoth sized amounts of increase in the biomedical waste that is being produced by hospitals that are treating COVID-infected patients is a fact that cannot be overlooked. However, it is not a revelation to assert that effectively managing medically contaminated trash contains the answer to controlling the coronavirus cycle. Hospitals guard human life, but, hazardous, COVID-led clinical waste created, particularly in the current pandemic crisis, continues to do irreparable harm (Reddy et al., 2023). Mismanagement of health care waste increases the number of airborne pathogenic infectious organisms, implying a growing hazard to the immediate environment of numerous pharmaceutical centres, research facilities, and public areas. While some of the top brains are contemplating finding a cure for the ongoing Corona virus, medical garbage contaminated with hazardous communicable pathogens has now turned into a distress for healthcare facilities as they can ramp up the curve of the affected cases. This book chapter focuses on the covivirus-infected trash handling system in the midst of the coronavirus pandemic. We systematically unravel the strata of knowledge between the surface of the panic caused by the recent coronavirus and the profundity of the dire need for the urgent management of biomedical waste, and everything in between. This book chapter has its objective to throw light on many topics that are of profound importance in the nightmare that this situation is turning into at the moment,, like the amendments made for treating the genomic misuse pre- and post-coronavirus upsurge, how the Internet of Medical Things is perceived as a boon to managing the healthcare sector, and the aggravating amounts of its subsequent trash. We move forward with special emphasis on the risks and repercussions of poor management of COVID-infected waste on scavengers and rag pickers,, and the front line warriors in particular in the medical field. We lay out some of the suggestions that might ameliorate the ongoing trepidation that the pandemic has pushed us towards (Boopathi, Khare, et al., 2023; S. et al., 2022).

The goal of this chapter is to manage current HWM procedures in the midst of the epidemic and identify the best strategy to control biomedical waste practises that are simple to implement using practical ways. For health care waste management and how to modify it, COVID-19 says waste management may necessitate particular measures, equipment, provision, operations, and administration during emergency scenarios.

BIOMEDICAL WASTE MANAGEMENT

A crystal-clear scenario is, with the growth of civilization, a larger population now has access to health care than ever before. Increased accessibility to healthcare facilities has not only enhanced the population's quality of life, but it has also put the community's health at risk due to the massive volume of biological waste produced. Because biomedical waste has both environmental and health dangers, it should be handled with care and properly disposed of. The regulation of biological waste is known as biomedical waste management. Multiple biological waste management failures have occurred in India, and the continuing covid-19 outbreak has made matters much more difficult. As an outcome, the governments and management have noticed an increase in biological waste as a result of the sudden development of a virus. Furthermore, the issue was compounded by weak infrastructure and a paucity of supplies. Management and the government have created a variety of SOPs as well as altered prevalent points to properly address this issue.

Levels And Analysis of Biomedical Waste

According to the BMW rules of 2016, each waste category is sub partitioned into a colour code of red, yellow, blue, and white. The most common source of BMW is formed in the handling or immunisation of covid-19, which is infectious when exposed to the elements. As a result, waste management is necessary for controlling and implementing awareness campaigns (Figure 1). The procedures by which garbage is managed and kept safe are referred to as levels.

Location of Root: According to the 2016 bio-medical waste management guidelines, garbage must be collected from sources such as hospitals, medical shops, and medical labs, and must be done under the supervision of a medical professional (in case of any misguide). Human or animal corpses should be eliminated of within 48 hours.

Figure 1. Levels of bio-medical waste

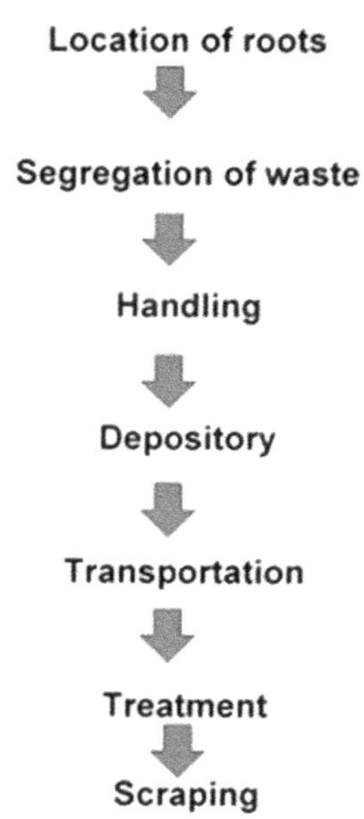

Classification: After the waste has been classified at the source level, it must be dispersed into different colour bags with correct labels, each with a specific capacity. Before coding, items like plastic, metals, and piercing objects must be removed.

Management and depository: After the colour bags have been coded, they must be precisely labelled with symbols, the date of entry, and the quantity, as well as the sender's and collector's information. Bags containing hazardous trash, such as radioactive and cytotoxic waste, that were dumped before 48 hours cannot be reused, and must not be stored after that time.

Transportation: According to BMWM,2018 rules, trucks or hospital facilities are employed to transport the tagged bags from adjacent hospitals that must be kept away from patients and visitors. The transport should be done on wheeled trolleys with no leaks or gas releases. The individual performing this procedure should be highly competent and aware of the risk of infection as well as the source of contamination in the event of a gas leak.

Treatment and scraping: The majority of the waste collected from medical facilities during COVID-19 is classified as general garbage. Those should be given to municipal employees, who will then separate them and dispose of them properly. Autoclaves, incineration, landfills, and deep burial are some of the treatment methods.

Some Effective Directives for Management of Biomedical Garbage

Many phases exist in the in the residual handling procedure that BMW management is fully reliant. Federal, sub-urban, and municipal governments will be able to draw a breath of healthcare/medical waste management policies. During this time of Covid-19, the waste policies in the municipal waste zone are beneficial to medical professionals. Those who did not have any kind of policy or strategy suffered a great deal during COVID-19. Municipalities or localities should include emergency preparation for unanticipated events, and ongoing pandemic COVID-19 waste management provocations can provide this knowledge (Figure 2).

The policy document should contour the logic for the legislation and national intentions(Anitha et al., 2023; Subha et al., 2023).

- Explanation of the health and safety issues that have arisen due to increased use of medical residue.
- The need for a safe environment for the introduction of HCWM at the outset of medical practise.

Following the Approved Disposal Procedures

- The use of damaging practises, such as disposal of hazardous medical waste in municipal areas, is discouraged.
- Costs of health-care waste management are being assessed.
- The following policy procedures should be outlined in technical guidelines: minimization, detachment, identification, management, treatment, and final ejection of trash.
- It should be digitally recorded or documented/written on paper
- Prepare before entering into the field.
- Governing the rules for the workers' safety and health

Biomedical waste management is based on the above-mentioned waste handling stages. As a result, in compliance with the Annual Health Survey, the (DLHS) has been confirmed to obtain advice on BMW arrangements. During the Covid-19 outbreak, the DLHS conducted a biological waste study across the country, which includes 30 states and union territories. The survey system consists of protocols that have been followed by collecting waste in different colour bins labelled as yellow for toxic waste and red for biohazard symbol. The survey system consists of protocols that have been followed by collecting waste in different colour bins labelled as yellow for toxic waste and red for biohazard symbol. Plastic ware is blue, sharps are white, and organs are black. Different techniques of biological waste treatment, such as autoclaves, chemical disinfectants, shedders, incinerators, and deep burial pits are used to dispose them. According to the facts provided above, many countries' problems include not just enforcing compliance but also formulating regulations and providing guidance. In times of political decision-making or government mobilisation, a medical waste management policy is required, and resources must be available to assist in policy implementation(Babu et al., 2023; Boopathi, Arigela, et al., 2023; Jeevanantham et al., 2023).

- **Regulations** stating medical residual organization as well as many sectors' transportation, training qualifications.
- **Updating**: Check the facts, resources, and explain the options to the authorities for waste employees and actions.
- **A Red alert plan**: We can declare that we have an emergency strategy in place for unforeseen situations and to ensure that all HCWM services are available. In such cases, an alternative plan

Figure 2. Social order of regulatory and organizational structure

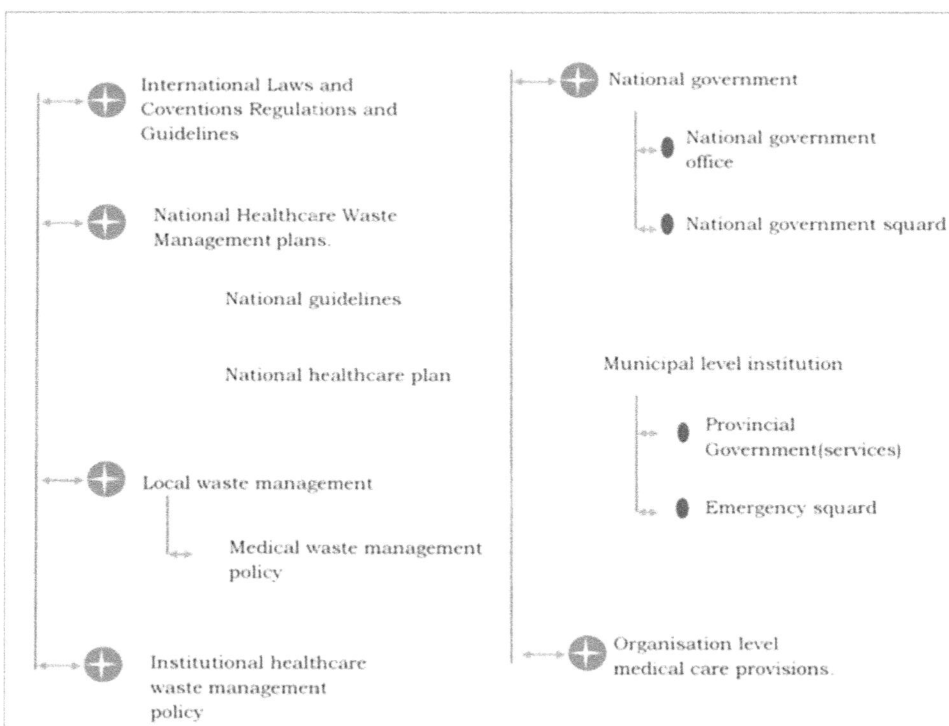

could include personal cars, disinfectant washing, and volunteer street cleaning services. If the environment is clear of diseases, one can feel safe.

Bio medical waste management depends on various steps of handling waste as mentioned above. consequently, the DLHS has been documented to procure advice on BMW according to the AHS. A survey was conducted regarding BMW all over our country 21. This survey was done to accumulate inputs from medical hospitals, households., The survey system consists of protocols followed by collecting the waste in different colour bins labelled as yellow for toxic waste, red for bio hazard symbol, blue for plastic ware and white for sharps and black for organs etc. They are disposed by different methods.

BIO-WASTE SEGREGATION, STORAGE, AND TRANSPORTATION OF COVID-19 DISPOSALS

Treatment And Disposal Methods

The current focus is on creating rapid assessments, therapies, and immunizations, as well as persuading the public to adopt control steps. Identifying and reacting to potential pandemic spread via different kinds of associated debris has received insufficient focus. While the regulations are being amended, a sudden shift in the categorization of certain rules may result in improper handling(Janardhana et al., 2023; Mohanty et al., 2023).

The NDMC gathered 3,599 kgs of BMW and 11,438 kgs of NBMW from many isolation centres and residences of patients. For this reason, the NDMC has partnered with other organizations. The refuse gathered is moved to designated hospitals/dispensaries for treatment and removal via a shared biological waste treatment centre.

Some low income countries are still leading a gruelling struggle against the topic in light. A few nations have used modern technology to steam-sterilize or chemically cleanse their medical waste, but these are the exception.

When dangerous and benign materials are mixed, the total refuse becomes toxic. Typically, discarded medical supplies and other kinds of medical debris are disposed in a clean dump or burned as part of a waste-to-energy reclamation program. However, in many developing countries, hospital refuses are discarded in the open or in badly administered dumps, where rubbish collectors and animals are regularly seen.

Regarding the pandemic's retreat, changes to the BMWM standards require that some bio-hazardous material be divided into well-labeled receptacles or sacks at the point of production before being kept, moved, treated, and disposed of. Only a specific vehicle held by a responsible authority, as specified by the government, may transport unprocessed medical refuse. Untreated medical refuse should not be kept or held for longer than 48 hours.The municipality will continue to gather and convey non-medical solid refuse produced in hospitals and care homes, as well as correctly cleaned medical waste, to a city dumpsite. Every year, each agency is required to file a report to the appropriate authorities documenting the kinds and quantities of medical refuse managed the previous year(Harikaran et al., 2023; Selvakumar et al., 2023; Vanitha et al., 2023). In this upcoming discussion, we lay out some of the safety directives following which will shield all the personnel irrespective of their sector of employment against the spread of covid 19 via the covid infected clinical waste handling.

Disposal Guidelines

- Follow national directives and WHO/WASH regulations.
- Separate coronavirus residues from other infectious trash at the source.
- Dangerous refuse should be discarded of in a colored foot-driven receptacle.
- Accumulation, transportation, and retention of the corona infected clinical waste should be carried out with extreme care.
- Utilize PPE and maintain hand cleanliness.
- Hybrid stream systems, microwaves, and ecological approachable handling techniques.
- Combined conductivity with sterilizer and radiation
- Use low-resource chlorine, an immersion sterilizer on the maximum level, sealing, and sterile interment for regular refuse receptacle decontamination and cleansing.
- The rag pickers or personnel associated to covid-led biomedical waste collection and handling, must fortify themselves against the risk of infection spread by wearing double layered, least penetration masks along with sturdy boots and gloves .
- The societies as a whole, must finance in waste disposal groundwork which will aid them in effectively managing the clinical or suspected biohazardous garbage.

Treatment Options

Due to the uncommon output of waste from both homes and clinics, there appears to be a garbage emergency due to the accumulation of gloves, robes, masks, and other protective apparel and equipment. Failure to correctly handle garbage produced by clinics may exacerbate the spread of the pandemic through secondary transmission. The potential rampant dumping, open combustion, and incineration could affect air quality and health outcomes due to toxin exposure. Here we present to you some of the most defined treatment options by which we can manage the healthcare waste:

- Hydrogen Peroxide (H_2O_2) >0.5% • Sustainable modifications with long-term improvements in system strengthening: steam-based treatment, autoclaving, or microwaving are chosen for waste treatment. Requirements.
- Improve tracking of the biomedical waste units which come out of the hospitals and reduce incineration.
- Waste may be sent for disposal after disinfection.
- Recycling should proceed as usual and dental should be held with care.

Challenges in Managing the Bio Medical Waste During Coronavirus

One of UNEP's main themes is to make full use of current waste management systems. "Disaster is the worst time for trading professional cards," as the saying goes in emergency management. It is also a bad time to build new medical WMS during an epidemic. In today's date, Covid 19 pandemic has spawned many challenges and setbacks in various walks of our lives(Boopathi, Khare, et al., 2023). Some of the challenges are explored below:

- At the moment, no appropriate drugs has been disclosed for addressing COVID-19; administration and companies are battling to discover operational medicines. The public are consuming medicines for curing themselves from the pandemic, which unnecessarily increases BMW.
- One of the most ignored aspects of health care in India is BMW regulation. Since the the administration made changes in 2016 by providing simpler classifications for division of separate BMWs, an update in 2018 was passed with the objective of increasing regulation compliance. With a marginal growth rate, appropriate classification, management, and removal remained a critical issue.
- People admit themselves out of fear and add to the BMW generation.

The Usage of PPE Kit

According to preliminary reports, conditions may worsen amid the epidemic, with large PPE supplies in hospitals. Despite WHO and Government of India guidelines on personal protective equipment (PPE), health care settings are experiencing high PPE demand due to infection fear. Fear frequently leads to PPE misuse, worsening the situation by causing BMW, which are difficult to keep and transport.

Unusual Demand of Safety Apparels

Medical supplies are in higher demand. Even in the community, the haphazard use and dumping of disposable medical shields adds to the danger. They are commonly discarded with inert trash from homes and domestic situations where colored BMW receptacles are scarce. Facial coverings have accumulated along the territory's environment routes and coastlines as a result of erroneous disposal in rivers and lakes, disturbing the aquatic ecosystem. Aquatic life suffers as a result of aquatic ecosystems mistaking inert polymers from facial shields for sustenance.

Expertise of Some Countries to Handle Covid-Led Clinical Waste

Production soared, but most nations were not prepared to manage such massive volumes of Bio hazardous trash. Today, most nations, particularly developing nations, suffer the same problem as Wuhan did in the past: insufficient medical waste treatment facilities. Since the established BMW facilities in developing countries lack the ability to handle a rise in medical waste during a epidemic, existing facilities must be extended. New treatment centers must be built.

Building Cleansing Employees' Capacity to Safeguard Their Own Opposing Contamination Progression

Clean-up employees, such as medical professionals and nurses, are a similarly important set of players engaged in effective COVID-19 epidemic control.

One of the most serious problems is that people who are subjected to biological garbage at different levels are completely ignorant of the risk. Raising knowledge and instruction of human resources engaged in BMW gathering and dumping are the primary areas for keeping safety and averting dangers. Non-technical hospital employees and college medical students lack adequate understanding of BMW administration. The workers should be educated on infection control measures, hand cleanliness, breathing decorum, social space, and the use of PPE, according to guidelines for BMW management.

ALTERNATIVE TECHNOLOGIES THAT HELP BIO MEDICAL WASTE MANAGEMENT

IOT

In terms of technology, the importance of healthcare waste management cannot be overstated. The IoT has the ability to revolutionize waste management. At all refuse sites, connected sensing devices and an organized IoT-based environment can be implemented. Smart garbage bins may be used to develop intelligent ways by combining the system, networks, and technology suppliers with IoT-based devices. In the BMW, intelligent receptacles with sufficient classification and upkeep can be used. In medical clinics, the traditional method of trash collection might expose personnel to harmful diseases and injuries. IoT sensors and devices can be used to transmit data related to significant variables involved in BMW. The implementation of IoT-based trash management also requires a comprehensive investigation of the problem from multiple perspectives. Implementing a solid refuse management framework helps to create a clean atmosphere as well as a safe community(Boopathi, Khare, et al., 2023).

The following some topics are in practice of medical waste management which are majorly contributed by IOT.

- In garbage management, the Internet of Things is revolting.
- Medical waste management research based on resources
- In the field of healthcare waste management, wearable technologies are being developed.
- Possibilities for reducing healthcare waste
- Smart technology has made a significant contribution to hospital waste management.
- A sophisticated healthcare waste monitoring system.
- Infectious and hazardous waste management systems have taken on new approaches in recent years.
- In healthcare, IoT-based scrap management

IoMT

The IoMT is a medical innovation. Social distancing norms, which have prompted the development of modern devices with internet of things (IoMT), such as using Wifi or Bluetooth to connect with patients without touching or standing nearby patients and allowing aid to be delivered remotely, have increased the need in the health-care system (Figure 3). If every individual in a large population country gets infected (for example, the Covid-19 pandemic), this technology will come in handy so that competent doctors can manage high-risk patients(Anitha et al., 2023; Subha et al., 2023).

This enables us to expand our networks while still safeguarding people and assuring business. The goal of this study was to look into places where we could use technological advances to significantly improve the client's treatment process despite increasing costs.

While IoMT and advances in medical treatment are wide subjects, this study concentrated on 9 particular applications and questioned ten physicians from the United States and India. One of the objectives of the initiative is to find applications that have a greater probability to be available for implementation at the moment rather than those that require greater hazards or additional time to develop.

Figure 3. IoMT

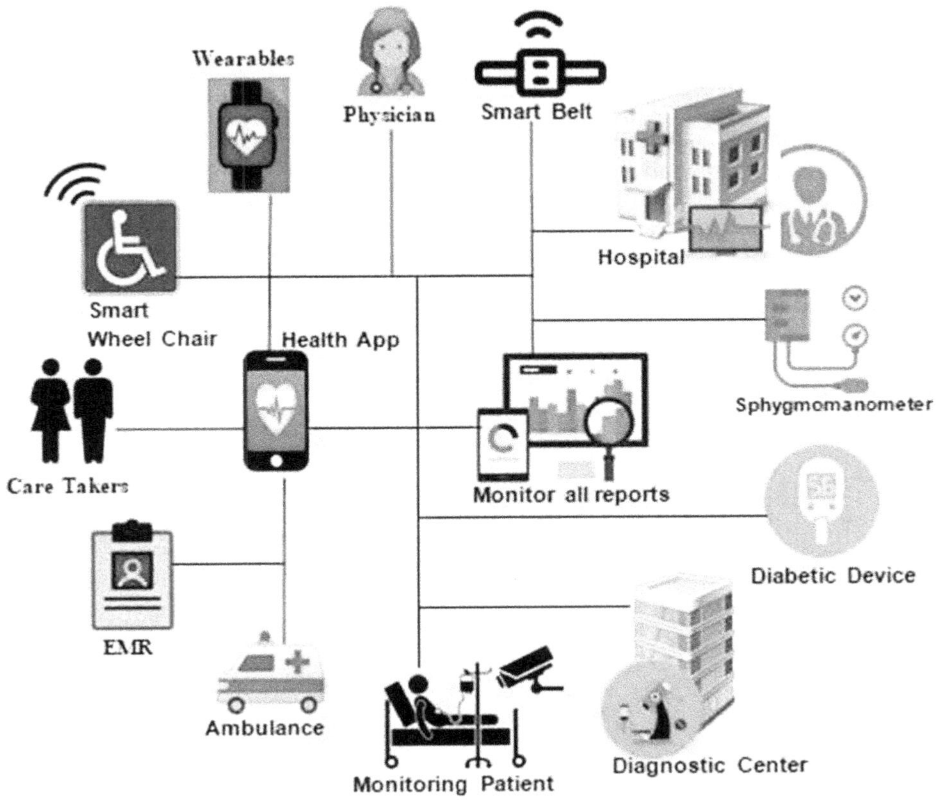

Among the most common applications were drone transportation, patient video tracking, in-patient monitoring, hospital trash management, and condition monitoring.

Waste Management Monitoring

Many hospitals' existing waste management systems entail inefficient rubbish collection since trash fluctuates over time. Trash administration tracking at medical facilities could notify healthcare facilities when containers achieve topping and avoiding redundant accumulations. Rubbish is currently collected on a schedule; for example, certain hospitals may only collect garbage once a week. This can lead to overflowing trash bins. Material administration observing should use detectors to alert healthcare facilities as inefficiency amounts grow above 75%, permitting the establishment to be more productive.

These innovations, corresponding to the medical professionals polled, are currently in use in many medical facilities, and the priority ought to be on employing recyclable materials, that could decrease both garbage and expenditure.

IOMT Helping in BMWM to Prevent the Spread of COVID-19

BMW arrangements comprise intelligent smart bins with correct waste material classification and maintenance. However, no one can handle the bins during the COVID-19 epidemic, therefore clever smart bins with automated lid opening come into play. Weight could be computed using IoMT, and signals sent to garbage collectors to pick up the rubbish could be sent automatically (Figure 4). Smart pill technology is used to meet pharmaceutical needs, as well as to categorise pharmaceuticals into two groups: helpful and non-essential for human usage, which helps with waste management.

Figure 4. Advantages of IoMT

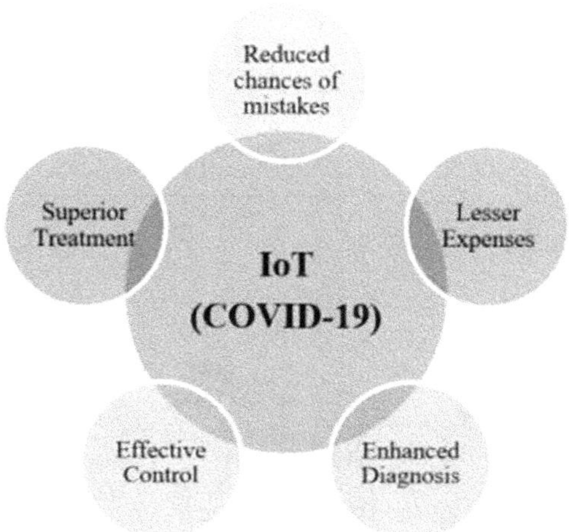

OTHER TECHNOLOGIES IN BIO MEDICAL WASTE MANAGEMENT

The projects mentioned below are only a few of the many that Health Care Without Harm and its partners are working on across the world(Philip et al., 2021; Raviteja & Supriya, 2020).

The Bagamayo project: An imaginative preliminary investigation employing a sterilizer for disinfecting infectious garbage is currently underway at HCWH where workers are instructed about discrimination. A sterilizer and splitter in collaboration with native non-governmental organizations were used. A waste administration group was formed.

The GEF Project: The HCWH is working on a global ecological equipment-assisted undertaking with the WHO and the UNO. The project's title is "Demonstrating and Promoting Best Techniques and Practices for Reducing Health Care Waste to Avoid Dioxin and Mercury Environmental Releases." It was primarily established to help poor nations. By encouraging best practises for reducing and managing health-care waste, the project's ultimate objective is to reduce dioxin and mercury emissions into the environment.

Advantages (Pros) of Waste Management in Bio Medical Field

Compliance with state and federal rules is the most fundamental advantage of effective hospital waste disposal. The WHO, CDC, OSHA, and USEPA have enacted rules to implement the cleanliness and waste regulation of medical facilities(Jeevanantham et al., 2023; Reddy et al., 2023; S. et al., 2022; Saha1 et al., 2022; Samikannu et al., 2023).

- Biomedical waste management may lead to a cleaner environment as there are so many diseases due to improper waste management like virus and many diseases. Used medical goods will be maintained in one location due to waste management. Water and wildlife habitats can be contaminated due to inadequate HWM. If not properly disposed of, toxic hospital waste can endanger humans, animals, and plants. Medical waste is sometimes placed in sewage, which flows to lakes and contaminates the water, as well as the organisms that live in the water and animals.
- Healthcare garbage handling is often implemented for accepted, but there are numerous benefits to having an effective healthcare garbage framework, like individuals as well as staff harmony of awareness and assurance. Poor hospital waste management exposes patients to infectious pathogens, needle sticks, and other harmful situations.
- Aids in the prevention of hospital-acquired illnesses by enhancing the preventive impact of good handwashing;
- Cuts infection cycles; and targets health-care personnel in a simple and cost-effective manner.
- When precise instructions and guidelines are followed at each step, waste management is extremely effective and efficient. The first stage should be to determine viable choices for BMWM based on the budget, technology, and preferences of the local community. When selecting the proper treatment technique for health care waste, several variables (such as volume, temperature, and composition) must be addressed. whether the waste is liquid or solid, hazardous or infectious). by determining these all we can be able to use precise technology and money to keep the hospitals labs clean
- One of the benefits of waste management is that it classifies garbage into distinct types of dry and wet trash, as well as different coloured trash. The hue development corresponds to a specific category of garbage and disposal alternatives. As a result, the procedure is simple.
- Many dangerous diseases have been halted as a result of biomedical waste management. This is due to the presence of hazardous miniature infectious agents like viruses, microbes, and bugs. If waste is not handled in a manner that destroys these infectious organisms, chemicals may penetrate the human system through cuts and other fluid layers in the oral cavity.

Limitations to Waste Management in Bio Medical Field

- Front line warriors have used many medical kits as well as surgical equipment. Yet, the problem comes after its usage which is a disadvantage. As public outcry grows in many nations, the US has seen an exponential surge in medical waste production. Every hour, 45 tonnes of garbage are created, necessitating biomedical waste management(Anitha et al., 2023; Subha et al., 2023).
- Many components or goods, such as syringes, pills, ppe kits, vaccinations, and so on, are non-biodegradable trash that takes many years to degrade, resulting in contamination of the environment and land use.

- Because much of the globe was impacted by the covid 19 epidemic, many pharmacies raised their output by 40 to 50 percent, resulting in a loss in natural resources.
- The total garbage from covid 19 unaffected individuals, including food conceals thrown by them, is three times that of confined individuals. As a result of the waste, employees may be afflicted and may infect others without their knowledge. Some countries, such as Indonesia (the Java area), suffer from poor garbage management.
- Burning waste management is less attractive when there is less of an emergency or a need for temporary disposal.
- In the event of impacted situations, transportation would be problematic.
- As the number of cases grew, so did operational capacity.
- Medical waste handles are at high risk during this pandemic in case of high injury or infection, especially from sharp objects.
- Most of the garbage is radioactive, rag pickers must use caution.

Waste management itself doesn't have any faults .But the types of BMWM have some disadvantages also such as:

Incineration: There appears to be an environmental crisis due to the accumulation of coverings, dresses, respirators, and other PPE owing to the unprecedented volume of trash from each household and healthcare establishments(Boopathi & Myilsamy, 2021; Haribalaji et al., 2022; Sampath et al., 2021; Vignesh et al., 2018). Inability to properly manage garbage generated by medical institutions as well as households could lead to supplementary COVID-19 transfer.In this sort of BMWM, the procedure is too costly. As a result of the process, various harmful gases are released, potentially polluting the air. The heavy metals that remain in the ash residue after the incineration are harmful. It's also a significant generator of dioxin and a colourless flammable poisonous liquid chemical.

Autoclave: The trash should be processed first and foremost using a drying system to dry any wet components that may be present. In the case of insufficient air supply, odours might be a concern near autoclaves. The procedure is confined to particular types of wastes and is not suited for all sorts of trash. We need to use a shredder to minimise the volume of garbage, and processed garbage can be maintained due to its physical appearance which requires further processing of final ejection.

Microwave: Some chemicals, such as volatile and semivolatile organic mixtures, mercury, and other hazardous chemical wastes, are not handled by microwave treatment, necessitating a different ultimate disposal technique, and volume and weight reduction are limited.

Future Work Extension

In the future study, the various parameters to control the medical waste pollution and management using various machine learning techniques(Babu et al., 2023; Boopathi, Khare, et al., 2023; Boopathi, Siva Kumar, et al., 2023; Boopathi, Arigela, et al., 2023; Domakonda et al., 2023; Jeevanantham et al., 2023; Kumara et al., 2023; Palaniappan et al., 2023; S. et al., 2022; Samikannu et al., 2023; Sampath et al., 2022; Senthil et al., 2023; Vanitha et al., 2023). The TOPSIS, VIKOR, NSGA-II and MOEA have been used to predict the result to minimize the medical waste reduction (Boopathi, 2022c, 2022a; Kannan et al., 2022; Myilsamy & Sampath, 2021; Sampath et al., 2022; Yupapin et al., 2023). The experiments have been performed using statistical approaches such as Taguchi, Response surface methodology and mixture methods(Boopathi, 2021, 2022b; Kalidas et al., 2012; Sampath et al., 2021; Saravanan et al., 2022; Trojovský et al., 2023).

RECOMMENDATIONS

Some Chinese regions bought portable rehabilitation automobiles as safeguards over the last ten years and acquired considerable expertise. As a backup plan for unforeseen circumstances, every nation and fundamental town should have some portable healthcare garbage implementation. amenities. While required, these establishments may help certain isolated regions and territories with the correct healthcare squander treatment.

- More abilities for portable amenities should be preserved, especially during epidemics, which can be critical for acquiring nations with limited healthcare garbage elimination facilities. Portable amenities are not only useful in a crisis, but they can also be used as a tactical protection ability for the nation in the years to come.
- Using societal mediums enhance belonging, worldwide, and worldwide consciousness.
- Choose the Basel convention regulations and have nations implement and tailor them to their specific circumstances.
- The societies as a whole, must finance in waste disposal groundwork and in the raising the cognizance in their residents by conducting online awareness sessions.

SUMMARY

Because of the world's overpopulation, demand for healthcare is uncertain. Improving healthcare effectiveness is the only way to reduce biological waste. This book chapter was a laborious effort to gather information and review all parameters related to the management of general medical waste on one hand and medical waste specific to the pandemic on the other. This is because, clinical waste requires special treatment due to its ramifications on the environment and the human race.

Furthermore, a clean human existence and a green environment are only achievable if one has a healthier environment thanks to a better BMWM system. It gives off a positive feeling about reducing occurrences caused by infectious illnesses, regulating the transmission of viruses and germs during the process, reusing, recycling, and reducing needless waste, and generating income via effective management in BMW. To summarise, each individual may contribute to the total eradication of the pandemic scenario caused by Covid 19 by obeying the government's as well as BMWM's laws and regulations to safeguard the environment and health quality.Many laws and regulations, such as adequate waste generation and segregation, a hospital-based biomedical waste management plant, and better health-care member measures, are adopted as emergency requirements and practised by every member of medical units in order to protect public health. Many organisations have been successful in supporting or maintaining the awareness programme and effort to organise knowledge for both health care personnel and covid 19 patients.

Owing to the strides in IOMT, medical equipment and use cases will be able to communicate with the patient and the emergency system workplace in the future, and with further analysis, it will be able to predict disease outcomes and prepare medicines in the event of a new disease outbreak, such as smart hospital data collection.

REFERENCES

Anitha, C., R, K. C., Vivekanand, C. V., Lalitha, S. D., Boopathi, S., & R, R. (2023, February). Artificial Intelligence driven security model for Internet of Medical Things ({IoMT}). *IEEE Explore*. doi:10.1109/ICIPTM57143.2023.10117713

Babu, B. S., Kamalakannan, J., Meenatchi, N., M, S. K. S., S, K., & Boopathi, S. (2023). Economic impacts and reliability evaluation of battery by adopting Electric Vehicle. *IEEE Explore*, 1–6. doi:10.1109/ICPECTS56089.2022.10046786

Boopathi, S. (2021). Improving of Green Sand-Mould Quality using Taguchi Technique. *Journal of Engineering Research*. doi:10.36909/jer.14079

Boopathi, S. (2022a). An Extensive Review on Sustainable Developments of Dry and Near-Dry Electrical Discharge Machining Processes. *Journal of Manufacturing Science and Engineering*, *144*(5), 50801. doi:10.1115/1.4052527

Boopathi, S. (2022b). Cryogenically treated and untreated stainless steel grade 317 in sustainable wire electrical discharge machining process: A comparative study. *Environmental Science and Pollution Research International*, 1–10. doi:10.100711356-022-22843-x PMID:36057706

Boopathi, S. (2022c). Experimental investigation and multi-objective optimization of cryogenic Friction-stir-welding of AA2014 and AZ31B alloys using MOORA technique. *Materials Today. Communications*, *33*, 104937. doi:10.1016/j.mtcomm.2022.104937

Boopathi, S., Arigela, S. H., Raman, R., Indhumathi, C., Kavitha, V., & Bhatt, B. C. (2023). Prominent Rule Control-based Internet of Things: Poultry Farm Management System. *IEEE Explore*, 1–6. doi:10.1109/ICPECTS56089.2022.10047039

Boopathi, S., Khare, R., Jaya Christiyan, K. G., Muni, T. V., & Khare, S. (2023). Additive manufacturing developments in the medical engineering field. In Development, Properties, and Industrial Applications of 3D Printed Polymer Composites (pp. 86–106). IGI Global. doi:10.4018/978-1-6684-6009-2.ch006

Boopathi, S., & Myilsamy, S. (2021). Material removal rate and surface roughness study on Near-dry wire electrical discharge Machining process. *Materials Today: Proceedings*, *45*(9), 8149–8156. doi:10.1016/j.matpr.2021.02.267

Boopathi, S., Siva Kumar, P. K., & Meena, R. S. J., S. I., P., S. K., & Sudhakar, M. (2023). Sustainable Developments of Modern Soil-Less Agro-Cultivation Systems. In Human Agro-Energy Optimization for Business and Industry (pp. 69–87). IGI Global. doi:10.4018/978-1-6684-4118-3.ch004

Domakonda, V. K., Farooq, S., Chinthamreddy, S., Puviarasi, R., Sudhakar, M., & Boopathi, S. (2023). Sustainable Developments of Hybrid Floating Solar Power Plants. In *Human Agro-Energy Optimization for Business and Industry* (pp. 148–167). IGI Global. doi:10.4018/978-1-6684-4118-3.ch008

Haribalaji, V., Venkatesan, G., Asif, M. M., Pandian, M., Subbiah, R., & Boopathi, S. (2022). Investigation on corrosion and tensile Characteristics: Friction stir welding of AA7075 and AA2014. *Materials Today: Proceedings*, *66*(3), 743–748. doi:10.1016/j.matpr.2022.04.037

Harikaran, M., Boopathi, S., Gokulakannan, S., & Poonguzhali, M. (2023). Study on the Source of E-Waste Management and Disposal Methods. In *Sustainable Approaches and Strategies for E-Waste Management and Utilization* (pp. 39–60). IGI Global. doi:10.4018/978-1-6684-7573-7.ch003

Janardhana, K., Singh, V., Singh, S. N., Babu, T. S. R., Bano, S., & Boopathi, S. (2023). Utilization Process for Electronic Waste in Eco-Friendly Concrete: Experimental Study. In Sustainable Approaches and Strategies for E-Waste Management and Utilization (pp. 204–223). IGI Global.

Jeevanantham, Y. A., A, S., V, V., J, S. I., Boopathi, S., & Kumar, D. P. (2023). Implementation of Internet-of Things (IoT) in Soil Irrigation System. *IEEE Explore*, 1–5. doi:10.1109/ICPECTS56089.2022.10047185

Kalidas, R., Boopathi, S., Sivakumar, K., & Mohankumar, P. (2012). Optimization of Machining Parameters of WEDM Process Based On the Taguchi Method. *Ijest, 6*(1\&2), 91.

Kannan, E., Trabelsi, Y., Boopathi, S., & Alagesan, S. (2022). Influences of cryogenically treated work material on near-dry wire-cut electrical discharge machining process. *Surface Topography : Metrology and Properties*, *10*(1), 15027. doi:10.1088/2051-672X/ac53e1

Kumara, V., Mohanaprakash, T. A., Fairooz, S., Jamal, K., Babu, T., & B., S. (2023). Experimental Study on a Reliable Smart Hydroponics System. In *Human Agro-Energy Optimization for Business and Industry* (pp. 27–45). IGI Global. doi:10.4018/978-1-6684-4118-3.ch002

Mohanty, A., Venkateswaran, N., Ranjit, P. S., Tripathi, M. A., & Boopathi, S. (2023). Innovative Strategy for Profitable Automobile Industries: Working Capital Management. In Handbook of Research on Designing Sustainable Supply Chains to Achieve a Circular Economy (pp. 412–428). IGI Global.

Myilsamy, S., & Sampath, B. (2021). Experimental comparison of near-dry and cryogenically cooled near-dry machining in wire-cut electrical discharge machining processes. *Surface Topography : Metrology and Properties*, *9*(3), 35015. doi:10.1088/2051-672X/ac15e0

Palaniappan, M., Tirlangi, S., Mohamed, M. J. S., Moorthy, R. M. S., Valeti, S. V., & Boopathi, S. (2023). Fused deposition modelling of polylactic acid (PLA)-based polymer composites: A case study. In Development, Properties, and Industrial Applications of 3D Printed Polymer Composites (pp. 66–85). IGI Global. doi:10.4018/978-1-6684-6009-2.ch005

Philip, N. Y., Rodrigues, J. J. P. C., Wang, H., Fong, S. J., & Chen, J. (2021). Internet of Things for In-Home Health Monitoring Systems: Current Advances, Challenges and Future Directions. *IEEE Journal on Selected Areas in Communications*, *39*(2), 300–310. doi:10.1109/JSAC.2020.3042421

Raviteja, K., & Supriya, M. (2020). Greenhouse Monitoring System Based on Internet of Things. *Lecture Notes in Electrical Engineering*, *637*, 581–591. doi:10.1007/978-981-15-2612-1_56

Reddy, M. A., Reddy, B. M., Mukund, C. S., Venneti, K., Preethi, D. M. D., & Boopathi, S. (2023). Social Health Protection During the COVID-Pandemic Using IoT. In *The COVID-19 Pandemic and the Digitalization of Diplomacy* (pp. 204–235). IGI Global. doi:10.4018/978-1-7998-8394-4.ch009

S., P. K., Sampath, B., R., S. K., Babu, B. H., & N., A. (2022). Hydroponics, Aeroponics, and Aquaponics Technologies in Modern Agricultural Cultivation. In *Trends, Paradigms, and Advances in Mechatronics Engineering* (pp. 223–241). IGI Global. doi:10.4018/978-1-6684-5887-7.ch012

Sahal, B. C., R, D., A, A., Thrinath, B. V. S., Boopathi, S., J. R., & Sudhakar, M. (2022). *IOT Based Smart Energy Meter For Smart Grid*. MDPI.

Samikannu, R., Koshariya, A. K., Poornima, E., Ramesh, S., Kumar, A., & Boopathi, S. (2023). Sustainable Development in Modern Aquaponics Cultivation Systems Using IoT Technologies. In *Human Agro-Energy Optimization for Business and Industry* (pp. 105–127). IGI Global. doi:10.4018/978-1-6684-4118-3.ch006

Sampath, B., Naveenkumar, N., Sampathkumar, P., Silambarasan, P., Venkadesh, A., & Sakthivel, M. (2021). Experimental comparative study of banana fiber composite with glass fiber composite material using Taguchi method. *Materials Today: Proceedings*, *49*(5), 1475–1480. doi:10.1016/j.matpr.2021.07.232

Sampath, B. C. S., & Myilsamy, S. (2022). Application of TOPSIS Optimization Technique in the Micro-Machining Process. In Trends, Paradigms, and Advances in Mechatronics Engineering (pp. 162–187). IGI Global. doi:10.4018/978-1-6684-5887-7.ch009

Saravanan, M., Vasanth, M., Boopathi, S., Sureshkumar, M., & Haribalaji, V. (2022). Optimization of Quench Polish Quench (QPQ) Coating Process Using Taguchi Method. *Key Engineering Materials*, *935*, 83–91. doi:10.4028/p-z569vy

Selvakumar, S., Adithe, S., Isaac, J. S., Pradhan, R., Venkatesh, V., & Sampath, B. (2023). A Study of the Printed Circuit Board (PCB) E-Waste Recycling Process. In Sustainable Approaches and Strategies for E-Waste Management and Utilization (pp. 159–184). IGI Global.

Senthil, T. S. R. Ohmsakthi vel, Puviyarasan, M., Babu, S. R., Surakasi, R., & Sampath, B. (2023). Industrial Robot-Integrated Fused Deposition Modelling for the 3D Printing Process. In Development, Properties, and Industrial Applications of 3D Printed Polymer Composites (pp. 188–210). IGI Global. doi:10.4018/978-1-6684-6009-2.ch011

Subha, S., Inbamalar, T. M., R, K. C., Suresh, L. R., Boopathi, S., & Alaskar, K. (2023, February). A Remote Health Care Monitoring system using internet of medical things ({IoMT}). *IEEE Explore*. doi:10.1109/ICIPTM57143.2023.10118103

Trojovský, P., Dhasarathan, V., & Boopathi, S. (2023). Experimental investigations on cryogenic friction-stir welding of similar ZE42 magnesium alloys. *Alexandria Engineering Journal*, *66*(1), 1–14. doi:10.1016/j.aej.2022.12.007

Vanitha, S. K. R., & Boopathi, S. (2023). Artificial Intelligence Techniques in Water Purification and Utilization. In *Human Agro-Energy Optimization for Business and Industry* (pp. 202–218). IGI Global. doi:10.4018/978-1-6684-4118-3.ch010

Vignesh, S., Arulshri, K. P., Syedsajith, S., Kathiresan, S., Boopathi, S., & Dinesh Babu, P. (2018). Design and development of ornithopter and experimental analysis of flapping rate under various operating conditions. *Materials Today: Proceedings*, *5*(11), 25185–25194. doi:10.1016/j.matpr.2018.10.320

Yupapin, P., Trabelsi, Y., Nattappan, A., & Boopathi, S. (2023). Performance Improvement of Wire-Cut Electrical Discharge Machining Process Using Cryogenically Treated Super-Conductive State of Monel-K500 Alloy. *Iranian Journal of Science and Technology. Transaction of Mechanical Engineering*, *47*(1), 267–283. doi:10.100740997-022-00513-0

Chapter 23
Biofuels and Health Hazards:
An Overview

Swapan Banerjee
Department of Nutrition, Seacom Skills University, India

Soumen Ghosh
Agricultural and Food Engineering Department, Indian Institute of Technology, Kharagpur, India

Gourav Dhar Bhowmick
Agricultural and Food Engineering Department, Indian Institute of Technology, Kharagpur, India

Digvijay Pandey
https://orcid.org/0000-0003-0353-174X
Department of Technical Education, IET, Dr. A.P.J. Abdul Kalam Technical University, Lucknow, India

Binay Kumar Pandey
https://orcid.org/0000-0002-4041-1213
Department of Information Technology, College of Technology, Govind Ballabh Pant University of Agriculture and Technology, Pantnagar, India

Pankaj Dadheech
https://orcid.org/0000-0001-5783-1989
Swami Keshvanand Institute of Technology, Management, and Gramothan, India

ABSTRACT

Biofuels, such as coal and oil, have been the most suitable and sufficient substitutes for other conventional fuels in the energy sector for the last few years. For decades, entire populations have suffered from deforestation, global warming, and carbon monoxide issues. Biofuel production is an eco-friendly process that still has health hazards for manufacturing unit workers. That is due to flammable and combustible raw materials and their chemical reactions. Bioethanol is a flammable biofuel prepared by cellulose items and the fermentation of grains. Biodiesel is another combustible liquid-based biofuel produced by alcohol and glyceride in vegetable oil duly catalyzed by strong alkaline like caustic soda. Biofuel plants or trees are the biggest isoprene producers, which meet the ozone layer in the earth's atmosphere. While combining with ozone always causes many health hazards for manufacturing units' employees despite several safety measures. One of the significant consequences is that some factory workers often have asthma, allergy, and lung disorders due to ozone attacks.

DOI: 10.4018/978-1-6684-8117-2.ch023

1. INTRODUCTION

Global energy supply primarily depends upon fossil fuels now. The need for alternative green technology in the explosive energy domain is well understood. The reason is that fossil fuel resources become scarce gradually, negatively impacting the environment and livelihood. Randomly using fossil fuels has become the primary cause of global warming, air& water pollution, and health-related issues. It creates a crisis in energy sectors due to the depletion of exhaustible energy resources. Biofuel is the future of fuel technology obtained from different biomasses and potential renewable energy sources. It includes organic matter derived from plants, animals& microorganisms. This will be a game-changing option in the energy industry with competitive, sustainable economic growth over traditional fossil fuel industries. Bio-based energy can also compete with other forms of renewable energy that are non-exhaustible, e.g., wind, tidal, geothermal, solar energy, inefficiency, and other parameters.

1.1. Biofuels and Biomass

As cetane and octane number is a matter of concern, biofuels can be used after blending with different fossil fuels to minimize carbon monoxide and sulfur oxide (SO_x) emissions (Chen, H. et al., 2015).The first initiates took in the late 19th century to produce bioethanol from corn and peanut oils. The alternative of fossil fuels has been prioritized due to the price hike at the beginning of the 20th century. Continuous research and development are being conducted to search for different biomass sources afterward. In 1980, bioethanol started to produce from sugarcane in Brazil. It has been observed that bioenergy production from various biomass sources encourages farmers to be involved in biomass production. Besides that, GHG (greenhouse gas) emissions are comparatively much lower in biofuels than in fossil fuels. As biofuels are derived from biomasses having closed carbon cycles, thus it does not have a significant contribution to the emission of GHGs (Van &Romijn, 2008).There are some disadvantages tothe synthesis of biofuels from organic resources. Especially in the 20th century, when the world fights against hunger, the ongoing competition between food and fuel has been aroused. Over the last few decades, the production of biofuels has become costly due to the need for more implementation of proper cutting-edge and sustainable technologies resulting in its expensive nature than fossil fuels.

1.2. Biofuels Production and Health Hazards

The industry persons involved in different stages and steps of biofuel production may have various health hazards, i.e., from asthma, allergy, lung disorders, for ozone attacks (Ajanovic & Haas, 2014).Several accidents during biodiesel production have been reported in many articles so far. Researchers are still developing a statistical and mathematical model to analyze the root cause and the severity of accidents classified into different groups.

Risk assessment is becoming essential before establishing a bioenergy plant as safety engineering is concerned (Jenkins, A. et al., 2013). Most accidents occur in biodiesel production plants due to a lack of experience in complex industrial and chemical engineering processes and management (Harper, S. W. et al., 2008). These trace gasses and aerosols harm air quality, negatively impacting human health (Verma, N., et al., 2021). So it is indisputable that bioprocess engineering principles involved or appropriate bioreactor designing for a controlled process and byproducts utilization & handling is a complex task (Fink &Medved, 2013; Gaffney & Marley, 2009).It has been observed that most of the accidents took

Table 1. Some terms used in biofuel production for risk assessment (Moreno & Cozzani, 2015)

Keyword	Definition
Bio-energy	The energy is produced & synthesized from living sources.
Biomass	The biodegradable waste and residues from meat, fish, agricultural sectors (including vegetable and animal substances), and industrial and municipal waste.
Bioliquids	Liquid fuel is used in different power generation units produced from biomass.
Biofuels	Gaseous or liquid fuel is produced from biomass and can be used in different engines after blending with other fossil fuels.
Biogas	A gas produced from the biodegradable waste fraction can be purified into natural gas or used as wood gas.

place due to a lack of statistical data analysis or risk interpretation. Safety training in such bioenergy industries should be mandatory to tackle unexpected accidents. Several studies revealed that hazardous materials are developed during biomass processing, but even production stages release some dangerous materials, resulting in severe health problems. The biomass produces many flue gasses during burning, which may cause severe respiratory troubles. Some carcinogenic compounds, such as carbonyl groups, are present in biomass, the precursor of major air pollutants. During biomass processing or extraction of gasses and oils, it emits ionizing radiation, which may become vulnerable to workers in the long run. The ozone layer depletion due to the biomass processing industry accelerates UV radiation (Naeher L. P. et al., 2007; Bakeas E. B. et al., 2003). The primary purpose of this book chapter is to entirely focus on the stats on the accidents that happened in bioprocess industries and the probabilistic human health hazard and suggest potential waste-based alternatives for hazardless biofuel production.

2. RESEARCH METHODS

This review was compiled from a search of open-access literature using medical subject heading words as keywords correlated with biofuels. These were biofuel, bioethanol, flammable, health hazard, isoprene, and ozone attack. Only articles were published and are accessible online and available in open-access databases like the Directory of Open Access Journals, PubMed, Google Scholar, Semantic Scholar, etc., between the years 2000 and 2019 considered for this article.

3. RESULTS AND DISCUSSIONS

3.1. Global Trends in the Synthesis of Bioenergy From Different Resources/Biomass

It is essential to understand different vital terms related to bioenergy and their risk assessment during production. Other critical words related to bioenergy synthesis are stated in Table 1.Various sources of biofuels have been reported in several papers and research articles. Some Tanzania villages adopted Jatropha cultivation for biofuel production as the expected yield ranges from 0.1 to 15 tons/ hectare/

Table 2. Estimating oil production and biodiesel yield from different crops in 24 MHa plots of the UK (Scott, S. A. et al., 2010; Rodolfi, L. et al., 2009)

Source	Oil content/tonne of biomass (wt% dry mass)	Oil production (t/ha/y)	Biodiesel yield (L/ha/y)
Nannochloropsis (Hill, J. et al.,2006)	Up to 50%	20.0–30.0	23000–34000
Oilseed rape (Stephenson, A.L. et al., 2008)	40–44% (of seed)	1.4	1560
Chlorella vulgaris (Stephenson, A.L., et al., 2010)	Up to 46%	7.2	8200
Soya (Upham, P., et al., 2009)	20% (of seed)	0.5	544
Jatropha (Kumar, A. et al., 2010)	30% (of seed)	2.4	2700

year (Heller, J., 1996), making it a promising source. The jatropha production chain is given in Fig 1. A detailed flowchart of processes will explain the intermediated steps themselves. Different research papers demonstrated the potentiality of microalgae cultivation, which could be used as a promising source of bioenergy production (Jones & Miller, 1992).This type of consideration is due to the diversity in an algal strain, which is diversified among 300,000 species compared to any land plant (Van &Romijn, 2008; Scott, S., A. et al.,2010). Some microalgae species can accumulate many triacylglycerides (TAGs), considered the bio refinery's significant biodiesel production feedstock. Regarding total yield, the choice of algal species is reported in several articles, as given in Table 2. From microalgal biomass, the lipid is extracted with extensive drying as water is present as an unbound moisture form.

Different drying techniques need comprehensive study and research to optimize lipid extraction. Generally, fatty acids and lipids are extracted from the algal biomass before processing biodiesel in the refinery (Kumar, A. et al., 2010).As found in research articles, two popular extraction methods, the Hexane method, and the Bligh-Dyer method, can be applied for the same (Demirbaş, A., 2008; Kanda, H. et al., 2013). Chemical solvent extraction methods have several limitations with the developed toxicity to humans and the atmosphere. After lipid extraction, TAG conversion into biodiesel through the transesterification process with methanol is the most suitable technology to produce methyl esters per industrial standards (Lam et al., 2010).Fig 2. Shows the esterification process of tags for synthesizing fatty acid methyl ester. The method of gasification can produce methanol and hydrogen from biomass. The production technique follows so many intermediate steps designed for the methanol-based reactors are given in Fig 3(Hamelinck & Faaij, 2002). A mixture of CO and H_2, known as syngas, is also needed to produce methanol or hydrogen.

In table 1,some terms used in biofuel production for risk assessment (Moreno &Cozzani, 2015) Some research based on agricultural, kitchen, waste frying oil, organic, and other food material wastes demonstrated the need for a sophisticated design procedure to convert them to biofuel easily. In this case, no such attention is needed to cultivate biomass for bioenergy purposes. Thus, the competition between food and fuel should be minimized, which is much more advantageous (Dutta K. et al., 2014).Different countries adopt biodiesel and bioethanol as substitutes for fossil fuels and gasoline. At the pilot and lab scales, it is possible to prepare refinery units for food, organic wastes, and cooking oil waste (Paladino &Neviani, 2018; Garcia, J., 2011).The Food Safety and Standards Authority of India (FSSAI) has made

Figure 1. Jatropha production chain
(Van &Romijn, 2008).

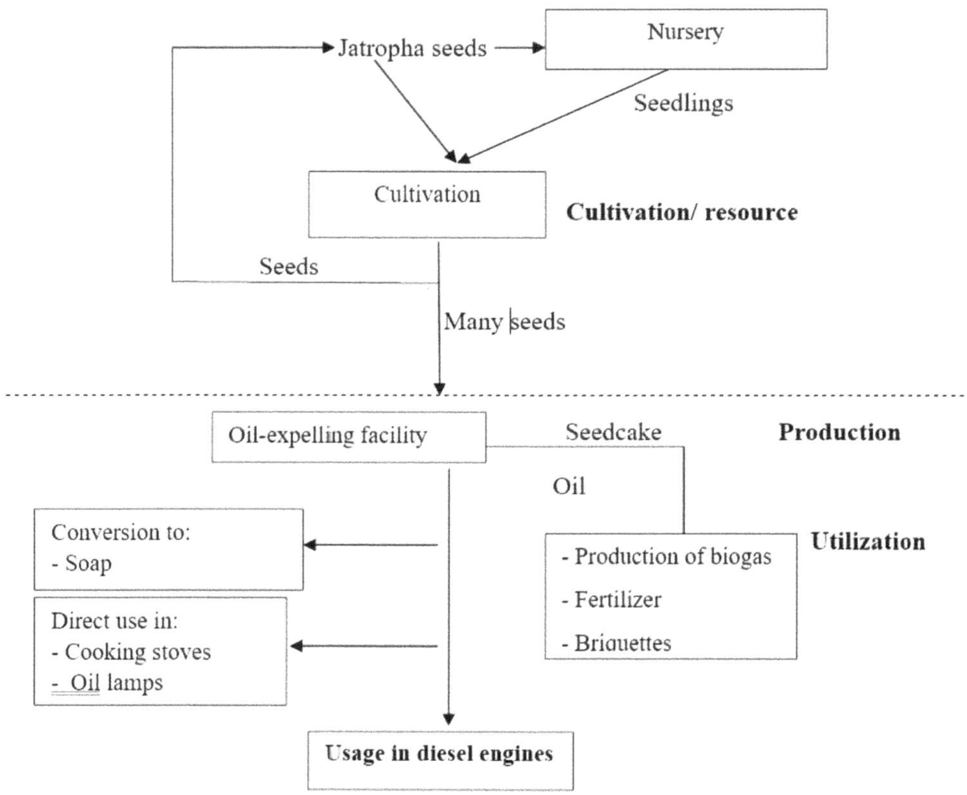

mandatory provisions and regulations regarding the repurposed use of cooking oil in India. It should be discarded after three times of usage. As reheating of cooking oils continuously degrades its quality, toxicity & viscosity increase, and carcinogenic polar compounds are generated due to the breakdown of fatty acids, rather than discarded, this used frying/cooking oil should be handed over to any biorefinery for other biofuels(Karmee& Lin, 2014). Generally, the combination of liquid biofuel from frying oil follows three stages. Gasification/pyrolysis, followed by combustion after waste treatment, produces a good syngas quality (Pham, T. P. T., et al., 2015). In the third stage, the transesterification of syngas produces biodiesel (Brennan & Owende, 2010). Due to the closed carbon cycle in this process, the carbon footprint is less. The complete integrated approach to bioenergy production is given in Fig 4.

Besides utilizing waste frying oil, much scope for bio-diesel production from food wastes has been studied. Food waste mainly includes peels or skins of fruits and vegetables and rotten or damaged agricultural produces. Fish and meat bones, skins, intestines, guts, eggshells, various dairy product wastes like cheese, ice creams, curds and byproducts like whey, any processed foods, ready-to-eat foods, ready-to-serve foods, noodles, restaurant wastes, fodder, and other animal foods, etc. Food waste is the most significant source of lipids, carbohydrates, and amino acids, which could be synthesized into biofuel (Karmee, S. K., 2016). Supercritical carbon dioxide ($scCO_2$) is a novel technique that gives maximum biofuel yield. This process needs to be optimized by setting pressure and temperature (Khan & Ghosh,

Figure 2. Esterification process of triacylglycerides for the synthesis of fatty acid methyl ester
(Scott, S. A. et al., 2010)

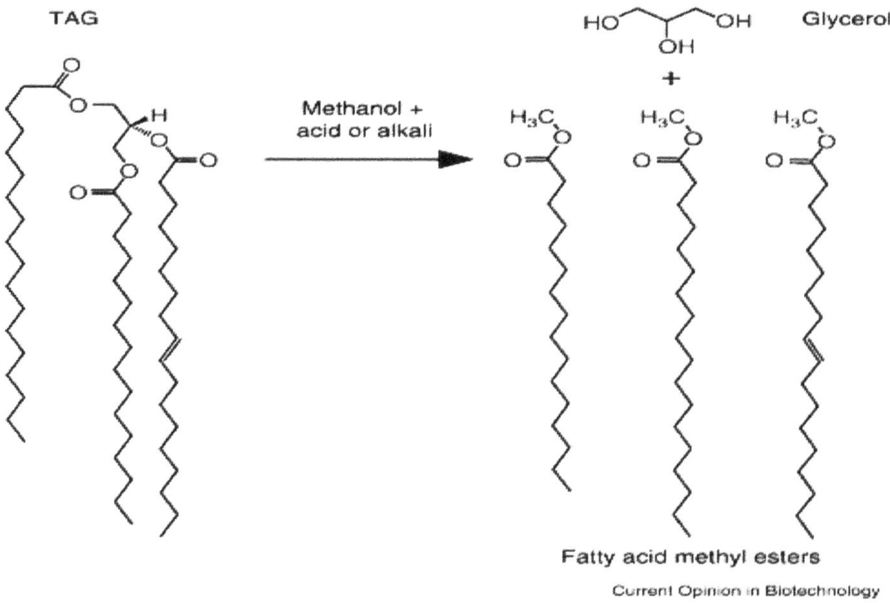

Figure 3. Major components for synthesizing methanol/hydrogen from biomass
(Hamelinck&Faaij, 2002).

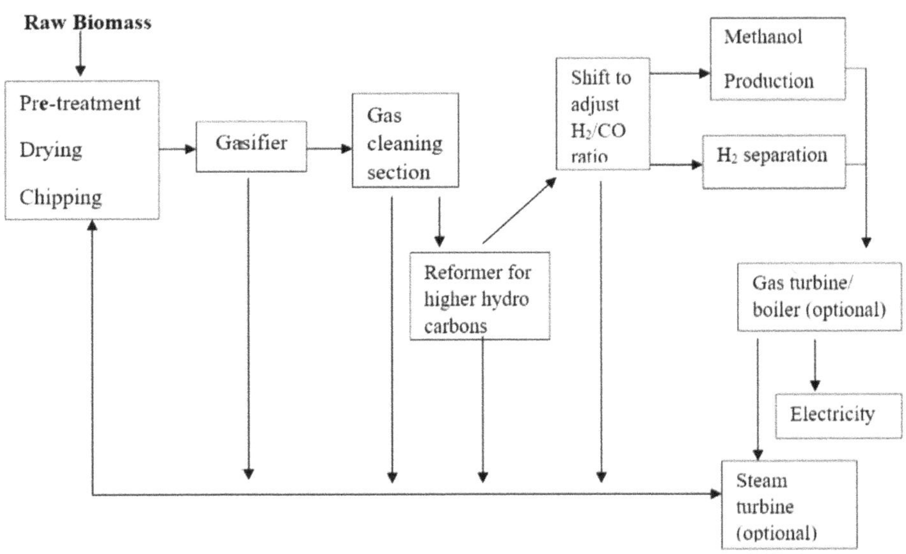

2017; Karmee S. K. et al., 2014). Carbohydrates in food wastes can be hydrolyzed to sucrose & fructose,

Figure 4. An integrated process for producing bioenergy from waste frying oil and organic wastes (Paladino & Neviani, 2018).

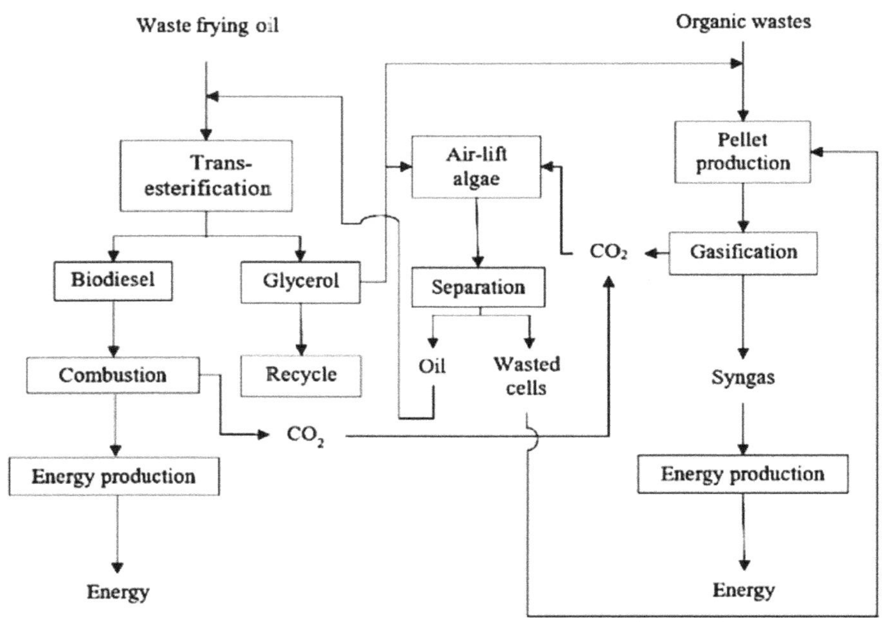

Figure 5. Production of bio-oil, bioethanol, and biodiesel from food waste (Karmee& Lin, 2014).

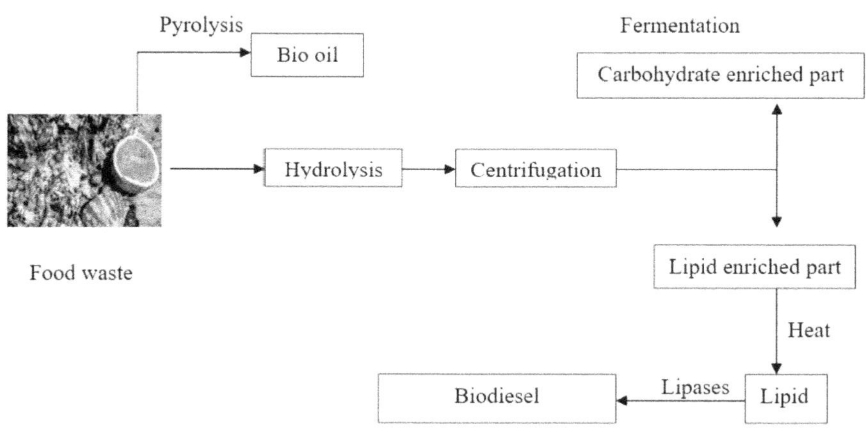

fermented to bioethanol. Researchers also found that food waste can produce bio-oil, ethanol, methane, and hydrogen (Kiran, E. U., et al., 2014). Simultaneous processes for bio-oil, biodiesel, and bioethanol productions are given in Fig 5. The synthesis of bioethanol from food wastes is shown in Table 3.

Table 3. Bioethanol synthesis by fermentation from different food waste types (Karmee, S. K., 2016).

Types of food waste	Microorganism	Method after fermentation
Mixed food	Saccharomyces cerevisiae H058	Hydrolysis (Bello, R. H., et al., 2014).
Banana	Saccharomyces cerevisiae	Enzymatic (Kim, J., et al., 2011)
Instant noodle	Saccharomyces cerevisiae K35	Saccharification (Yang, X. et al., 2014).
Household food	Saccharomyces cerevisiae	Enzymatic liquefaction/ saccharification (Matsakas, L. et al., 2014).
Potato peel	Saccharomyces cerevisiae var. Bayanus	Acidic hydrolysis or enzymatic hydrolysis (Yan S. et al., 2013).

3.2. Health Hazard in Biorefineries and Accidental Risk Assessment

Many researchers investigated the database on bio-refinery accidents using different statistical models to determine the significant risk and severity of accidents to implement proper safety engineering for bio-refineries. Production of biomass involves intermediated steps like distillation, and the fermentation process requires fossil fuel, which may cause a health hazard. A complete study was done on DALYs, i.e., total disability-adjusted life years for each unit mass (kg) produced biofuel (Arapoglou D. et al., 2010). It has been observed that a lower concentration of carcinogenic compound emissions in biofuel compared to fossil fuels and fewer emissions of carbonyl compounds shows statistically significant lower emissions. However, carbonyl compounds are the precursors of ozone and PAN (Grosjean D. et al., 2001). However, researchers found a significant increase in carbonyl compounds in warm engines (Pre-harvest biomass burning led to carbon monoxide and ozone. In Brazil, the burning of sugar cane led to air pollution, and as a result, several cases of respiratory attack and deaths were found (Pang X. et

*Figure 6. Disability-adjusted life years (DALY); statistically significant from fossil diesel (*p < 0.05, **p < 0.001) or fossil petrol (#p < 0.05, ##p <0.001)*
(Fink &Medved, 2013).

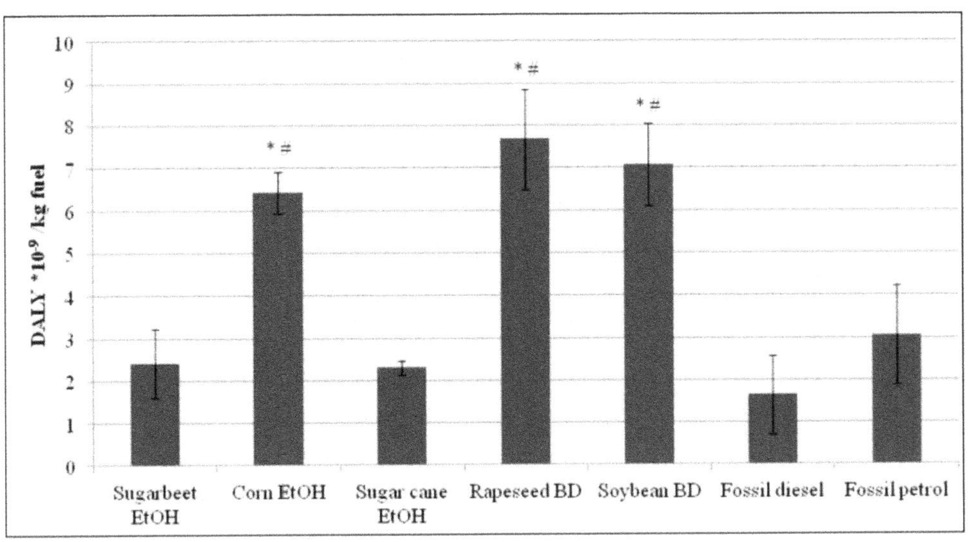

Table 4. A case history of the accidental database

Sl No.	Nation	Year	Bioenergy	Site	Scenario	Inferences
1	Italy	1997	BG	Production	Explosion	2 fatalities + 1 injury
2	Denmark	1998	BM	Storage	Fire	Asset damage
3	USA	2000	BB	Feedstock	-Do-	2 injuries
4	Sweden	2004	BM	Storage	-Do-	Asset damage
5	-Do-	2005	-Do-	-Do-	-Do-	-Do-
6	USA	2005	BB	Feedstock	Explosion	1 fatality
7	Germany	2005	BG	-Do-	-Do-	4 fatalities
8	-Do-	2005	-Do-	-Do-	Toxic Release	4 fatalities
9	USA	2006	BM	Conversion	Fire	2 injuries
10	-Do-	2006	-Do-	-Do-	-Do-	Asset damage
11	-Do-	2006	BB	Feedstock	-Do-	1 fatality
12	Netherlands	2007	BM	Conversion	Explosion and fire	Asset damage
13	USA	2007	-Do-	-Do-	Fire	-Do-
14	Sweden	2007	-Do-	Storage	-Do-	-Do-
15	France	2007	BB	Feedstock	Dust explosion	2 injuries
16	Germany	2007	BG	-Do-	Explosion	-Do-
17	France	2008	BB	-Do-	Fire	1 injury
18	-Do-	2008	BG	-Do-	Explosion	2 injuries
19	Philippines	2008	-Do-	-Do-	Toxic Release	4 fatalities
20	India	2009	-Do-	-Do-	Explosion	4 fatalities + 3 injuries
21	USA	2009	BM	Conversion	-Do-	Asset damage
22	-Do-	2009	-Do-	-Do-	Fire	-Do-
23	France	2010	BG	Feedstock	Toxic Release	-Do-
24	USA	2010	BM	Conversion	Fire	-Do-
25	Germany	2010	-Do-	Storage	Explosion	3 fatalities
26	France	2010	-Do-	Conversion	Fire	Asset damage
27	USA	2010	-Do-	-Do-	-Do-	-Do-
28	Netherland	2010	-Do-	-Do-	Explosion and fire	-Do-
29	Norway	2010	BM	Storage	Fire	-Do-
30	Germany	2010	BG	Feedstock	Explosion and fire	3 injuries
31	Sweden	2010	BM	Storage	Fire	Asset damage
32	Germany	2010	BG	Feedstock	Explosion	2 injuries
33	France	2011	BM	Conversion	Fire	1 injury
34	UK	2011	-Do-	-Do-	-Do-	Asset damage
35	France	2011	-Do-	Feedstock	-Do-	-Do-
36	USA	2011	-Do-	-Do-	Explosion	-Do-

continued on following page

Table 4. Continued

Sl No.	Nation	Year	Bioenergy	Site	Scenario	Inferences
37	Germany	2011	BG	-Do-	-Do-	2 injuries
38	-Do-	2011	-Do-	-Do-	-Do-	2 injuries
39	USA	2011	BM	Conversion	Fire	Asset damage
40	France	2012	-Do-	-Do-	Explosion	-Do-
41	Netherland	2012	-Do-	-Do-	-Do-	-Do-
42	USA	2012a	-Do-	-Do-	Fire	-Do-
43	-Do-	2012b	-Do-	-Do-	-Do-	-Do-
44	Denmark	2012	-Do-	-Do-	-Do-	-Do-
45	-Do-	2012a	-Do-	-Do-	Dust explosion	3 injuries
46	Denmark	2012b	-Do-	-Do-	Fire	Asset damage
47	UK	2012	-Do-	-Do-	-Do-	Asset damage
48	USA	2012	-Do-	-Do-	-Do-	2 injuries
49	-Do-	2012	-Do-	-Do-	Explosion	Asset damage
50	-Do-	2012	-Do-	-Do-	Fire	Asset damage
51	-Do-	2012	-Do-	-Do-	Explosion	1 injury
52	-Do-	2012	BB	Feedstock	Explosion and fire	-Do-
53	France	2013	BM	-Do-	Fire	Asset damage
54	UK	2013	-Do-	Conversion	-Do-	-Do-
55	USA	2013	-Do-	-Do-	boiler	2 injuries (1 serious)
56	-Do-	2013	-Do-	-Do-	Explosion and fire	Asset damage
57	UK	2013	-Do-	-Do-	Fire	-Do-

al., 2006; Cançado, J. E. et al.2006). Researchers found that the reason behind this respiratory illness is the presence of different complex, volatile organic compounds in biofuel, which are significantly higher than fossil fuels, as shown in Fig. 6

Several investigations on biofuel productions from different agricultural products have been considered to create a valuable database for further research with the help of statistical modeling to find out the significant health hazards caused due to some carcinogenic byproducts, inorganic, organic compounds, ultra-violet and ionizing radiation (Fink &Medved, 2013). The effect of radiation from biofuel-based heat and power plants has also been reported elsewhere. In Wood industries, concentrated radioactive elements have been found in their waste products, ash, liquors, and sludge. This ash and mud, when dumped into the landfills, emit radiation. Researchers have found several radioactive elements, namely ^{134}Cs, ^{137}Cs, ^{40}K, and ^{90}Sr. Before dumping them, workers need to take appropriate precautions in those cases to avoid unnecessary health hazards (Ravila& Holm, 1994).

Table 5. Chemical compounds (by-products of biofuels) and their health hazards (Li, Z. et al., 2013).

Chemical Compound	Chemical Structure	Average-ppm	Possible Health Hazards
Methane	CH_4	50,000 -1,50,00ppm	Potentially explosive, acute pulmonary toxicity causes asphyxiation, a drowsy mentality, and hypoxemia.
		>1,50,000 ppm	Causes Pneumonitis, in toxication, loss of consciousness or asphyxiation, and acute pulmonary toxicity.
Hydrogen Sulfide	H_2S	1-10 ppm	Nausea, vomiting, tears, headache, mild irritation, breathing trouble.
		10-150 ppm	Irritation of lungs and eyes. Skin rashes.
		150-750 ppm	Multiple organ failure and death may occur.
		>750	Confirmed death
Carbon Dioxide	CO_2	2,000-5,000 ppm:	Suffocation due to crisis of air, Incapacitation, and no consciousness. Poor concentration, headache, vertigo, double vision, tinnitus, and seizures.
		>5,000 ppm:	Breathing difficulties, seizures, toxicity or oxygen deprivation, cardiovascular diseases.
		>40,000 ppm:	Coma, stroke, and maybe death or severe nervous disorders.
Nitrous Oxide	N_2O	> 25 ppm	Loss of blood pressure, death by hypoxia. This may lead to a vitamin B12 deficiency, so neurological issues as high risks.
Fluorocarbons or Perfluorocarbons	C_xF_y	1,000 ppm. -10,000	Severe outbreaks of respiratory syndromes and bronchitis, characterized by dyspnea, cough, and chest pain. Respiratory failure, acute respiratory distress syndrome (ARDS), and even death may occur.
Ozone	O_3	0.10ppm for 8 hrs. for workers	Respiratory tract syndrome, bronchitis, emphysema, asthma, chest pain, coughing, throat irritation, and airway inflammation, asthma

A few researchers have reported some accidental stats or scenarios globally to make a robust database for analyzing them statistically, as shown in Table 4. This data will fulfill the need for appropriate technology to reduce human and environmental hazards (Moreno & Cozzani, 2015). According to researchers, there is an utmost need for comprehensive technology to develop a single reaction system and optimized separation and byproduct utilization (Sharma, M et al.2023) as cost-effectiveness & process efficacy are crucial concerns. Fig. 7 indicates the increasing trends of accidents in bio refineries from 1997. Table 4 shows a case history of the accidental database (BB: bio liquids/biofuels; BG: biogas; BM: biomass); sites where feed stocks are converted to bioenergy (Moreno &Cozzani, 2015).

3.3. Biofuels and Health Hazards Through Byproducts

Biofuels are natural energy and fuel that we can get from biomass (Ravila& Holm, 1994). Biofuels are essential for our modern life, which we can get from various agricultural fields, private enterprises, household areas, and manufacturing refuse (for organic refuses sources). Biofuels are the same as multiple edible oils, bio-ethers, solid biofuels, biogas, biodiesel, etc Madhumathy, P., & Pandey, D. (2022). Our modern civilization is in massive biofuels demand from regions like China, India, Brazil, and Europe (Bikis, A. et al., 2023).

3.3.1. Sources and History of Biofuel (Various Sources of Biofuel)

First-generation Biofuels: are produced using food crops developed on arable land and are unequivocally designed for fuel creation. Sugars, starch, vegetable oil, or similar products received from the harvests are changed into ethanol or biodiesel, utilizing trans-esterification or yeast maturation found in available literature (Singh & Singh, 2010).

Second-generation: biofuels are derived from different biomass varieties, an agricultural residue. Second-generation biofuels are processed from some crops, such as paddy, corn, and wheat. Also, biofuels generated used raw materials as a wood form of cellulose and produced ethanol (C_2H_5OH) (Swanson, K. J., 2007).

Third-generation biofuels: These biofuels, which are used for producing biofuels such as Algae, it has the possible scope of modern generation high potential biomass (Sanchez & Cardona, 2008).

Fourth-generation: 4th generation biofuels belong to electro fuels, also called photo biological solar fuels. They are comparatively environmentally friendly and less hazardous. (Briggs, M ., 2004)

3.3.2. Production and Processing

The below-mentioned fuels are procured following the generations as 1st, 2nd, 3rd, or 4th generation as per the manufacturing process of biofuels (Aro, E.M., 2016).

Biogas: Biogas can be manufactured easily from biodegradable waste materials from recycling industries, landfill waste products, and various crops. We get from this project as a by-product such as bio fertilizers and biofuels (Janu, Neha, et al. (2022)). This biogas is a potent greenhouse gas. Farmers can quickly produce biogas from manure from their cattle by using an anaerobic digester. Undoubtedly this is very economical and easy (Douglas, C. C. et al., 2008).

Syngas: This gas combines carbon monoxide, hydrogen, and a few more hydrocarbons generated through partly biomass combustion. Combustion produces oxygen but cannot convert biomass into $CO_2$and water (Evans, G., 2008).

Hydrogen and Dimethyl: ME, *i.e.,* Dimethyl, either gets converted by the Fischer-Tropsch process to substitute a blending of alcohols or diesel that may be added with gasoline. Gasification happens at more than 700°C.DME has many fuel properties for the best use in diesel engines. ***Ethanol Ethyl alcohol:*** *Ethanol* is one of the popular biofuels worldwide. The sugar fermentation process prepares ethyl biofuels to obtain substances like corn, beetroot, sugar cane, molasses, fruit waste, potato, and starch or sugar. This is very familiar to biofuel in Brazil, the United Kingdom, and other European countries (Evans, G., 2008).

***Biodiesel*:** It comprises a combination of fatty acid ethyl or methyl esters. Biodiesel is produced from vegetables, animal fats, rapeseed oil, soy, mahua, mustard, sunflower, palm, Pongamia pinnata plant, and algae. Australian Scientist CSIRO 2020 has observed safflower oil, an engine lubricant (Lee, Tim., 2020). Biodiesel is associated with O_2 fuel, which is a reduced amount of carbon and excessive oxygen (O_2), and hydrogen (H_2) (Nylund&Koponen, 2013). Biodiesel is very attentive to manipulation and conveys because it is nonpoisonous or biodegradable, having a flashpoint of 126°F (53°C) (Swanson, K. J., 2007).In the United States of America, around 82% of city buses and Lorrie's ride using biodiesel. So the US biodiesel market value has increased since 2004 (Traviss, N., 2012).

Green diesel is manufactured by hydrocracking organic oil live feed stocks, e.g., animal fats and veg-sourced oils. Hydrocracking is a cleaning procedure that utilizes raised- pressure and temperatures in the existence of catalysts to disintegrate big particles. Green diesel constitutes various vegetable oils into small hydrocarbon bonds used in diesel engines (Knothe, G., 2010). Green diesel, also called renewable diesel, is hydro-treated vegetable oil (HVO fuel). Its existence has been enlarged in Singapore and Louisiana by many largest companies like Conoco Philips, Valero, UOP-Dynamic Fuels, Honeywell, Neste Oil, and Preem. Gothenburg city in Sweden is generating rising Swedish petroleum-biofuels or green diesel. Lubricant, along with low-quality oil used in these grounds, has been processed into biodiesel usually (Gupta, Sunita, et al.(2022)). Fat, along with lubricants, could be hydrogenated to replace diesel. This bio-diesel is compatible with an engine or motor vehicle to maintain low temperatures during driving or transportation(Albrecht &Hallen, 2011).

3.3.3. Safety and Health Risks Associated With a Biogas Plant (Rapport, J. et al., 2008)

The various types of risks include

- Fire and explosion
- Risk of Asphyxiation
- Confined space hazards
- Risk of gas poisoning (H_2S, NH_3)
- The risk associated with rotating mechanical equipment
- The risk of high-pressureliquid leaks or gas.
- Different risks correlated with pathogens(diseases)

***Fire and explosion*: -** Methane gas, about 60% of biogas. So biogas, when mixed in the oxygen air, create an explosion. Biogas quickly dilutes between 10-30% of the atmosphere of O2, causing fire and explosion hazards. Hydrogen sulfide and ammonia explosions cannot always open flames. So the strict prohibition of smoke in places associated with biogas equipment and linked lines must be followed.

***Risk of Asphyxiation*:** Health Risk of Biogas is asphyxiation; showed one case of three people died from asphyxiation generated by swine manure gas in the surrounding area. Cowpats can give rise to methane gas at a research-based amount to extend the manure and provide the space O_2 deficiently.

***Confined space hazards*:** There are no improperly sized tanks, vessels, or different types of containers, bunds, pits, cellars, and chambers. These may cause a toxic environment due to unwanted chemicals and low oxygen levels (Traviss, N., 2012).

***Risk of gas poisoning* (H_2S, NH_3);** Hydrogen Sulfide (H_2S) is an exceedingly poisonous gas that is weighty than air. This gas, at minimal volume, scents such as rotten eggs. Also, H_2S can affect eye irritation. It is hazardous for our human health levels; it causes respiratory paralysis and eliminates the sensation of smell. So, there is no smell to alert its existence at threatening and mortal levels, where one can drop dead. Ammonia (NH_3) Ammonia is a gas that is bright than air. It has a pungent smell and can aggravate the respiratory tract and the eyes.NH_3canreplace oxygen in the blood flow. Precautions Manufacturer Warnings: Failure to take note of producer caution may influence severe injury or death outcomes. Communicate with the producer for services, demand, and maintenance, and favor opportunity (Lee, Tim., 2020).

***The risk associated with rotating mechanical equipment*:** There should be a thorough check-up and verification of machinery parts and all types of equipment. To avoid hazards, machine faults are to be identified, followed by preventive measures cum maintenance. There are three ways to control risks: i) Limitation of time near machinery, ii) Must identify what machine is defective and what types of hazards iii) Preventive maintenance pathways.

***Risk of high-pressure gas or liquid leaks*:** Some unadulterated compacted gases are synthetically insecure. When presented with a slight temperature, weight increments, or mechanical stun, they can promptly experience particular kinds of compound responses, such as polymerization or decay. These responses may get rough, bringing about fire or blast. Some perilously responsive gases have synthetic substances called inhibitors to forestall these dangerous responses that are always related to the biogas industry. Standard perilously receptive gases are acetylene, vinyl chloride, 1, 3-butadiene, methyl acetylene, and tetrafluoroethylene (Evans, G., 2008).

Risks associated with pathogens (Disease): Natural manure carries viruses, bacteria, and viable parasites. The anaerobic assimilation of manure develops biogas. The bacteria in animal misuse can generate contamination when they touch substances, movement suitable hazardous by utilizing PPE to avoidmanure communication. After completion of work, clean thoroughly.It primarily advocates washing hands before drinking and eating and affecting the eyes or other mucous membranes commonly prone to contagious diseases (Douglas, C. C. et al., 2008).

3.3.4. Component of Biogas

Biogas comprises 40% carbon dioxide (CO_2) and 60% methane gas with little hydrogen sulfide and other gases. All these gases can replace oxygen (O2) during biogas production (Li Z. et al., 2019).

Methane: Methane gas is lighter than air and will gather concerning the overlying capacity of the construction. It is inflammable at 5% - 15% of attentiveness. The gas is not toxic but replaces ventilation in a constricted capacity; it generates a lack of oxygen in the airspace (Zhang X. et al., 2016).

***Carbon Dioxide*:** A non-odorous gas that is more weighty than air; carbon dioxide can be covered close by the ground in an inactive space. Carbon dioxide increases the level of respiratory rate and heart rate. High volume replaces oxygen in the blood vessels, which can cause unconsciousness or death (Li Z. et al., 2019).Safety Walk-Through safety can assist in controlling prospective risk and preventive estimates. Cornel University grows a self-evaluation comprehensive instruction for farmers. It is intentional to be used by farm staff responsible for the performance and farm managers or farm owners, a continuation of anaerobic digesters, and they are associated with this operation in all developed countries. Gas sensors' suffocation, explosion, and toxic gas risk may be recognized using gas sensors. These sensors incorporate both replaceable and electronic sensors. It requires testing regularly (Dogiwal, Sanwta Ram, et al (2022)). A certified person should be allowed to maintain these sensors to control an area for safety.

3.3.5. Personal-Protective-Equipment (PPE):

The region of manure should never go in without any suitable personal protective equipment (PPE), which may be associated with a self–contained breathing apparatus (SCBA). The operator should be certified in its use with medical permission and fit device testing in this field. Biogas is composed of 61% average methane. According to the environmental report on the impact of greenhouse gas emissions in the United States of America, there are four specific greenhouse gases. The specific gases are methane, hydrogen sulfide, carbon dioxide, nitrous oxide, and fluorinated gas. Ozone is another vital gas that is good and evil from different perspectives, including industrial use (Bikis, A., & Pandey, D. (2021)). According to various environmental studies, methane in the atmosphere has doubled for centuries, reaching approximately 1,900 ppb or more. Apart from methane, carbon dioxide, nitrous oxide, and other greenhouse gases have been elevating their level and influencing global atmospheric concentrations in the last few decades. After the Industrial Revolution, nitrous oxide concentrations in the atmosphere exceeded 280 ppb to approximately 300 ppb in 1920. In 2015, various environmental authorities reported a level of around 328 ppb. Further, concentrations of halogenated gases were found to be almost nil early in the 1900 century but had been rising quickly as per industrial product developments. Ozone contains three oxygen atoms and is available in upper atmospheres and ground levels. Good ozone is naturally found in the upper atmosphere and forms a preventive layer protecting human society from harmful ultraviolet rays. However, many chemicals and industrial products have partially destroyed the generous ozone layer in the last few decades. Ozone is present at ground level as a harmful air pollutant, also abundantly found in smog (Jo J. Y.et al.,2013).

Figure 7. Accidents (%) related to the type of bioenergy production involved for all 57 records

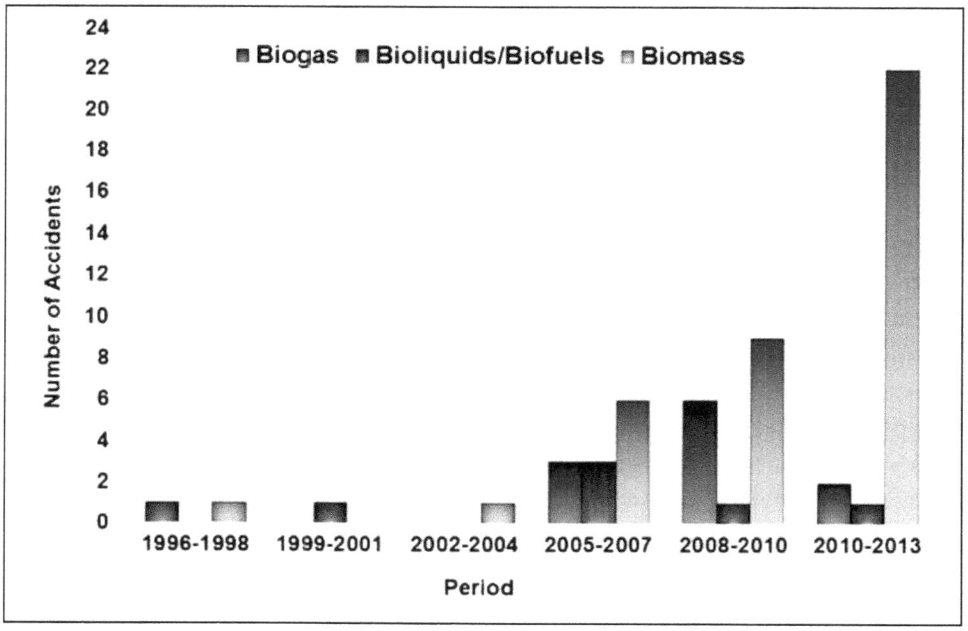

3.4. Chemical Hazards Affect Entire Lifestyles

Health hazards also significantly impact sedentary people living in big cities and highly populated countries like India, China, and the United States of America. High-calorie food consumption and no such exercises are the significant causes of obesity, diabetes, renal disorders, cardiovascular diseases, liver diseases, and cancer (Kumar, M. S.. 2021) Many people suffer from endocrine conditions like type-II diabetes, hypothyroidism, polycystic ovarian diseases, etc. Besides modern medicines, diet as therapy and yoga may help provide at-home relief from these diseases (Banerjee, S., 2020). As per recent studies, many chemicals are treated as endocrine disruptors called thyroid disruptors. Hence chemical hazards should be identified, and isolated person-wise to avoid long-termhypersensitizes to the skin and other organs (Shahzad, K. et al. 2020). These may affect and infect the human body's respiratory, endocrine, lymphatic, and other systems. We have provided an easy table to understand various chemical compounds with their average presence measuring parts per million (ppm) that often cause different health hazards (.Dadheech, Pankaj, et al.(2022)). While concerning health safety, we can recommend prevention is better than cure. Preventions can be done by raising immunity through a balanced diet, daily exercises, and at the same time, identifying the hazards that are affecting our lifestyles (Massart F. et al., 2007; Banerjee, S., 2019).

4. CONCLUSION

In general, the biofuel production process is eco-friendly; however, it may cause serious health hazards to those working in the production units. That is due to the lack of experience in complex industrial & chemical engineering processes and management practices and the volatile chemical constituents of the raw materials used for biofuel production. The probabilistic human health hazard analysis is thus to be made mandatory before establishing the biofuel production units. Replacing conventional biomass with waste-based alternatives like food waste or used oil can be a better alternative for hazardless biofuel production. Carbon dioxide, methane, hydrogen sulfide, and fluorocarbon emissions, including harmful ozone during biofuel manufacturing, may harm the human body, even death. Therefore, if manufacturers can control much below-average production of these unwanted chemicals and any other byproducts in-unit parts per million (ppm), then there is a chance of eco-friendly Production of Biofuels without less affecting the living world.

5. REFERENCES

Ajanovic, A., & Haas, R. (2014). On the prospects and limits of biofuels in Brazil, the US, and the EU. *Applied Energy*, *135*, 730–737. doi:10.1016/j.apenergy.2014.07.001

Albrecht, K. O., & Hallen, R. T. (2011). A Brief Literature Overview of Various Routes to Biorenewable Fuels from Lipids for the National Alliance of Advanced Biofuels and Bio-products NAAB Consortium. US Department of Energy. doi:10.2172/1019208

Arapoglou, D., Varzakas, T., Vlyssides, A., & Israilides, C. (2010). Ethanol production from potato peel waste (PPW). *Waste Management (New York, N.Y.)*, *30*(10), 1898–1902. doi:10.1016/j.wasman.2010.04.017 PMID:20471817

Aro, E. M. (2016). From first-generation biofuels to advanced solar biofuels. *Ambio*, *45*(S1, Suppl 1), S24–S31. doi:10.100713280-015-0730-0 PMID:26667057

Bakeas, E. B., Argyris, D. I., & Siskos, P. A. (2003). Carbonyl compounds in the urban environment of Athens, Greece. *Chemosphere*, *52*(5), 805–813. doi:10.1016/S0045-6535(03)00257-1 PMID:12757781

Banerjee, S. (2019). Study on Yoga Intervention and Diet on Hypothyroidism Associated with Obesity among Sedentary Working Women in West Bengal. *International Journal of Yoga and Allied Sciences*, *8*(1), 18–23.

Banerjee, S. (2020). The Essence of Indian Indigenous Knowledge in the perspective of Ayurveda, Nutrition, and Yoga. *Research & Reviews in Biotechnology & Biosciences*, *7*(2), 20–27.

Bello, R. H., Linzmeyer, P., Franco, C. M. B., Souza, O., Sellin, N., Medeiros, S. H. W., & Marangoni, C. (2014). Pervaporation of ethanol produced from banana waste. *Waste Management (New York, N.Y.)*, *34*(8), 1501–1509. doi:10.1016/j.wasman.2014.04.013 PMID:24834817

Bikis, A., & Pandey, D. (2021). Air quality at public transportation stations/stops: contribution of light rail transit to reduce air pollution. *Aerosol Science and Engineering*, 1-16.

Bikis, A., & Pandey, D. (2023). Squatter settlement and informal urbanization: Causes and consequences. *Environmental Science and Pollution Research International*, *30*(9), 23276–23294. doi:10.100711356-022-23778-z PMID:36322351

Brennan, L., & Owende, P. (2010). Biofuels from microalgae- a review of technologies for production, processing, and extract biofuels and co-products. *Renewable & Sustainable Energy Reviews*, *14*(2), 557–577. doi:10.1016/j.rser.2009.10.009

Briggs, M. (2004). *Widescale Biodiesel Production from Algae*. UNH Biodiesel Group (University of New Hampshire).

Cançado, J. E., Saldiva, P. H., Pereira, L. A., Lara, L. B., Artaxo, P., Martinelli, L. A., & Braga, A. L. (2006). The impact of sugar cane–burning emissions on the respiratory system of children and the elderly. *Environmental Health Perspectives*, *114*(5), 725–729. doi:10.1289/ehp.8485 PMID:16675427

Chen, H., Qiu, T., Rong, J., He, C., & Wang, Q. (2015). Microalgal biofuel revisited: An informatics-based analysis of developments to date and future prospects. *Applied Energy*, *155*, 585–598. doi:10.1016/j.apenergy.2015.06.055

Dadheech, P. (2022). A Compressive Review on Internet of Things in Healthcare: Opportunity, Application and Challenges. *Emerging Technologies in Computer Engineering: Cognitive Computing and Intelligent IoT: 5th International Conference, ICETCE 2022*. Cham: Springer International Publishing. 10.1007/978-3-031-07012-9_20

Demirbaş, A. (2008). Production of biodiesel from algae oils. *Energy Sources. Part A, Recovery, Utilization, and Environmental Effects*, *31*(2), 163–168. doi:10.1080/15567030701521775

Dogiwal, S. R. (2022). Internet of Things Based Real-Time Monitoring System for Grid Data. *Emerging Technologies in Computer Engineering: Cognitive Computing and Intelligent IoT: 5th International Conference, ICETCE 2022*. Cham: Springer International Publishing. 10.1007/978-3-031-07012-9_21

Douglas, C. C., Lawrence, J. C., Bush, N. C., & Oster, R. a, Gower, B. A., & Darnell, B. E. (2008). Ability of the Harris Benedict to predict energy requirements differs with weight history and ethnicicty. *Obesity Research*, *27*(4), 194–199.

Dutta, K., Daverey, A., & Lin, J. G. (2014). Evolution retrospective for alternative fuels: First to fourth generation. *Renewable Energy*, *69*, 114–122. doi:10.1016/j.renene.2014.02.044

Evans, G. (2008). *Liquid Transport Biofuels -Technology Status Report*. National Non-Food Crops Centre.

Fink, R., & Medved, S. (2013). Health impact assessment of liquid biofuel production. *International Journal of Environmental Health Research*, *23*(1), 66–75. doi:10.1080/09603123.2012.699030 PMID:22774773

Gaffney, J. S., & Marley, N. A. (2009). The impacts of combustion emissions on air quality and climate–From coal to biofuels and beyond. *Atmospheric Environment*, *43*(1), 23–36. doi:10.1016/j.atmosenv.2008.09.016

Garcia, J. (2011). *Food waste recycled as biofuel*. Stanford University.

Grosjean, D., Grosjean, E., & Gertler, A. W. (2001). On-road emissions of carbonyls from light-duty and heavy-duty vehicles. *Environmental Science & Technology*, *35*(1), 45–53. doi:10.1021/es001326a PMID:11352025

Gupta, S., Nawal, M., Janu, N., & Goyal, D. (2022). IoT, Enabling Technologies, and Sensor Node Deployment Pattern in WSN. *ECS Transactions*, *107*(1), 7441–7455. doi:10.1149/10701.7441ecst

Hamelinck, C. N., & Faaij, A. P. (2002). Future prospects for the production of methanol and hydrogen from biomass. *Journal of Power Sources*, *111*(1), 1–22. doi:10.1016/S0378-7753(02)00220-3

Harper, S. W., Etchells, J. C., Summerfield, A. J., & Cockton, A. (2008). Health and safety in biodiesel manufacture. In *Institution of Chemical Engineers Symposium Series, 154*(943), 1-9. Institution of Chemical Engineers.

Heller, J. (1996). *Physic nut, Jatropha curcas L (Vol. 1)*. Bioversity international.

Hill, J., Nelson, E., Tilman, D., Polasky, S., & Tiffany, D. (2006). Environmental, economic, and energetic costs and benefits of biodiesel and ethanol biofuels. *Proceedings of the National Academy of Sciences of the United States of America*, *103*(30), 11206–11210. doi:10.1073/pnas.0604600103 PMID:16837571

Janu, N. (2022). *Cross-lingual Word Sense Disambiguation Using Multilingual Co-occurrence Graphs*. Graph Learning and Network Science for Natural Language Processing. doi:10.1201/9781003272649-9

Janu, N. (2022). Query-Based Image Retrieval Using SVM. *Emerging Technologies in Computer Engineering: Cognitive Computing and Intelligent IoT: 5th International Conference, ICETCE 2022*. Cham: Springer International Publishing.

Jenkins, A., Gornall, L., & Cripps, H. (2013). Lessons for safe design and operation of anaerobic digesters. *Loss Prevention Bulletin*, 229.

Jo, J. Y., Kwon, Y. S., Lee, J. W., Park, J. S., Rho, B. H., & Choi, W. I. (2013). Acute respiratory distress due to methane inhalation. *Tuberculosis and Respiratory Diseases*, *74*(3), 120–123. doi:10.4046/trd.2013.74.3.120 PMID:23579434

Jones, N., & Miller, J. H. (1992). Jatropha curcas: A multipurpose species for problematic sites. *Scientific Research Publishing.*

Kanda, H., Li, P., Yoshimura, T., & Okada, S. (2013). Wet extraction of hydrocarbons from Botryococcus-braunii by dimethyl ether as compared with dry extraction by hexane. *Fuel*, *105*, 535–539. doi:10.1016/j.fuel.2012.08.032

Karmee, S. K. (2016). Liquid biofuels from food waste: Current trends, prospects, and limitations. *Renewable & Sustainable Energy Reviews*, *53*, 945–953. doi:10.1016/j.rser.2015.09.041

Karmee, S. K., & Lin, C. S. K. (2014). Valorization of food waste to biofuel: Current trends and technological challenges. *Sustainable Chemical Processes*, *2*(1), 22. doi:10.118640508-014-0022-1

Karmee, S. K., Niemeijer, B., Casiraghi, L., Mlambo, B., Lapkin, A., & Greiner, L. (2014). Facile biocatalytic synthesis of a macrocyclic lactone in sub-and supercritical solvents. *Biocatalysis and Biotransformation*, *32*(2), 125–131. doi:10.3109/10242422.2014.893579

Khan, S., & Ghosh, S. (2017). Dense phase carbon dioxide: an emerging non-thermal technology in food processing. *Physical Science International Journal*, 1-7.

Kim, J. H., Lee, J. C., & Pak, D. (2011). Feasibility of producing ethanol from food waste. *Waste Management (New York, N.Y.)*, *31*(9-10), 2121–2125. doi:10.1016/j.wasman.2011.04.011 PMID:21596551

Kiran, E. U., Trzcinski, A. P., Ng, W. J., & Liu, Y. (2014). Bioconversion of food waste to energy: A review. *Fuel*, *134*, 389–399. doi:10.1016/j.fuel.2014.05.074

Knothe, G. (2010). Biodiesel and renewable diesel: A comparison. *Progress in Energy and Combustion Science*, *36*(3), 364–373. doi:10.1016/j.pecs.2009.11.004

Kumar, A., Ergas, S., Yuan, X., Sahu, A., Zhang, Q., Dewulf, J., & Van Langenhove, H. (2010). Enhanced CO2 fixation and biofuel production via microalgae: Recent developments and future directions. *Trends in Biotechnology*, *28*(7), 371–380. doi:10.1016/j.tibtech.2010.04.004 PMID:20541270

Kumar, M. S., Sankar, S., Nassa, V. K., Pandey, D., Pandey, B. K., & Enbeyle, W. (2021). Innovation and creativity for data mining using computational statistics. In *Methodologies and Applications of Computational Statistics for Machine Intelligence* (pp. 223–240). IGI Global. doi:10.4018/978-1-7998-7701-1.ch012

Lam, M. K., Lee, K. T., & Mohamed, A. R. (2010). Homogeneous, heterogeneous, and enzymatic catalysis for transesterification of high free fatty acid oil (waste cooking oil) to biodiesel: A review. *Biotechnology Advances*, *28*(4), 500–518. doi:10.1016/j.biotechadv.2010.03.002 PMID:20362044

Lee, T. (2020). *ABC News. Landline*. Australian Broadcasting Corporation.

Li, Z., Jia, Y., Feng, Y., Cui, R., Wang, Z., Qu, K., Liu, C., & Zhang, J. (2019). Methane-Rich Saline Protects Against Sepsis-Induced Liver Damage by Regulating the PPAR-γ/NF-κB Signaling Pathway. *Shock (Augusta, Ga.)*, *52*(6), e163–e172. doi:10.1097/SHK.0000000000001310 PMID:30601406

Madhumathy, P., & Pandey, D. (2022). Deep learning based photo acoustic imaging for non-invasive imaging. *Multimedia Tools and Applications*, *81*(5), 7501–7518. doi:10.100711042-022-11903-6

Massart, F., & Meucci, V. (2007). Environmental thyroid toxicants and child endocrine health. *Pediatric Endocrinology Reviews; PER*, *5*(1), 500–509. PMID:17925791

Matsakas, L., Kekos, D., Loizidou, M., & Christakopoulos, P. (2014). Utilization of household food waste for the production of ethanol at high dry material content. *Biotechnology for Biofuels*, *7*(1), 1–9. doi:10.1186/1754-6834-7-4 PMID:24401142

Moreno, V. C., & Cozzani, V. (2015). Major accident hazard in bioenergy production. *Journal of Loss Prevention in the Process Industries*, *35*, 135–144. doi:10.1016/j.jlp.2015.04.004

Naeher, L. P., Brauer, M., Lipsett, M., Zelikoff, J. T., Simpson, C. D., Koenig, J. Q., & Smith, K. R. (2007). Woodsmoke health effects: A review. *Inhalation Toxicology*, *19*(1), 67–106. doi:10.1080/08958370600985875 PMID:17127644

Nylund, N. & Koponen, K. (2013). *Fuel and Technology Alternatives for Buses. Overall Energy Efficiency and Emission Performance.* IEA Bioenergy Task 46.

Paladino, O., & Neviani, M. (2018). A closed-loop biowaste to biofuel integrated process fed with waste frying oil, organic waste, and algal biomass: Feasibility at pilot scale. *Renewable Energy*, *124*, 61–74. doi:10.1016/j.renene.2017.08.027

Pang, X., Shi, X., Mu, Y., He, H., Shuai, S., Chen, H., & Li, R. (2006). Characteristics of carbonyl compound emission from a diesel engine using biodiesel–ethanol–diesel as fuel. *Atmospheric Environment*, *40*(36), 7057–7065. doi:10.1016/j.atmosenv.2006.06.010

Pham, T. P. T., Kaushik, R., Parshetti, G. K., Mahmood, R., & Balasubramanian, R. (2015). Food waste-to-energy conversion technologies: Current status and future directions. *Waste Management (New York, N.Y.)*, *38*, 399–408. doi:10.1016/j.wasman.2014.12.004 PMID:25555663

Rapport, J., Zhang, R., Jenkins, B.M., & Willams, R.B. (2008). *Current anaerobic digestion technologies used for treatment of municipal organic solid waste*. Semantic Scholar.

Ravila, A., & Holm, E. (1994). Radioactive elements in the forest industry. *The Science of the Total Environment*, *157*, 339–356. doi:10.1016/0048-9697(94)90598-3

Rodolfi, L., Chini Zittelli, G., Bassi, N., Padovani, G., Biondi, N., Bonini, G., & Tredici, M. R. (2009). Microalgae for oil: Strain selection, induction of lipid synthesis, and outdoor mass cultivation in a low-cost photobioreactor. *Biotechnology and Bioengineering*, *102*(1), 100–112. doi:10.1002/bit.22033 PMID:18683258

Sanchez, O. J., & Cardona, C. A. (2008). Trends in biotechnological production of fuel ethanol from different feedstocks. *Bioresource Technology*, *99*(13), 5270–5295. doi:10.1016/j.biortech.2007.11.013 PMID:18158236

Scott, S. A., Davey, M. P., Dennis, J. S., Horst, I., Howe, C. J., Lea-Smith, D. J., & Smith, A. G. (2010). Biodiesel from algae: Challenges and prospects. *Current Opinion in Biotechnology*, *21*(3), 277–286. doi:10.1016/j.copbio.2010.03.005 PMID:20399634

Shahzad, K., Abbas, F., Pandey, D., Ajmal, S., Khadim, M., & Tahir, M. U. (2020). Synthesis, characterization and biological evaluation of novel tetrasubsituted Imidazole compounds. *ChemRxiV.*

Sharma, M., Sharma, B., Gupta, A. K., & Pandey, D. (2023). Recent developments of image processing to improve explosive detection methodologies and spectroscopic imaging techniques for explosive and drug detection. *Multimedia Tools and Applications*, *82*(5), 6849–6865. doi:10.100711042-022-13578-5

Singh, S. P., & Singh, D. (2010). Biodiesel production through the use of different sources and characterization of oils and their esters as the substitute of diesel. *Renewable & Sustainable Energy Reviews*, *14*(1), 200–216. doi:10.1016/j.rser.2009.07.017

Stephenson, A. L., Dennis, J. S., Howe, C. J., Scott, S. A., & Smith, A. G. (2010). Influence of nitrogen-limitation regime on the Production by Chlorella vulgaris of lipids for biodiesel feedstocks. *Biofuels*, *1*(1), 47–58. doi:10.4155/bfs.09.1

Stephenson, A. L., Dennis, J. S., & Scott, S. A. (2008). Improving the sustainability of the production of biodiesel from oilseed rape in the UK. *Process Safety and Environmental Protection*, *86*(6), 427–440. doi:10.1016/j.psep.2008.06.005

Swanson, K. J., Madden, M. C., & Ghio, A. J. (2007). Biodiesel exhaust: The need for health effects research. *Environmental Health Perspectives*, *115*(4), 496–499. doi:10.1289/ehp.9631 PMID:17450214

Traviss, N. (2012). Breathing easier? The known impacts of biodiesel on air quality. *Biofuels*, *3*(3), 285–291. doi:10.4155/bfs.12.22 PMID:23585814

Upham, P., Thornley, P., Tomei, J., & Boucher, P. (2009). Substitutable biodiesel feedstocks for the UK: A review of sustainability issues with reference to the UK RTFO. *Journal of Cleaner Production*, *17*, S37–S45. doi:10.1016/j.jclepro.2009.04.014

Van Eijck, J., & Romijn, H. (2008). Prospects for Jatropha biofuels in Tanzania: An analysis with strategic niche management. *Energy Policy*, *36*(1), 311–325. doi:10.1016/j.enpol.2007.09.016

Van Eijck, J., & Romijn, H. (2008). Prospects for Jatropha biofuels in Tanzania: An analysis with strategic niche management. *Energy Policy*, *36*(1), 311–325. doi:10.1016/j.enpol.2007.09.016

Verma, N., Sehrawat, K. D., Sehrawat, A. R., & Pandey, D. (2021). Effective green synthesis, characterization and antibacterial efficacy of silver nanoparticles from seaweed treated sprouts of Moth Bean (Vigna aconitifolia Jacq.). *Regenerative Engineering and Translational Medicine*, 1–14.

Yan, S., Chen, X., Wu, J., & Wang, P. (2013). Pilot-scale production of fuel ethanol from concentrated food waste hydrolysates using Saccharomyces cerevisiae H058. *Bioprocess and Biosystems Engineering*, *36*(7), 937–946. doi:10.100700449-012-0827-9 PMID:23010725

Yang, X., Lee, S. J., Yoo, H. Y., Choi, H. S., Park, C., & Kim, S. W. (2014). Biorefinery of instant noodle waste to biofuels. *Bioresource Technology*, *159*, 17–23. doi:10.1016/j.biortech.2014.02.068 PMID:24632436

Zhang, X., Li, N., Shao, H., Meng, Y., Wang, L., Wu, Q., Yao, Y., Li, J., Bian, J., Zhang, Y., & Deng, X. (2016). Methane limit LPS-induced NF-κB/MAPKs signal in macrophages and suppress immune response in mice by enhancing PI3K/AKT/GSK-3β-mediated IL-10 expression. *Scientific Reports*, *6*(1), 29359. doi:10.1038rep29359 PMID:27405597

Compilation of References

Abanades, S., Abbaspour, H., Ahmadi, A., Das, B., Ehyaei, M. A., Esmaeilion, F., El Haj Assad, M., Hajilounezhad, T., Hmida, A., Rosen, M. A., Safari, S., Shabi, M. A., & Silveira, J. L. (2022). A conceptual review of sustainable electrical power generation from biogas. *Energy Science & Engineering, 10*(2), 630–655. doi:10.1002/ese3.1030

Abbaszadeh, S., Zagreus, L., Lehrer, D., & Huizenga, C. (2006). Occupant satisfaction with indoor environmental quality in green buildings. Proceedings of Healthy Buildings, Lisbon, Vol. III, 365-370.

Abd Wahab, M. H., Kadir, A. A., Tomari, M. R., & Jabbar, M. H. (2014). Smart recycle bin: a conceptual approach of smart waste management with integrated web based system. In 2014 international conference on IT convergence and security (ICITCS),1-4. IEEE.

Abdel-Shafy, H. I., & Mansour, M. S. (2018). Solid waste issue: Sources, composition, disposal, recycling, and valorization. *Egyptian journal of petroleum, 27*(4), 1275-1290.

Abdulsahib, J. S., Eneizan, B., & Alabboodi, A. S. (2019). Environmental concern, health consciousness and purchase intention of green products: An application of extended theory of planned behavior. *The Journal of Social Sciences Research, 5*(4), 868–880. doi:10.32861/jssr.54.1203.1215

Abid, N., Khan, A. M., Shujait, S., Chaudhary, K., Ikram, M., Imran, M., Haider, J., Khan, M., Khan, Q., & Maqbool, M. (2022). Synthesis of nanomaterials using various top-down and bottom-up approaches, influencing factors, advantages, and disadvantages: A review. *Advances in Colloid and Interface Science, 300*, 102597–102597. doi:10.1016/j.cis.2021.102597 PMID:34979471

Abuabdou, S. M., Ahmad, W., Aun, N. C., & Bashir, M. J. (2020). A review of anaerobic membrane bioreactors (AnMBR) for the treatment of highly contaminated landfill leachate and biogas production: Effectiveness, limitations and future perspectives. *Journal of Cleaner Production, 255*, 120215. doi:10.1016/j.jclepro.2020.120215

Abuga, D., & Raghava, N. S. (2021). Real-time smart garbage bin mechanism for solid waste management in smart cities. *Sustainable Cities and Society, 75*, 103347. doi:10.1016/j.scs.2021.103347

Ádám, B., Göen, T., Scheepers, P. T., Adliene, D., Batinic, B., Budnik, L. T., & Au, W. W. (2021). From inequitable to sustainable e-waste processing for reduction of impact on human health and the environment. *Environmental Research, 194*, 110728. doi:10.1016/j.envres.2021.110728 PMID:33444608

Adanu, S. K., Gbedemah, S. F., & Attah, M. K. (2020). Challenges of adopting sustainable technologies in e-waste management at Agbogbloshie, Ghana. *Heliyon, 6*(8), e04548. doi:10.1016/j.heliyon.2020.e04548 PMID:32802976

Adesina, A. O. G. (2000). *Integrated Waste Management System.* CORE.

Adrados, A., de Marco, I., Caballero, B. M., López, A., Laresgoiti, M. F., & Torres, A. (2012). Pyrolysis of plastic packaging waste: A comparison of plastic residuals from material recovery facilities with simulated plastic waste. *Waste Management (New York, N.Y.)*, *32*(5), 826–832. doi:10.1016/j.wasman.2011.06.016 PMID:21795037

AffairsCloud. (2023, February 28). *Current Affairs*. AffairsCloud. https://affairscloud.com/current-affairs-28-february-2023/

Aggarwal, P., Rana, M., & Akhai, S. (2022). Briefings on e-waste hazard until COVID era in India. *Materials Today: Proceedings*, *71*(2), 389–393. doi:10.1016/j.matpr.2022.09.507

Ahirwar, R., & Tripathi, A. K. (2021). E-waste management: A review of recycling process, environmental and occupational health hazards, and potential solutions. *Environmental Nanotechnology, Monitoring and Management*, *15*(November 2020), 100409. doi:10.1016/j.enmm.2020.100409

Ahmed, M. J. (2017). Adsorption of non-steroidal anti-inflammatory drugs from aqueous solution using activated carbons [Review]. *Journal of Environmental Management*, *190*, 274–282. doi:10.1016/j.jenvman.2016.12.073 PMID:28063293

Ajanovic, A., & Haas, R. (2014). On the prospects and limits of biofuels in Brazil, the US, and the EU. *Applied Energy*, *135*, 730–737. doi:10.1016/j.apenergy.2014.07.001

Ajayan, P. M., Schadler, L. S., & Braun, P. V. (2006). *Nanocomposite science and technology*. John Wiley & Sons.

Ajay, C. M., Mohan, S., & Dinesha, P. (2021). Decentralized energy from portable biogas digesters using domestic kitchen waste: A review. *Waste Management (New York, N.Y.)*, *125*, 10–26. doi:10.1016/j.wasman.2021.02.031 PMID:33667979

Ajwang, S. O., Abila, J. O., & Tajero-Dakay, I. (2021). Adoption conceptual model for intelligent waste management in smart cities: Theoretical review. *International Journal of Computer Science Research*, *5*(1), 426–440. doi:10.25147/ijcsr.2017.001.1.50

Akhai, S., Mala, S., & Jerin, A. A. (2020). Apprehending Air Conditioning Systems in Context to COVID-19 and Human Health: A Brief Communication. *International Journal of Healthcare Education & Medical Informatics*, *7*(1&2), 28-30.

Akhai, S., Mala, S., & Jerin, A. A. (2021). Understanding whether air filtration from air conditioners reduces the probability of virus transmission in the environment. *Journal of Advanced Research in Medical Science & Technology*, *8*(1), 36-41.

Akhai, S., Singh, V. P., & John, S. (2016). Human performance in industrial design centers with small unit air conditioning systems. *Journal of Advanced Research in Production Industrial Engineering*, *3*(2), 5–11.

Akhai, S., Singh, V. P., & John, S. (2016). Investigating Indoor Air Quality for the Split-Type Air Conditioners in an Office Environment and Its Effect on Human Performance. *Journal of Mechanical Civil Engineering*, *13*(6), 113–118.

Akhai, S., Srivastava, P., & Sharma, S. (2020). Developments in Horizontal Axis Wind Turbines - A Brief Review. *Journal of Critical Reviews*, *7*(19), 255–260.

Akhai, S., Srivastava, P., Sharma, V., & Bhatia, A. (2021). Investigating weld strength of AA8011-6062 alloys joined via friction-stir welding using the RSM approach. []. IOP Publishing.]. *Journal of Physics: Conference Series*, *1950*(1), 012016. doi:10.1088/1742-6596/1950/1/012016

Akhai, S., Thareja, P., & Singh, V. P. (2017). Assessment of Indoor Environment Health Sustenance in Air Conditioned Class Rooms. *Journal of Advanced Research in Civil and Environmental Engineering*, *4*(1&2), 1–9. doi:10.24321/2393.8307.201701

Akshay, D., Gaikwad, S., & Mendhe, A. (2021). AUTOMATIC DRAINAGE CLEANING MACHINE. *International Journal of Innovations in Engineering Research and Technology*, 1–3. https://repo.ijiert.org/index.php/ijiert/article/view/1730

Alami, H., Gagnon, M. P., Wootton, R., Fortin, J. P., & Zanaboni, P. (2017). 'Exploring factors associated with the uneven utilization of telemedicine in Norway : A mixed methods study', *BMC Medical Informatics and Decision Making. BMC Medical Informatics and Decision Making*, *17*(180), 1–15. doi:10.118612911-017-0576-4 PMID:29282048

Alam, P., & Ahmade, K. (2013). Impact of solid waste on health and the environment. [IJSDGE]. *International Journal of Sustainable Development and Green Economics*, *2*(1), 165–168.

Albayrakoglu, M. (2020). *Supply Chain Management In The Textile Industry : A Supplier.* Semantic Scholar.

Albino, V., Berardi, U., Dangelico, R.M. (2015). Smart cities: definitions, dimensions, performance, and initiatives. *Journal of Urban Technology, 22* (1), 3e21.

Albrecht, K. O., & Hallen, R. T. (2011). A Brief Literature Overview of Various Routes to Biorenewable Fuels from Lipids for the National Alliance of Advanced Biofuels and Bio-products NAAB Consortium. US Department of Energy. doi:10.2172/1019208

Ali, G., Nitivattananon, V., Abbas, S., & Sabir, M. (2012). Green waste to biogas: Renewable energy possibilities for Thailand's green markets. *Renewable & Sustainable Energy Reviews*, *16*(7), 5423–5429. doi:10.1016/j.rser.2012.05.021

Ali, I., Alharbi, O. M. L., Alothman, Z. A., Al-Mohaimeed, A. M., & Alwarthan, A. (2019). Modeling of fenuron pesticide adsorption on CNTs for mechanistic insight and removal in water. *Environmental Research*, *170*, 389–397. doi:10.1016/j.envres.2018.12.066 PMID:30623886

Al-Khatib, I. A., Monou, M., Abu Zahra, A. S., Shaheen, H. Q., & Kassinos, D. (2010). Solid waste characterization, quantification and management practices in developing countries. A case study: Nablus District – Palestine. *Journal of Environmental Management*, *91*(5), 1131–1138. doi:10.1016/j.jenvman.2010.01.003 PMID:20116162

AlKhidir, T., & Zailani, S. (2009). Going green in supply chain towards environmental sustainability. *Global Journal of Environmental Research*, *3*(3), 246–251.

Al-Mansour, F., Sucic, B., & Pusnik, M. (2014). Challenges and prospects of electricity production from renewable energy sources in Slovenia. *Energy*, *77*, 73–81. doi:10.1016/j.energy.2014.04.106

Ament, S. M. C., Gillissen, F., Maessen, J. M. C., Dirksen, C. D., van der Weijden, T., & von Meyenfeldt, M. F. (2012). Sustainability of healthcare innovations (SUSHI): Long term effects of two implemented surgical care programmes (protocol). *BMC Health Services Research*, *12*(1), 423. Advance online publication. doi:10.1186/1472-6963-12-423 PMID:23174024

Amoah, J. O., Britwum, A. O., Essaw, D. W., & Mensah, J. (2023). Solid waste management and gender dynamics: Evidence from rural Ghana. *Research in Globalization*, *6*, 100111. doi:10.1016/j.resglo.2023.100111

Anand, A., & Singh, S. (2014). Membrane technique for leachate treatment: A literature review. *International Journal of Environmental Research and Development*, *4*(1), 33–36.

Andeobu, L., Wibowo, S., & Grandhi, S. (2021). A systematic review of E-waste generation and environmental management of Asia Pacific countries. *International Journal of Environmental Research and Public Health*, *18*(17), 9051. doi:10.3390/ijerph18179051 PMID:34501640

Andeobu, L., Wibowo, S., & Grandhi, S. (2022). Artificial intelligence applications for sustainable solid waste management practices in Australia: A systematic review. *The Science of the Total Environment*, *834*, 155389. doi:10.1016/j.scitotenv.2022.155389 PMID:35460765

Andreou, V. G., & Clonis, Y. D. (2002). A portable fiber-optic pesticide biosensor based on immobilized cholinesterase and sol–gel entrapped bromcresol purple for in-field use. *Biosensors & Bioelectronics*, *17*(1-2), 61–69. doi:10.1016/S0956-5663(01)00261-5 PMID:11742736

Andreou, V. G., & Clonis, Y. D. (2002). Novel fiber-optic biosensor based on immobilized glutathione S-transferase and sol–gel entrapped bromcresol green for the determination of atrazine. *Analytica Chimica Acta*, *460*(2), 151–161. doi:10.1016/S0003-2670(02)00250-7

Anitha, C., R, K. C., Vivekanand, C. V., Lalitha, S. D., Boopathi, S., & R, R. (2023, February). Artificial Intelligence driven security model for Internet of Medical Things ({IoMT}). *IEEE Explore*. doi:10.1109/ICIPTM57143.2023.10117713

Annamalai, J. (2015). Occupational health hazards relate to informal recycling of e-waste in India: An overview. *Indian Journal of Occupational and Environmental Medicine*, *19*(1), 61–65. http://www.ijoem. com/article.asp?issn=00195278;year=2015;volume=19;issue=1;spage=61;epage=65;aulast=Annamalai. doi:10.4103/0019-5278.157013 PMID:26023273

Ansar Ali, S. K., Kumar, P., Sharma, V. K., Akhai, S., Kumar, V., & Joshi, R. S. (2022). Determination of friction factor and heat transfer rate in laminar flow in an agitated vessel. *Proceedings of the Institution of Mechanical Engineers, Part E: Journal of Process Mechanical Engineering*.

Antonio, A. (2015). 'Lean in Healthcare: A comprehensive review', *Health policy. Health Policy (Amsterdam)*, *9*(1), 197–209. doi:10.1016/j.healthpol.2015.02.002

Anuar, A., Saad, R., & Yusoff, R. Z. (2018). 'Sustainability through Lean Healthcare and Operational Performance in the Private Hospitals : A Proposed Framework', *Int. J Sup. Chain. Mgt*, *7*(5), 221–227.

Arapoglou, D., Varzakas, T., Vlyssides, A., & Israilides, C. (2010). Ethanol production from potato peel waste (PPW). *Waste Management (New York, N.Y.)*, *30*(10), 1898–1902. doi:10.1016/j.wasman.2010.04.017 PMID:20471817

Ariza-Tarazona, M. C., Villarreal-Chiu, J. F., Hernández-López, J. M., Rivera De la Rosa, J., Barbieri, V., Siligardi, C., & Cedillo-González, E. I. (2020). Microplastic pollution reduction by a carbon and nitrogen-doped TiO2: Effect of pH and temperature in the photocatalytic degradation process. *Journal of Hazardous Materials*, *395*, 122632–122632. doi:10.1016/j.jhazmat.2020.122632 PMID:32315794

Arni, S. S. A., & Elwaheidi, M. M. (2020). *Concise Handbook of Waste Treatment Technologies*. CRC Press.

Aro, E. M. (2016). From first-generation biofuels to advanced solar biofuels. *Ambio*, *45*(S1, Suppl 1), S24–S31. doi:10.100713280-015-0730-0 PMID:26667057

Arvola, J., Belt, P., Harkonen, J., Kess, P., & Imppola, R. (2012). Biogas as an option for industrial applications. *International Journal of Sustainable Economy*, *4*(1), 71–88. doi:10.1504/IJSE.2012.043996

Arya, S., & Kumar, S. (2020). Bioleaching: Urban mining option to curb the menace of E-waste challenge. *Bioengineered*, *11*(1), 640–660. doi:10.1080/21655979.2020.1775988 PMID:32538256

Arya, S., & Kumar, S. (2020). E-waste in India at a glance: Current trends, regulations, challenges and management strategies. *Journal of Cleaner Production*, *271*, 122707. doi:10.1016/j.jclepro.2020.122707

Ashikuzzaman, M., & Howlader, H. M. (2019). *Sustainable Solid Waste Management in Bangladesh: Issues and Challenges*. IGI Global., doi:10.4018/978-1-7998-0198-6.ch002

Ashodian, J. (2016). *Industry 4. 0 Is Transforming Supply Chains*. SICK. https://scg-mmh.s3.amazonaws.com/pdfs/sick_wp_industry4.0_100616.pdf

Assamoi, B., & Lawryshyn, Y. (2012). The environmental comparison of landfilling vs. incineration of MSW accounting for waste diversion. *Waste Management (New York, N.Y.)*, *32*(5), 1019–1030. doi:10.1016/j.wasman.2011.10.023 PMID:22099926

Awan, A. (2020). Batteries: An essential technology to electrify road transport. In Global EV Outlook 2020: Entering the decade of electric drive? (pp. 185-219): International Energy Agency.

Awasthi, A. K., Wang, M., Wang, Z., Awasthi, M. K., & Li, J. (2018). E-waste management in India: A mini-review. *Waste Management & Research*, *36*(5), 408–414. doi:10.1177/0734242X18767038 PMID:29732961

Awasthi, A. K., Zeng, X., & Li, J. (2016). Integrated bioleaching of copper metal from waste printed circuit board—A comprehensive review of approaches and challenges. *Environmental Science and Pollution Research International*, *23*(21), 21141–21156. doi:10.100711356-016-7529-9 PMID:27678000

Ayodeji, I. (2010). Exploring secondary school students' understanding and practices of waste management in Ogun State, Nigeria. *International Journal of Environmental and Science Education*, *5*(2), 201–215. https://files.eric.ed.gov/fulltext/EJ884419.pdf

Azeh Engwa, G., Udoka Ferdinand, P., Nweke Nwalo, F., & Unachukwu, N. M. (2019). Mechanism and Health Effects of Heavy Metal Toxicity in Humans. In Poisoning in the Modern World - New Tricks for an Old Dog?: IntechOpen.

Babu, B. S., Kamalakannan, J., Meenatchi, N., Karthik, S., & Boopathi, S. (2022). Economic impacts and reliability evaluation of battery by adopting Electric Vehicle. *IEEE Explore*, (pp. 1–6). IEEE.

Babu, B. S., Kamalakannan, J., Meenatchi, N., Karthik, S., & Boopathi, S. (2022). Economic impacts and reliability evaluation of battery by adopting Electric Vehicle. *IEEE Explore*, 1–6.

Babu, B. S., Kamalakannan, J., Meenatchi, N., M, S. K. S., S, K., & Boopathi, S. (2023). Economic impacts and reliability evaluation of battery by adopting Electric Vehicle. *IEEE Explore*, 1–6. doi:10.1109/ICPECTS56089.2022.10046786

Babu, B.S., Kamalakannan, J., Meenatchi, N., Karthik, S., Boopathi, S. (2022). Economic impacts and reliability evaluation of battery by adopting Electric Vehicle. *IEEE Explor.* 1–6.

Babu, S. Z. D., Pandey, D., Naidu, G. T., Sumathi, S., Gupta, A., Bader Alazzam, M., & Pandey, B. K. (2022, July). Analysation of Big Data in Smart Healthcare. In *Artificial Intelligence on Medical Data: Proceedings of International Symposium, ISCMM 2021* (pp. 243-251). Singapore: Springer Nature Singapore.

Bachmann, N., la Cour Jansen, J., Bochmann, G., & Montpart, N. (2015). *Sustainable biogas production in municipal wastewater treatment plants*. IEA Bioenergy.

Badrnezhad, R., & Mirza, B. (2014, March). Modeling and optimization of cross-flow ultrafiltration using hybrid neural network-genetic algorithm approach. *Journal of Industrial and Engineering Chemistry*, *20*(2), 528–543. doi:10.1016/j.jiec.2013.05.012

Baeyens, J., Appels, L., Peng, L. Y., & Dewil, R. (2016). Biogas as a renewable fuel. *Renewable Energy, 96*(Part B), 1055-1152.

Bagherzadeh, F., Mehrani, M. J., Basirifard, M., & Roostaei, J. (2021, June). Comparative study on total nitrogen prediction in wastewater treatment plant and effect of various feature selection methods on machine learning algorithms performance. *Journal of Water Process Engineering*, *41*, 102033. doi:10.1016/j.jwpe.2021.102033

Bag, S., Wood, L. C., Mangla, S. K., & Luthra, S. (2019, September). Procurement 4.0 and Its Implications on Business Process Performance in a Circular Economy. *Resources, Conservation and Recycling*, *152*, 104502. doi:10.1016/j.resconrec.2019.104502

Bahauddin, K. M., & Uddin, M. H. (2012). Prospect of Solid Waste Situation and an Approach of Environmental Management Measure (EMM) Model for Sustainable Solid Waste Management: Case Study of Dhaka City. *Journal of Environmental Science & Natural Resources*, *5*(1), 99–111. doi:10.3329/jesnr.v5i1.11601

Bains, M., Yarker, J., Amir, Z., Wynn, P., & Munir, F. (2012). Helping Cancer Survivors Return to Work : What Providers Tell Us About the Challenges in Assisting Cancer Patients with Work Questions. *Journal of Occupational Rehabilitation*, *22*(1), 71–77. doi:10.100710926-011-9330-4 PMID:21870065

Bakeas, E. B., Argyris, D. I., & Siskos, P. A. (2003). Carbonyl compounds in the urban environment of Athens, Greece. *Chemosphere*, *52*(5), 805–813. doi:10.1016/S0045-6535(03)00257-1 PMID:12757781

Bakker, T., Van Asselt, K., Bontsema, J., Müller, J., & van Straten, G. (2010). A path following algorithm for mobile robots. *Autonomous Robots*, *29*(1), 85–97. doi:10.100710514-010-9182-3

Baldé, C. P., Wang, F., Kuehr, R., & Huisman, J. (2015). *The global e-waste monitor–2014, United Nations University, IAS–SCYCLE.* ISBN Print.

Baldo, D., Mecocci, A., Parrino, S., Peruzzi, G., & Pozzebon, A. (2021). A multi-layer lorawan infrastructure for smart waste management. *Sensors (Basel)*, *21*(8), 2600. doi:10.339021082600 PMID:33917255

Banerjee, S. (2019). Study on Yoga Intervention and Diet on Hypothyroidism Associated with Obesity among Sedentary Working Women in West Bengal. *International Journal of Yoga and Allied Sciences*, *8*(1), 18–23.

Banerjee, S. (2020). The Essence of Indian Indigenous Knowledge in the perspective of Ayurveda, Nutrition, and Yoga. *Research & Reviews in Biotechnology & Biosciences*, *7*(2), 20–27.

Barceló, D., & Farré, M. (2012). *Analysis and risk of nanomaterials in environmental and food samples* (Vol. 59). Newnes. doi:10.1016/B978-0-444-56328-6.09999-0

Bayda, S., Adeel, M., Tuccinardi, T., Cordani, M., & Rizzolio, F. (2019). The History of Nanoscience and Nanotechnology: From Chemical–Physical Applications to Nanomedicine. *Molecules (Basel, Switzerland)*, *25*(1), 112–112. doi:10.3390/molecules25010112 PMID:31892180

Belloni, F., Kütahyali, C., Rondinella, V. V., Carbol, P., Wiss, T., & Mangione, A. (2009). Can carbon nanotubes play a role in the field of nuclear waste management? *Environmental Science & Technology*, *43*(5), 1250–1255. doi:10.1021/es802764g PMID:19350887

Bello, R. H., Linzmeyer, P., Franco, C. M. B., Souza, O., Sellin, N., Medeiros, S. H. W., & Marangoni, C. (2014). Pervaporation of ethanol produced from banana waste. *Waste Management (New York, N.Y.)*, *34*(8), 1501–1509. doi:10.1016/j.wasman.2014.04.013 PMID:24834817

Benjamin, E. J., Virani, S. S., Callaway, C. W., Chamberlain, A. M., Chang, A. R., Cheng, S., Chiuve, S. E., Cushman, M., Delling, F. N., Deo, R., de Ferranti, S. D., Ferguson, J. F., Fornage, M., Gillespie, C., Isasi, C. R., Jiménez, M. C., Jordan, L. C., Judd, S. E., Lackland, D., & Muntner, P. (2018). Heart disease and stroke statistics—2018 update: A report from the American Heart Association. *Circulation*, *137*(12), e67–e492. doi:10.1161/CIR.0000000000000558 PMID:29386200

Berthel, S., Folke, C., Cloding, J. (2010). *Social–ecological memory in urban gardens Retaining the capacity for management of ecosystem services.* Project: SUPER Sustainable Urban Planning for Ecosystem Services and Resilience. . doi:10.1016/j.gloenvcha.2010.01.001

Bharathiraja, B., Sudharsanaa, T., Bharghavi, A., Jayamuthunagai, J., & Praveenkumar, R. (2016). Biohydrogen and Biogas–An overview on feedstocks and enhancement process. *Fuel*, *185*, 810–828. doi:10.1016/j.fuel.2016.08.030

Bhatia, R., & Baruah, P. (2020). Exclusive talent management and its consequences: A review of literature. *Asian Journal of Business Ethics*, *9*(2), 193–209. doi:10.100713520-020-00105-8

Bhatt, A. H., & Tao, L. (2020). Economic perspectives of biogas production via anaerobic digestion. *Bioengineering (Basel, Switzerland)*, *7*(3), 74. doi:10.3390/bioengineering7030074 PMID:32674480

Bhunia, P., Chatterjee, S., Rudra, P., & De, S. (2018). Chelating polyacrylonitrile beads for removal of lead and cadmium from wastewater. *Separation and Purification Technology*, *193*, 202–213. doi:10.1016/j.seppur.2017.11.001

Bhutta, M. K. S., Omar, A., & Yang, X. (2011). Electronic Waste: A Growing Concern in Today's Environment. *Economic Research International*, *2011*, 1–8. doi:10.1155/2011/474230

Bick, I. A., Bardhan, R., & Beaubois, T. (2018). Applying fuzzy logic to open data for sustainable development decision-making: A case study of the planned city Amaravati. *Natural Hazards*, *91*(3), 1317–1339. doi:10.100711069-018-3186-2

Bienhaus, F., & Haddud, A. (2018). Procurement 4.0: Factors Influencing the Digitisation of Procurement and Supply Chains. *Business Process Management Journal*, *24*(4), 965–984. doi:10.1108/BPMJ-06-2017-0139

Bikis, A., & Pandey, D. (2021). Air quality at public transportation stations/stops: contribution of light rail transit to reduce air pollution. *Aerosol Science and Engineering*, 1-16.

Bikis, A., & Pandey, D. (2023). Squatter settlement and informal urbanization: Causes and consequences. *Environmental Science and Pollution Research International*, *30*(9), 23276–23294. doi:10.100711356-022-23778-z PMID:36322351

Bimir, M. N. (2020). Revisiting e-waste management practices in selected African countries. *Journal of the Air & Waste Management Association*, *70*(7), 659–669. doi:10.1080/10962247.2020.1769769 PMID:32401631

Biogas Forum India. (2015). *E–News Letter, 6*(2). Biogas Forum. http://www.biogasforum.in/newsletter/BFI_Newsletter_Jan-Jun_2015.pdf

Blinová, L., & Godovčin, P. (2021). Importance of Recycling the Waste-Cables Containing Copper and PVC. *Res. Pap. Fac. Mater. Sci. Technol. Slovak Univ. Technol.*, *29*(48), 1–21. doi:10.2478/rput-2021-0001

BMC. (2018). *Swachchta Sarvekshan –2018*. Bhopal Municipal Corporation. https://cdn.cseindia.org/docs/photogallery/slideshows/02_20171212_BHOPAL_SBM_PPT.pdf

Bocken, N. M. P., de Pauw, I., Bakker, C., & van der Grinten, B. (2016). Product design and business model strategies for a circular economy. *J. Ind. Prod. Eng.*, *2016*(33), 308–320. doi:10.1080/21681015.2016.1172124

Bommareddy, S., Khan, J. A., & Pandey, D. (2022). IoT implementation and challenges in healthcare industries. In *Networking Technologies in Smart Healthcare* (pp. 211–230). CRC Press.

Bong, C. P. C., Ho, W. S., Hashim, H., Lim, J. S., Ho, C. S., Tan, W. S. P., & Lee, C. T. (2017). Review on the renewable energy and solid waste management policies towards biogas development in Malaysia. *Renewable & Sustainable Energy Reviews*, *70*, 988–998. doi:10.1016/j.rser.2016.12.004

Boopathi, S. (2013). Experimental Study and Multi- Objective Optimization of Near-Dry Wire-Cut Electrical Discharge Machining Process Faculty of Mechanical Engineering. In *Shodhganga@INFLIBNET* (Issue August)http://hdl.handle.net/10603/16933

Boopathi, S. (2021). Improving of Green Sand-Mould Quality using Taguchi Technique. *Journal of Engineering Research*, 1–12. doi:10.36909/jer.14079

Boopathi, S., & Khare, R. K.G., J.C., Muni, T.V., & Khare, S. (2023). Additive Manufacturing Developments in the Medical Engineering Field. In Development, Properties, and Industrial Applications of 3D Printed Polymer Composites (pp. 86–106). IGI Global

Boopathi, S., & Khare, R. KG, J. C., Muni, T. V., & Khare, S. (2023). Additive Manufacturing Developments in the Medical Engineering Field. In Development, Properties, and Industrial Applications of 3D Printed Polymer Composites (pp. 86–106). IGI Global.

Boopathi, S., Arigela, S. H., Raman, R., Indhumathi, C., Kavitha, V., & Bhatt, B. C. (2022). Prominent Rule Control-based Internet of Things: Poultry Farm Management System. *IEEE Explore*, (pp. 1–6). IEEE.

Boopathi, S., Arigela, S. H., Raman, R., Indhumathi, C., Kavitha, V., & Bhatt, B. C. (2022). Prominent Rule Control-based Internet of Things: Poultry Farm Management System. *IEEE Explore*, 1–6.

Boopathi, S., Arigela, S. H., Raman, R., Indhumathi, C., Kavitha, V., & Bhatt, B. C. (2023). Prominent Rule Control-based Internet of Things: Poultry Farm Management System. *IEEE Explore*, 1–6. doi:10.1109/ICPECTS56089.2022.10047039

Boopathi, S., Arigela, S.H., Raman, R., Indhumathi, C., Kavitha, V., & Bhatt, B.C. (2022). Prominent Rule Control-based Internet of Things: Poultry Farm Management System. *IEEE Explor.* 1–6.

Boopathi, S., Khare, R., Jaya Christiyan, K. G., Muni, T. V., & Khare, S. (2023). Additive manufacturing developments in the medical engineering field. In Development, Properties, and Industrial Applications of 3D Printed Polymer Composites (pp. 86–106). IGI Global. doi:10.4018/978-1-6684-6009-2.ch006

Boopathi, S., Siva Kumar, P. K., & Meena, R. S. J., S. I., P., S. K., & Sudhakar, M. (2023). Sustainable Developments of Modern Soil-Less Agro-Cultivation Systems. In Human Agro-Energy Optimization for Business and Industry (pp. 69–87). IGI Global. doi:10.4018/978-1-6684-4118-3.ch004

Boopathi, S. (2019). Experimental investigation and parameter analysis of LPG refrigeration system using Taguchi method. *SN Applied Sciences*, *1*(8), 892. doi:10.100742452-019-0925-2

Boopathi, S. (2022a). An experimental investigation of Quench Polish Quench (QPQ) coating on AISI 4150 steel. *Eng. Res. Express*, *4*(4), 045009. doi:10.1088/2631-8695/ac9ddd

Boopathi, S. (2022a). An Extensive Review on Sustainable Developments of Dry and Near-Dry Electrical Discharge Machining Processes. *Journal of Manufacturing Science and Engineering*, *144*(5), 1–37. doi:10.1115/1.4052527

Boopathi, S. (2022b). (in press). Cryogenically treated and untreated stainless steel grade 317 in sustainable wire electrical discharge machining process: A comparative study. *Environmental Science and Pollution Research International*, 1–10. doi:10.100711356-022-22843-x PMID:36057706

Boopathi, S. (2022c). Experimental investigation and multi-objective optimization of cryogenic Friction-stir-welding of AA2014 and AZ31B alloys using MOORA technique. *Materials Today. Communications*, *33*, 104937. doi:10.1016/j.mtcomm.2022.104937

Boopathi, S., Balasubramani, V., & Sanjeev Kumar, R. (2023). Influences of various natural fibers on the mechanical and drilling characteristics of coir-fiber-based hybrid epoxy composites. *Eng. Res. Express*, *5*(1), 15002–15020. doi:10.1088/2631-8695/acb132

Boopathi, S., & Myilsamy, S. (2021). Material removal rate and surface roughness study on Near-dry wire electrical discharge Machining process. *Materials Today: Proceedings*, *45*(9), 8149–8156. doi:10.1016/j.matpr.2021.02.267

BoopathiS.MyilsamyS.SukkasamyS. (2021). Experimental Investigation and Multi-Objective Optimization of Cryogenically Cooled Near-Dry Wire-Cut EDM Using TOPSIS Technique. *Ijamt Preprint*. doi:10.21203/rs.3.rs-254117/v1

Boopathi, S., & Sivakumar, K. (2013). Experimental investigation and parameter optimization of near-dry wire-cut electrical discharge machining using multi-objective evolutionary algorithm. *International Journal of Advanced Manufacturing Technology*, *67*(9–12), 2639–2655. doi:10.100700170-012-4680-4

Boopathi, S., & Sivakumar, K. (2016). Optimal parameter prediction of oxygen-mist near-dry Wire-cut EDM. *Int. J. Manuf. Technol. Manag.*, *30*(3–4), 164–178. doi:10.1504/IJMTM.2016.077812

Boopathi, S., Sivakumar, K., & Kalidas, R. (2012). Parametric Study of Dry WEDM Using Taguchi Method. *International Journal of Engineering Research and Development*, 2(4), 1–6.

Boopathi, S., Thillaivanan, A., Azeem, M. A., Shanmugam, P., & Pramod, V. R. (2022). Experimental investigation on abrasive water jet machining of neem wood plastic composite. *Funct. Compos. Struct.*, *4*(2), 025001. doi:10.1088/2631-6331/ac6152

Borbón, L. O. P. (2011). *Computational studies of transition metal nanoalloys*. Springer Science & Business Media. doi:10.1007/978-3-642-18012-5

Borthakur, A., & Govind, M. (2018). Public understandings of E-waste and its disposal in urban India: From a review towards a conceptual framework. *Journal of Cleaner Production*, *172*, 1053–1066. doi:10.1016/j.jclepro.2017.10.218

Bratby, J. (2016). *Coagulation and flocculation in water and wastewater treatment*. IWA publishing. doi:10.2166/9781780407500

Brennan, L., & Owende, P. (2010). Biofuels from microalgae- a review of technologies for production, processing, and extract biofuels and co-products. *Renewable & Sustainable Energy Reviews*, *14*(2), 557–577. doi:10.1016/j.rser.2009.10.009

Briggs, M. (2004). *Widescale Biodiesel Production from Algae*. UNH Biodiesel Group (University of New Hampshire).

Brito, L. B. Q., Agrawal, P., de Mélo, T. J. A., Brito, G. de F., & Morais, C. R. da S. (2023). Recycling of waste glass for the production of hollow blocks using the kiln-casting process. *Cleaner Waste Systems*, *4*, 100079. doi:10.1016/j.clwas.2023.100079

Brunner, P. H., & Rechberger, H. (2015). Waste to energy–key element for sustainable waste management. *Waste Management (New York, N.Y.)*, *37*, 3–12. doi:10.1016/j.wasman.2014.02.003 PMID:24630214

Bui, T. D., Tseng, J. W., Tseng, M. L., & Lim, M. K. (2022). Opportunities and challenges for solid waste reuse and recycling in emerging economies: A hybrid analysis. *Resources, Conservation and Recycling*, *177*, 105968. doi:10.1016/j.resconrec.2021.105968

Bui, T. D., Tseng, J. W., Tseng, M. L., Wu, K. J., & Lim, M. K. (2023). Municipal solid waste management technological barriers: A hierarchical structure approach in Taiwan. *Resources, Conservation and Recycling*, *190*, 106842. Advance online publication. doi:10.1016/j.resconrec.2022.106842

Bundgaard, S. S., Kofoed-Wiuff, A., Herrmann, I. T., & Karlsson, K. B. (2014). Experiences with biogas in Denmark. *DTU Management Engineering*.

Bundhoo, Z. M. (2018). Solid waste management in least developed countries: Current status and challenges faced. *Journal of Material Cycles and Waste Management*, *20*(3), 1867–1877. doi:10.100710163-018-0728-3

Calvo, F., Moreno, B., Zamorano, M., & Szanto, M. (2005). Environmental Diagnosis Methodology for Municipal Waste Dumping sites. *Waste Management (New York, N.Y.)*, *25*(8), 768–779. doi:10.1016/j.wasman.2005.02.019 PMID:15905084

Cançado, J. E., Saldiva, P. H., Pereira, L. A., Lara, L. B., Artaxo, P., Martinelli, L. A., & Braga, A. L. (2006). The impact of sugar cane–burning emissions on the respiratory system of children and the elderly. *Environmental Health Perspectives, 114*(5), 725–729. doi:10.1289/ehp.8485 PMID:16675427

Cao, D., Chan, M., & Ng, S. (2023, February 17). Modeling and Forecasting of nanoFeCu Treated Sewage Quality Using Recurrent Neural Network (RNN). *Computation (Basel, Switzerland), 11*(2), 39. doi:10.3390/computation11020039

Cao, S. (2020). Evaluation of Smart Humanity Systems and Novel UV-Oriented Solution for Integration, Resilience, Inclusiveness and Sustainability. *5th International Conference on Universal Village, UV 2020*, (21). 10.1109/UV50937.2020.9426226

Capolongo, S., Bottero, M. C., & Lettieri, E. (2015) Chapter 1 Healthcare Sustainability Challenge. *International Academic Conference, 6*(3), 0–9. 10.1007/978-3-319-14036-0

Caruso, M. C., Braghieri, A., Capece, A., Napolitano, F., Romano, P., Galgano, F., Altieri, G., & Genovese, F. (2019). Recent updates on the use of agro-food waste for biogas production. *Applied Sciences (Basel, Switzerland), 9*(6), 1217. doi:10.3390/app9061217

Castagnoli, R., Büchi, G., Coeurderoy, R., & Cugno, M. (2022). Evolution of Industry 4.0 and International Business: A Systematic Literature Review and a Research Agenda. *European Management Journal, 40*(4), 572–589. doi:10.1016/j.emj.2021.09.002

Castro, M. de F., Mateus, R., & Bragança, L. (2017). Healthcare Building Sustainability Assessment tool - Sustainable Effective Design criteria in the Portuguese context. *Environmental Impact Assessment Review, 67*(August), 49–60. doi:10.1016/j.eiar.2017.08.005

Cavicchi, C., Oppi, C., & Vagnoni, E. (2022). Back and Forth on Sustainable Development: A Focus on Healthcare Organisations. *Sustainability (Basel), 14*(9), 1–15. doi:10.3390u14094958

Census of India. (2011). *PCA_AY_2011_Revised (1)*. Government of India.

Central Public Health and Environmental Engineering Organisation (CPHEEO). (2016). *SWACHH BHARAT MISSION Municipal Solid Waste Management Manual Part II: The Manual.* Ministry of Urban Development. http://moud.gov.in/sites/default/files/SBM-II-2016.pdf

Chaianong, A., & Pharino, C. (2022). How to design an area-based prioritization of biogas production from organic municipal solid waste? Evidence from Thailand. *Waste Management (New York, N.Y.), 138*, 243–252. doi:10.1016/j.wasman.2021.11.042 PMID:34906780

Chakravarthi, P. K., Yuvaraj, D., & Venkataramanan, V. (2022). IoT-based smart energy meter for smart grids. *ICDCS 2022 - 2022 6th International Conference on Devices, Circuits and Systems*, (pp. 360–363). IEEE. 10.1109/ICDCS54290.2022.9780714

Chalmeta, R. & Santos-dele, N. (2020). *Sustainable Supply Chain in the Era of Industry 4 . 0 and Big Data : A Systematic Analysis of Literature and Research.* MDPI.

Chang, P., Bao, X., Meng, F., & Lu, R. (2023, April). Multi-objective Pigeon-inspired Optimized feature enhancement soft-sensing model of Wastewater Treatment Process. *Expert Systems with Applications, 215*, 119193. doi:10.1016/j.eswa.2022.119193

Chen, W. E., Wang, Y. H., Huang, P. C., Huang, Y. Y., & Tsai, M. Y. (2018). A smart IoT system for waste management. In *2018 1st International cognitive cities conference (IC3)*, 202-203. IEEE. 10.1109/IC3.2018.00-24

Chen, B., Xiong, R., Li, H., Sun, Q., & Yang, J. (2019). Pathways for Sustainable Energy Transition. *Journal of Cleaner Production, 228*, 1564–1571. doi:10.1016/j.jclepro.2019.04.372

Chen, C., Shih, H.-S., Shyur, H.-J., & Wu, K.-S. (2012). A business strategy selection of green supply chain management via an. *Computers & Mathematics with Applications (Oxford, England), 64*(1), 2544–2546. doi:10.1016/j.camwa.2012.06.013

Chen, G., Wu, G., Li, N., Lu, X., Zhao, J., He, M., Yan, B., Zhang, H., Duan, X., & Wang, S. (2021). Landfill leachate treatment by persulphate related advanced oxidation technologies. *Journal of Hazardous Materials, 418*, 126355–126355. doi:10.1016/j.jhazmat.2021.126355 PMID:34329014

Cheng, H., & Hu, Y. (2010). Analysis of the Spatial Distribution of Lead Concentration in the Soil of Anka, Zamfara State, Nigeria. *Journal of Environment Pollution and Human Health., 4*(1), 16–23. doi:10.12691/jephh-4-1-3

Chen, H., Qiu, T., Rong, J., He, C., & Wang, Q. (2015). Microalgal biofuel revisited: An informatics-based analysis of developments to date and future prospects. *Applied Energy, 155*, 585–598. doi:10.1016/j.apenergy.2015.06.055

Chen, W. T., Wardill, J. T., Sun, Y., Pulver, S. R., Renninger, L. S., Baohan, A., Schreiter, E. R., Kerr, R. A., Orger, M. B., Jayaraman, V., Looger, L. L., Svoboda, K., & Kim, D. S. (2013). Ultrasensitive fluorescent proteins for imaging neuronal activity. *Nature, 499*(7458), 295–300. doi:10.1038/nature12354 PMID:23868258

Chen, W., Gu, Z., Ran, G., & Li, Q. (2021). Application of membrane separation technology in the treatment of leachate in China: A review. *Waste Management (New York, N.Y.), 121*, 127–140. doi:10.1016/j.wasman.2020.12.002 PMID:33360812

Chen, Y., Chen, M., Li, Y., Wang, B., Chen, S., & Xu, Z. (2018). Impact of technological innovation and regulation development on e-waste toxicity: A case study of waste mobile phones. *Scientific Reports, 8*(1), 1–9. doi:10.103841598-018-25400-0 PMID:29740013

Cherian, J., & Jacob, J. (2012). Green marketing: A study of consumers' attitude towards environment friendly products. *Asian Social Science, 8*(12), 117–126. doi:10.5539/ass.v8n12p117

Cherukuri, I., Sultana, N., & Podila, S. P. (2018). Status Of E-Waste In India- A Review. *IOSR Journal of Environmental Science, Toxicology and Food Technology, 12*(11), 8–16. doi:10.9790/2402-1211010816

Chiu, S. L., & Lo, I. M. (2016). Reviewing the anaerobic digestion and co-digestion process of food waste from the perspectives on biogas production performance and environmental impacts. *Environmental Science and Pollution Research International, 23*(24), 24435–24450. doi:10.100711356-016-7159-2 PMID:27380183

Choi, E., Bhandari, T. R., & Shrestha, N. (2020). Social inequality, noise pollution, and quality of life of slum dwellers in Pokhara, Nepal. *Archives of Environmental & Occupational Health, 0*(0), 1–12. doi:10.1080/19338244.2020.1860880 PMID:33355030

Choi, J.-W., Kim, Y.-K., Oh, B.-K., Song, S.-Y., & Lee, W. H. (2003). Optical biosensor for simultaneous detection of captan and organophosphorus compounds. *Biosensors & Bioelectronics, 18*(5-6), 591–597. doi:10.1016/S0956-5663(03)00016-2 PMID:12706567

Cloete, T. E. (2010). *Nanotechnology in water treatment applications*. Caister Academic Press.

Constantinescu, A., Platon, V., Surugiu, M., Frone, S., Antonescu, D., & Mazilescu, R. (2022). The Influence of Eco-Investment on E-Waste Recycling-Evidence From EU Countries. *Frontiers in Environmental Science, 10*(July), 1–15. doi:10.3389/fenvs.2022.928955

Copat, C., Cristaldi, A., Fiore, M., Grasso, A., Zuccarello, P., Santo Signorelli, S., Conti, G. O., & Ferrante, M. (2020). The role of air pollution (PM and NO2) in COVID-19 spread and lethality: A systematic review. *Environmental Research, 191*, 110129. doi:10.1016/j.envres.2020.110129 PMID:32853663

Cremiato, R., Mastellone, M. L., Tagliaferri, C., Zaccariello, L., & Lettieri, P. (2018). Environmental impact of municipal solid waste management using life cycle assessment: The effect of anaerobic digestion, materials recovery and secondary fuels production. *Renewable Energy*, *124*, 180–188. doi:10.1016/j.renene.2017.06.033

Cronin, J. J. Jr, Smith, J. S., Gleim, M. R., Ramirez, E., & Martinez, J. D. (2011). Green marketing strategies: An examination of stakeholders and the opportunities they present. *Journal of the Academy of Marketing Science*, *39*(1), 158–174. doi:10.100711747-010-0227-0

Cui, J., & Zhang, L. (2008). Metallurgical recovery of metals from electronic waste: A review. *Journal of Hazardous Materials*, *158*(2-3), 228–256. doi:10.1016/j.jhazmat.2008.02.001 PMID:18359555

Cullen, K. (2005). Workplace-Based Return-to-Work Interventions. *Journal of Occupational Rehabilitation*, *15*(4), 607–631. doi:10.100710926-005-8038-8 PMID:16254759

Czekała, W. (2022). Biogas as a Sustainable and Renewable Energy Source. *Clean Fuels for Mobility*, 201-214.

D'Adamo, I., Gastaldi, M., & Rosa, P. (2020). Recycling of end-of-life vehicles: Assessing trends and performances in Europe. *Technological Forecasting and Social Change*, *152*, 119887. doi:10.1016/j.techfore.2019.119887

da Silva, R. F. B., Batistella, M., & Moran, E. F. (2017). Socioeconomic changes and environmental policies as dimensions of regional land transitions in the Atlantic Forest, Brazil. *Environmental Science \& Policy, 74*, 14–22.

Dabija, D. C., Bejan, B. M., & Grant, D. B. (2018). The impact of consumer green behaviour on green loyalty among retail formats. *Moravian geographical reports,*26(3),173-185.

Dadheech, P. (2022). A Compressive Review on Internet of Things in Healthcare: Opportunity, Application and Challenges. *Emerging Technologies in Computer Engineering: Cognitive Computing and Intelligent IoT: 5th International Conference, ICETCE 2022*. Cham: Springer International Publishing. 10.1007/978-3-031-07012-9_20

Dadheech, P., Dogiwal, S. R., Kumar, A., Raja, L., Varshney, N., & Janu, N. (2022, May). A Compressive Review on Internet of Things in Healthcare: Opportunity, Application and Challenges. In *Emerging Technologies in Computer Engineering: Cognitive Computing and Intelligent IoT: 5th International Conference, ICETCE 2022*, (pp. 226-235). Cham: Springer International Publishing.

Damodharan, U. (2013). Bioaccumulation of heavy metals in contaminated river water-Uppanar, Cuddalore South East Coast of India. *Intech Open*. . doi:10.5772/5334

Dandge, A., & Valsson, S. (2020). Conservation of Material, Technology and Practice in Heritage Structure and its Relevance in Today's Context. Encyclopedia of Renewable and Sustainable Materials, 39–49. doi:10.1016/B978-0-12-803581-8.10723-4

Daniels, T., Williams, I., Robinson, S., & Spence, K. (2013). Tackling disinvestment in health care services The views of resource allocators. *Journal of Health Organization and Management*, *27*(6), 762–780. doi:10.1108/JHOM-11-2012-0225 PMID:24422258

Das, A. K. (2013). E-waste management in India current scenario. *International Journal of Management, Technology and Engineering, 6*(6), 1–41. https://www.epa.gov/sites/default/files/2014-05/documents/india.pdf

David, S., Duraipandian, K., Chandrasekaran, D., Pandey, D., Sindhwani, N., & Pandey, B. K. (2023). Impact of blockchain in healthcare system. In *Unleashing the Potentials of Blockchain Technology for Healthcare Industries* (pp. 37–57). Academic Press. doi:10.1016/B978-0-323-99481-1.00004-3

De Feo, G., Antoniou, G., Fardin, H. F., El-Gohary, F., Zheng, X. Y., Reklaityte, I., Butler, D., Yannopoulos, S., & Angelakis, A. N. (2014). The historical development of sewers worldwide. *Sustainability (Basel)*, *6*(6), 3936–3974. doi:10.3390u6063936

Debnath, B., Roychoudhuri, R., & Ghosh, S. K. (2016). E-waste management–a potential route to green computing. *Procedia Environmental Sciences*, *35*, 669–675. doi:10.1016/j.proenv.2016.07.063

Debrah, J. K., Vidal, D. G., & Dinis, M. A. P. (2021). Raising awareness on solid waste management through formal education for sustainability: A developing countries evidence review. *Recycling*, *6*(1), 6. doi:10.3390/recycling6010006

Demirbaş, A. (2008). Production of biodiesel from algae oils. *Energy Sources. Part A, Recovery, Utilization, and Environmental Effects*, *31*(2), 163–168. doi:10.1080/15567030701521775

Dermatas, D., Mpouras, T., & Panagiotakis, I. (2018). Application of Nanotechnology for waste management: Challenges and limitations. *Waste Management & Research*, *36*(3), 197–199. doi:10.1177/0734242X18758820 PMID:29466938

Desmond, S. (2018). Health service planning and sustainable development : Considering what, where and how care is delivered through a pro-environmental lens. *Australian Health Review*, *42*(2), 140–145. doi:10.1071/AH16217 PMID:28248634

Devarakonda, S., Sevusu, P., Liu, H., Liu, R., Iftode, L., & Nath, B. (2013). Real-time air quality monitoring through mobile sensing in metropolitan areas. *Proceedings of the 2nd ACM SIGKDD International Workshop on Urban Computing*, (pp. 1–8). ACM. 10.1145/2505821.2505834

Dhawan, A., Singh, S., Kumar, A., & Shanker, R. (2018). *Nanobiotechnology: Human Health and the Environment*. CRC Press. doi:10.1201/9781351031585

Di Maria, F., Mastrantonio, M., & Uccelli, R. (2021). The Life Cycle Approach for assessing the impact of municipal solid waste incineration on the environment and on human health. *The Science of the Total Environment*, *776*, 145785. doi:10.1016/j.scitotenv.2021.145785 PMID:33647657

Dimitrov, R., & Zlateva, P. (2019, September). Possible Uses of Biogas for Power Purposes. In *2019 11th Electrical Engineering Faculty Conference (BulEF)* (pp. 1-4). IEEE. 10.1109/BulEF48056.2019.9030756

Dogiwal, S. R., Dadheech, P., Kumar, A., Raja, L., Kumar Singh, M., & Janu, N. (2022, May). Internet of Things Based Real-Time Monitoring System for Grid Data. In *Emerging Technologies in Computer Engineering: Cognitive Computing and Intelligent IoT: 5th International Conference, ICETCE 2022*, (pp. 236-247). Cham: Springer International Publishing. 10.1007/978-3-031-07012-9_21

Domakonda, V. K., Farooq, S., Chinthamreddy, S., Puviarasi, R., Sudhakar, M., & Boopathi, S. (2022). Sustainable Developments of Hybrid Floating Solar Power Plants: Photovoltaic System. In Human Agro-Energy Optimization for Business and Industry (pp. 148–167). IGI Global.

Domakonda, V. K., Farooq, S., Chinthamreddy, S., Puviarasi, R., Sudhakar, M., & Boopathi, S. (2023). Sustainable Developments of Hybrid Floating Solar Power Plants. In *Human Agro-Energy Optimization for Business and Industry* (pp. 148–167). IGI Global. doi:10.4018/978-1-6684-4118-3.ch008

Douglas, C. C., Lawrence, J. C., Bush, N. C., & Oster, R. a, Gower, B. A., & Darnell, B. E. (2008). Ability of the Harris Benedict to predict energy requirements differs with weight history and ethnicicty. *Obesity Research*, *27*(4), 194–199.

Duffus, J. H. (2002). Heavy metals" a meaningless term?(IUPAC Technical Report). *Pure and Applied Chemistry*, *74*(5), 793–807. doi:10.1351/pac200274050793

Dutta, K., Daverey, A., & Lin, J. G. (2014). Evolution retrospective for alternative fuels: First to fourth generation. *Renewable Energy*, *69*, 114–122. doi:10.1016/j.renene.2014.02.044

El Essawy, N. A., Ali, S. M., Farag, H. A., Konsowa, A. H., Elnouby, M., & Hamad, H. A. (2017). Green synthesis of graphene from recycled PET bottle wastes for use in the adsorption of dyes in aqueous solution. *Ecotoxicology and Environmental Safety*, *145*, 57–68. doi:10.1016/j.ecoenv.2017.07.014 PMID:28708982

Elleuch, B., Bouhamed, F., Elloussaief, M., & Jaghbir, M. (2018). Environmental sustainability and pollution prevention. *Environmental Science and Pollution Research International*, *25*(19), 18223–18225. doi:10.100711356-017-0619-5 PMID:29971753

Elsheekh, K. M., Kamel, R. R., Elsherif, D. M., & Shalaby, A. M. (2021). Achieving Sustainable Development Goals from the perspective of solid waste management plans. *Journal of Engineering and Applied Sciences (Asian Research Publishing Network)*, *68*(1), 9. Advance online publication. doi:10.118644147-021-00009-9

Eltayeb, T., & Zailani, S. (2014). Going green through green supply chain initiatives toward environmental sustainability. *Operations and Supply Chain Management: an International Journal*, *2*(2), 93–110. doi:10.31387/oscm040019

Eneizan, B. M., Abd-Wahab, K., & Obaid, T. F. (2016). Effects of green marketing strategy on the financial and non-financial performance of firms: A conceptual paper. *Oman Chapter of Arabian Journal of Business and Management Review*, *5*(12), 14–27. doi:10.12816/0031457

EPA. U. (2020). Advancing sustainable materials management: 2018 tables and figures. Assessing Trends in Materials Generation and Management in the United States. United States Environmental Protection Agency.

Erdogan, O., & Kara, M. (2019). Analytical approach to the waste management of nanomaterials in developing countries. *Front. Drug Chem. Clin. Res*, *2*, 1–5.

Esayas Lelisho, M., Akessa, G. M., Kifle Demissie, D., Fikadu Yermosa, S., Andargie, S. A., Tareke, S. A., & Pandey, D. (2023). Application of Parametric Shared Frailty Models to Analyze Time-to-Death of Gastric Cancer Patients. *Journal of Gastrointestinal Cancer*, *54*(1), 104–116. doi:10.100712029-021-00775-y PMID:35064523

Esmaeili, A., Heydari Shalmani, S., Ebrahimian Pirbazari, A., Esmaeili Khalil Saraei, F., Tabatabai-Yazdi, F. S., Ebrahimian Pirbazari, A., & Derakhshesh, A. (2023, January 24). Pharmaceutical wastewater treatment using TiO2 nanosheets deposited by cobalt co-catalyst as hybrid photocatalysts: Combined experimental study and artificial intelligence modeling. *Chemical Product and Process Modeling*, *0*(0). doi:10.1515/cppm-2022-0070

Esmaeilian, B., Wang, B., Lewis, K., Duarte, F., Ratti, C., & Behdad, S. (2018). The future of waste management in smart and sustainable cities: A review and concept paper. *Waste Management (New York, N.Y.)*, *81*, 177–195. doi:10.1016/j.wasman.2018.09.047 PMID:30527034

European Commission Communication. COM (2015) 614/2: Closing the Loop–an EU Action Plan for the Circular Economy; European Commission Communication: Brussels, Belgium, 2015

Evans, G. (2008). *Liquid Transport Biofuels -Technology Status Report*. National Non-Food Crops Centre.

Falagas, M. E., Pitsouni, E. I., Malietzis, G. A., & Pappas, G. (2008). Comparison of PubMed, Scopus, Web of Science, and Google Scholar : Strengths and weaknesses. *The FASEB Journal*, *22*(338–342), 338–342. Advance online publication. doi:10.1096/fj.07-9492LSF PMID:17884971

Fallavi, K. N., Kumar, V. R., & Chaithra, B. M. (2017). Smart waste management using Internet of Things: A survey. In *2017 International Conference on I-SMAC (IoT in Social, Mobile, Analytics and Cloud)(I-SMAC)*, (pp. 60-64). IEEE. 10.1109/I-SMAC.2017.8058247

Fan, M., Chang, X., Meng, Q., Wan, L. J., & Guo, Y. G. (2021). Progress in the sustainable recycling of spent lithium-ion batteries. *SusMat, 1*(2), 241–254. doi:10.1002us2.16

Farah, M. J. (2019). Challenges of Solid Waste Management and factors influencing its effectiveness: A case study in Burao Municipality.

Fatorachian, H., & Kazemi, H. (2020). Impact of Industry 4.0 on Supply Chain Performance. *Production Planning and Control, 32*(1), 63–81. doi:10.1080/09537287.2020.1712487

Fatouh, M., & El Kafafy, M. (2006) Experimental evaluation of a domestic refrigerator working with LPG. In Applied Thermal Engineering, 26(14–15), 1593–1603. doi:10.1016/j.applthermaleng.2005.11.026

Faunce, T., & Kolodziejczyk, B. (2017). Nanowaste: Need for disposal and recycling standards. *G20 Insights, 148*, 202-213.

Ferronato, N., & Torretta, V. (2019). Waste Mismanagement in Developing Countries: A Review of Global Issues. *International Journal of Environmental Research and Public Health, 16*(6), 1060. doi:10.3390/ijerph16061060 PMID:30909625

Fink, R., & Medved, S. (2013). Health impact assessment of liquid biofuel production. *International Journal of Environmental Health Research, 23*(1), 66–75. doi:10.1080/09603123.2012.699030 PMID:22774773

Flores-Albornoz, J., Asis, E. H., Asnate-Salazar, E. A. S., Espinoza-Rodríguez, H., Ramirez-Asis, N., Pandey, D., & Dadheech, P. (2023). Metaverse: The Development of Human Culture in the Future and Cyberspace. In Applications of Neuromarketing in the Metaverse (pp. 52-61). IGI Global.

Folianto, F., Low, Y. S., & Yeow, W. L. (2015). Smartbin: Smart waste management system. In *2015 IEEE Tenth International Conference on Intelligent Sensors, Sensor Networks and Information Processing (ISSNIP)*, (pp. 1-2). IEEE.

Forti, V., Balde, C., & Kuehr, R. (2018) Guidelines on Classification, Reporting and Indicators. *E-Waste Stat. 2*. https://ewastemonitor.info/

Forti, V., Baldé, C. P., Kuehr, R., & Bel, G. (2020). *The global e-waste monitor 2020. United Nations University (UNU), International Telecommunication Union (ITU) & International Solid Waste Association*. ISWA.

Fuss, M., Barros, R. T. V., & Poganietz, W. R. (2018). Designing a framework for municipal solid waste management towards sustainability in emerging economy countries-An application to a case study in Belo Horizonte (Brazil). *Journal of Cleaner Production, 178*, 655–664. doi:10.1016/j.jclepro.2018.01.051

Gabay, M. (2012). Resource Conservation and Recovery Act: Management of Pharmaceutical Waste. *Hospital Pharmacy, 47*(10), 769–770. doi:10.1310/hpj4710-769

Gabriel, M., & Pessl, E. (2016). Industry 4.0 and Sustainability Impacts: Critical Discussion of Sustainability Aspects With a Special Focus on Future of Work and Ecological Consequences. *Annals of the Faculty of Engineering Hunedoara, 14*(2), 131.

Gabrys, J. (2016). *Program earth: Environmental sensing technology and the making of a computational planet* (Vol. 49). U of Minnesota Press.

Gaffney, J. S., & Marley, N. A. (2009). The impacts of combustion emissions on air quality and climate–From coal to biofuels and beyond. *Atmospheric Environment, 43*(1), 23–36. doi:10.1016/j.atmosenv.2008.09.016

Gaidajis, G., Angelakoglou, K., & Aktsoglou, D. (2010). E-waste: Environmental problems and current management. *Journal of Engineering Science and Technology Review, 3*(1), 193–199. doi:10.25103/jestr.031.32

Ganesh Saratale, R., Cho, S.-K., Dattatraya Saratale, G., Kadam, A. A., Ghodake, G. S., Kumar, M., Naresh Bharagava, R., Kumar, G., Su Kim, D., Mulla, S. I., & Seung Shin, H. (2021). A comprehensive overview and recent advances on polyhydroxyalkanoates (PHA) production using various organic waste streams. *Bioresource Technology*, *325*, 124685–124685. doi:10.1016/j.biortech.2021.124685 PMID:33508681

Ganguly, R. (2016). E-waste management in India-An overview. *International Journal of Earth Sciences and Engineering*, *9*(2), 574–588.

Garay-Rondero, C. L., Martinez-Flores, J. L., Smith, N. R., Santiago, O. C. M., & Aldrette-Malacara, A. (2019). Digital Supply Chain Model in Industry 4.0. *Journal of Manufacturing Technology Management*, *31*(5), 887–933. doi:10.1108/JMTM-08-2018-0280

Garcia, C., & Hora, G. (2017). State-of-the-art of waste wood supply chain in Germany and selected European countries. *Waste Management (New York, N.Y.)*, *70*, 189–197. doi:10.1016/j.wasman.2017.09.025 PMID:28951150

Garcia, J. (2011). *Food waste recycled as biofuel*. Stanford University.

Gardiner, B. (2014). Biogas, a low-tech fuel with a big payoff. *The International New York Times. Nova Iorque, 30.*

Garg, N., & Adhana, D. (2019). E-waste management in India: A study of the current scenario, International Journal of Management. *Technology and Engineering*, *9*(1), 2791–2803.

Garud, K. S., Jayaraj, S., & Lee, M. (2020, June 25). A review on modeling of solar photovoltaic systems using artificial neural networks, fuzzy logic, genetic algorithm and hybrid models. *International Journal of Energy Research*, *45*(1), 6–35. doi:10.1002/er.5608

Gavrilita, P. (2006). *Environmental Systems Analysis of Municipal Solid Waste Management in Chisinau.*

George, A. H., Fernando, M., George, A. S., Baskar, T., & Pandey, D. (2021). Metaverse: The next stage of human culture and the internet. [IJARTET]. *International Journal of Advanced Research Trends in Engineering and Technology*, *8*(12), 1–10.

Ghadge, A., Kara, M. E., Moradlou, H., & Goswami, M. (2020). The Impact of Industry 4.0 Implementation on Supply Chains. *Journal of Manufacturing Technology Management*, *31*(4), 669–686. doi:10.1108/JMTM-10-2019-0368

Ghatge, K. (2014). Waste to energy: A way forward for sustainable solid waste management in India. *iCrier.* https://icrier.org/Urbanisation/events/30-5-14/Waste_Energy_Ketaki%20Ghatge.pdf

Ghosh, P., Shah, G., Sahota, S., Singh, L., & Vijay, V. K. (2020). Biogas production from waste: Technical overview, progress, and challenges. *Bioreactors*, 89-104.

Giammar, D. E., Greene, D. M., Mishrra, A., Rao, N., Sperling, J. B., Talmadge, M., Miara, A., Sitterley, K. A., Wilson, A., Akar, S., Kurup, P., Stokes-Draut, J. R., & Coughlin, K. (2021, December 6). Cost and Energy Metrics for Municipal Water Reuse. *ACS ES&T Engineering*, *2*(3), 489–507. doi:10.1021/acsestengg.1c00351

Giang, N. V., Kochanek, K., Vu, N. T., & Duan, N. B. (2018). Landfill leachate assessment by Hydrological and geophysical data: Case study namson, Hanoi, Vietnam. *Journal of Material Cycles and Waste Management*, *20*(3), 1648–1662. doi:10.100710163-018-0732-7

Gillpatrick, T. (2019). Understanding The Role Of Consumer Behavior In Forecasting The Impact Of Industry 4 . 0 And The Wave Of Digital. *Accenture.*

Giunipero, L. C., Hooker, R. E., & Denslow, D. (2012). Purchasing and supply management sustainability: Drivers and barriers. *Journal of Purchasing and Supply Management*, *18*(4), 258–269. doi:10.1016/j.pursup.2012.06.003

Godbole, N. S., & Lamb, J. (2018) 'Research into Making Healthcare Green with Cloud, Green IT, and Data Science to Reduce Healthcare Costs and Combat Climate Change', *2018 9th IEEE Annual Ubiquitous Computing, Electronics and Mobile Communication Conference, UEMCON 2018*. IEEE, pp. 189–195. 10.1109/UEMCON.2018.8796529

Goh, C. Y., & Marimuthu, M. (2016) 'The Path towards Healthcare Sustainability: The Role of Organisational Commitment', Procedia - Social and Behavioral Sciences. The Author(s), 224(August 2015), pp. 587–592. doi:10.1016/j.sbspro.2016.05.445

Gong, J., Liu, J., Wen, X., Jiang, Z., Chen, X., Mijowska, E., & Tang, T. (2014). Upcycling waste polypropylene into graphene flakes on organically modified montmorillonite. *Industrial & Engineering Chemistry Research*, *53*(11), 4173–4181. doi:10.1021/ie4043246

Goyal, S., Pandey, D., Singh, H., Singh, J., Kakkar, R., & Srinivasu, P. N. (2022). Mathematical modelling for prediction of spread of corona virus and artificial intelligence/machine learning-based technique to detect COVID-19 via smartphone sensors. *International Journal of Modelling Identification and Control*, *41*(1-2), 43–52. doi:10.1504/IJMIC.2022.127096

Graham, A. L., Burke, M. V., Jacobs, M. A., Cha, S., Croghan, I. T., Schroeder, D. R., Moriarty, J. P., Borah, B. J., Rasmussen, D. F., Brookover, M. J., Suesse, D. B., Midthun, D. E., & Hays, J. T. (2017). 'An integrated digital / clinical approach to smoking cessation in lung cancer screening : Study protocol for a randomized controlled trial', *BioMed Central. Trials*, *18*(568), 1–14. doi:10.118613063-017-2312-x PMID:29179734

Grant, K., Goldizen, F. C., Sly, P. D., Brune, M.-N., Neira, M., Berg, M. V. D., & Norman, R. E. (2013). Health consequences of exposure to e-waste: A systematic review. *The Lancet. Global Health*, *1*(6), 350–361. http://www.sciencedirect.com/ science/article/pii/S2214109X13701013. doi:10.1016/S2214-109X(13)70101-3 PMID:25104600

Greenhalgh, T., MacFarlane, F., Barton-Sweeney, C., & Woodard, F. (2012). "If we build it, will it stay?" A case study of the sustainability of whole-system change in London. *The Milbank Quarterly*, *90*(3), 516–547. doi:10.1111/j.1468-0009.2012.00673.x PMID:22985280

Greenhalgh, T., Robert, G., MacFarlane, F., Bate, P., & Kyriakidou, O. (2004). Diffusion of innovations in service organizations: Systematic review and recommendations. *The Milbank Quarterly*, *82*(4), 581–629. doi:10.1111/j.0887-378X.2004.00325.x PMID:15595944

Grosjean, D., Grosjean, E., & Gertler, A. W. (2001). On-road emissions of carbonyls from light-duty and heavy-duty vehicles. *Environmental Science & Technology*, *35*(1), 45–53. doi:10.1021/es001326a PMID:11352025

Gruen, R. L. et al. (2008) 'Sustainability science: an integrated approach for health-programme planning', *The Lancet*. Elsevier Ltd, 372(9649), pp. 1579–1589. . doi:10.1016/S0140-6736(08)61659-1

Guerrero, L. A., Maas, G., & Hogland, W. (2013). Solid waste management challenges for cities in developing countries. *Waste Management (New York, N.Y.)*, *33*(1), 220–232. doi:10.1016/j.wasman.2012.09.008 PMID:23098815

Gu, F., Summers, P. A., & Hall, P. (2019). Recovering materials from waste mobile phones: Recent technological developments. *Journal of Cleaner Production*, *237*, 117657. doi:10.1016/j.jclepro.2019.117657

Guidolin, M., & Alpcan, T. (2019). Transition to Sustainable Energy Generation in Australia: Interplay between Coal, Gas and Renewables. *Renewable Energy*, *139*, 359–367. doi:10.1016/j.renene.2019.02.045

Gunasekaran, K., Boopathi, S., & Sureshkumar, M. (2022). Analysis of a Cryogenically Cooled Near-Dry Wedm Process Using Different Dielectrics. *Materiali in Tehnologije*, *56*(2), 179–186. doi:10.17222/mit.2022.397

Gunatilake, S. (2015). Methods of removing heavy metals from industrial wastewater. *Methods (San Diego, Calif.)*, *1*(1), 14.

Gundupalli, S. P., Hait, S., & Thakur, A. (2017). A review on automated sorting of source-separated municipal solid waste for recycling. *Waste Management (New York, N.Y.)*, *60*, 56–74. doi:10.1016/j.wasman.2016.09.015 PMID:27663707

Guo, Z., Sun, Y., Pan, S. Y., & Chiang, P. C. (2019). Integration of green energy and advanced energy-efficient technologies for municipal wastewater treatment plants. *International Journal of Environmental Research and Public Health*, *16*(7), 1282. doi:10.3390/ijerph16071282 PMID:30974807

Gupta, M., Sharma, S., & Bansal, S. (2022, April). Neuromarketing: An Emerging Domain in the Formal Education System. In *2022 3rd International Conference on Intelligent Engineering and Management (ICIEM)* (pp. 53-58). IEEE.

Gupta, R., & Pierdzioch, C. (2022). Climate risks and forecastability of the realized volatility of gold and other metal prices. *Resour. Pol., 77.*

Gupta, S. (2011). E-waste management: Teaching how to reduce, reuse and recycle for sustainable development-need of some educational strategies. *Journal of Education and Practice*, *2*(3), 2222–1735.

Gupta, S., Nawal, M., Janu, N., & Goyal, D. (2022). IoT, Enabling Technologies, and Sensor Node Deployment Pattern in WSN. *ECS Transactions*, *107*(1), 7441–7455. doi:10.1149/10701.7441ecst

Guthi, V. R., Arepalli, S., & Ganapa, P. (2016). Study of socio demographic factors among persons affected by leprosy in Kurnool division of Kurnool district, Andhra Pradesh, India. [Internet]. *International Journal of Community Medicine and Public Health*, *3*(12), 3548–3555. doi:10.18203/2394-6040.ijcmph20164290

Habib, M. A., Ahmed, M. M., Aziz, M., Beg, M. R. A., & Hoque, M. E. (2021). Municipal solid waste management and waste-to-energy potential from rajshahi city corporation in Bangladesh. *Applied Sciences (Basel, Switzerland)*, *11*(9), 3744. doi:10.3390/app11093744

Hagos, K., Zong, J., Li, D., Liu, C., & Lu, X. (2017). Anaerobic co-digestion process for biogas production: Progress, challenges and perspectives. *Renewable & Sustainable Energy Reviews*, *76*, 1485–1496. doi:10.1016/j.rser.2016.11.184

Haider, M. Z., & Riaz, M. R. (2021). Municipal Solid Waste Management In Bangladesh: A Study Of X Municipality. *Khulna University Studies*, *18*(1), 37–43. doi:10.53808/KUS.2021.18.01.2101-S

Halder, M., Rahman, T., Mahmud, A., Jim, S. A., Akbor, M. A., Siddique, M. A. B., & Joardar, J. C. (2022). Are the vegetables grown in the soil of municipal solid waste dumping sites safe for human health? An assessment from trace elements contamination and associated health risks. *Environmental Nanotechnology, Monitoring & Management*, *18*, 100731. doi:10.1016/j.enmm.2022.100731

Halim, L., & Suharyanti, Y. (2019). E-Waste: Current Research and Future Perspective on Developing Countries. *International Journal of Industrial Engineering and Engineering Management*, *1*(2), 25–42. doi:10.24002/ijieem.v1i2.3214

Hamelinck, C. N., & Faaij, A. P. (2002). Future prospects for the production of methanol and hydrogen from biomass. *Journal of Power Sources*, *111*(1), 1–22. doi:10.1016/S0378-7753(02)00220-3

Handfield, R., Jeong, S., & Choi, T. (2019). Emerging Procurement Technology: Data Analytics and Cognitive Analytics. *International Journal of Physical Distribution & Logistics Management*, *49*(10), 972–1002. doi:10.1108/IJPDLM-11-2017-0348

Hannan, M.A., Abdulla Al Mamun, M., Hussain, A., Basri, H., Begum, R.A. (2015). A review on technologies and their usage in solid waste monitoring and management systems: issues and challenges. *Waste Management. 43*, 509e523.

Haribalaji, V., Venkatesan, G., Asif, M. M., Pandian, M., Subbiah, R., & Boopathi, S. (2022). Investigation on corrosion and tensile Characteristics: Friction stir welding of AA7075 and AA2014. *Materials Today: Proceedings*, *66*(3), 743–748. doi:10.1016/j.matpr.2022.04.037

Harikaran, M., Boopathi, S., Gokulakannan, S., & Poonguzhali, M. (2023). Study on the Source of E-Waste Management and Disposal Methods. In *Sustainable Approaches and Strategies for E-Waste Management and Utilization* (pp. 39–60). IGI Global. doi:10.4018/978-1-6684-7573-7.ch003

Harper, G., Sommerville, R., Kendrick, E., Driscoll, L., Slater, P., Stolkin, R., . . . Lambert, S. (2019). Recycling lithium-ion batteries from electric vehicles. *Nature, 575*(7781), 75-86.

Harper, S. W., Etchells, J. C., Summerfield, A. J., & Cockton, A. (2008). Health and safety in biodiesel manufacture. In *Institution of Chemical Engineers Symposium Series, 154*(943), 1-9. Institution of Chemical Engineers.

Harris, C., Allen, K., Waller, C., & Brooke, V. (2017). 'Sustainability in health care by allocating resources effectively (SHARE) 3 : Examining how resource allocation decisions are made, implemented and evaluated in a local healthcare setting', *BMC Health Services Research. BMC Health Services Research*, *17*(340), 1–21. doi:10.118612913-017-2207-2 PMID:28486953

Harris, C., Green, S., Ramsey, W., Allen, K., & King, R. (2017). 'Sustainability in Health care by allocating resources effectively (SHARE) 1 : Introducing a series of papers reporting an investigation of disinvestment in a local healthcare setting', *BMC Health Services Research. BMC Health Services Research*, *17*(323), 1–13. doi:10.118612913-017-2210-7 PMID:28472962

Hasan, S. (1998). *Problems of municipal waste management in Bangladesh: An inquiry into its nature.* Habitat International. . doi:10.1016/S0197-3975(97)00039-8

Heller, J. (1996). *Physic nut, Jatropha curcas L (Vol. 1).* Bioversity international.

Herat, S. (2021). E-waste management in Asia Pacific region: Review of issues, challenges and solutions. *Nature Environment and Pollution Technology*, *20*(1), 45–53. doi:10.46488/NEPT.2021.v20i01.005

Hettiarachchi, H., Meegoda, J. N., & Ryu, S. (2018). Organic waste buyback as a viable method to enhance sustainable municipal solid waste management in developing countries. *International Journal of Environmental Research and Public Health*, *15*(11), 2483. doi:10.3390/ijerph15112483 PMID:30405058

He, Z. L., Yang, X. E., & Stoffella, P. J. (2005). Trace elements in agroecosystems and impacts on the environment. *Journal of Trace Elements in Medicine and Biology*, *19*(2-3), 125–140. doi:10.1016/j.jtemb.2005.02.010 PMID:16325528

Hillie, T., & Hlophe, M. (2007). Nanotechnology and the challenge of clean water. *Nature Nanotechnology*, *2*(11), 663–664. doi:10.1038/nnano.2007.350 PMID:18654395

Hill, J., Nelson, E., Tilman, D., Polasky, S., & Tiffany, D. (2006). Environmental, economic, and energetic costs and benefits of biodiesel and ethanol biofuels. *Proceedings of the National Academy of Sciences of the United States of America*, *103*(30), 11206–11210. doi:10.1073/pnas.0604600103 PMID:16837571

Hira, M., Yadav, S., Morthekai, P., Linda, A., Kumar, S., & Sharma, A. (2018). Mobile Phones—An asset or a liability: A study based on characterization and assessment of metals in waste mobile phone components using leaching tests. *Journal of Hazardous Materials*, *342*, 29–40. doi:10.1016/j.jhazmat.2017.08.008 PMID:28822247

Hirpe, L., & Yeom, C. (2021). Municipal solid waste management policies, practices, and challenges in Ethiopia: A systematic review. *Sustainability (Basel)*, *13*(20), 11241. doi:10.3390u132011241

Hiscock, K. M., & Grischek, T. (2002). Attenuation of groundwater pollution by bank? ltration. *Journal of Hydrology (Amsterdam)*, *266*(139–144).

Hoornweg, D., & Thomas, L. (1999). *What a waste: solid waste management in Asia.* The World Bank.

Hornak, J. (2021). Synthesis, Properties, and Selected Technical Applications of Magnesium Oxide Nanoparticles: A Review. *International Journal of Molecular Sciences, 22*(23), 12752–12752. doi:10.3390/ijms222312752 PMID:34884556

Hovlid, E., Bukve, O., Haug, K., Aslaksen, A. B., & von Plessen, C. (2012). Sustainability of healthcare improvement : What can we learn from learning theory? *BMC Health Services Research, 12*(235), 2–13. doi:10.1186/1472-6963-12-235 PMID:22863199

Hubau, A., Minier, M., Chagnes, A., Joulian, C., Silvente, C., & Guezennec, A. G. (2020). Recovery of metals in a double-stage continuous bioreactor for acidic bioleaching of printed circuit boards (PCBs). *Separation and Purification Technology, 238*, 116481. doi:10.1016/j.seppur.2019.116481

Hu, J. S., Zhong, L. S., Song, W. G., & Wan, L. J. (2008). Synthesis of hierarchically structured metal oxides and their application in heavy metal ion removal. *Advanced Materials, 20*(15), 2977–2982. doi:10.1002/adma.200800623

Hull, M. S., Vikesland, P. J., & Schultz, I. R. (2013). Uptake and retention of metallic nanoparticles in the Mediterranean mussel (Mytilus galloprovincialis). *Aquatic Toxicology (Amsterdam, Netherlands), 140-141*, 89–97. doi:10.1016/j.aquatox.2013.05.005 PMID:23765031

Hussain, M. et al. (2018) 'Exploration of Social Sustainability in Healthcare Supply Chain', *Journal of Cleaner Production*. Elsevier B.V. . doi:10.1016/j.jclepro.2018.08.157

Hu, W., Zhou, Y., Min, X., Liu, J., Li, X., Luo, L., Zhang, J., Mao, Q., Chai, L., & Zhou, Y. (2018). The study of a pilot-scale aerobic/Fenton/anoxic/aerobic process system for the treatment of landfill leachate. *Environmental Technology, 39*(15), 1926–1936. doi:10.1080/09593330.2017.1344325 PMID:28625109

IBEF. (2019). *Indian economy growth rate & statistics*. IBEF. https://www.ibef. org/economy/indian-economy-overview

Igliński, B., Buczkowski, R., & Cichosz, M. (2015). Biogas production in Poland—Current state, potential and perspectives. *Renewable & Sustainable Energy Reviews, 50*, 686–695. doi:10.1016/j.rser.2015.05.013

Ihsanullah, I., Alam, G., Jamal, A., & Shaik, F. (2022). Recent advances in applications of artificial intelligence in solid waste management: A review. *Chemosphere, 309*, 136631. doi:10.1016/j.chemosphere.2022.136631 PMID:36183887

Ikumapayi, O. M., Okokpujie, I. P., Afolalu, S. A., Ajayi, O. O., Akilabi, E. T., & Bodunde, O. P. (2018). Effects of quenchants on impact strength of single-vee butt welded joint of mild steel. *IOP Conference Series. Materials Science and Engineering, 391*(1), 12007. doi:10.1088/1757-899X/391/1/012007

Ilankoon, I. M. S. K., Ghorbani, Y., Chong, M. N., Herath, G., Moyo, T., & Petersen, J. (2018). E-waste in the international context–A review of trade flows, regulations, hazards, waste management strategies and technologies for value recovery. *Waste Management (New York, N.Y.), 82*, 258–275. doi:10.1016/j.wasman.2018.10.018 PMID:30509588

Imani, M. M., & Safaei, M. (2019). Optimized Synthesis of Magnesium Oxide Nanoparticles as Bactericidal Agents. *Journal of Nanotechnology, 2019*, 1–6. doi:10.1155/2019/6063832

IRENA. (2018). *Renewable Energy Market Analysis: Southeast Asia*. International Renewable Energy Agency.

Isernia, R., Passaro, R., Quinto, I., & Thomas, A. (2019). The reverse supply chain of the e-waste management processes in a circular economy framework: Evidence from Italy. Sustainability (Switzerland) https://dx.doi.org/ doi:10.3390/su11082430

Işıldar, A., van de Vossenberg, J., Rene, E. R., van Hullebusch, E. D., & Lens, P. N. L. (2016). Two-step bioleaching of copper and gold from discarded printed circuit boards (PCB). *Waste Management (New York, N.Y.), 57*, 149–157. doi:10.1016/j.wasman.2015.11.033 PMID:26704063

Islam, A., Ahmed, T., Awual, M. R., Rahman, A., Sultana, M., Abd Aziz, A., & Hasan, M. (2020). Advances in sustainable approaches to recover metals from e-waste-A review. *Journal of Cleaner Production, 244*, 118815. doi:10.1016/j.jclepro.2019.118815

Islam, F. S. (2015). The engineers role in climate change mitigation. *Journal of Modern Science and Technology, 3*(1), 117–124.

Islam, M. T., Huda, N., Baumber, A., Shumon, R., Zaman, A., Ali, F., Hossain, R., & Sahajwalla, V. (2021). A global review of consumer behavior towards e-waste and implications for the circular economy. *Journal of Cleaner Production, 316*, 128297. doi:10.1016/j.jclepro.2021.128297

Ivanov, D., Dolgui, A., & Sokolov, B. (2019). The Impact of Digital Technology and Industry 4.0 on the Ripple Effect and Supply Chain Risk Analytics. *International Journal of Production Research, 57*(3), 829–846. doi:10.1080/00207543.2018.1488086

Iwuozor, K. O., Chizitere Emenike, E., Ighalo, J. O., Omoarukhe, F. O., Omuku, P. E., & George Adeniyi, A. (2022). A review on the thermochemical conversion of sugarcane bagasse into biochar. *Cleaner Materials, 6*, 100162. doi:10.1016/j.clema.2022.100162

Iyyanar, P., Anand, R., Shanthi, T., Nassa, V. K., Pandey, B. K., George, A. S., & Pandey, D. (2023). A Real-Time Smart Sewage Cleaning UAV Assistance System Using IoT. In *Handbook of Research on Data-Driven Mathematical Modeling in Smart Cities* (pp. 24–39). IGI Global.

Jacobsen, B. H., Laugesen, F. M., & Dubgaard, A. (2014). The economics of biogas in Denmark: a farm and socioeconomic perspective. *International Journal of Agricultural Management, 3.*

Janardhana, K., Singh, V., Singh, S. N., Babu, T. S. R., Bano, S., & Boopathi, S. (2023). Utilization Process for Electronic Waste in Eco-Friendly Concrete: Experimental Study. In Sustainable Approaches and Strategies for E-Waste Management and Utilization (pp. 204–223). IGI Global.

Jangre, J., Prasad, K., & Patel, D. (2022). Analysis of barriers in e-waste management in developing economies: An integrated multiple-criteria decision-making approach. *Environmental Science and Pollution Research International, 29*(48), 72294–72308. doi:10.100711356-022-21363-y PMID:35696062

Janu, N. (2022). Query-Based Image Retrieval Using SVM. *Emerging Technologies in Computer Engineering: Cognitive Computing and Intelligent IoT: 5th International Conference, ICETCE 2022.* Cham: Springer International Publishing, 2022.

Janu, N. (2022). Query-Based Image Retrieval Using SVM. *Emerging Technologies in Computer Engineering: Cognitive Computing and Intelligent IoT: 5th International Conference, ICETCE 2022.* Cham: Springer International Publishing.

Janu, N., & Kumar, A. (2021). Automated fruit grading system using image fusion. In Smart Agricultural Services Using Deep Learning, Big Data, and IoT (pp. 32-45). IGI Global. doi:10.4018/978-1-7998-5003-8.ch002

Janu, N., ... “Query-Based Image Retrieval Using SVM.” Emerging Technologies in Computer Engineering: Cognitive Computing and Intelligent IoT: 5th International Conference, ICETCE 2022, Jaipur, India, February 4–5, 2022, Revised Selected Papers. Cham: Springer International Publishing, 2022.

Janu, N. (2022). *Cross-lingual Word Sense Disambiguation Using Multilingual Co-occurrence Graphs.* Graph Learning and Network Science for Natural Language Processing. doi:10.1201/9781003272649-9

Janu, N., Kumar, A., Dadheech, P., Sharma, G., Kumar, A., & Raja, L. (2021, April). Multiple Watermarking Scheme for Video & Image for Authentication & Copyright Protection. []. IOP Publishing.]. *IOP Conference Series. Materials Science and Engineering, 1131*(1), 012020. doi:10.1088/1757-899X/1131/1/012020

Jawad, J., Hawari, A. H., & Javaid Zaidi, S. (2021, September). Artificial neural network modeling of wastewater treatment and desalination using membrane processes: A review. *Chemical Engineering Journal*, *419*, 129540. doi:10.1016/j.cej.2021.129540

Jayakrishnan, T., Jeeja, M. C., & Bhaskar, R. (2013). Occupational Health Problems of Municipal Solid Waste Management Workers in India. *International Journal of Environmental Health Engineering*, *2*(1), 42. doi:10.4103/2277-9183.122430

Jeevanantham, Y. A., A, S., V, V., J, S. I., Boopathi, S., & Kumar, D. P. (2023). Implementation of Internet-of Things (IoT) in Soil Irrigation System. *IEEE Explore*, 1–5. doi:10.1109/ICPECTS56089.2022.10047185

Jeevanantham, Y. A., Saravanan, A., Vanitha, V., Boopathi, S., & Kumar, D. P. (2022). Implementation of Internet-of Things (IoT) in Soil Irrigation System. *IEEE Explore*, 1–5.

Jeevanantham, Y.A., Saravanan, A., Vanitha, V., Boopathi, S., & Kumar, D.P. (2022). Implementation of Internet-of Things (IoT) in Soil Irrigation System. *IEEE Explor.* 1–5.

Jenkins, A., Gornall, L., & Cripps, H. (2013). Lessons for safe design and operation of anaerobic digesters. *Loss Prevention Bulletin*, 229.

Jerath, N. (2021). Awareness and attitude of students towards municipal solid waste management to achieve Sustainable Development Goals - A Case Study. *International Journal of Plant And Environment*, *7*(01), 78–85. doi:10.18811/ijpen.v7i01.9

Jerin, D. T., Sara, H. H., Radia, M. A., Hema, P. S., Hasan, S., Urme, S. A., Audia, C., Hasan, M. T., & Quayyum, Z. (2022). An overview of progress towards implementation of solid waste management policies in Dhaka, Bangladesh. *Heliyon*, *8*(2), e08918. doi:10.1016/j.heliyon.2022.e08918 PMID:35243053

Jeyaraj, P. (2021). Management of E-waste in India – Challenges and recommendations. *World J. Adv. Res. Rev.*, *11*(2), 193–218. doi:10.30574/wjarr.2021.11.2.0329

Jha, R., Rao, M. D., Meshram, A., Verma, H. R., & Singh, K. K. (2020). Potential of polymer inclusion membrane process for selective recovery of metal values from waste printed circuit boards: A review. *Journal of Cleaner Production*, *265*, 121621. doi:10.1016/j.jclepro.2020.121621

Jin & Shin. (2020). ScienceDirect Changing the Game to Compete : Innovations in the Fashion Retail Industry from the Disruptive Business Model. *Business Horizons*, *7*(2), 23–36. . doi:10.1016/j.bushor.2020.01.004

Ji, X., Yang, M., Wan, A., Yu, S., & Yao, Z. (2022). Bioleaching of Typical Electronic Waste—Printed Circuit Boards (WPCBs): A Short Review. *International Journal of Environmental Research and Public Health*, *19*(12), 7508. doi:10.3390/ijerph19127508 PMID:35742757

Johnston, R. B. (2016) *Transforming Our World: The 2030 Agenda For Sustainable Development, Global Sustainability - Proceedings of the 6th International Congress on Arsenic in the Environment, AS 2016*. 10.1201/b20466-7

Jo, J. Y., Kwon, Y. S., Lee, J. W., Park, J. S., Rho, B. H., & Choi, W. I. (2013). Acute respiratory distress due to methane inhalation. *Tuberculosis and Respiratory Diseases*, *74*(3), 120–123. doi:10.4046/trd.2013.74.3.120 PMID:23579434

Jones, N., & Miller, J. H. (1992). Jatropha curcas: A multipurpose species for problematic sites. *Scientific Research Publishing*.

Joon, V., Shahrawat, R., & Kapahi, M. (2017). The emerging environmental and public health problem of electronic waste in India. *Journal of Health & Pollution*, *7*(15), 1–7. Advance online publication. doi:10.5696/2156-9614-7.15.1 PMID:30524825

Joshi, R., & Ahmed, S. (2016). Status and challenges of municipal solid waste management in India: A review. *Cogent Environmental Science*, *2*(1), 1139434. doi:10.1080/23311843.2016.1139434

Judd, S. (2010). *The MBR book: principles and applications of membrane bioreactors for water and wastewater treatment*. Elsevier.

Julander, A., Lundgren, L., Skare, L., Grandér, M., Palm, B., Vahter, M., & Lidén, C. (2014). Formal recycling of e-waste leads to increased exposure to toxic metals: AN occupational exposure study from Sweden. *Environment International*, *73*, 243–251. doi:10.1016/j.envint.2014.07.006 PMID:25300751

Juneja, V., Singh, S., Jain, V., Pandey, K. K., Dhabliya, D., Gupta, A., & Pandey, D. (2023). Optimization-Based Data Science for an IoT Service Applicable in Smart Cities. In *Handbook of Research on Data-Driven Mathematical Modeling in Smart Cities* (pp. 300–321). IGI Global. doi:10.4018/978-1-6684-6408-3.ch016

Kabeyi, M. J. B., & Olanrewaju, O. A. (2022). Biogas production and applications in the sustainable energy transition. *Journal of Energy*. doi:10.1155/2022/8750221

Kahhat, R., Kim, J., Xu, M., Allenby, B., Williams, E., & Zhang, P. (2008). Exploring e-waste management systems in the United States. *Resources, Conservation and Recycling*, *52*(7), 955–964. doi:10.1016/j.resconrec.2008.03.002

Kalaiselvan, N., Glivin, G., Bakthavatsalam, A. K., Mariappan, V., Premalatha, M., Raveendran, P. S., Jayaraj, S., & Sekhar, S. J. (2022). A waste to energy technology for Enrichment of biomethane generation: A review on operating parameters, types of biodigesters, solar assisted heating systems, socio economic benefits and challenges. *Chemosphere*, *293*, 133486. doi:10.1016/j.chemosphere.2021.133486 PMID:35016951

Kalidas, R., Boopathi, S., Sivakumar, K., & Mohankumar, P. (2012). Optimization of Machining Parameters of WEDM Process Based On the Taguchi Method. *Ijest, 6*(1\&2), 91.

Kalidas, R., Boopathi, S., Sivakumar, K., & Mohankumar, P. (2012). Optimization of Machining Parameters of WEDM Process Based On the Taguchi Method. *IJEST, 6*(1\&2), 91.

Kamali, M., Appels, L., Yu, X., Aminabhavi, T. M., & Dewil, R. (2021, August). Artificial intelligence as a sustainable tool in wastewater treatment using membrane bioreactors. *Chemical Engineering Journal*, *417*, 128070. doi:10.1016/j.cej.2020.128070

Kanda, H., Li, P., Yoshimura, T., & Okada, S. (2013). Wet extraction of hydrocarbons from Botryococcusbraunii by dimethyl ether as compared with dry extraction by hexane. *Fuel*, *105*, 535–539. doi:10.1016/j.fuel.2012.08.032

Kannan, E., Trabelsi, Y., Boopathi, S., & Alagesan, S. (2022). Influences of cryogenically treated work material on near-dry wire-cut electrical discharge machining process. *Surface Topography : Metrology and Properties*, *10*(1), 015027. doi:10.1088/2051-672X/ac53e1

Kantabutra, S., & Punnakitikashem, P. (2020). Exploring the Process Toward Corporate Sustainability at a Thai SME. *Sustainability (Basel)*, *12*(1), 2–19. doi:10.3390u12219204

Kanwar, V. S., Shetty, A., & Mehta, T. (2022). Water audit of Chitkara University, Himachal Pradesh–A zero discharge campus. *Materials Today: Proceedings*, *69*, 383–386. doi:10.1016/j.matpr.2022.09.057

Karak, T., Bhagat, R. M., & Bhattacharyya, P. (2012). Municipal solid waste generation, composition, and management: The world scenario. *Critical Reviews in Environmental Science and Technology*, *42*(15), 1509–1630. doi:10.1080/10643389.2011.569871

Karmee, S. K. (2016). Liquid biofuels from food waste: Current trends, prospects, and limitations. *Renewable & Sustainable Energy Reviews*, *53*, 945–953. doi:10.1016/j.rser.2015.09.041

Karmee, S. K., & Lin, C. S. K. (2014). Valorization of food waste to biofuel: Current trends and technological challenges. *Sustainable Chemical Processes*, *2*(1), 22. doi:10.118640508-014-0022-1

Karmee, S. K., Niemeijer, B., Casiraghi, L., Mlambo, B., Lapkin, A., & Greiner, L. (2014). Facile biocatalytic synthesis of a macrocyclic lactone in sub-and supercritical solvents. *Biocatalysis and Biotransformation*, *32*(2), 125–131. doi:10.3109/10242422.2014.893579

Karthik, R., Chen, S.-M., Elangovan, A., Muthukrishnan, P., Shanmugam, R., & Lou, B.-S. (2016). Phyto mediated biogenic synthesis of gold nanoparticles using Cerasus serrulata and its utility in detecting hydrazine, microbial activity and DFT studies. *Journal of Colloid and Interface Science*, *468*, 163–175. doi:10.1016/j.jcis.2016.01.046 PMID:26835586

Kasar, T., Barlas, P., Adam, S., Chatelain, C., & Paquet, T. (2013). Learning to detect tables in scanned document images using line information. *2013 12th International Conference on Document Analysis and Recognition*, (pp. 1185–1189). ACM.

Kashyap, A. (2012) 'Building Sustainable Healthcare Knowledge Systems by Harnessing Efficiencies from Biomedical Linked Open Data', in IEEE, pp. 16–18. doi:10.1109/INDCON.2011.6139343

Kasper, A. C., Bernardes, A. M., & Veit, H. M. (2011). Characterization and recovery of polymers from mobile phone scrap. *Waste Management & Research*, *29*(7), 714–726. doi:10.1177/0734242X10391528 PMID:21382879

Kassaw, C., & Pandey, D. (2021). COVID-19 pandemic related to anxiety disorder among communities using public transport at Addis Ababa, Ethiopia, March 2020: cross-sectional study design. *Human Arenas*, 1-10.

Kassinis, G., & Vafeas, N. (2006). Stakeholder pressures and environmental performance. *Academy of Management Journal*, *49*(1), 145–159. doi:10.5465/amj.2006.20785799

Kaufman, S. M., & Themelis, N. J. (2009). Using a Direct Method to Characterize and Measure Flows of Municipal Solid Waste in the United States. *Journal of the Air & Waste Management Association*, *59*(12), 1386–1390. doi:10.3155/1047-3289.59.12.1386 PMID:20066903

Kaur, A., & Deswal, S. (2019). Sustainable solid waste management in Indian cities. In *Recycled Waste Materials: Proceedings of EGRWSE 2018* (pp. 239-251). Research Gate.

Kaushik, M. V. (2018). E- Waste and Its Management. *International Journal for Research in Applied Science and Engineering Technology*, *6*(6), 170–178. doi:10.22214/ijraset.2018.6031

Kaushik, P. R., & Herat, S. (2020). Current state of e-waste management in India. *International Journal of Environment and Waste Management*, *25*(3), 322. doi:10.1504/IJEWM.2020.106293

Kazancoglu, Y., Mangla, S. K., & Ram, M. (2020). Risk assessment for sustainability in e-waste recycling in circular economy. *Clean Technologies and Environmental Policy*, *24*(4), 1145–1157. doi:10.100710098-020-01901-3

Kaza, S., Yao, L. C., Bhada-Tata, P., & Van Woerden, F. (2018). *What a Waste 2.0: A Global Snapshot of Solid Waste Management to 2050*. World Bank. doi:10.1596/978-1-4648-1329-0

Keegan, D., Kretschmer, B., Elbersen, B., & Panoutsou, C. (2013). Cascading use: A systematic approach to biomass beyond the energy sector. *Biofuels, Bioproducts & Biorefining*, *7*(2), 193–206. doi:10.1002/bbb.1351

Kenny, C., & Priyadarshini, A. (2021). Review of Current Healthcare Waste Management Methods and Their Effect on Global Health. *Health Care*, *9*(284), 2–17. PMID:33807606

Kesikoğlu, F., & Yıldırım, E. (2014). The Causal Effect of Shifting Oil to Natural Gas Consumption on Current Account Balance and Economic Growth in 11 OECD Countries: Evidence from Bootstrap-corrected Panel Causality Test. *Procedia: Social and Behavioral Sciences*, *143*, 1064–1069. doi:10.1016/j.sbspro.2014.07.556

Kevin, M. (2020). Camp, David Mead, Stephen B. Reed, Christopher Sitter, and Derek Wasilewski, "From the barrel to the pump: The impact of the COVID-19 pandemic on prices for petroleum products," Monthly Labor Review. *U.S. Bureau of Labor Statistics*, (October). doi:10.21916/mlr.2020

Khalil, M., Berawi, M. A., Heryanto, R., & Rizalie, A. (2019). Waste to energy technology: The potential of sustainable biogas production from animal waste in Indonesia. *Renewable & Sustainable Energy Reviews*, *105*, 323–331. doi:10.1016/j.rser.2019.02.011

Khan, S., & Ghosh, S. (2017). Dense phase carbon dioxide: an emerging non-thermal technology in food processing. *Physical Science International Journal*, 1-7.

Khan, A. H., Sharholy, M., Alam, P., Al-Mansour, A. I., Ahmad, K., Kamal, M. A., Alam, S., Pervez, M. N., & Naddeo, V. (2022). Evaluation of cost benefit analysis of municipal solid waste management systems. *Journal of King Saud University. Science*, *34*(4), 101997. doi:10.1016/j.jksus.2022.101997

Khan, B., Hasan, A., Pandey, D., Ventayen, R. J. M., Pandey, B. K., & Gowwrii, G. (2021). Fusion of datamining and artificial intelligence in prediction of hazardous road accidents. In *Machine learning and iot for intelligent systems and smart applications* (pp. 201–223). CRC Press. doi:10.1201/9781003194415-12

Khan, H., Veeraiah, V., Jain, V., Rajkumar, A., Gupta, A., & Pandey, D. (2023). Integrating Deep Learning in an IoT Model to Build Smart Applications for Sustainable Cities. In *Handbook of Research on Data-Driven Mathematical Modeling in Smart Cities* (pp. 238–261). IGI Global. doi:10.4018/978-1-6684-6408-3.ch013

Khan, H., Veeraiah, V., Jain, V., Rajkumar, A., Gupta, A., & Pandey, D. (2023). Integrating Deep Learning in an IoT Model to Build Smart Applications for Sustainable Cities. In Handbook of Research on Data-Driven Mathematical Modeling in Smart Cities (pp. 238-261). IGI Global.22.Bansod, S. J., &Thakre, S. (2014). Near Infrared Spectroscopy based Soil Nitrogen measurement-A Review. *International Journal of Current Engineering and Technology*, *4*(1), 268–272.

Khateeb, M.A. (2013). *Landfill leachate impact on groundwater quality report.* Institute of water and environmental science solid and hazardous waste management.

Kiddee, P., Naidu, R., & Wong, M. H. (2013). Electronic waste management approaches: An overview. *Waste management, 33*(5), 1237-1250.

Kiddee, P., Naidu, R., & Wong, M. H. (2013). Electronic waste management approaches: An overview. *Waste Management (New York, N.Y.)*, *33*(5), 1237–1250. doi:10.1016/j.wasman.2013.01.006 PMID:23402807

Kiel, D. (2017). What Do We Know about Industry 4.0" so Far? *International Association for Management of Technology 3*, 9–22. https://www.researchgate.net/publication/316684847

Kim, J. H., Lee, J. C., & Pak, D. (2011). Feasibility of producing ethanol from food waste. *Waste Management (New York, N.Y.)*, *31*(9-10), 2121–2125. doi:10.1016/j.wasman.2011.04.011 PMID:21596551

Kiran, E. U., Trzcinski, A. P., Ng, W. J., & Liu, Y. (2014). Bioconversion of food waste to energy: A review. *Fuel*, *134*, 389–399. doi:10.1016/j.fuel.2014.05.074

Klassen, V., Blifernez-Klassen, O., Wobbe, L., Schlüter, A., Kruse, O., & Mussgnug, J. H. (2016). Efficiency and biotechnological aspects of biogas production from microalgal substrates. *Journal of Biotechnology*, *234*, 7–26. doi:10.1016/j.jbiotec.2016.07.015 PMID:27449486

Knothe, G. (2010). Biodiesel and renewable diesel: A comparison. *Progress in Energy and Combustion Science*, *36*(3), 364–373. doi:10.1016/j.pecs.2009.11.004

Kocak, E., & Baglitas, H. H. (2022). The path to sustainable municipal solid waste management: Do human development, energy efficiency, and income inequality matter? *Sustainable Development (Bradford)*, *30*(6), 1947–1962. doi:10.1002d.2361

Kolhe, N. P., & Dhote, K. K. (2019). Spatial assessment of rurban slums: A new perspective of urban villages. *International Journal of Recent Technology and Engineering*, *8*(3), 8688–8693. doi:10.35940/ijrte.A9215.098319

Kollikkathara, N., Feng, H., & Stern, E. (2009). A purview of waste management evolution: Special emphasis on USA. *Waste Management (New York, N.Y.)*, *29*(2), 974–985. doi:10.1016/j.wasman.2008.06.032 PMID:18796347

Kosuge, T., Imato, K., Goseki, R., & Otsuka, H. (2016). Polymer–inorganic composites with dynamic covalent mechanochromophore. *Macromolecules*, *49*(16), 5903–5911. doi:10.1021/acs.macromol.6b01333

Kress. (2019). *Desalination technologies. Marine Impacts of Seawater Desalination*. Elsevier.

Krishna, R., & Saha, S. (2022). *Study Paper On e-waste management.*

Kshirsagar, P. R., Chippalkatti, P. P., & Karve, S. M. (2018). Performance optimization of neural net- work using GA incorporated PSO. *Journal of Advanced Research in Dynamical and Control Systems*, *10*(4), 156–169.

Kulikowska, D., & Klimiuk, E. (2008). The effect of dumping site age on municipal leachate composition. *Bioresource Technology*, *99*(13), 5981–5985. doi:10.1016/j.biortech.2007.10.015 PMID:18060769

Kulkarni, B. N. (2020). Environmental sustainability assessment of land disposal of municipal solid waste generated in Indian cities–A review. *Environmental Development*, *33*, 100490. doi:10.1016/j.envdev.2019.100490

Kulshrestha, D., Tiwari, M.K., Shalender, K., & Sharma, S. (2022). Consumer Acatalepsy Towards Buying Behaviour for Need-Based Goods for Sustainability During the COVID-19 Pandemic. *Indian Journal of Marketing*, *52*(10), 50–63, doi: 10.17010/ijom/2022/v52/i10/172347

Kumar, R., & Karishma. (2016). Current Scenario of e-waste management in India:issues and strategies. *International Journal of Scientific and Research Publications*, *6*(1), 424–430. www.ijsrp.org

Kumar, R., Neerudu, U. K., Gothalwal, R., Mohapatra, S., Deshpande, P. K., Vani, M. M., & Merugu, R. (2021). Bioprocess parameters for thermophilic and mesophilic biogas production: recent trends and challenges. *Bioenergy Research: Basic and Advanced Concepts*, 225-256.

Kumara, V., Mohanaprakash, T. A., Fairooz, S., Jamal, K., Babu, T., & B., S. (2023). Experimental Study on a Reliable Smart Hydroponics System. In *Human Agro-Energy Optimization for Business and Industry* (pp. 27–45). IGI Global. doi:10.4018/978-1-6684-4118-3.ch002

Kumar, A. (2018). Recent Advances in Solid Waste Minimization using Nanotechnology. *International Journal for Research in Applied Science and Engineering Technology*, *6*(6), 759–764. doi:10.22214/ijraset.2018.6118

Kumar, A., & Agrawal, A. (2020). Recent trends in solid waste management status, challenges, and potential for the future Indian cities—A review. *Current Research in Environmental Sustainability*, *2020*(2), 100011. doi:10.1016/j.crsust.2020.100011

Kumar, A., Dadheech, P., Goyal, D., Patidar, P. K., Dogiwal, S. R., & Janu, N. (2021). A novel scheme for prevention and detection of black hole & gray hole attack in VANET network. *Recent Patents on Engineering*, *15*(2), 263–274. doi:10.2174/1872212114999200512120211

Kumar, A., Ergas, S., Yuan, X., Sahu, A., Zhang, Q., Dewulf, J., & Van Langenhove, H. (2010). Enhanced CO2 fixation and biofuel production via microalgae: Recent developments and future directions. *Trends in Biotechnology*, *28*(7), 371–380. doi:10.1016/j.tibtech.2010.04.004 PMID:20541270

Kumar, M. S., Sankar, S., Nassa, V. K., Pandey, D., Pandey, B. K., & Enbeyle, W. (2021). Innovation and creativity for data mining using computational statistics. In *Methodologies and Applications of Computational Statistics for Machine Intelligence* (pp. 223–240). IGI Global. doi:10.4018/978-1-7998-7701-1.ch012

Kumar, M., Kichambare, P. D., Sharon, M., Ando, Y., & Zhao, X. (1999). Synthesis of conducting fibers, nanotubes, and thin films of carbon from commercial kerosene. *Materials Research Bulletin*, *34*(5), 791–801. doi:10.1016/S0025-5408(99)00066-5

Kumar, P., & Akhai, S. (2022). Effective energy management in smart buildings using VRV/VRF systems. *Additive Manufacturing in Industry*, *4*(0), 27–35. doi:10.1201/9781003360001-2

Kumar, R., Kumar, R., & Kaur, G. (2021). *New frontiers of nanomaterials in environmental science*. Springer. doi:10.1007/978-981-15-9239-3

Kumar, S., Bhattacharyya, J. K., Vaidya, A. N., Chakrabarti, T., Devotta, S., & Akolkar, A. B. (2009). Assessment of the status of Municipal Solid Waste Management in Metro cities, state capitals, class I cities, and class II towns in India: An insight. *Waste Management (New York, N.Y.)*, *29*(2), 883–895. doi:10.1016/j.wasman.2008.04.011 PMID:18595684

Kumar, S., Kwon, H. T., Choi, K. H., Lim, W., Cho, J. H., Tak, K., & Moon, I. (2011). I. Moon LNG: An eco-friendly cryogenic fuel for sustainable development. *Applied Energy*, *88*(12), 4264–4273. doi:10.1016/j.apenergy.2011.06.035

Kumar, S., Mukherjee, A., & Dutta, J. (2020). Chitosan based nanocomposite films and coatings: Emerging antimicrobial food packaging alternatives. *Trends in Food Science & Technology*, *97*, 196–209. doi:10.1016/j.tifs.2020.01.002

Kumar, S., Pradhan, A. K., Tiwari, A. K., & Kang, S. H. (2019). Correlations and volatility spillovers between oil, natural gas, and stock prices in India. *Resources Policy*, *62*, 282–291. doi:10.1016/j.resourpol.2019.04.004

Kumar, S., & Preetha, G. S. (2012). Health promotion: An effective tool for global health. *Indian Journal of Community Medicine*, *37*(1), 5–12. doi:10.4103/0970-0218.94009 PMID:22529532

Kumar, V. V., Shastri, Y., & Hoadley, A. (2020). A consequence analysis study of natural gas consumption in a developing country: Case of India. *Energy Policy*, *145*, 111675. doi:10.1016/j.enpol.2020.111675

Kyung, H., Lee, J., & Choi, W. (2005). Simultaneous and Synergistic Conversion of Dyes and Heavy Metal Ions in Aqueous TiO ₂ Suspensions under Visible-Light Illumination. *Environmental Science & Technology*, *39*(7), 2376–2382. doi:10.1021/es0492788 PMID:15871279

Lagaros, N. D. (2023, January 14). Artificial Neural Networks Applied in Civil Engineering. *Applied Sciences (Basel, Switzerland)*, *13*(2), 1131. doi:10.3390/app13021131

Lamb, J. J. (2020). *Biogas and the Energy Sector*. Anaerob. Dig. From Biomass to Biogas. doi:10.48216/9788269203325ch1

Lam, M. K., Lee, K. T., & Mohamed, A. R. (2010). Homogeneous, heterogeneous, and enzymatic catalysis for transesterification of high free fatty acid oil (waste cooking oil) to biodiesel: A review. *Biotechnology Advances*, *28*(4), 500–518. doi:10.1016/j.biotechadv.2010.03.002 PMID:20362044

Lazaro, K., Malempati, S., & Varela, S. (2015). *Analysis of the production and usage of biogas at the San Francisco agricultural school. An Interactive Qualifying Project*. Worcester Polytechnic Institute. https://web.wpi.edu/Pubs/E-project/Available/E-project-051515-102329/unrestricted/Analysis_of_the_Production_and_Usage_of_Biogas_at_the_San_Francisco_Agricultural_School.pdf

Leape, L. (2009). *Transforming healthcare : a safety imperative*. Harvard School of Public Health., doi:10.1136/qshc.2009.036954

Lee, C. C., & Huffman, G. L. (1989). Incineration of solid waste. *Environmental Progress*, *8*(3), 143–151. doi:10.1002/ep.3300080312

Lee, G. H., & Yoo, J. Y. (2000). Performance analysis and simulation of automobile air conditioning system. *International Journal of Refrigeration*, *23*(3), 243–254. doi:10.1016/S0140-7007(99)00047-X

Lee, M. Y., Albright, S., O'Leary, L., Terkla, D. G., & Wilson, N. (2008). Expanding the reach of health sciences education and empowering others : The OpenCourseWare initiative at Tufts University. *Medical Teacher*, *30*(1), 159–163. doi:10.1080/01421590701881665 PMID:18464140

Lee, T. (2020). *ABC News. Landline*. Australian Broadcasting Corporation.

Lee, Y., & Hu, J. (2018). Integrated Approaches for Business Sustainability : The Perspective of Corporate Social Responsibility. *Sustainability (Basel)*, *10*(2), 2–20. doi:10.3390u10072318

Lelisho, M. E., Pandey, D., Alemu, B. D., Pandey, B. K., & Tareke, S. A. (2023). The negative impact of social media during COVID-19 pandemic. *Trends in Psychology*, *31*(1), 123–142. doi:10.100743076-022-00192-5

Liang, G., Mo, Y., & Zhou, Q. (2010). Novel strategies of bioleaching metals from printed circuit boards (PCBs) in mixed cultivation of two acidophiles. *Enzyme and Microbial Technology*, *47*(7), 322–326. doi:10.1016/j.enzmictec.2010.08.002

Liao, C.-H., Huang, C.-W., & Wu, J. C. S. (2012). Hydrogen Production from Semiconductor-based Photocatalysis via Water Splitting. *Catalysts*, *2*(4), 490–516. doi:10.3390/catal2040490

Li, J., Zhao, L., Qin, L., Tian, X., Wang, A., Zhou, Y., Meng, L., & Chen, Y. (2016). Removal of refractory organics in nanofiltration concentrates of municipal solid waste leachate treatment plants by combined Fenton oxidative-coagulation with photo – Fenton processes. *Chemosphere*, *146*, 442–449. doi:10.1016/j.chemosphere.2015.12.069 PMID:26741550

Li, L. (2020). Education Supply Chain in the Era of Industry 4.0. *Systems Research and Behavioral Science*, *37*(4), 579–592. doi:10.1002res.2702

Li, L., Rong, S., Wang, R., & Yu, S. (2021, February). Recent advances in artificial intelligence and machine learning for nonlinear relationship analysis and process control in drinking water treatment: A review. *Chemical Engineering Journal*, *405*, 126673. doi:10.1016/j.cej.2020.126673

Lin, D., Lee, C., Lau, H., & Yang, Y. (2017). Industrial Management & Data Systems Strategic Response to Industry 4.0: An Empirical Investigation on The Chinese Automotive Industry Article Information. *Industrial Management & Data Systems 118* (3), 0–18.

Liu, A., Ren, F., Lin, W. Y., & Wang, J.-Y. (2015). A review of municipal solid waste environmental standards with a focus on incinerator residues. *International Journal of Sustainable Built Environment*, *4*(2), 165–188. doi:10.1016/j.ijsbe.2015.11.002

Liu, C., Liu, Y., Liu, J., Yang, G., Zhao, X., & Quan, W. (2016). Electro-hydraulic servo plate-inclined plunger hydraulic transformer. *IEEE Access : Practical Innovations, Open Solutions*, *4*(1), 8608–8616. doi:10.1109/ACCESS.2016.2628355

Liu, M., Chen, C., Hu, J., Wu, X., & Wang, X. (2011). Synthesis of magnetite/graphene oxide composite and application for cobalt (II) removal. *The Journal of Physical Chemistry C*, *115*(51), 25234–25240. doi:10.1021/jp208575m

Liu, M., Wei, Y., & Leng, X. (2021). Improving biogas production using additives in anaerobic digestion: A review. *Journal of Cleaner Production*, *297*, 126666. doi:10.1016/j.jclepro.2021.126666

Liu, R., Li, J., & Ge, Z. (2016). Review on Chromobacterium Violaceum for Gold Bioleaching from E-waste. *Procedia Environmental Sciences*, *31*, 947–953. doi:10.1016/j.proenv.2016.02.119

Liu, R., Wei, Y., Zheng, J., Zhang, H., & Sheng, Q. (2013). A Hydrogen Peroxide Sensor Based on Silver Nanoparticles Biosynthesized by <i>Bacillus subtilis</i>. *Chinese Journal of Chemistry*, *31*(12), 1519–1525. doi:10.1002/cjoc.201300487

Liu, S., Zheng, Z., & Li, X. (2013). Advances in pesticide biosensors: Current status, challenges, and future perspectives. *Analytical and Bioanalytical Chemistry*, *405*(1), 63–90. doi:10.100700216-012-6299-6 PMID:22892800

Li, Z., Jia, Y., Feng, Y., Cui, R., Wang, Z., Qu, K., Liu, C., & Zhang, J. (2019). Methane-Rich Saline Protects Against Sepsis-Induced Liver Damage by Regulating the PPAR-γ/NF-κB Signaling Pathway. *Shock (Augusta, Ga.)*, *52*(6), e163–e172. doi:10.1097/SHK.0000000000001310 PMID:30601406

Lo, H. et al. (2018) 'An integrated model for solving problems in green', *Journal of Cleaner Production*. Elsevier B.V. . doi:10.1016/j.jclepro.2018.04.105

Louie, S. M., Dale, A. L., Casman, E. A., & Lowry, G. V. (2016). Challenges facing the environmental nanotechnology research enterprise. *Engineered Nanoparticles and the Environment: Biophysicochemical Processes and Toxicity: Biophysicochemical Processes and Toxicity*, 1-19.

Loukissas, Y. A. (2016). A place for Big Data: Close and distant readings of accessions data from the Arnold Arboretum. *Big Data \& Society, 3*(2), 2053951716661365.

Lowe, M., Qin, R., & Mao, X. (2022, April 24). A Review on Machine Learning, Artificial Intelligence, and Smart Technology in Water Treatment and Monitoring. *Water (Basel)*, *14*(9), 1384. doi:10.3390/w14091384

Lowry, G. V., Gregory, K. B., Apte, S. C., & Lead, J. R. (2012). Transformations of Nanomaterials in the Environment. *Environmental Science & Technology*, *46*(13), 6893–6899. doi:10.1021/es300839e PMID:22582927

Lu, C., Zhang, L., Zhong, Y., Ren, W., Tobias, M., Mu, Z., Ma, Z., Geng, Y., & Xue, B. (2015). An overview of e-waste management in China. *Journal of Material Cycles and Waste Management*, *17*(1), 1–12. doi:10.100710163-014-0256-8

Lu, J. W., Chang, N. B., Liao, L., & Liao, M. Y. (2015). Smart and green urban solid waste collection systems: Advances, challenges, and perspectives. *IEEE Systems Journal*, *11*(4), 2804–2817. doi:10.1109/JSYST.2015.2469544

Luthra, S., Mangla, S. K., & Kharb, R. K. (2015). Sustainable assessment in energy planning and management in Indian perspective. *Renewable & Sustainable Energy Reviews*, *47*, 58–73. doi:10.1016/j.rser.2015.03.007

M., G., B., R., S., S., & Boopathi, S. (2018). Fabrication of Solar Irrigation System. *International Research Journal of Engineering and Technology (IRJET), 5*(2), 2106–2108.

Ma, G. A., Diaby, V., & Xiao, H. (2014) 'Application of multicriteria decision analysis in health care : a systematic review and bibliometric analysis', health Epectations, 18, pp. 1894–1905. doi:10.1111/hex.12287

Maamoun, I., Rushdi, M. A., Falyouna, O., Eljamal, R., & Eljamal, O. (2023, March). Insights into machine-learning modeling for Cr (VI) removal from contaminated water using nano-nickel hydroxide. *Separation and Purification Technology*, *308*, 122863. doi:10.1016/j.seppur.2022.122863

Machado, C. M. L. (2015) 'Sustainability at the healthcare organizations: An analysis of the impact on the environment, society, and economy', Chemical Engineering Transactions, 45(June 2017), pp. 727–732. doi:10.3303/CET1545122

Machado, C. G., Winroth, M. P., & da Silva, E. H. D. R. (2020). Sustainable Manufacturing in Industry 4.0: An Emerging Research Agenda. *International Journal of Production Research*, *58*(5), 1462–1484. doi:10.1080/00207543.2019.1652777

Madhumathy, P., & Pandey, D. (2022). Deep learning based photo acoustic imaging for non-invasive imaging. *Multimedia Tools and Applications*, *81*(5), 7501–7518. doi:10.100711042-022-11903-6

Mæng, H., Lund, H., & Hvelplund, F. (1999). Biogas plants in Denmark: Technological and economic developments. *Applied Energy*, *64*(1-4), 195–206. doi:10.1016/S0306-2619(99)00067-7

Mahapatra, I., Clark, J. R. A., Dobson, P. J., Owen, R., Lynch, I., & Lead, J. R. (2018). Expert perspectives on potential environmental risks from nanomedicines and adequacy of the current guideline on environmental risk assessment. *Environmental Science. Nano*, *5*(8), 1873–1889. doi:10.1039/C8EN00053K

Mahmudul, H. M., Rasul, M. G., Akbar, D., Narayanan, R., & Mofijur, M. (2021). A comprehensive review of the recent development and challenges of a solar-assisted biodigester system. *The Science of the Total Environment*, *753*, 141920. doi:10.1016/j.scitotenv.2020.141920 PMID:32889316

Mäkinen, J., Bachér, J., Kaartinen, T., Wahlström, M., & Salminen, J. (2015). The effect of flotation and parameters for bioleaching of printed circuit boards. *Minerals Engineering*, *75*, 26–31. doi:10.1016/j.mineng.2015.01.009

Malche, T., & Maheshwary, P. (2015). Harnessing the Internet of things (IoT): A review. *International Journal (Toronto, Ont.)*, *5*(8).

Malhotra, M., & Singh, J. (2013). Factors affecting the adoption of online shopping in youngsters-an empirical study. *SSRN*, *2*(1), 44–54. doi:10.2139srn.2384552

Malinauskaite, J., Jouhara, H., Czajczyńska, D., Stanchev, P., Katsou, E., Rostkowski, P., Thorne, R. J., Colón, J., Ponsá, S., Al-Mansour, F., Anguilano, L., Krzyżyńska, R., López, I. C., Vlasopoulos, A., & Spencer, N. (2017). Municipal solid waste management and waste-to-energy in the context of a circular economy and energy recycling in Europe. *Energy*, *141*, 2013–2044. doi:10.1016/j.energy.2017.11.128

Mallikarjunaiah, S., Pattabhiramaiah, M., & Metikurki, B. (2020). Application of nanotechnology in the bioremediation of heavy metals and wastewater management. *Nanotechnology for food, agriculture, and environment*, 297-321. Springer.

Mandal, D., Mishra, S., & Singh, R. K. (2018). Green Synthesized Nanoparticles as Potential Nanosensors. In (pp. 137-164). doi:10.1007/978-981-10-7751-7_7

Mandal, P., Dubey, B. K., & Gupta, A. K. (2017). Review on landfill leachate treatment by electrochemical oxidation: Drawbacks, challenges and future scope. *Waste Management (New York, N.Y.)*, *69*, 250–273. doi:10.1016/j.wasman.2017.08.034 PMID:28865908

Mao, C., Feng, Y., Wang, X., & Ren, G. (2015). Review on research achievements of biogas from anaerobic digestion. *Renewable & Sustainable Energy Reviews*, *45*, 540–555. doi:10.1016/j.rser.2015.02.032

Marampoutis, I., Vinot, M., & Trilling, L. (2022). Multi-objective vehicle routing problem with flexible scheduling for the collection of refillable glass bottles: A case study. *EURO Journal on Decision Processes*, *10*, 100011. doi:10.1016/j.ejdp.2021.100011

Mariyate, J., & Bera, A. (2023). Paradigm shift towards the sustainability in upstream oil industry for enhanced recovery - A state-of-art review. In *Journal of Cleaner Production* (Vol. 386). Elsevier Ltd., doi:10.1016/j.jclepro.2022.135784

Maskey, B. (2018). *Municipal solid waste management in Nepal: A case study of Gorkha Municipality*. Sl: sn.

Massart, F., & Meucci, V. (2007). Environmental thyroid toxicants and child endocrine health. *Pediatric Endocrinology Reviews; PER*, *5*(1), 500–509. PMID:17925791

Masud, M. H., Akram, W., Ahmed, A., Ananno, A. A., Mourshed, M., Hasan, M., & Joardder, M. U. H. (2019). Towards the effective E-waste management in Bangladesh: A review. *Environmental Science and Pollution Research International*, *26*(2), 1250–1276. doi:10.100711356-018-3626-2 PMID:30456610

Mathews, J. A., & Tan, H. (2016). Circular economy: Lessons from China. *Nature*, *531*(7595), 440–442. doi:10.1038/531440a PMID:27008954

Matsakas, L., Kekos, D., Loizidou, M., & Christakopoulos, P. (2014). Utilization of household food waste for the production of ethanol at high dry material content. *Biotechnology for Biofuels*, *7*(1), 1–9. doi:10.1186/1754-6834-7-4 PMID:24401142

Matt, D. T., Modrák, V., & Zsifkovits, H. (2020). *Industry 4.0 for Smes: Challenges, Opportunities and Requirements. Industry 4.0 for SMEs: Challenges*. Opportunities and Requirements. doi:10.1007/978-3-030-25425-4

Maurer-Jones, M. A., Gunsolus, I. L., Murphy, C. J., & Haynes, C. L. (2013). Toxicity of Engineered Nanoparticles in the Environment. *Analytical Chemistry*, *85*(6), 3036–3049. doi:10.1021/ac303636s PMID:23427995

Maxwell, J. C. (1873). *A treatise on electricity and magnetism* (Vol. 1). Clarendon Press.

McCullough, B. P., Pfahl, M. E., & Nguyen, S. N. (2016). The green waves of environmental sustainability in sport. *Sport in Society*, *19*(7), 1040–1065. doi:10.1080/17430437.2015.1096251

Meenakshi, P. (2005). Elements of Environmental Science and Engineering. Prentice Hall of India Private limited, India.

Meneguette, R. I. (2016). A vehicular cloud-based framework for the intelligent transport management of big cities. *International Journal of Distributed Sensor Networks*, *12*(5), 8198597. doi:10.1155/2016/8198597

Menon, V., Pandey, D., Khosla, D., Kaur, M., Vashishtha, H. K., George, A. S., & Pandey, B. K. (2022). A Study on COVID–19, Its Origin, Phenomenon, Variants, and IoT-Based Framework to Detect the Presence of Coronavirus. In *IoT Based Smart Applications* (pp. 1–13). Springer International Publishing.

Meslie, Y., Enbeyle, W., Pandey, B. K., Pramanik, S., Pandey, D., Dadeech, P., & Saini, A. (2021). Machine intelligence-based trend analysis of COVID-19 for total daily confirmed cases in Asia and Africa. In *Methodologies and Applications of Computational Statistics for Machine Intelligence* (pp. 164–185). IGI Global. doi:10.4018/978-1-7998-7701-1.ch009

Ministry of Housing and Urban Affairs. (2015). Swachh Bharat Mission (Urban). In *Handbook for Swachh Bharat Mission (Urban)* (Vol. 2). MoHUA. https://mohua.gov.in/upload/uploadfiles/files/Part2.pdf

Ministry of petroleum and natural gas. (2019). *About Natural Gas*. MPNG. http://petroleum.nic.in/natural-gas/about-natural-gas.

Min, J., Kim, Y., Lee, S., Tae-won, J., Kim, I., & Song, J. (2019). The Fourth Industrial Revolution and Its Impact on Occupational Health and Safety, Worker ' s Compensation and Labor Conditions. *Safety and Health at Work*, *10*(4), 400–408. doi:10.1016/j.shaw.2019.09.005 PMID:31890322

Mohamed, H. M. (2020). Sensors and biosensors for environment contaminants. *Nanosensor technologies for environmental monitoring*, 109-134. NIH.

Mohanty, A., Venkateswaran, N., Ranjit, P. S., Tripathi, M. A., & Boopathi, S. (2023). Innovative Strategy for Profitable Automobile Industries: Working Capital Management. In Handbook of Research on Designing Sustainable Supply Chains to Achieve a Circular Economy (pp. 412–428). IGI Global.

Moher, D. (2009). Academia and Clinic Annals of Internal Medicine Preferred Reporting Items for Systematic Reviews and Meta-Analyses. *Annals of Internal Medicine*, *151*(4), 264–269. doi:10.7326/0003-4819-151-4-200908180-00135 PMID:19622511

Mohsin,M.; Chinyama,A. (2016) Impact of Solid Waste Management practices on Environment and Public Health: A Case of Bahawalpur City, Pakistan. *Journal of Environmental and Agricultural Sciences*.

Mohsin, M., & Chinyama, A. (2016). Impacts of solid waste management practices on environment and public health: A case of Bahawalpur city, Pakistan. *Journal of Environmental and Agricultural Sciences*, *9*, 69–79.

Mongeon, P., & Paul-Hus, A. (2016). The journal coverage of Web of Science and Scopus: A comparative analysis. *Scientometrics*, *106*(1), 213–228. doi:10.100711192-015-1765-5

Monika, & Kishore, J. (2010). E-waste management: As a challenge to public health in India. *Indian Journal of Community Medicine*, *35*(3), 382–385. doi:10.4103/0970-0218.69251

Montour, A., Baumann, A., Blythe, J., & Hunsberger, M. (2009). The changing nature of nursing work in rural and small community hospitals. *Rural and Remote Health*, *9*(1), 1089–1092. http://usblogs.pwc.com/industrialinsights/2017/05/09/megatrends-to-watch-in-the-paper-and-packaging-industry/. doi:10.22605/RRH1089 PMID:19199373

Moreda, I. L. (2016). The potential of biogas production in Uruguay. *Renewable & Sustainable Energy Reviews*, *54*, 1580–1591. doi:10.1016/j.rser.2015.10.099

More, K. S., & Wolkersdorfer, C. (2023, January 18). Intelligent Mine Water Management Tools—eMetsi and Machine Learning GUI. *Mine Water and the Environment*, *42*(1), 111–120. doi:10.100710230-023-00917-7

Moreno, V. C., & Cozzani, V. (2015). Major accident hazard in bioenergy production. *Journal of Loss Prevention in the Process Industries*, *35*, 135–144. doi:10.1016/j.jlp.2015.04.004

Mor, R. S., Sangwan, K. S., Singh, S., Singh, A., & Kharub, M. (2021). E-waste Management for Environmental Sustainability: An Exploratory Study. *Procedia CIRP*, *98*, 193–198. doi:10.1016/j.procir.2021.01.029

Morrissey, A. J., & Browne, J. (2004). Waste management models and their application to sustainable waste management. *Waste Management (New York, N.Y.)*, *24*(3), 297–308. doi:10.1016/j.wasman.2003.09.005 PMID:15016418

Mor, S., Ravindra, K., Dahiya, R. P., & Chandra, A. (2006). Leachate characterization and assessment of groundwater pollution near municipal solid waste dumping site. *Environmental Monitoring and Assessment*, *118*(1-3), 435–456. doi:10.100710661-006-1505-7 PMID:16897556

Mubarak, M. F., & Petraite, M. (2020). Industry 4.0 technologies, digital trust and technological orientation: What matters in open innovation? *Technological Forecasting and Social Change*, *161*, 120332. doi:10.1016/j.techfore.2020.120332

Müller, J. M., & Voigt, K. (2018). Sustainable Industrial Value Creation in SMEs: A Comparison between Industry 4.0 and Made in China 2025." International Journal of Precision Engineering and Manufacturing - Green Technology 5 (5): 659–70. . doi:10.1007/s40684-018-0056-z

Müller, F. M., de Oliveira, D., & Michels, C. (2023, April). Current status, gaps and challenges of rendering industries wastewater. *Journal of Water Process Engineering*, *52*, 103480. doi:10.1016/j.jwpe.2022.103480

Mundada, M. N., Kumar, S., & Shekdar, A. V. (2004). E-waste: A new challenge for waste management in India. *The International Journal of Environmental Studies*, *61*(3), 265–279. doi:10.1080/0020723042000176060

Mundhe, N., Jaybhaye, R., & Dorik, B. (2014). Assessment of municipal solid waste management of Pune city using geospatial tools. *International Journal of Computer Applications*, *100*(10), 24–32. doi:10.5120/17562-8184

Municipal solid wastes (management and handling) rules, 2000. (n.d.). TSPCB. https://tspcb.cgg.gov.in/GOs/Municipal%20Solid%20Waste%20Rules.pdf

Murata, C., Pant, N., Lyer, S., Veitch, J., & Campbell, M. (2020). *Seizing the Trade 4. 0 Opportunity About the Authors*. Deloitte Insights.

Murthy, V., & Ramakrishna, S. (2022). A review on global E-waste management: Urban mining towards a sustainable future and circular economy. *Sustainability (Basel), 14*(2), 647. doi:10.3390u14020647

Mustafa, H. M., Hayder, G., & Mustapa, S. I. (2022, April 22). Circular Economy Framework for Energy Recovery in Phytoremediation of Domestic Wastewater. *Energies, 15*(9), 3075. doi:10.3390/en15093075

Mustapha, M. A., Manan, Z. A., & Alwi, S. R. W. (2017). Sustainable Green Management System (SGMS) - an integrated approach towards organisational sustainability. *Journal of Cleaner Production, 146*, 158–172. doi:10.1016/j.jclepro.2016.06.033

Myilsamy, S., & Sampath, B. (2021). Experimental comparison of near-dry and cryogenically cooled near-dry machining in wire-cut electrical discharge machining processes. *Surface Topography : Metrology and Properties, 9*(3), 035015. doi:10.1088/2051-672X/ac15e0

Nabegu, A. B. (2010). An analysis of Municipal Solid Waste in Kano Metropolis, Nigeria. *Journal of Human Ecology (Delhi, India), 31*(2), 111–119. doi:10.1080/09709274.2010.11906301

Naeher, L. P., Brauer, M., Lipsett, M., Zelikoff, J. T., Simpson, C. D., Koenig, J. Q., & Smith, K. R. (2007). Woodsmoke health effects: A review. *Inhalation Toxicology, 19*(1), 67–106. doi:10.1080/08958370600985875 PMID:17127644

Nagi, D. S., & Kurien, D. J. V. (2019). International Journal of Advanced in Management, Technology and Engineering Sciences ISSN NO : 2249-7455. *International Journal of Management. Technology And Engineering, 9*(IV), 2854.

Nanda, S., & Berruti, F. (2020). Thermochemical conversion of plastic waste to fuels: A Review. *Environmental Chemistry Letters, 19*(1), 123–148. doi:10.100710311-020-01094-7

Narasaiah, B., & Mandal, B. (2019). Bio-fabricated CuO NPs as green catalyst towards remediation of environmental pollutants. *Lett. Appl. NanoBioSci, 8*, 597–603.

Narayana, T. (2009). Municipal solid waste management in India: From waste disposal to recovery of resources? *Waste Management (New York, N.Y.), 29*(3), 1163–1166. doi:10.1016/j.wasman.2008.06.038 PMID:18829290

Nath, D. (2018). Nanomaterial for the Management of Radioactive Waste. In *Handbook of Ecomaterials* (pp. 1–18). Springer International Publishing. doi:10.1007/978-3-319-48281-1_49-1

Needhidasan, S., Samuel, M., & Chidambaram, R. (2014) Electronic waste – an emerging threat to the environment of urban India. *J. Environ. Heal. Sci. Eng.* 1–9.

Ngo, H.-H., Guo, W., & Xing, W. (2009). Applied technologies in municipal solid waste landfill leachate treatment. *Waste Water Treat Technol, 2*, 199.

Ngulube, P., Pandey, B. K., Pandey, D., Nassa, V. K., George, A. S., Gupta, M. K., & Dadheech, P. (2022). Evaluation of the SARS-COV-2 Omicron variant's features, prospective consequences, and upcoming prospects. An Interdisciplinary Approach in the Post-COVID-19 Pandemic Era, 63-76.

Ngulube, P., Pandey, B. K., Pandey, D., Nassa, V. K., George, A. S., Gupta, M. K., & Dadheech, P. (2022). Evaluation of the SARS-COV-2 Omicron variant's features, prospective consequences, and upcoming prospects. *An Interdisciplinary Approach in the Post-COVID-19 Pandemic Era*, 63-76. Nova Science Publishers.

Nielsen, J. B. H., & Oleskowicz-Popiel, P. (2008, December). Biogas–a promising renewable energy source for Europe. In *AEBIOM Workshop-European Parliament, Brussels.* Scientific Research Publishing.

Nitti, M., Pilloni, V., Giusto, D., & Popescu, V. (2017). IoT Architecture for a sustainable tourism application in a smart city environment. *Mobile Information Systems, 2017*, 2017. doi:10.1155/2017/9201640

Niu, C., Li, X., Dai, R., & Wang, Z. (2022, June). Artificial intelligence-incorporated membrane fouling prediction for membrane-based processes in the past 20 years: A critical review. *Water Research*, *216*, 118299. doi:10.1016/j.watres.2022.118299 PMID:35325824

Njoku, P. O., Edokpayi, J. N., & Odiyo, J. O. (2019). Health and environmental risks of residents living close to a landfill: A case study of thohoyandou landfill, Limpopo Province, South Africa. *International Journal of Environmental Research and Public Health*, *16*(12), 2125. doi:10.3390/ijerph16122125 PMID:31208082

Njuguna, J., Pielichowski, K., & Zhu, H. (2021). *Health and environmental safety of nanomaterials: polymer nanocomposites and other materials containing nanoparticles*. Woodhead Publishing.

Nkuna, R., Roopnarain, A., Rashama, C., & Adeleke, R. (2022). Insights into organic loading rates of anaerobic digestion for biogas production: A review. *Critical Reviews in Biotechnology*, *42*(4), 487–507. PMID:34315294

Nnorom, Innocent C., & Odeyingbo, O. A. (2019). *Electronic waste management practices in Nigeria. Handbook of Electronic Waste Management: International Best Practices and Case Studies*. INC. doi:10.1016/B978-0-12-817030-4.00014-0

Nnorom, I. C., & Osibanjo, O. (2008). Overview of electronic waste (e-waste) management practices and legislations, and their poor applications in the developing countries. *Resources, Conservation and Recycling*, *52*(6), 843–858. doi:10.1016/j.resconrec.2008.01.004

Nnorom, I. C., & Osibanjo, O. (2009a). Heavy metal characterization of waste portable rechargeable batteries used in mobile phones. *International Journal of Environmental Science and Technology*, *6*(4), 641–650. doi:10.1007/BF03326105

Nnorom, I. C., & Osibanjo, O. (2009b). Toxicity characterization of waste mobile phone plastics. *Journal of Hazardous Materials*, *161*(1), 183–188. doi:10.1016/j.jhazmat.2008.03.067 PMID:18448243

Norsyahida, M., Mohamad, I. W. W., Indati, M. S., & Victor, A. B. (2021). Natural Gas as a Key Alternative Energy Source in Sustainable Renewable Energy Transition: A Mini Review. *Frontiers in Energy Research*, *9*. https://www.frontiersin.org/articles/10.3389/fenrg.2021.625023

Nourani, V., Asghari, P., & Sharghi, E. (2021, April). Artificial intelligence based ensemble modeling of wastewater treatment plant using jittered data. *Journal of Cleaner Production*, *291*, 125772. doi:10.1016/j.jclepro.2020.125772

Nowakowski, P. (2016). The influence of residents' behaviour on waste electrical and electronic equipment collection effectiveness. Waste Management &amp. *Waste Management & Research*, *34*(11), 1126–1135. doi:10.1177/0734242X16669997 PMID:27705882

Nowakowski, P., & Pamuła, T. (2020). Application of deep learning object classifier to improve e-waste collection planning. *Waste Management (New York, N.Y.)*, *109*, 1–9. doi:10.1016/j.wasman.2020.04.041 PMID:32361385

Nsair, A., Onen Cinar, S., Alassali, A., Abu Qdais, H., & Kuchta, K. (2020). Operational parameters of biogas plants: A review and evaluation study. *Energies*, *13*(15), 3761. doi:10.3390/en13153761

Nylund, N. & Koponen, K. (2013). *Fuel and Technology Alternatives for Buses. Overall Energy Efficiency and Emission Performance*. IEA Bioenergy Task 46.

O'Connor, M. P., Coulthard, R. M., & Plata, D. L. (2018). Electrochemical deposition for the separation and recovery of metals using carbon nanotube-enabled filters. *Environmental Science. Water Research & Technology*, *4*(1), 58–66. doi:10.1039/C7EW00187H

Oakes, T. W., & Shank, K. E. (1979). *Brief review of the Resource Conservation and Recovery Act (RCRA) 1976*. OSTI.

Okot-Okumu, J., & Nyenje, R. (2011). Municipal solid waste management under decentralisation in Uganda. *Habitat International*, *35*(4), 537–543. doi:10.1016/j.habitatint.2011.03.003

Olapiriyakul, S., & Caudill, R. J. (2008, 19-22 May 2008). A framework for risk management and end-of-life (EOL) analysis for nanotechnology products: A case study in lithium-ion batteries. Paper presented at the *2008 IEEE International Symposium on Electronics and the Environment*. IEEE. 10.1109/ISEE.2008.4562877

Olson, C., Lenzmann, F. (2016). The social and economic consequences of the fossil fuel supply chain. *MRS Energy & Sustainability, 3.*

Omar, M. F., Termizi, A. A. A., Zainal, D., Wahap, N. A., Ismail, N. M., & Ahmad, N. (2016). Implementation of spatial smart waste management system in Malaysia. [IOP Publishing.]. *IOP Conference Series. Earth and Environmental Science*, *37*(1), 012059. doi:10.1088/1755-1315/37/1/012059

Osibanjo, O., & Nnorom, I. C. (2007). The challenge of electronic waste (e-waste) management in developing countries. *Waste Management & Research*, *25*(6), 489–501. doi:10.1177/0734242X07082028 PMID:18229743

Othman, S. A., & Fadzil, N. F. (2021). Implementation of Nanofiller for Bioplastic Reinforcement: A Review. *International Journal of Advanced Research in Technology and Innovation*, *3*(2), 19–27.

Paladino, O., & Neviani, M. (2018). A closed-loop biowaste to biofuel integrated process fed with waste frying oil, organic waste, and algal biomass: Feasibility at pilot scale. *Renewable Energy*, *124*, 61–74. doi:10.1016/j.renene.2017.08.027

Palaniappan, M., Tirlangi, S., Mohamed, M. J. S., Moorthy, R. M. S., Valeti, S. V., & Boopathi, S. (2023). Fused Deposition Modelling of Polylactic Acid (PLA)-Based Polymer Composites: A Case Study. In Development, Properties, and Industrial Applications of 3D Printed Polymer Composites (pp. 66–85). IGI Global

Palaniappan, M., Tirlangi, S., Mohamed, M. J. S., Moorthy, R. M. S., Valeti, S. V., & Boopathi, S. (2023). Fused Deposition Modelling of Polylactic Acid (PLA)-Based Polymer Composites: A Case Study. In Development, Properties, and Industrial Applications of 3D Printed Polymer Composites (pp. 66–85). IGI Global.

Palaniappan, M., Tirlangi, S., Mohamed, M. J. S., Moorthy, R. M. S., Valeti, S. V., & Boopathi, S. (2023). Fused deposition modelling of polylactic acid (PLA)-based polymer composites: A case study. In Development, Properties, and Industrial Applications of 3D Printed Polymer Composites (pp. 66–85). IGI Global. doi:10.4018/978-1-6684-6009-2.ch005

Pandey, D., & Wairya, S. (2022). Perfomance Analysis of Text Extraction from Complex Degraded Image Using Fusion of DNN, Steganography, and AGSO. In *VLSI, Microwave and Wireless Technologies: Select Proceedings of ICVMWT 2021* (pp. 195-203). Singapore: Springer Nature Singapore.

Pandey, D., Aswari, A., Taufiqurakman, M., Khalim, A., & Azahrah, F. F. (2021). System Of Education Changes Due to Covid-19 Pandemic. *Asian Journal of Advances in Research*, 10-15.

Pandey, D., Babu, S. Z. D., Noibi, T. O., Pandey, B. K., Dey, P. T. D., Kassaw, C., & Canete, J. J. O. (2020). A study on novel covid-19 epidemic threat for universal physical situation in 2020. *Asian Journal of Advances in Research*, 27-33.

Pandey, D., Gul, R., Canete, J. J. O., Rocha, I. C. N., Gowwrii, G., Pandey, B. K., & Peter, S. N. (2021). Mental Stress In Online Learning During The Pandemic: An Assessment Of Learners' perception. *Asian Journal of Advances in Research*, 37-49.

Pandey, D., Hasan, A., Pandey, B. K., Lelisho, M. E., George, A. H., & Shahul, A. (2023). COVID-19 epidemic anxiety, mental stress, and sleep disorders in developing country university students. CSI Transactions on ICT, 1-9.

Pandey, D., Hasan, A., Pandey, B. K., Lelisho, M. E., George, A. H., & Shahul, A. (2023). COVID-19 epidemic anxiety, mental stress, and sleep disorders in developing country university students. *CSI Transactions on ICT*, 1-9. NIH.

Pandey, D., Islam, T., & Malik, M. A. (2021). Novel coronavirus disease (SARS-COV-2): an overview. *Asian Journal of Advances in Medical Science*, 8-12.

Pandey, D., Lelisho, M. E., Hasan, A., Rajput, N., Kumar, B., George, A. H., & Shahul, A. (2023). Factors Associated With Sleep Disorders Among Indian and Ethiopian Students in the Digital World in the Midst of the COVID-19 Pandemic. In Applications of Neuromarketing in the Metaverse (pp. 137-154). IGI Global.

Pandey, D., Pandey, B. K., Noibi, T. O., Babu, S., Patra, P. M., Kassaw, C., & Canete, J. J. O. (2020). Covid-19: Unlock 1.0 risk, test, transmission, incubation and infectious periods and reproduction of novel Covid-19 pandemic. *Asian Journal of Advances in Medical Science*, 23-28.

Pandey, D., Wairya, S., Sharma, M., Gupta, A. K., Kakkar, R., & Pandey, B. K. (2022). An approach for object tracking, categorization, and autopilot guidance for passive homing missiles. *Aerospace Systems*, 1-14.

Pandey, J. K., Jain, R., Dilip, R., Kumbhkar, M., Jaiswal, S., Pandey, B. K., & Pandey, D. (2022). Investigating Role of IoT in the Development of Smart Application for Security Enhancement. In *IoT Based Smart Applications* (pp. 219-243). Cham: Springer International Publishing.

Pandey, B. K., Pandey, D., & Agarwal, A. (2022). Encrypted Information Transmission by Enhanced Steganography and Image Transformation. [IJDAI]. *International Journal of Distributed Artificial Intelligence*, *14*(1), 1–14. doi:10.4018/IJDAI.297110

Pandey, B. K., Pandey, D., Wariya, S., Aggarwal, G., & Rastogi, R. (2021). Deep learning and particle swarm optimisation-based techniques for visually impaired humans' text recognition and identification. *Augmented Human Research*, *6*(1), 1–14. doi:10.100741133-021-00051-5

Pandey, D., Islam, T., Magray, J. A., Gulzar, A., & Zargar, S. A. (2021). Use of statistical analysis to monitor novel coronavirus-19 cases in Jammu and Kashmir, India. *European Journal of Biological Research*, *11*(3), 274–282.

Pandey, D., & Pandey, B. K. (2022). An Efficient Deep Neural Network with Adaptive Galactic Swarm Optimization for Complex Image Text Extraction. In *Process Mining Techniques for Pattern Recognition* (pp. 121–137). CRC Press. doi:10.1201/9781003169550-10

Pandey, D., Pandey, B. K., & Wairya, S. (2021). Hybrid deep neural network with adaptive galactic swarm optimization for text extraction from scene images. *Soft Computing*, *25*(2), 1563–1580. doi:10.100700500-020-05245-4

Pandey, D., Wairya, S., Pradhan, B., & Wangmo. (2022). Understanding COVID-19 response by twitter users: A text analysis approach. *Heliyon*, *8*(8), e09994. doi:10.1016/j.heliyon.2022.e09994 PMID:35873536

Pandey, R. U., Surjan, A., & Kapshe, M. (2018). Exploring linkages between sustainable consumption and prevailing green practices in reuse and recycling of household waste: Case of Bhopal city in India. *Journal of Cleaner Production*, *173*, 49–59. doi:10.1016/j.jclepro.2017.03.227

Pandey, S., Goswami, G. K., & Nanda, K. K. (2013). Green synthesis of polysaccharide/gold nanoparticle nanocomposite: An efficient ammonia sensor. *Carbohydrate Polymers*, *94*(1), 229–234. doi:10.1016/j.carbpol.2013.01.009 PMID:23544533

Pandyaswargo, A. H., Jagath Dickella Gamaralalage, P., Liu, C., Knaus, M., Onoda, H., Mahichi, F., & Guo, Y. (2019). Challenges and an implementation framework for sustainable municipal organic waste management using biogas technology in emerging Asian Countries. *Sustainability (Basel)*, *11*(22), 6331. doi:10.3390u11226331

Pang, X., Shi, X., Mu, Y., He, H., Shuai, S., Chen, H., & Li, R. (2006). Characteristics of carbonyl compound emission from a diesel engine using biodiesel–ethanol–diesel as fuel. *Atmospheric Environment*, *40*(36), 7057–7065. doi:10.1016/j.atmosenv.2006.06.010

Paredes, E. M. A., Murillo, M. F. M., Ramírez, V. B. S., & Pandey, D. (2023). Lighting Evaluation and Implementation of an Energy Efficiency Plan in the Company Mantransve. In *Applications of Neuromarketing in the Metaverse* (pp. 62–77). IGI Global. doi:10.4018/978-1-6684-8150-9.ch005

Pargal, S., Gilligan, D., & Huq, M. (2002). Does social capital increase participation in voluntary solid waste management? Evidence from Dhaka, Bangladesh. *The role of social capital in development*, 188-212. Research Gate.

Part, F., Berge, N., Baran, P., Stringfellow, A., Sun, W., Bartelt-Hunt, S., Mitrano, D., Li, L., Hennebert, P., Quicker, P., Bolyard, S. C., & Huber-Humer, M. (2018). A review of the fate of engineered nanomaterials in municipal solid waste streams. *Waste Management (New York, N.Y.)*, *75*, 427–449. doi:10.1016/j.wasman.2018.02.012 PMID:29477652

Parthasarathy, M. (2021). Challenges and emerging trends in toner waste recycling: A review. *Recycling*, *6*(3), 57. doi:10.3390/recycling6030057

Patel, K. (2021). *Understanding the current E-waste management system in India: A study on strengthening the integration of E-waste management sectors in India under.* MDPI.

Patel, N., Pathak, P., Rai, D., & Chaudhary, V. K. (2020). Biosensors used for monitoring of environmental contaminants. *Nanosensor technologies for environmental monitoring*, 69-83. NIH.

Pathak, P., Srivastava, R. R., & Ojasvi. (2017). Assessment of legislation and practices for the sustainable management of Waste Electrical and electronic equipment in India. *Renewable & Sustainable Energy Reviews*, *78*, 220–232. doi:10.1016/j.rser.2017.04.062

Patil, D. A. (2020). The study of industry 4 . 0 and its impact on supply chain management. *International Research Journal of Engineering and Technology.*

Patil, G. S., Pawar, R. A., Borole, M. D., Ahire, S. G., Krishnani, A. L., & Karwande, A. H. (2018). *Review Paper on Drainage Water Cleaner Machine. International Research Journal of Engineering and Technology* (Vol. 5). IRJET.

Patil, S. A., & Sharma, N. M. (2013). Electronic Waste - A Literature Review. *International Journal of Scientific Research*, *4*(4), 2319–7064. https://www.niir.org/information/content.phtml?content=

Patil, Y., & Rao, P. (2014). Industrial waste management in the era of climate change: a smart sustainable model based on utilization of passive biomass. *Patil, Yogesh and Rao, Prakash (2014) Industrial Waste Management in the Era of Climate Change: A Smart Sustainable Model Based on Utilization of Passive Biomass.* In W. L. Filho (Ed.), *Handbook of Climate Change Adaptation* (pp. 1–13). Springer-Verlag Berlin Heidelberg. doi:10.1007/978-3-642-40455-9_49-1

Patinvoh, R. J., & Taherzadeh, M. J. (2019). Challenges of biogas implementation in developing countries. *Current Opinion in Environmental Science & Health*, *12*, 30–37. doi:10.1016/j.coesh.2019.09.006

Pavoski, G., Junior, A. B. B., Chaves, R. M., Maraschin, T., Oviedo, L. R., Martins, T. A. G., & Espinosa, D. C. R. (2022). Nanotechnology and recycling, remanufacturing, and reusing battery. In Nano Technology for Battery Recycling, Remanufacturing, and Reusing (pp. 53-78): Elsevier. doi:10.1016/B978-0-323-91134-4.00022-4

Peattie, K. (2001). Golden goose or wild goose? The hunt for the green consumer. *Business Strategy and the Environment*, *10*(4), 187–199. doi:10.1002/bse.292

Perkins, D. N., Drisse, M.-N. B., Nxele, T., & Sly, P. D. (2014). E-waste: A global hazard. *Annals of Global Health*, *80*(4). Advance online publication. http://www.annalsofglobalhealth.org/article/S2214- 9996(14)00320-8/pdf. doi:10.1016/j.aogh.2014.10.001 PMID:25459330

Pham, T. P. T., Kaushik, R., Parshetti, G. K., Mahmood, R., & Balasubramanian, R. (2015). Food waste-to-energy conversion technologies: Current status and future directions. *Waste Management (New York, N.Y.)*, *38*, 399–408. doi:10.1016/j.wasman.2014.12.004 PMID:25555663

Phan Hoang, T. T., & Kato, T. (2016). Measuring the effect of environmental education for sustainable development at elementary schools: A case study in Da Nang city, Vietnam. *Sustainable Environment Research*, *26*(6), 274–286. doi:10.1016/j.serj.2016.08.005

Philip, N. Y., Rodrigues, J. J. P. C., Wang, H., Fong, S. J., & Chen, J. (2021). Internet of Things for In-Home Health Monitoring Systems: Current Advances, Challenges and Future Directions. *IEEE Journal on Selected Areas in Communications*, *39*(2), 300–310. doi:10.1109/JSAC.2020.3042421

Pinzone, M., Lettieri, E. and Masella, C. (2012) 'Sustainability in Healthcare : Combining Organizational and Architectural Levers Regular Paper', *Intech open science*, 4(4), pp. 1–9. . doi:10.5772/54841

Plano Clark, V. L., Ivankova, N. V., & Yang, N. (2023). *Frameworks for conceptualizing mixed methods research* (4th ed.). International Encyclopedia of Education. doi:10.1016/B978-0-12-818630-5.11038-3

Poje, T. (2016). Situation and development trends of agricultural biogas plants in Slovenia. *51. hrvatski i 11. međunarodni simpozij agronoma, 15.-18. veljače 2016. godine, Opatija, Hrvatska. Zbornik radova*, 502-506.

Poje, T. (2010). Slovenian legislation-driving power and obstacle to the development of agricultural biogas plants. *Novi izzivi v poljedelstvu 2010. Zbornik simpozija, Rogaška. Slatina*, *2-3*(December), 277–283.

Polonsky, M. J. (2011). Transformative green marketing: Impediments and opportunities. *Journal of Business Research*, *64*(12), 1311–1319. doi:10.1016/j.jbusres.2011.01.016

Pop, M., Bucur, I., Zoldan, D., Imre, K., Nichita, I., Cristina, G., Tîrziu, A., & Tîrziu, E. (2022). Chemical and Microbiological Air Quality in a Municipal Solid Waste Landfill and Its Surroundings, in South-Eastern Romania. *Sustainability (Basel)*, *14*(1), 156. doi:10.3390u14010156

Pöschl, M., Ward, S., & Owende, P. (2010). Evaluation of energy efficiency of various biogas production and utilization pathways. *Applied Energy*, *87*(11), 3305–3321. doi:10.1016/j.apenergy.2010.05.011

Pradeep, M.D. & Akhilesh Suresh A Kuckian (2016). Green Marketing to meet consumer demands and sustainable development-Challenges and Opportunities. International journal of Advanced Trends in Engineering and Technology (IJATET), 1(1), 34-41.

Prajapati, K. K., Yadav, M., Singh, R. M., Parikh, P., Pareek, N., & Vivekanand, V. (2021). An overview of municipal solid waste management in Jaipur city, India - Current status, challenges and recommendations. *Renewable & Sustainable Energy Reviews*, *152*(August), 111703. doi:10.1016/j.rser.2021.111703

Preindl, R., Nikolopoulos, K., & Litsiou, K. (2020). Transformation Strategies for the Supply Chain: The Impact of Industry 4.0 and Digital Transformation. *Supply Chain Forum, 21* (1), 26–34. 10.1080/16258312.2020.1716633

Prema, K., Kumar, N. S., & Sunitha, K. A. (2009, June). Online temperature control based on virtual instrumentation. In *2009 International Conference on Control, Automation, Communication and Energy Conservation* (pp. 1-4). IEEE.

Priti, & Mandal, K. (2019). Review on evolution of Municipal Solid Waste Management in India: Practices, challenges and policy implications. *Journal of Material Cycles and Waste Management, 21*(6), 1263–1279. doi:10.1007/s10163-019-00880-y

Priya, A., & Hait, S. (2017). Comparative assessment of metallurgical recovery of metals from electronic waste with special emphasis on bioleaching. *Environmental Science and Pollution Research International, 24*(8), 6989–7008. doi:10.100711356-016-8313-6 PMID:28091997

Proctor, E., Luke, D., Calhoun, A., McMillen, C., Brownson, R., McCrary, S., & Padek, M. (2015). 'Sustainability of evidence-based healthcare : Research agenda, methodological advances, and infrastructure support', *Implementation Science. Implementation Science : IS, 10*(88), 1–13. doi:10.118613012-015-0274-5 PMID:26062907

Propris, L., & Bailey, D. (2020). *Industry 4.0 and Regional Transformations*. Routledge. . doi:10.4324/9780429057984

Pujara, Y., Pathak, P., Sharma, A., & Govani, J. (2019). Review on Indian Municipal Solid Waste Management practices for reduction of environmental impacts to achieve sustainable development goals. *Journal of Environmental Management, 248*, 109238. doi:10.1016/j.jenvman.2019.07.009 PMID:31319199

Punnakitikashem, P., & Hallinger, P. (2020). Bibliometric review of the knowledge base on healthcare management for sustainability, 1994-2018. *Sustainability (Basel), 12*(1), 1–17. doi:10.3390u12010205

Pushparaj, K., Liu, W.-C., Meyyazhagan, A., Orlacchio, A., Pappusamy, M., Vadivalagan, C., Robert, A. A., Arumugam, V. A., Kamyab, H., Klemeš, J. J., Khademi, T., Mesbah, M., Chelliapan, S., & Balasubramanian, B. (2022). Nano- from nature to nurture: A comprehensive review on facets, trends, perspectives and sustainability of nanotechnology in the food sector. *Energy, 240*, 122732–122732. doi:10.1016/j.energy.2021.122732

Quazi, B., Hassan, K. M., & Haque, E. M. (2012). *Solid waste recycling in Rajshahi city of Bangladesh.* ELSIEVER Journal., doi:10.1016/j.wasman.2012.05.036

Qureshi, M. I., Qayyum, S., Nassani, A. A., Aldakhil, A. M., Qazi Abro, M. M., & Zaman, K. (2019). Management of various socio-economic factors under the United Nations sustainable development agenda. *Resources Policy, 64*(October), 101515. doi:10.1016/j.resourpol.2019.101515

Qu, X., Alvarez, P. J. J., & Li, Q. (2013). Applications of nanotechnology in water and wastewater treatment. *Water Research, 47*(12), 3931–3946. doi:10.1016/j.watres.2012.09.058 PMID:23571110

Qu, Y., Wang, W., Liu, Y., & Zhu, Q. (2019). Understanding residents' preferences for e-waste collection in China - A case study of waste mobile phones. *Journal of Cleaner Production, 228*, 52–62. doi:10.1016/j.jclepro.2019.04.216

Rabanimotlagh, A., & … . (2014) 'Healthcare facilities: Planning for increasing extreme heat events', *IIE Annual Conference and Expo 2014*, pp. 1810–1816.

Rada, E. C., Ragazzi, M., & Fedrizzi, P. (2013). Web-GIS oriented systems viability for municipal solid waste selective collection optimization in developed and transient economies. *Waste Management (New York, N.Y.), 33*(4), 785–792. doi:10.1016/j.wasman.2013.01.002 PMID:23402896

Radwan, E., Alattar, E., Radwan, A., Radwan, W., Alajez, M., Pandey, D., & Jebreen, K. (2023). Knowledge, attitudes, and practices of palestinian students toward COVID-19: a cross-sectional study during the first wave of the COVID-19 pandemic. *Human Arenas*, 1-22. NIH.

Rahman, M. M., & Jakia, T. (2015). *Solid Waste Management of Rajshahi City in Bangladesh and Its Impacts on Human Health and Environment*. NTUA.

Rajkovic, S., Bornhöft, N. A., van der Weijden, R., Nowack, B., & Adam, V. (2020). Dynamic probabilistic material flow analysis of engineered nanomaterials in European waste treatment systems. *Waste Management (New York, N.Y.), 113*, 118–131. doi:10.1016/j.wasman.2020.05.032 PMID:32531660

Rajpurohit, S. S., & Khan, T. (2018). A Review Paper on Blockchain and Cryptocurrency Based Transaction Systems. *International Journal on Latest Technology in Engineering, management, and Applied Sciences, 7*(3), 241–242.

Raman, N., & Sathiya,, D. (2008). Impact of solid waste effect on ground water and soil quality nearer to pallavaram solid waste landfill site in chennai. *Rasayan Journal of Chemistry, 1*(4), 828–836.

Ramdas, B. Y., & Galande, S. G. (2014). Green Growth Management by Using Arm Controller. *International Journal of Engineering Research and Applications, 4*(3), 360–363.

Ramirez, B., Oetjen, R. M., & Malvey, D. (2011). Sustainability and the health care manager: Part i. *The Health Care Manager, 30*(2), 133–138. doi:10.1097/HCM.0b013e318216f4e5 PMID:21537134

Ramirez, B., West, D. J., & Costell, M. M. (2013). Development of a culture of sustainability in health care organizations. *Journal of Health Organization and Management, 27*(5), 665–672. doi:10.1108/JHOM-11-2012-0226 PMID:24341182

Rana, R., Ganguly, R., & Gupta, A. (2014). *Solid Waste Management in Chandigarh – A Case Study*. Research Gate.

Rao, K. C., Parthan, B., & Doshi, K. (2018). *Power from municipal solid waste at Pune Municipal Corporation (Pune, Maharashtra, India)*. CGIAR.

Rapport, J., Zhang, R., Jenkins, B.M., & Willams, R.B. (2008). *Current anaerobic digestion technologies used for treatment of municipal organic solid waste*. Semantic Scholar.

Rarotra, S., Sahu, S., Kumar, P., Kim, K. H., Tsang, Y. F., Kumar, V., Kumar, P., Srinivasan, M., Veksha, A., & Lisak, G. (2020). Progress and challenges on battery waste management: A critical review. *ChemistrySelect, 5*(20), 6182–6193. doi:10.1002lct.202000618

Rasheed, R., Rizwan, A., Javed, H., Sharif, F., Yasar, A., Tabinda, A. B., Mahfooz, Y., Ahmed, S. R., & Su, Y. (2022). Analysis of environmental sustainability of e-waste in developing countries—A case study from Pakistan. *Environmental Science and Pollution Research International, 29*(24), 36721–36739. doi:10.100711356-022-18691-4 PMID:35064515

Rautela, R., Arya, S., Vishwakarma, S., Lee, J., Kim, K. H., & Kumar, S. (2021). E-waste management and its effects on the environment and human health. *The Science of the Total Environment, 773*, 145623. doi:10.1016/j.scitotenv.2021.145623 PMID:33592459

Raven, R. P., & Gregersen, K. H. (2007). Biogas plants in Denmark: Successes and setbacks. *Renewable & Sustainable Energy Reviews, 11*(1), 116–132. doi:10.1016/j.rser.2004.12.002

Ravila, A., & Holm, E. (1994). Radioactive elements in the forest industry. *The Science of the Total Environment, 157*, 339–356. doi:10.1016/0048-9697(94)90598-3

Ravindra, K., & Mor, S. (2019). E-waste generation and management practices in Chandigarh, India and economic evaluation for sustainable recycling. *Journal of Cleaner Production, 221*, 286–294. doi:10.1016/j.jclepro.2019.02.158

Raviteja, K., & Supriya, M. (2020). Greenhouse Monitoring System Based on Internet of Things. *Lecture Notes in Electrical Engineering, 637*, 581–591. doi:10.1007/978-981-15-2612-1_56

Ray, S. and Ray, B. (2018) 'Business Group Affiliation and Corporate Sustainability Strategies of Firms : An Investigation of Firms in India', *Journal of Business Ethics*. Springer Netherlands, 0(0), p. 0. . doi:10.1007/s10551-018-3917-z

Reddy, P. J. (2011). Municipal solid waste management. The Netherlands: CRC Press/Balkema..

Reddy, M. A., Reddy, B. M., Mukund, C. S., Venneti, K., Preethi, D. M. D., & Boopathi, S. (2023). Social Health Protection During the COVID-Pandemic Using IoT. In *The COVID-19 Pandemic and the Digitalization of Diplomacy* (pp. 204–235). IGI Global. doi:10.4018/978-1-7998-8394-4.ch009

Reshadi, M. A. M., Bazargan, A., & McKay, G. (2020). A review of the application of adsorbents for landfill leachate treatment: Focus on magnetic adsorption. *The Science of the Total Environment*, *731*, 138863–138863. doi:10.1016/j.scitotenv.2020.138863 PMID:32446150

Reza, M. (2020b). Relationship with Industry 4.0 and Supply Chain Management System of FMCG. *Journal of Business Management Studies.*

Reza, M. (2020a). Relationship with Industry 4.0 and Supply Chain Management System of FMCG. *Journal of Business and Management Studies*, *2*(2), 24–31.

Riboni, D., Bettini, C., Civitarese, G., Janjua, Z. H., & Helaoui, R. (2016). 'Artificial Intelligence in Medicine SmartFABER : Recognizing fine-grained abnormal behaviors for early detection of mild cognitive impairment', *Artificial Intelligence In Medicine. Artificial Intelligence in Medicine*, *67*, 57–74. doi:10.1016/j.artmed.2015.12.001 PMID:26809483

Richardson, E. A., & Mitchell, R. (2010). Gender differences in relationships between urban green space and health in the United Kingdom. *Social Science & Medicine*, *71*(3), 568–575. doi:10.1016/j.socscimed.2010.04.015 PMID:20621750

Rizal, N. N. M., & Hayder, G. (2022, November 19). Forecasting Effluent Biochemical Oxygen Demand in Sewage Treatment Plants Using Machine Learning and User-Friendly Interface. *International Journal of Environmental Research*, *17*(1), 4. doi:10.100741742-022-00493-8

Robinson, S., Williams, I., Dickinson, H., Freeman, T., & Rumbold, B. (2012). Social Science & Medicine Priority-setting and rationing in healthcare : Evidence from the English experience. *Social Science & Medicine*, *75*(12), 2386–2393. doi:10.1016/j.socscimed.2012.09.014 PMID:23083894

Robinson, T. (2007). Membrane bioreactors: Nanotechnology improves landfill leachate quality. *Filtration & Separation*, *44*(9), 38–39. doi:10.1016/S0015-1882(07)70288-4

Robles, A., Roxana, S., & Hinostroza, M. L. (2015). *Guidebook for the development of Nationally Appropriate Mitigation Actions (NAMAs).* CORE.

Rodić, L., & Wilson, D. (2017). Resolving governance issues to achieve priority sustainable development goals related to solid waste management in developing countries. *Sustainability (Basel)*, *9*(3), 404. doi:10.3390u9030404

Rodolfi, L., Chini Zittelli, G., Bassi, N., Padovani, G., Biondi, N., Bonini, G., & Tredici, M. R. (2009). Microalgae for oil: Strain selection, induction of lipid synthesis, and outdoor mass cultivation in a low-cost photobioreactor. *Biotechnology and Bioengineering*, *102*(1), 100–112. doi:10.1002/bit.22033 PMID:18683258

Rodriguez, R., Svensson, G., & Eriksson, D. (2021). Priorities determining future directions of sustainable development in business models of the healthcare industry—Findings and framework. *Sustainability (Basel)*, *13*(11), 1–11. doi:10.3390u13116507

Rodríguez-Rángel, H., Arias, D. M., Morales-Rosales, L. A., Gonzalez-Huitron, V., Valenzuela Partida, M., & García, J. (2022, March 29). Machine Learning Methods Modeling Carbohydrate-Enriched Cyanobacteria Biomass Production in Wastewater Treatment Systems. *Energies*, *15*(7), 2500. doi:10.3390/en15072500

Rosenzweig, C., Solecki, W. D., Romero-Lankao, P., Mehrotra, S., Dhakal, S., & Ibrahim, S. A. (2018). *Climate change and cities: Second assessment report of the urban climate change research network.* Cambridge University Press. doi:10.1017/9781316563878

Roubík, H., Mazancová, J., Le Dinh, P., Dinh Van, D., & Banout, J. (2018). Biogas quality across small-scale biogas plants: A case of central Vietnam. *Energies*, *11*(7), 1794. doi:10.3390/en11071794

Ruggeri, K., & … . (2020). 'Well-being is more than happiness and life satisfaction: A multidimensional analysis of 21 countries. Health and Quality of Life OutcomesHealth and Quality of Life Outcomes [revista en Internet] 2020 [acceso 4 de julio de 2021]; 18(1): 1-16.', *Health and Quality of Life Outcomes. Health and Quality of Life Outcomes*, *1*(2), 1–16. https://hqlo.biomedcentral.com/track/pdf/10.1186/s12955-020-01423-y.pdf

S., P. K., Sampath, B., R., S. K., Babu, B. H., & N., A. (2022). Hydroponics, Aeroponics, and Aquaponics Technologies in Modern Agricultural Cultivation. In *Trends, Paradigms, and Advances in Mechatronics Engineering* (pp. 223–241). IGI Global. doi:10.4018/978-1-6684-5887-7.ch012

Safaei, M., Taran, M., & Imani, M. M. (2019). Preparation, structural characterization, thermal properties and antifungal activity of alginate-CuO bionanocomposite. *Materials Science and Engineering C*, *101*, 323–329. doi:10.1016/j.msec.2019.03.108 PMID:31029325

Safari, A., Das, N., Langhelle, O., Roy, J., & Assadi, M. (2019). Natural gas: A transition fuel for sustainable energy system transformation? *Energy Science & Engineering*, *7*(4), 1075–1094. doi:10.1002/ese3.380

Saha1, B. C., R, D., A, A., Thrinath, B. V. S., Boopathi, S., J. R., & Sudhakar, M. (2022). *IOT Based Smart Energy Meter For Smart Grid.* MDPI.

Saha, H. N., Auddy, S., Pal, S., Kumar, S., Pandey, S., Singh, R., Singh, A. K., Banerjee, S., Ghosh, D., & Saha, S. (2017). Waste management using Internet of Things (IoT). In *8th annual industrial automation and electromechanicalengineering conference*, Bangkok, Thailand. 10.1109/IEMECON.2017.8079623

Sahoo, S. et al. (2022) *Blockchain for sustainable supply chain management: trends and ways forward, Electronic Commerce Research*. Springer US. . doi:10.1007/s10660-022-09569-1

Sahoo, N. G., Rana, S., Cho, J. W., Li, L., & Chan, S. H. (2010). Polymer nanocomposites based on functionalized carbon nanotubes. *Progress in Polymer Science*, *35*(7), 837–867. doi:10.1016/j.progpolymsci.2010.03.002

Saini, Ritesh, & Siddiqui, Nihal & Bahukhandi, kanchan Deoli & J.Ahuja, Neelu. (2018). Study of effects of Municipal solid waste on water quality of Dehradun City. *Journal of Environmental Science and Health, Part A: Environmental Science and Engineering*, *7*, 409–416.

Salam, M. A. (2019). Analyzing Manufacturing Strategies and Industry 4.0 Supplier Performance Relationships from a Resource-Based Perspective. *Benchmarking*. doi:10.1108/BIJ-12-2018-0428

Salehi, M. (2022, January). Global water shortage and potable water safety; Today's concern and tomorrow's crisis. *Environment International*, *158*, 106936. doi:10.1016/j.envint.2021.106936 PMID:34655888

Saleh, T. A., Mustaqeem, M., & Khaled, M. (2022, May). Water treatment technologies in removing heavy metal ions from wastewater: A review. *Environmental Nanotechnology, Monitoring & Management*, *17*, 100617. doi:10.1016/j.enmm.2021.100617

Salem, M. A., Shawtari, F. A., Shamsudin, M. F., & Hussain, H. I. (2016). 'The relation between stakeholders ' integration and environmental competitiveness'. *Social Responsibility Journal*, *12*(4), 755–768. doi:10.1108/SRJ-12-2015-0189

Salgot, M., & Folch, M. (2018, April). Wastewater treatment and water reuse. *Current Opinion in Environmental Science & Health*, *2*, 64–74. doi:10.1016/j.coesh.2018.03.005

Salisu, A. A., Ndako, U. B., & Vo, X. V. (2023). Transition risk, physical risk, and the realized volatility of oil and natural gas prices. *Resources Policy*, *81*, 103383. doi:10.1016/j.resourpol.2023.103383

Samans, R. (2019). *Globalization 4.0: Shaping a New Global Architecture in the Age of the Fourth Industrial Revolution*. World Economic Forum. https://www3.weforum.org/docs/WEF_Globalization_4.0_Call_for_Engagement.pdf

Sami, K., & Edith, N. (2019). Real-time air pollution monitoring with sensors on city bus. *Digital Communications and Networks*.

Samikannu, R., Koshariya, A. K., Poornima, E., Ramesh, S., Kumar, A., & Boopathi, S. (2022). Sustainable Development in Modern Aquaponics Cultivation Systems Using IoT Technologies. In *Human Agro-Energy Optimization for Business and Industry* (pp. 105–127). IGI Global.

Sampath, B. C. S., & Myilsamy, S. (2022). Application of TOPSIS Optimization Technique in the Micro-Machining Process. In Trends, Paradigms, and Advances in Mechatronics Engineering (pp. 162–187). IGI Global. doi:10.4018/978-1-6684-5887-7.ch009

Sampath, B., & Myilsamy, S. (2021). Experimental investigation of a cryogenically cooled oxygen-mist near-dry wire-cut electrical discharge machining process. *Strojniski Vestnik. Jixie Gongcheng Xuebao*, *67*(6), 322–330. doi:10.5545v-jme.2021.7161

Sampath, B., Naveenkumar, N., Sampathkumar, P., Silambarasan, P., Venkadesh, A., & Sakthivel, M. (2021). Experimental comparative study of banana fiber composite with glass fiber composite material using Taguchi method. *Materials Today: Proceedings*, *49*(5), 1475–1480. doi:10.1016/j.matpr.2021.07.232

Sampath, B., Pandian, M., Deepa, D., & Subbiah, R. (2022). Operating parameters prediction of liquefied petroleum gas refrigerator using simulated annealing algorithm. *AIP Conference Proceedings*, *2460*(1), 070003. doi:10.1063/5.0095601

Sanchez, O. J., & Cardona, C. A. (2008). Trends in biotechnological production of fuel ethanol from different feedstocks. *Bioresource Technology*, *99*(13), 5270–5295. doi:10.1016/j.biortech.2007.11.013 PMID:18158236

Sandhu, K. (2014). Historical trajectory of waste management; an analysis using the health belief model. *Management of Environmental Quality*, *25*(5), 615–630. doi:10.1108/MEQ-09-2013-0098

Santoso, L., Singh, B., Rajest, S., Regin, R., & Kadhim, K. (2018, July 13). A Genetic Programming Approach to Binary Classification Problem. *EAI Endorsed Transactions on Energy Web*, *165523*, 165523. doi:10.4108/eai.13-7-2018.165523

Saoji, A. (2012). E-WASTE MANAGEMENT: AN EMERGING ENVIRONMENTAL AND HEALTH ISSUE. *National Journal of Medical Research*, *2*(01), 107–110.

Sarath, P., Bonda, S., Mohanty, S., & Nayak, S. K. (2017). Identification and thermomechanical characterization of polymers recovered from mobile phone waste. *Journal of Material Cycles and Waste Management*, *19*(4), 1391–1399. doi:10.100710163-016-0525-9

Saravanamurthy, D., E, R. R. C. V, Sarangan, B., & Bharathkumar, R. (2018). *E-Waste Management, 6*(02).

Saravanan, M., Vasanth, M., Boopathi, S., Sureshkumar, M., & Haribalaji, V. (2022). Optimization of Quench Polish Quench (QPQ) Coating Process Using Taguchi Method. *Key Engineering Materials*, *935*, 83–91. doi:10.4028/p-z569vy

Saur, G., & Milbrandt, A. (2014). Renewable hydrogen potential from biogas in the United States (No. NREL/TP-5400-60283). National Renewable Energy Lab (NREL).

Sayed, E. T., Rezk, H., Abdelkareem, M. A., & Olabi, A. (2023, January). Artificial neural network based modelling and optimization of microalgae microbial fuel cell. *International Journal of Hydrogen Energy*. doi:10.1016/j.ijhydene.2022.12.081

Scanlon, N. L. (2007). An analysis and assessment of environmental operating practices in hotel and resortproperties. *International Journal of Hospitality Management*, *26*(3), 711–723. doi:10.1016/j.ijhm.2006.07.003

Schafer, B. (2014). D-waste: Data disposal as challenge for waste management in the Internet of Things. *International Journal of Information Ethics*, *22*, 101–107. doi:10.29173/irie131

Scheirer, M. A. (2005). Is sustainability possible? A review and commentary on empirical studies of program sustainability. *The American Journal of Evaluation*, *26*(3), 320–347. doi:10.1177/1098214005278752

Schneider, P., Castell, N., Vogt, M., Dauge, F. R., Lahoz, W. A., & Bartonova, A. (2017). Mapping urban air quality in near real-time using observations from low-cost sensors and model information. *Environment International*, *106*, 234–247. doi:10.1016/j.envint.2017.05.005 PMID:28668173

Schoolman, E. D., Shriberg, M., Schwimmer, S., & Tysman, M. (2016). Green cities and ivory towers: How do higher education sustainability initiatives shape millennials' consumption practices? *Journal of Environmental Studies and Sciences*, *6*(3), 490–502. doi:10.100713412-014-0190-z

Scott, S. A., Davey, M. P., Dennis, J. S., Horst, I., Howe, C. J., Lea-Smith, D. J., & Smith, A. G. (2010). Biodiesel from algae: Challenges and prospects. *Current Opinion in Biotechnology*, *21*(3), 277–286. doi:10.1016/j.copbio.2010.03.005 PMID:20399634

Selvakumar, S., Adithe, S., Isaac, J. S., Pradhan, R., Venkatesh, V., & Sampath, B. (2023). A Study of the Printed Circuit Board (PCB) E-Waste Recycling Process. In Sustainable Approaches and Strategies for E-Waste Management and Utilization (pp. 159–184). IGI Global.

Sen, M. (2020). Nanocomposite materials. *Nanotechnology and the Environment*, 1-12. InTech Open.

Sengupta, D., Ilankoon, I. M. S. K., Dean Kang, K., & Nan Chong, M. (2022). Circular economy and household e-waste management in India: Integration of formal and informal sectors. *Minerals Engineering*, *184*(May), 107661. doi:10.1016/j.mineng.2022.107661

Senthil, T. S. R. Ohmsakthi vel, Puviyarasan, M., Babu, S. R., Surakasi, R., & Sampath, B. (2023). Industrial Robot-Integrated Fused Deposition Modelling for the 3D Printing Process. In Development, Properties, and Industrial Applications of 3D Printed Polymer Composites (pp. 188–210). IGI Global. doi:10.4018/978-1-6684-6009-2.ch011

Senthil, T. S., Puviyarasan, M., Babu, S. R., Surakasi, R., & Sampath, B. (2023). Industrial Robot-Integrated Fused Deposition Modelling for the 3D Printing Process. In Development, Properties, and Industrial Applications of 3D Printed Polymer Composites (pp. 188–210). IGI Global

Senthil, T. S., Puviyarasan, M., Babu, S. R., Surakasi, R., Sampath, B., & Associates. (2023). Industrial Robot-Integrated Fused Deposition Modelling for the 3D Printing Process. In Development, Properties, and Industrial Applications of 3D Printed Polymer Composites (pp. 188–210). IGI Global.

Shah, A. V., Srivastava, V. K., Mohanty, S. S., & Varjani, S. (2021). Municipal solid waste as a sustainable resource for energy production: State-of-the-art review. *Journal of Environmental Chemical Engineering*, *9*(4), 105717. doi:10.1016/j.jece.2021.105717

Shah, S., Mehta, M., & Mukherjee, M. D. (2001). *Occupational Health Hazards Encountered at Health Care Facility and Medical College in India*. American Industrial Hygiene Association. doi:10.3320/1.2765786

Shahzad, K., Abbas, F., Pandey, D., Ajmal, S., Khadim, M., & Tahir, M. U. (2020). Synthesis, characterization and biological evaluation of novel tetrasubsituted Imidazole compounds. *ChemRxiV.*

Shalender, K., & Yadav, R. K. (2019). Strategic Flexibility, Manager Personality, and Firm Performance: The Case of Indian Automobile Industry. *Global Journal of Flexible Systems Managment*, *20*(1), 77–90. doi:10.100740171-018-0204-x

Shankar, Y. S., & Khandelwal, R. (2017). Sustainable waste management strategy for a campus: A case study of JUET, Guna. *Management of Environmental Quality*, *28*(5), 6. doi:10.1108/MEQ-01-2016-0008

Sharholy, M., Ahmad, K., Mahmood, G., & Trivedi, R. C. (2008). Municipal solid waste management in Indian cities–A review. *Waste Management (New York, N.Y.)*, *28*(2), 459–467. doi:10.1016/j.wasman.2007.02.008 PMID:17433664

Sharma, B. K., & Chandel, M. K. (2021). Life cycle cost analysis of municipal solid waste management scenarios for Mumbai, India. *Waste Management (New York, N.Y.)*, *124*, 293–302. doi:10.1016/j.wasman.2021.02.002 PMID:33640669

Sharma, H. B., Vanapalli, K. R., Samal, B., Cheela, V. R. S., Dubey, B. K., & Bhattacharya, J. (2021). Circular economy approach in solid waste management system to achieve UN-SDGs: Solutions for post-COVID recovery. *The Science of the Total Environment*, *800*, 149605. doi:10.1016/j.scitotenv.2021.149605 PMID:34426367

Sharma, M., Joshi, S., Kannan, D., Govindan, K., Singh, R., & Purohit, H. C. (2020). Internet of Things (IoT) adoption barriers of smart cities' waste management: An Indian context. *Journal of Cleaner Production*, *270*, 122047. doi:10.1016/j.jclepro.2020.122047

Sharma, M., Sharma, B., Gupta, A. K., & Pandey, D. (2023). Recent developments of image processing to improve explosive detection methodologies and spectroscopic imaging techniques for explosive and drug detection. *Multimedia Tools and Applications*, *82*(5), 6849–6865. doi:10.100711042-022-13578-5

Sharon, M., Mukhopadhyay, K., Yase, K., Iijima, S., Ando, Y., & Zhao, X. (1998). Spongy carbon nanobeads—A new material. *Carbon*, *36*(5-6), 507–511. doi:10.1016/S0008-6223(98)00060-8

Sharon, M., Pradhan, D., Ando, Y., & Zhao, X. (2002). Electrical and optical properties of semiconducting Camphoric Carbon Films. *Current Applied Physics*, *2*(6), 445–450. doi:10.1016/S1567-1739(02)00154-2

Shazwin, T. M., & Nakagoshi, N. (2010). Sustainable waste management through international cooperation: Review of comprehensive waste management technique and training course. *J. Int. Dev. Coop.*, *16*(1), 23–33.

Shediac-rizkallah, M. C., & Bone, L. R. (1998). Planning for the sustainability of community-based health programs : Conceptual frameworks and future directions for research, practice and policy. *Health Education Research*, *13*(1), 87–108. doi:10.1093/her/13.1.87 PMID:10178339

Sheet, I., Kabbani, A., & Holail, H. (2014). Removal of heavy metals using nanostructured graphite oxide, silica nanoparticles and silica/graphite oxide composite. *Energy Procedia*, *50*, 130–138. doi:10.1016/j.egypro.2014.06.016

Shekdar, A. V. (2009). Sustainable solid waste management: An integrated approach for Asian countries. *Waste Management (New York, N.Y.)*, *29*(4), 1438–1448. doi:10.1016/j.wasman.2008.08.025 PMID:19081236

Shen, Y., Linville, J. L., Urgun-Demirtas, M., Mintz, M. M., & Snyder, S. W. (2015). An overview of biogas production and utilization at full-scale wastewater treatment plants (WWTPs) in the United States: Challenges and opportunities towards energy-neutral WWTPs. *Renewable & Sustainable Energy Reviews*, *50*, 346–362. doi:10.1016/j.rser.2015.04.129

Sherazi, H. H. R., Khan, Z. A., Iqbal, R., Rizwan, S., Imran, M. A., & Awan, K. (2019). A heterogeneous IoV architecture for data forwarding in vehicle to infrastructure communication. *Mobile Information Systems*, *2019*, 2019. doi:10.1155/2019/3101276

Sheth, A., ... (2008) 'Reviewing the sustainability of existing healthcare facilities', Loughborough university, pp. 1193–1202.

Shittu, O. S., Williams, I. D., & Shaw, P. J. (2021). Global E-waste management: Can WEEE make a difference? A review of e-waste trends, legislation, contemporary issues and future challenges. *Waste Management (New York, N.Y.)*, *120*, 549–563. doi:10.1016/j.wasman.2020.10.016 PMID:33308953

Shi, W. Q., Yuan, L. Y., Li, Z. J., Lan, J. H., Zhao, Y. L., & Chai, Z. F. (2012). Nanomaterials and nanotechnologies in nuclear energy chemistry. *Radiochimica Acta*, *100*(8-9), 727–736. doi:10.1524/ract.2012.1961

Shukla, R. (2016). Smart waste management. *Asian Journal of Pharm Educational Research*, *5*(2), 48–54.

Shumon, M. R. H., Ahmed, S., & Islam, M. T. (2014). Electronic waste: Present status and future perspectives of sustainable management practices in Malaysia. *Environmental Earth Sciences*, *72*(7), 2239–2249. doi:10.100712665-014-3129-5

Sichiweza, E. (2017). Participation of households in solid waste management and circular economy towards sustainability: a case study of Kabwe Town, Central Province of Zambia [Master's thesis, University of Twente].

Silva, A., Lebron, Y., Moreira, V., Ribeiro, L., Koch, K., & Amaral, M. (2023, March). High-retention membrane bioreactors for sugarcane vinasse treatment: Opportunities for environmental impact reduction and wastewater valorization. *Journal of Environmental Management*, *329*, 117001. doi:10.1016/j.jenvman.2022.117001 PMID:36565496

Silva, C. E., Hoppe, A. E., Ravanello, M. M., & Mello, N. (2005). Medical waste management in the south of Brazil. *Waste Management (New York, N.Y.)*, *25*(6), 600–605. doi:10.1016/j.wasman.2004.03.002 PMID:15993344

Singh, D., Chaturvedi, P., Dhawan, G., Singh, A., & Aggarwal, R. (2022, February). WBMS: Waste Bin Management System for densely populated urban areas. In *2022 2nd International Conference on Innovative Practices in Technology and Management (ICIPTM)* (*Vol. 2*, pp. 770-774). IEEE.

Singh, H., Pandey, B. K., George, S., Pandey, D., Anand, R., Sindhwani, N., & Dadheech, P. (2022, July). Effective Overview of Different ML Models Used for Prediction of COVID-19 Patients. In *Artificial Intelligence on Medical Data: Proceedings of International Symposium, ISCMM 2021* (pp. 185-192). Singapore: Springer Nature Singapore.

Singh, I.B., and Sarma, R.K. (1996) Hospital Waste Disposal System and Technology. *Journal of Academy of Hospital Administration, 8*(2), pp. 44-48.

Singh, K. R., & Singh, R. P. (2021). *Utility of nanobiosensors in environmental analysis and monitoring*. Research Gate.

Singh, S., Chakma, S., Alawa, B., Kalyanasundaram, M., & Diwan, V. (2023). Hazardous Materials Advances Identification, characterization, and implications of microplastics in soil – A case study of Bhopal, central India. *Journal of Hazardous Materials Advances, 9*, 100225. doi:10.1016/j.hazadv.2022.100225

Singh, A. (2022). Indicators and ICTs application for municipal waste management. *Waste Management & Research*, *40*(1), 24–33. doi:10.1177/0734242X211010367 PMID:33836633

Singhal, S., Deepak, A., & Marwaha, V. (2018). Green Initiatives Practices in Indian Hotels. *IOSR Journal of Business and Management*, *20*(8), 10–13.

Singh, N., Duan, H., Yin, F., Song, Q., & Li, J. (2018). Characterizing the Materials Composition and Recovery Potential from Waste Mobile Phones: A Comparative Evaluation of Cellular and Smart Phones. *ACS Sustainable Chemistry & Engineering*, *6*(10), 13016–13024. doi:10.1021/acssuschemeng.8b02516

Singh, S. P., & Singh, D. (2010). Biodiesel production through the use of different sources and characterization of oils and their esters as the substitute of diesel. *Renewable & Sustainable Energy Reviews*, *14*(1), 200–216. doi:10.1016/j.rser.2009.07.017

Singh, V. V. (2022). Green nanotechnology for environmental remediation. In *Sustainable Nanotechnology for Environmental Remediation* (pp. 31–61). Elsevier. doi:10.1016/B978-0-12-824547-7.00017-5

Sitharaman, B. (2016). *Nanobiomaterials handbook*. CRC Press. doi:10.1201/b10970

Sit, M., Demiray, B. Z., Xiang, Z., Ewing, G. J., Sermet, Y., & Demir, I. (2020, August 5). A comprehensive review of deep learning applications in hydrology and water resources. *Water Science and Technology*, *82*(12), 2635–2670. doi:10.2166/wst.2020.369 PMID:33341760

Sivaramanan, S. (2013). E-Waste Management, Disposal and Its Impacts on the Environment. *Universal Journal of Environmental Research and Technology*, *3*(5), 531–537. Www.Environmentaljournal.Org. doi:10.13140/2.1.2978.0489

Skoulidas, A. I., Sholl, D. S., & Johnson, J. K. (2006). Adsorption and diffusion of carbon dioxide and nitrogen through single-walled carbon nanotube membranes. *The Journal of Chemical Physics*, *124*(5), 054708–054708. doi:10.1063/1.2151173 PMID:16468902

Smetana, G., & Grosser, A. (2023, January 3). The Oxygenic Photogranules—Current Progress on the Technology and Perspectives in Wastewater Treatment: A Review. *Energies*, *16*(1), 523. doi:10.3390/en16010523

Smith. (2010). What is Solid Waste Management? *Greek.com* greek.com/ what-is-solid-waste-managment.htm)

Smit, S., Musango, J. K., Kovacic, Z., & Brent, A. C. (2017). Conceptualising slum in an urban African context. *Cities (London, England)*, *62*, 107–119. doi:10.1016/j.cities.2016.12.018

Sobral, T., Galvão, T., & Borges, J. (2019). Visualization of urban mobility data from intelligent transportation systems. *Sensors (Basel)*, *19*(2), 332. doi:10.339019020332 PMID:30650641

Solano, R., Patiño-Ruiz, D., Tejeda-Benitez, L., & Herrera, A. (2021). Metal- and metal/oxide-based engineered nanoparticles and nanostructures: A review on the applications, nanotoxicological effects, and risk control strategies. *Environmental Science and Pollution Research International*, *28*(14), 16962–16981. doi:10.100711356-021-12996-6 PMID:33638785

Somani, P., Navaneethan, R. D., & Thangaiyan, S. (2021, May). Integrated solid waste management in urban India: A mini review. [). IOP Publishing.]. *Journal of Physics: Conference Series*, *1913*(1), 012084. doi:10.1088/1742-6596/1913/1/012084

Song, W., & Yu, H. (2018). Green innovation strategy and green innovation: The roles of green creativity and green organizational identity. *Corporate Social Responsibility and Environmental Management*, *25*(2), 135–150. doi:10.1002/csr.1445

Sovacool, B. K. (2009). Energy Policy and Cooperation in Southeast Asia: The History, Challenges, and Implications of the Trans-ASEAN Gas Pipeline (TAGP) Network. *Energy Policy*, *37*(6), 2356–2367. doi:10.1016/j.enpol.2009.02.014

Spellman, F. R. (2018). *The Science of Water*. CRC Press. doi:10.1201/b17484

Spitalsky, Z., Tasis, D., Papagelis, K., & Galiotis, C. (2010). Carbon Nanotube–Polymer Composites: Chemistry, processing, mechanical and electrical properties. *Progress in Polymer Science*, *35*(3), 357–401. doi:10.1016/j.progpolymsci.2009.09.003

Staley, B. F., & Barlaz, M. A. (2009). Composition of Municipal Solid Waste in the United States and Implications for Carbon Sequestration and Methane Yield. *Journal of Environmental Engineering*, *135*(10), 901–909. doi:10.1061/(ASCE)EE.1943-7870.0000032

Štangar, U. L., & Kranjčevič, E. (2008). *Renewable energy sources in slovenia: facts and plans*. Springer Netherlands.

Stankovic, M., Gupta, R., & Figueroa, J. (2017). *Industry 4.0 - Opportunities behind the Challenge. UNIDO Background Paper*, 1–56. United Nations Industrial Development Organization. https://www.unido.org/sites/default/files/files/2017-11/UNIDO Background Paper on Industry 4.0_27112017.pdf

Stephenson, A. L., Dennis, J. S., Howe, C. J., Scott, S. A., & Smith, A. G. (2010). Influence of nitrogen-limitation regime on the Production by Chlorella vulgaris of lipids for biodiesel feedstocks. *Biofuels*, *1*(1), 47–58. doi:10.4155/bfs.09.1

Stephenson, A. L., Dennis, J. S., & Scott, S. A. (2008). Improving the sustainability of the production of biodiesel from oilseed rape in the UK. *Process Safety and Environmental Protection*, *86*(6), 427–440. doi:10.1016/j.psep.2008.06.005

Stirman, S. W., & … . (2012). *The sustainability of new programs and innovations : a review of the empirical literature and recommendations for future research*. Implmentation Science.

Stock, T., & Seliger, G. (2016). Opportunities of Sustainable Manufacturing in Industry 4.0. *Procedia CIRP 40* (Icc): 536–41. . doi:10.1016/j.procir.2016.01.129

Subha, S., Inbamalar, T. M., R, K. C., Suresh, L. R., Boopathi, S., & Alaskar, K. (2023, February). A Remote Health Care Monitoring system using internet of medical things ({IoMT}). *IEEE Explore*. doi:10.1109/ICIPTM57143.2023.10118103

Subramanian, L., Thomas, S., & Koshy, O. (2017). Green synthesis of silver nanoparticles using aqueous plant extracts and its application as optical sensor. *International Journal of Biosensors & Bioelectronics*, *2*(3), 82–85.

Sugania, S. S., Vishalishwaran, D. S., & Kumar, J. V. (2018). Automated System for Road Safety Enhancement using Big. *Data Report*.

Sun, H. (2019). Grand Challenges in Environmental Nanotechnology. *Frontiers in Nanotechnology*, *1*, 2. doi:10.3389/fnano.2019.00002

Sunil, M., Narayan, A., Bhat, V., & Vinay, S. (2013). Smart biogas plant. *International Journal of Innovative Technology and Exploring Engineering*, *3*(3), 62–66.

Suresh, D. S., Jyothi Prakash, K. V., & Rajendra, C. J. (2013). Automated soil testing device. *ITSI Transactions on Electrical & Electronics Engineering*, (pp. 2320-8945). Semantic Scholar.

Swanson, K. J., Madden, M. C., & Ghio, A. J. (2007). Biodiesel exhaust: The need for health effects research. *Environmental Health Perspectives*, *115*(4), 496–499. doi:10.1289/ehp.9631 PMID:17450214

Tabatabaei, M., Aghbashlo, M., Valijanian, E., Panahi, H. K. S., Nizami, A. S., Ghanavati, H., & Karimi, K. (2020). A comprehensive review on recent biological innovations to improve biogas production, part 2: Mainstream and downstream strategies. *Renewable Energy*, *146*, 1392–1407. doi:10.1016/j.renene.2019.07.047

Tabatabaei, M., & Ghanavati, H. (Eds.). (2018). *Biogas: fundamentals, process, and operation* (Vol. 6). Springer. doi:10.1007/978-3-319-77335-3_6

Tałałaj, I. A., Biedka, P., & Bartkowska, I. (2019). Treatment of landfill leachates with biological pretreatments and reverse osmosis. *Environmental Chemistry Letters*, *17*(3), 1177–1193. doi:10.100710311-019-00860-6

Tan, J., Zhang, M., Li, H., Guo, W., & Wang, L. (2018). A Bidirectional Green Wave Band Method under Asymmetric Phase Sequence Mode based on Mobile Navigation data. *2018 IEEE 7th Data Driven Control and Learning Systems Conference (DDCLS)*, (pp. 1191–1196). IEEE.

Tania, F. (2014). Solid waste management of Dhaka City: A socio-economic analysis. *Banglavision*, *13*(1), 91–100.

Tan, Q., Yang, L., Wei, F., Chen, Y., & Li, J. (2023). Is reusable packaging an environmentally friendly alternative to the single-use plastic bag? A case study of express delivery packaging in China. *Resources, Conservation and Recycling*, *190*, 106863. doi:10.1016/j.resconrec.2022.106863

Tanwar, N., & Akhai, S. (2017). Survey Analysis for Quality Control Comfort Management in Air Conditioned Classroom. *Journal of Advanced Research in Civil and Environmental Engineering*, *4*(1&2), 20–23. doi:10.24321/2393.8307.201702

Taran, M., Safaei, M., Karimi, N., & Almasi, A. (2021). Benefits and application of nanotechnology in environmental science: An overview. *Biointerface Research in Applied Chemistry*, *11*(1), 7860–7870.

Tauseef, S. M., Premalatha, M., Abbasi, T., & Abbasi, S. A. (2013). Methane capture from livestock manure. *Journal of Environmental Management, 117*, 187–207. doi:10.1016/j.jenvman.2012.12.022 PMID:23376302

Tchounwou, P. B., Yedjou, C. G., Patlolla, A. K., & Sutton, D. J. (2012). Heavy metal toxicity and the environment. *Molecular, clinical and environmental toxicology: volume 3: environmental toxicology*, 133-164. NIH.

Tesfaye, F., Lindberg, D., Hamuyuni, J., Taskinen, P., & Hupa, L. (2017). Improving urban mining practices for optimal recovery of resources from E-waste. *Minerals Engineering, 111*, 209–221. doi:10.1016/j.mineng.2017.06.018

Thakare, S., & Nandi, S. (2016). Study on potential of gasification technology for municipal solid waste (MSW) in Pune city. *Energy Procedia, 90*, 509–517. doi:10.1016/j.egypro.2016.11.218

The Federal Institute for Geosciences and Natural Resources (BGR). (2011). *Energy Study 2012*. BGR. http://www.bgr.bund.de/DE/Gemeinsames/Produkte/Downloads/DERA_Rohstoffinformationen/rohstoffinformatio nen-15e.pdf?__blob=publicationFile&v=3\

Thiruselvi, D., Kumar, P. S., Kumar, M. A., Lay, C. H., Aathika, S., Mani, Y., Jagadiswary, D., Dhanasekaran, A., Shanmugam, P., Sivanesan, S., & Show, P. L. (2021). A critical review on global trends in biogas scenario with its up-gradation techniques for fuel cell and future perspectives. *International Journal of Hydrogen Energy, 46*(31), 16734–16750. doi:10.1016/j.ijhydene.2020.10.023

Thomas, P., & Soren, N. (2020). An overview of municipal solid waste-to-energy application in Indian scenario. *Environment, Development and Sustainability, 22*(2), 575–592. doi:10.100710668-018-0235-7

Tian, H., Wang, X., Lim, E. Y., Lee, J. T., Ee, A. W., Zhang, J., & Tong, Y. W. (2021). Life cycle assessment of food waste to energy and resources: Centralized and decentralized anaerobic digestion with different downstream biogas utilization. *Renewable & Sustainable Energy Reviews, 150*, 111489. doi:10.1016/j.rser.2021.111489

Tirabeni, L., De Bernardi, P., Forliano, C., & Franco, M. (2019). How can organisations and business models lead to a more sustainable society? A framework from a systematic review of the industry 4.0. *Sustainability (Basel), 11*(22), 6363. doi:10.3390u11226363

Tiruvalam, R. C., Pritchard, J. C., Dimitratos, N., Lopez-Sanchez, J. A., Edwards, J. K., Carley, A. F., Hutchings, G. J., & Kiely, C. J. (2011). Aberration corrected analytical electron microscopy studies of sol-immobilized Au + Pd, Au{Pd} and Pd{Au} catalysts used for benzyl alcohol oxidation and hydrogen peroxide production. *Faraday Discussions, 152*, 63–63. doi:10.1039/c1fd00020a PMID:22455039

Tjahjono, B., Esplugues, C., Ares, E., & Pelaez, G. (2017). What Does Industry 4.0 Mean to Supply Chain? *Procedia Manufacturing, 13*, 1175–1182. doi:10.1016/j.promfg.2017.09.191

Tofa, T. S., Kunjali, K. L., Paul, S., & Dutta, J. (2019). Visible light photocatalytic degradation of microplastic residues with zinc oxide nanorods. *Environmental Chemistry Letters, 17*(3), 1341–1346. doi:10.100710311-019-00859-z

Tomar, R. S., Jyoti, A., & Kaushik, S. (2020). Nanobiotechnology: Concepts and Applications in Health. *Agriculture, and Environment*.

Tomić, T., & Schneider, D. R. (2020). Circular economy in waste management – Socio-economic effect of changes in waste management system structure. *Journal of Environmental Management, 267*(April), 110564. doi:10.1016/j.jenvman.2020.110564 PMID:32421664

Tonelli, F. M. P., Goulart, V. A. M., Gomes, K. N., Ladeira, M. S., Santos, A. K., Lorençon, E., Ladeira, L. O., & Resende, R. R. (2015). Graphene-based nanomaterials: Biological and medical applications and toxicity. *Nanomedicine (London), 10*(15), 2423–2450. doi:10.2217/nnm.15.65 PMID:26244905

Tong, H., & Kang, J. (2021). Characteristics of noise complaints and the associations with urban morphology: A comparison across densities. *Environmental Research*, *197*(March), 111045. doi:10.1016/j.envres.2021.111045 PMID:33745930

Topmiller, J. L., & Dunn, K. H. (2013). *Current strategies for engineering controls in nanomaterial production and downstream handling processes*. Retrieved from Usman, M., Cheema, S. A., & Farooq, M. (2020). Heterogeneous Fenton and persulfate oxidation for treatment of landfill leachate: A review supplement. *Journal of Cleaner Production*, *256*, 120448–120448. doi:10.1016/j.jclepro.2020.120448

Torkayesh, A. E. et al. (2021) 'An integrated BWM-LBWA-CoCoSo framework for evaluation of healthcare sectors in Eastern Europe', *Socio-Economic Planning Sciences*. Elsevier Ltd, 78(September 2020), p. 101052. . doi:10.1016/j.seps.2021.101052

Traviss, N. (2012). Breathing easier? The known impacts of biodiesel on air quality. *Biofuels*, *3*(3), 285–291. doi:10.4155/bfs.12.22 PMID:23585814

Tripathi, S., & Gupta, M. (2020). A Framework for Procurement Process Re-Engineering in Industry 4.0. *Business Process Management Journal*, *27*(2), 439–458. doi:10.1108/BPMJ-07-2020-0321

Trojovský, P., Dhasarathan, V., & Boopathi, S. (2023). Experimental investigations on cryogenic friction-stir welding of similar ZE42 magnesium alloys. *Alexandria Engineering Journal*, *66*(1), 1–14. doi:10.1016/j.aej.2022.12.007

Tsui, T. H., & Wong, J. W. (2019). A critical review: Emerging bioeconomy and waste-to-energy technologies for sustainable municipal solid waste management. *Waste Disposal & Sustainable Energy*, *1*(3), 151–167. doi:10.100742768-019-00013-z

Umap, V. S., & Deshmukh, P. H. (n.d.). Intelligent Transportation System—Recent Trends in Transportation System. *RE:view*.

UNEP. (2005). *Solid Waste Management*. United Nations Environment Programme - Division of Technology, Industry, and Economics.

Upham, P., Thornley, P., Tomei, J., & Boucher, P. (2009). Substitutable biodiesel feedstocks for the UK: A review of sustainability issues with reference to the UK RTFO. *Journal of Cleaner Production*, *17*, S37–S45. doi:10.1016/j.jclepro.2009.04.014

US IEA. (2013). *Key World Energy Statistics 2013*. US IEA. http://www.iea.org/publications/freepublications/publication/KeyWorld2 013.pdf https://www.eia.gov/energyexplained/oil-and-petroleum-products/ https://education.seattlepi.com/causes-negative-effects-oil-pollution-4661.html

Vaani, I., Sushil, S. J., Kunjamma, U. V., Ramachandran, A., Bai, V. T., & Thyla, B. (2017, May). BhrtyArtana (A pipe cleaning and inspection robot). In *2017 Third International Conference on Sensing, Signal Processing and Security (ICSSS)* (pp. 422-425). IEEE. 10.1109/SSPS.2017.8071632

Vaidya, H., & Chatterji, T. (2020). *SDG 11 Sustainable Cities and Communities*, 173–185. Springer. doi:10.1007/978-981-32-9927-6_12

Valkenburg, C., Walton, C. W., Thompson, B. L., Gerber, M. A., Jones, S. B., & Stevens, D. J. (2008). *Municipal solid waste (MSW) to Liquid Fuels Synthesis, volume 1: Availability of feedstock and Technology*. OSTI. doi:10.2172/962858

van Eck, N. J., & Waltman, L. (2017). Citation-based clustering of publications using CitNetExplorer and VOSviewer. *Scientometrics*, *111*(2), 1053–1070. doi:10.100711192-017-2300-7 PMID:28490825

Van Eijck, J., & Romijn, H. (2008). Prospects for Jatropha biofuels in Tanzania: An analysis with strategic niche management. *Energy Policy*, *36*(1), 311–325. doi:10.1016/j.enpol.2007.09.016

Van, D. P., Fujiwara, T., Tho, B. L., Toan, P. P. S., & Minh, G. H. (2020). A review of anaerobic digestion systems for biodegradable waste: Configurations, operating parameters, and current trends. *Environmental Engineering Research, 25*(1), 1–17. doi:10.4491/eer.2018.334

Vanitha, S. K. R., & Boopathi, S. (2023). Artificial Intelligence Techniques in Water Purification and Utilization. In *Human Agro-Energy Optimization for Business and Industry* (pp. 202–218). IGI Global. doi:10.4018/978-1-6684-4118-3.ch010

Vasanthi, P., Kaliappan, S., & Srinivasaraghavan, R. (2008). Impact of poor solid waste management on ground water. *Environmental Monitoring and Assessment, 143*(1-3), 227–238. doi:10.100710661-007-9971-0 PMID:17999155

Vashisht, S., & Rakshit, D. (2021). Recent advances and sustainable solutions in automobile air conditioning systems. *Journal of Cleaner Production, 329*, 129754. doi:10.1016/j.jclepro.2021.129754

Venkiteela, L. K. (2020). Status and challenges of solid waste management in Tirupati city. *Materials Today: Proceedings, 33*, 470–474. doi:10.1016/j.matpr.2020.05.044

Vennila, T., Karuna, M. S., Srivastava, B. K., Venugopal, J., Surakasi, R., & Sampath, B. (2022). New Strategies in Treatment and Enzymatic Processes: Ethanol Production From Sugarcane Bagasse. In Human Agro-Energy Optimization for Business and Industry (pp. 219–240). IGI Global.

Vergara-Araya, M., Lehn, H., & Poganietz, W. R. (2020). Integrated water, waste and energy management systems–A case study from Curauma, Chile. *Resources, Conservation and Recycling, 156*, 104725. doi:10.1016/j.resconrec.2020.104725

Verma, T., Aggarwal, R., Bhardwaj, G., Verma, V., & Agarwal, S. (2022, February). Geyser Time Controller for Effective Electricity and Water Management. In *2022 2nd International Conference on Innovative Practices in Technology and Management (ICIPTM)* (*Vol. 2*, pp. 378-382). IEEE. 10.1109/ICIPTM54933.2022.9753816

Verma, V., Chauhan, H., Gupta, D., & Gupta, S. (2020). Different Convolution Neural Network Based Models for Garbage Detection: A Review. In *2020 IEEE International Women in Engineering (WIE) Conference on Electrical and Computer Engineering (WIECON-ECE)* (pp. 117-121). IEEE. 10.1109/WIECON-ECE52138.2020.9397961

Verma, J. (2020). A holistic approach of quality in higher education system vis-a-vis sustainable development. [IJM]. *International Journal of Management, 11*(10), 433–440.

Verma, N., Sehrawat, K. D., Sehrawat, A. R., & Pandey, D. (2021). Effective green synthesis, characterization and antibacterial efficacy of silver nanoparticles from seaweed treated sprouts of Moth Bean (Vigna aconitifolia Jacq.). *Regenerative Engineering and Translational Medicine*, 1–14.

Vest, J. R., & Gamm, L. D. (2009). 'A critical review of the research literature on Six Sigma, Lean and StuderGroup ' s Hardwiring Excellence in the United States : The need to demonstrate and communicate the effectiveness of transformation strategies in healthcare'. *Implementation Science : IS, 4*(35), 1–9. doi:10.1186/1748-5908-4-35 PMID:19570218

Viet, N. D., & Jang, A. (2023, February). Machine learning-based real-time prediction of micropollutant behaviour in forward osmosis membrane (waste) water treatment. *Journal of Cleaner Production, 389*, 136023. doi:10.1016/j.jclepro.2023.136023

Vignesh, S., Arulshri, K. P., Syedsajith, S., Kathiresan, S., Boopathi, S., & Dinesh Babu, P. (2018). Design and development of ornithopter and experimental analysis of flapping rate under various operating conditions. *Materials Today: Proceedings, 5*(11), 25185–25194. doi:10.1016/j.matpr.2018.10.320

Vijayan, R. V., Krishnan, M. M., Parayitam, S., Anantharaman Duraisami, S. P., & Saravanaselvan, N. R. (2023). Exploring e-waste recycling behaviour intention among the households: Evidence from India. *Cleaner Materials, 7*, 100174. doi:10.1016/j.clema.2023.100174

Vinodhini, V., Kumar, M. S., Sankar, S., Pandey, D., Pandey, B. K., & Nassa, V. K. (2022). IoT-based early forest fire detection using MLP and AROC method. *International Journal of Global Warming*, *27*(1), 55–70. doi:10.1504/IJGW.2022.122794

Vyas, S., Prajapati, P., Shah, A. V., & Varjani, S. (2022). Municipal solid waste management: Dynamics, risk assessment, ecological influence, advancements, constraints and perspectives. *The Science of the Total Environment*, *814*, 152802. doi:10.1016/j.scitotenv.2021.152802 PMID:34982993

Vyas, V., Mehta, K., & Sharma, R. (2023). The nexus between toxic-air pollution, health expenditure, and economic growth: An empirical study using ARDL. *International Review of Economics and Finance*, *84*, 154–166. doi 10.1016/j.iref.2022.11.017

Walser, T., Limbach, L. K., Brogioli, R., Erismann, E., Flamigni, L., Hattendorf, B., Juchli, M., Krumeich, F., Ludwig, C., Prikopsky, K., Rossier, M., Saner, D., Sigg, A., Hellweg, S., Günther, D., & Stark, W. J. (2012). Persistence of engineered nanoparticles in a municipal solid-waste incineration plant. *Nature Nanotechnology*, *7*(8), 520–524. doi:10.1038/nnano.2012.64 PMID:22609690

Wan. (2007). *International Journal of Health Care Quality Assurance*, *20*(3), 253–263.

Wang, H., Yi, H., Zhu, C., Wang, X., & Jin Fan, H. (2015). Functionalized highly porous graphitic carbon fibers for high-rate supercapacitive electrodes. *Nano Energy*, *13*, 658–669. doi:10.1016/j.nanoen.2015.03.033

Wang, K., Li, L., Tan, F., & Wu, D. (2018). Treatment of Landfill Leachate Using Activated Sludge Technology: A Review. *Archaea (Vancouver, B.C.)*, *2018*, 1–10. doi:10.1155/2018/1039453 PMID:30254508

Wang, X. (2019). System Simulation Optimization of Resource and Environmental Effects of Circular Economy. *Nature Environment & Pollution Technology*, *18*(5), 1573–1578.

Wang, X., Huang, K., Chen, Y., Liu, J., Chen, S., Cao, J., Mei, S., Zhou, Y., & Jing, T. (2018). Preparation of dumbbell manganese dioxide/gelatin composites and their application in the removal of lead and cadmium ions. *Journal of Hazardous Materials*, *350*, 46–54. doi:10.1016/j.jhazmat.2018.02.020 PMID:29448213

Waqas, S., Harun, N. Y., Sambudi, N. S., Arshad, U., Nordin, N. A. H. M., Bilad, M. R., Saeed, A. A. H., & Malik, A. A. (2022, August 23). SVM and ANN Modelling Approach for the Optimization of Membrane Permeability of a Membrane Rotating Biological Contactor for Wastewater Treatment. *Membranes*, *12*(9), 821. doi:10.3390/membranes12090821 PMID:36135840

Wath, S. B., Dutt, P. S., & Chakrabarti, T. (2011). E-waste scenario in India, its management and implications. *Environmental Monitoring and Assessment*, *172*(1–4), 249–262. doi:10.100710661-010-1331-9 PMID:20151189

Wath, S. B., Vaidya, A. N., Dutt, P. S., & Chakrabarti, T. (2010). A roadmap for development of sustainable E-waste management system in India. *The Science of the Total Environment*, *409*(1), 19–32. doi:10.1016/j.scitotenv.2010.09.030 PMID:20951410

Weber, S., Weber, O., Habib, K., & Dias, G. M. (2023). Textile waste in Ontario, Canada: Opportunities for reuse and recycling. *Resources, Conservation and Recycling*, *190*, 106835. doi:10.1016/j.resconrec.2022.106835

White, H. D., & McCain, K. W. (1998). Visualizing a discipline: An author co-citation analysis of information science, 1972–1995. *Journal of the American Society for Information Science*, *49*(4), 327–355.

WHO. (2006). *Protecting Groundwater for Health: Managing the Quality of Drinkingwater Sources* (O. Schmoll, G. Howard, J. Chilton, & I. Chorus, Eds.). Published by IWA Publishing.

Widmer, R., Oswald-Krapf, H., Sinha-Khetriwal, D., Schnellmann, M., & Böni, H. (2005). Global perspectives on e-waste. *Environmental Impact Assessment Review*, *25*(5), 436–458. doi:10.1016/j.eiar.2005.04.001

Wijaya, J., & Oh, S. (2023, April). Machine learning reveals the complex ecological interplay of microbiome in a full-scale membrane bioreactor wastewater treatment plant. *Environmental Research*, *222*, 115366. doi:10.1016/j.envres.2023.115366 PMID:36706897

Willner, M. R., & Vikesland, P. J. (2018). Nanomaterial enabled sensors for environmental contaminants. *Journal of Nanobiotechnology*, *16*(1), 95–95. doi:10.118612951-018-0419-1 PMID:30466465

Wilson, D. C., & Velis, C. A. (2015). Waste management–still a global challenge in the 21st century: An evidence-based call for action. *Waste Management & Research*, *33*(12), 1049–1051. doi:10.1177/0734242X15616055 PMID:26574579

Wilson, D. G. (1976). A brief history of solid-waste management. *The International Journal of Environmental Studies*, *9*(2), 123–129. doi:10.1080/00207237608737618

Wolf, J. (2011). Sustainable supply chain management integration: A qualitative analysis of the German manufacturing industry. *Journal of Business Ethics*, *102*(2), 221–235. doi:10.100710551-011-0806-0

Wong, S., Ngadi, N., Inuwa, I. M., & Hassan, O. (2018). Recent advances in applications of activated carbon from biowaste for wastewater treatment: A short review. *Journal of Cleaner Production*, *175*, 361–375. doi:10.1016/j.jclepro.2017.12.059

Woolridge, A. C., Ward, G. D., Phillips, P. S., Collins, M., & Gandy, S. (2006). Life cycle assessment for reuse/recycling of donated waste textiles compared to use of virgin material: An UK energy saving perspective. *Resources, Conservation and Recycling*, *46*(1), 94–103. doi:10.1016/j.resconrec.2005.06.006

WWAP (World Water Assessment Programme). (2012). *The United Nations World Water Development Report 4: Managing Water under Uncertainty and Risk*. UNESCO.

Xavier, L. H., Giese, E. C., Ribeiro-Duthie, A. C., & Lins, F. A. F. (2021). Sustainability and the circular economy: A theoretical approach focused on e-waste urban mining. *Resources Policy*, *74*, 101467. doi:10.1016/j.resourpol.2019.101467

Xavier, L. H., Ottoni, M., & Abreu, L. P. P. (2023). A comprehensive review of urban mining and the value recovery from e-waste materials. *Resources, Conservation and Recycling*, *190*, 106840. doi:10.1016/j.resconrec.2022.106840

Xiang, Y., Wu, P., Zhu, N., Zhang, T., Liu, W., Wu, J., & Li, P. (2010). Bioleaching of copper from waste printed circuit boards by bacterial consortium enriched from acid mine drainage. *Journal of Hazardous Materials*, *184*(1–3), 812–818. doi:10.1016/j.jhazmat.2010.08.113 PMID:20869807

Xie, X., Mai, Y., & Zhou, X.XIE. (2005). Dispersion and alignment of carbon nanotubes in polymer matrix: A Review. *Materials Science and Engineering R Reports*, *49*(4), 89–112. doi:10.1016/j.mser.2005.04.002

Xiong, G., Zhu, F., Fan, H., Dong, X., Kang, W., & Teng, T. (2014). Novel ITS based on space-air-ground collected big-data. *17th International IEEE Conference on Intelligent Transportation Systems (ITSC)*, (pp. 1509–1514). IEEE. 10.1109/ITSC.2014.6957647

Yaacob, M., Ibrahim, M. H., & Nasir, Z. M. (2019). Solid Waste Management in Malaysia: The Perspectives of Non-governmental Organizations (NGOs) and Youths on Consumers'. *Commitment*, 27–41. doi:10.1007/978-981-13-9173-6_3

Yaashikaa, P. R., Priyanka, B., Senthil Kumar, P., Karishma, S., Jeevanantham, S., & Indraganti, S. (2022). A review on recent advancements in recovery of valuable and toxic metals from e-waste using bioleaching approach. *Chemosphere*, *287*, 132230. doi:10.1016/j.chemosphere.2021.132230 PMID:34826922

Yacob, P., Wong, L. S., & Khor, S. C. (2018). An empirical investigation of green initiatives and environmental sustainability for manufacturing SMEs. *Journal of Manufacturing Technology Management*, *30*(1), 2–25. doi:10.1108/JMTM-08-2017-0153

Yadav, H., Soni, U., & Kumar, G. (2023). Analysing challenges to smart waste management for a sustainable circular economy in developing countries: A fuzzy DEMATEL study. *Smart and Sustainable Built Environment*, *12*(2), 361–384. doi:10.1108/SASBE-06-2021-0097

Yadav, I. C., & Devi, N. L. (2009). Studies on municipal solid waste management in Mysore city — A case study. *Report and Opinion*, *1*(3), 15–21.

Yadeta, A. E., & Pandey, D. (2020). Analysis of the global impact of the pandemic (COVID-19) on construction industry: Possible scenarios. *Current Trends in Civil & Structural Engineering*, *6*(4), 1–8. doi:10.33552/CTCSE.2020.06.000641

Yang, T., Xu, Z., Wen, J., & Yang, L. (2009). Factors influencing bioleaching copper from waste printed circuit boards by Acidithiobacillus ferrooxidans. *Hydrometallurgy*, *97*(1–2), 29–32. doi:10.1016/j.hydromet.2008.12.011

Yang, X., Lee, S. J., Yoo, H. Y., Choi, H. S., Park, C., & Kim, S. W. (2014). Biorefinery of instant noodle waste to biofuels. *Bioresource Technology*, *159*, 17–23. doi:10.1016/j.biortech.2014.02.068 PMID:24632436

Yan, S., Chen, X., Wu, J., & Wang, P. (2013). Pilot-scale production of fuel ethanol from concentrated food waste hydrolysates using Saccharomyces cerevisiae H058. *Bioprocess and Biosystems Engineering*, *36*(7), 937–946. doi:10.100700449-012-0827-9 PMID:23010725

Yasmin Begum, R. N., & Shamshuddin, R. N. (2014). Green Marketing: Challenges and Strategies for Indian Companies in 21st Century. *International Journal of Research in Business Management*, *2*(5), 91–104.

Ying, Y., Pourrahimi, A. M., Sofer, Z., Matějková, S., & Pumera, M. (2019). Radioactive Uranium Preconcentration <i>via</i> Self-Propelled Autonomous Microrobots Based on Metal–Organic Frameworks. *ACS Nano*, *13*(10), 11477–11487. doi:10.1021/acsnano.9b04960 PMID:31592633

Yin, S., Wang, L., Kabwe, E., Chen, X., Yan, R., An, K., Zhang, L., & Wu, A. (2018). Copper bioleaching in China: Review and prospect. *Minerals (Basel)*, *8*(2), 32. doi:10.3390/min8020032

Yi, S., Li, C., & Li, Q. (2015). A survey of fog computing: concepts, applications and issues. *Proceedings of the 2015 Workshop on Mobile Big Data*, (pp. 37–42). ACM. 10.1145/2757384.2757397

Yosif, Z. M., Mahmood, B. S., & Saeed, S. Z. (2022, June 30). Artificial Techniques Based on Neural Network and Fuzzy Logic Combination Approach for Avoiding Dynamic Obstacles. *Journal Européen des Systèmes Automatisés*, *55*(3), 339–348. doi:10.18280/jesa.550306

Yosufi, A. (2019). Poor Municipal Waste Management and Its Health Implication: A Case Study of Kabul City. *Education Quarterly Eeview*, *2*(2). doi:10.31014/aior.1993.02.02.75

Yousuf, T. B. (2014). 3R (Reduce, Reuse and Recycle) in Bangladesh. In *Municipal Solid Waste Management in Asia and the Pacific Islands* (pp. 61–75). Springer. doi:10.1007/978-981-4451-73-4_4

Yupapin, P., Trabelsi, Y., Nattappan, A., & Boopathi, S. (2022). Performance Improvement of Wire-Cut Electrical Discharge Machining Process Using Cryogenically Treated Super-Conductive State of Monel-K500 Alloy. *Iranian Journal of Science and Technology - Transactions of Mechanical Engineering*, 1–17.

Yupapin, P., Trabelsi, Y., Nattappan, A., & Boopathi, S. (2022). Performance Improvement of Wire-Cut Electrical Discharge Machining Process Using Cryogenically Treated Super-Conductive State of Monel-K500 Alloy. *Iranian Journal of Science and Technology - Transactions of Mechanical Engineering*, 1–17. doi:10.1007/s40997-022-00513-0

Zaghloul, M. S., & Achari, G. (2022, June). Application of machine learning techniques to model a full-scale wastewater treatment plant with biological nutrient removal. *Journal of Environmental Chemical Engineering*, *10*(3), 107430. doi:10.1016/j.jece.2022.107430

Zahur, M. (2007). *Solid waste management of Dhaka city: public private community partnership*. Research Gate.

Zaidi Ab, G., Yusoff, M. S., & Andas, J. (2015). Review on Applications of Nanoparticles in Landfill Leachate Treatment. *Applied Mechanics and Materials*, *802*, 525–530. doi:10.4028/www.scientific.net/AMM.802.525

Zantalis, F., Koulouras, G., Karabetsos, S., & Kandris, D. (2019). A review of machine learning and IoT in smart transportation. *Future Internet*, *11*(4), 94. doi:10.3390/fi11040094

Zaree, T., & Honarvar, A. R. (2018). Improvement of air pollution prediction in a smart city and its correlation with weather conditions using metrological big data. *Turkish Journal of Electrical Engineering and Computer Sciences*, *26*(3), 1302–1313.

Zhang, A., Venkatesh, V. G., Liu, Y., Wan, M., Qu, T., & Huisingh, D. (2019). Barriers to smart waste management for a circular economy in China. *Journal of Cleaner Production*, *240*, 118198. doi:10.1016/j.jclepro.2019.118198

Zhang, H., He, J., Shi, X., Hong, Q., Bao, J., & Xue, S. (2020). Technology characteristics, stakeholder pressure, social influence, and green innovation: Empirical evidence from Chinese express companies. *Sustainability (Basel)*, *12*(7), 2891. doi:10.3390u12072891

Zhang, K., Schnoor, J. L., & Zeng, E. Y. (2012). E-waste recycling: Where does it go from here? Environmental Science &amp. *Environmental Science & Technology*, *46*(20), 10861–10867. doi:10.1021/es303166s PMID:22998401

Zhang, P., Dong, G., Sun, B., Zhang, L., Chen, X., Ma, N., Yu, F., Guo, H., Huang, H., Lee, Y. L., Tang, N., & Chen, J. (2011). Long-term exposure to ambient air pollution and mortality due to cardiovascular disease and cerebrovascular disease in Shenyang, China. *PLoS One*, *6*(6), e20827. doi:10.1371/journal.pone.0020827 PMID:21695220

Zhang, X., Li, N., Shao, H., Meng, Y., Wang, L., Wu, Q., Yao, Y., Li, J., Bian, J., Zhang, Y., & Deng, X. (2016). Methane limit LPS-induced NF-κB/MAPKs signal in macrophages and suppress immune response in mice by enhancing PI3K/AKT/GSK-3β-mediated IL-10 expression. *Scientific Reports*, *6*(1), 29359. doi:10.1038rep29359 PMID:27405597

Zhang, X., & Liu, Y. (2020). Nanomaterials for radioactive wastewater decontamination. *Environmental Science. Nano*, *7*(4), 1008–1040. doi:10.1039/C9EN01341E

Zhao, G., Huang, X., Tang, Z., Huang, Q., Niu, F., & Wang, X. (2018). Polymer-based nanocomposites for heavy metal ions removal from aqueous solution: A review. *Polymer Chemistry*, *9*(26), 3562–3582. doi:10.1039/C8PY00484F

Zhao, Y., Jia, G., Shang, Y., Zhao, P., Cui, X., & Guo, Q. (2022). Chlorine migration during hydrothermal carbonization of recycled paper wastes and fuel performance of hydrochar. *Process Safety and Environmental Protection*, *158*, 495–502. doi:10.1016/j.psep.2021.12.041

Zheng, T., Ardolino, M., Bacchetti, A., & Perona, M. (2020). The Applications of Industry 4.0 Technologies in Manufacturing Context: A Systematic Literature Review. *International Journal of Production Research*, *0*(0), 1–33. doi:10.1080/00207543.2020.1824085

Zheng, Y., Mao, C., Zhang, Z., Liu, R., & Yan, B. (2022). Nanotechnology for the Remediation of Plastic Wastes. In *Sustainable Nanotechnology* (pp. 117–143). Royal Society of Chemistry. doi:10.1039/9781839165771-00117

Zhou, Z., & Zhang, L. (2022). Sustainable waste management and waste to energy: Valuation of energy potential of MSW in the Greater Bay Area of China. *Energy Policy*, *163*, 112857. doi:10.1016/j.enpol.2022.112857

Zhu, J., Wei, S., Chen, M., Gu, H., Rapole, S. B., Pallavkar, S., Ho, T. C., Hopper, J., & Guo, Z. (2013). Magnetic nanocomposites for environmental remediation. *Advanced Powder Technology*, *24*(2), 459–467. doi:10.1016/j.apt.2012.10.012

Zhu, M., Wang, J., Yang, X., Zhang, Y., Zhang, L., Ren, H., Wu, B., & Ye, L. (2022, June). A review of the application of machine learning in water quality evaluation. *Eco-Environment & Health*, *1*(2), 107–116. doi:10.1016/j.eehl.2022.06.001

Zupic, I., & Čater, T. (2015). Bibliometric Methods in Management and Organization. *Organizational Research Methods*, *18*(3), 429–472. doi:10.1177/1094428114562629

About the Contributors

Prateek Srivastava is currently working as Associate Professor in Chitkara University Innovation and Research Network, Chitkara University, Punjab. In addition, has additional charge of Manager-Technology Enabling Center (A Govt. of India funded project) at Chitkara University, Punjab India. He received his Ph.D. in the field of Advanced Air Conditioning System (Energy Efficiency) from Malaviya National Institute of Technology, Jaipur, India. He completed his M.Tech. (Thermal Engineering) from National Institute of Technology, Silchar, India. He completed his B.Tech. (Mechanical Engineering) from Rajasthan Technical University, Kota. His areas of research include: Heating Ventilation Air Conditioning, Building Energy Efficiency, Green Buildings, Energy Recovery, Renewable Energy, Water recycling. He has over 25 publications in international journals of repute including SCOPUS, Web of Science and SCI indexed database and refereed international conferences. He has 4 granted patents to his credit. He has a total of 3 years of experience in teaching and research. He was winner of prestigious Building Energy Efficiency Higher and Advanced Network (BHAVAN) Fellowship supported by the Department of Science and Technology, Govt. of India, and the Indo-U.S. Science and Technology Forum (IUSSTF). He is certified Zero Effect Zero Defect (ZED) consultant, ZED is a process rating system for MSME companies implemented by Quality Council of India (QCI). He is an active member of Radiant Cooling System Technical Group of Indian Society of Heating Refrigeration, Air Conditioning Engineers (ISHRAE). He is Chair-Student Activities and BOG member of ASHRAE Chandigarh Chapter for SY 2021-2022. He is member of energy management group responsible for energy audit, energy conservation and green building initiatives, started energy conservation club for students at Chitkara University. He has organized various Faculty Development Programmes, International Conferences, Workshops, and Seminar for researchers, Ph.D., UG and PG level students.

Durgaprasad Ramteke is currently a researcher in Fiber and particle Engineering at the University of Oulu in Finland. Prior to his recent appointment at the UOULU, he was a visiting researcher at the University of Padova, Italy. Dr. Ramteke received his M.Sc. degree in Physics from University of Nagpur and his Ph.D. in Material Science from Visvesvaraya National Institute of Technology (VNIT, Nagpur). Dr. Ramteke published several papers in preferred Journals and participated in a range of forums on material science. He also presented various academic as well as research-based papers at several national and international conferences in the USA, South Africa, and Europe. He was also part of the HORIZON 2020 program. His research work mainly focused on glasses, glass ceramics, ceramics, inorganic waste management, foam glass and geopolymers for direct applications.

Ankush Kumar Bedyal is currently working as an Assistant Professor at the Department of Physics, Cluster University of Jammu, Jammu, India. Formerly he worked as a Research Associate at the Department of Physics and Astronomical Sciences, Central University of Jammu, Jammu, India. He also worked as Assistant Professor at Chitkara University, Punjab since, October 2019. Before that, he was Post Doctorate fellow at the Department of Physics, University of the Free State, Bloemfontein, South Africa for 3 years and 4 months. He did his Ph.D. in Physics from the School of Physics, Shri Mata Vaishno Devi University, Katra, Jammu & Kashmir, India in 2016. Dr. Bedyal has more than 12 years of experience in research. He has worked in the field of luminescence properties of phosphors particularly, Photoluminescence, Thermoluminescence, and Cathodoluminescence for their applications in lighting, White Light Emitting Diodes, Displays, Dosimetry and many more. He worked on the modification of luminescence, optical and structural properties of phosphors by Swift Heavy Ion irradiation. He also worked on the surface properties of the materials. Dr. Bedyal has outstanding research credentials. He has published more than 50 research papers in reputed journals and proceedings of International and National Conferences. He has presented his work at many national and international conferences in India and Abroad. He has delivered a couple of invited lectures at national conferences in India. His current focus is to synthesize the nanophosphors for light emitting diode applications, CTR phosphors, material modification using swift heavy ion irradiation and radiation dosimetry and to study the energy transfer mechanism among the activator ions in the phosphor material. Currently, he is also working on the ZnO/nanophosphor composites to enhance the UV/visible light photocatalytic activity of ZnO.

Monika Gupta joined Chitkara University as Associate Professor in November 2021. She is Ph.D. (Management), MBA (Marketing), UGC-NET (Management), B.Sc. (General), B.Ed. with about thirteen years of academic experience and four years of corporate experience. She has five papers publications in SCOPUS, one paper in ABDC 'C' Listed journal and one book review in ABDC Listed journal (out of thirty nine papers in national and international journals of repute). She is Editorial Board Member of International Journal of Economics and Financial Issues (IJEFI), ABDC 'C' Listed journal. She received Cash honorarium from AIMA for paper publication. She remained actively involved in organizing National and International Conferences. She has attended more than twenty National and International Conferences. She also served as HOD (MBA, BBA) for three years. Her areas of interest are Marketing Management, Sales and Retail Management, Consumer Behavior, Integrated Marketing Communications, Services Marketing, Consumer Relationship Management, Strategic Management and Management Fundamentals.

Jasminder Kaur Sandhu is working at Deptt. of CSE, Chandigarh University, Punjab. She also served ass Assistant Professor, Chitkara University Research and Innovation Network (CURIN), Chitkara University, Punjab. She is having an additional charge of First Information officer (FIO) – CURIN and Joint Editor of the Newsletter "Res Novae". She is Ph.D. in Computer Science and Engineering from Thapar Institute of Engineering and Technology, Patiala. She completed her M.Tech. (CSE) securing top merit position at Punjabi University, Patiala. She is having approximately 10 years of research and teaching experience. She is an active reviewer of many reputed journals such as Springer International Journal of Machine Learning and Cybernetics, IEEE Access. She has to her credit more than 25 publications in reputed SCI-Indexed Journals and International Conferences. Her research interests include e-Waste Management, Machine Learning, Ensemble Modelling Artificial Intelligence, Soft Computing, Dependability Evaluation, Quality of Service, Wireless Sensor Networks, and Ad-Hoc Networks.

* * *

Kanimozhi is with the Department of Computer Science and Engineering, Saveetha School of Engineering, Saveetha Institute of Medical and Technical Sciences, Saveetha University, Chennai, India.

Rakesh A. Afre completed his M.Sc (Physics with specialization in Energy Studies) from North Maharashtra University, Jalgaon, Maharashtra and PhD (Nanotechnology) from Nagoya Institute of Technology (NIT), Nagoya, Japan under guidance of Prof. Tetsuo Soga. After completion of M.Sc he joined as a JRF at IIT Mumbai to work on DST sponsored project entitled "Carbon nanomaterials for solar cell" under guidance of Prof. Maheshwar Sharon, Dept. of Chemistry. Later he moved to Japan to pursue his PhD degree under Japan Government Scholarship (Monbukagakusho). His doctoral studies were focused on "Synthesis of carbon nanotubes by renewable natural precursors" In 2007; he has been awarded with PhD degree and continued his further research on organic photovoltaic (OPV) devices at NIT. Then he moved to KRICT, South Korea for his next assignment to continue research on DSSCs and small molecule OPVs. Furthermore, he moved to Australia to continue his research on fabrication of transparent electrodes using nanomaterials for OPVs and then to Ethiopia as a Professor. He came back to India and joined Walchand Center for Research in Nanotechnology and Bionanotechnology (wcRnb), WCAS, Solapur as a HOD of Nanotechnology department under eminent directors Dr. Madhuri Sharon and Prof. Maheshwar Sharon. Presently he is working as Professor and Deputy Director (Innovations) at Assam down town University, Guwahati, Assam. Having 15+ years of experience in research and development and published more than 25 high quality research papers and 2 international patents.

J. Ananth is with the department of Marine Engineering, AMET Deemed to be University, Chennai, India.

Manohar. B.S. is with the Department of Mechanical Engineering, Panimalar Engineering College, Chennai, India.

Kalyan Bamane is with the Department of Information Technology, D Y Patil College of Engineering Akurdi, Pune, India.

Esha Bansal is a PGT teacher at Red Roses Public School, Gurugram and is company secretary.

Sunandan Baruah is currently the Director, Research and Innovation at Assam down town University, Guwahati. He also holds additional responsibilities of Director, Centre of Excellence in Nanotechnology and Director, Technology, Innovation, Transfer and Entrepreneurship Centre and President of the Institution's Innovation Council. He is also the CEO of Globinno Engineering Solutions, Guwahati. He is a social entrepreneur and his company uses Nanotechnology based solutions to solve societal problems. Previously, he was the Director, Center of Excellence in Nanotechnology at Assam Don Bosco University. He also worked as a Research Faculty at the Angstrom Laboratoriet, Uppsala University, Sweden and as a Research Specialist at the Asian Institute of Technology, Thailand. Prof. Baruah's research interest is in Environmental Nanotechnology. He has 112 publications in international peer reviewed journals and conference proceedings and 13 book chapters. Prof. Baruah has innovated many nanotechnology-based systems and has published 11 patents. Prof. Baruah is also a Consultant to the Ministry of MSME and is involved in upgradation of traditional industries of India through inclusion of modern technology. He is also a mentor to the budding entrepreneurs of the North East through the OctaNE program of Software Technology Parks of India. He is a Rotarian from Rotary Club of Guwahati.

Archana Bhaskar is with the Department of School of CSA, REVA University, Rukmini Knowledge Park, Bengaluru, Karnataka 560064, India.

Devanshri Bhatt is a research assistant as well as a postgraduate Master of Commerce student at Jain University. Her research and writing is rooted in multiple fields and interest areas. She has an interest in public speaking and academic culture as well. She previously worked as the Vice President of the placement department at Jain University and has handled 20+ company recruitment drives. She is also a swimmer and is passionate about psychology.

Sumanta Bhattacharya is with the Department of textile technology, MAKAUT Howrah, West Bengal.

Sampath Boopathi completed his undergraduate in Mechanical Engineering and postgraduate in the field of Computer-Aided Design. He completed his Ph.D. from Anna University and his field of research includes Manufacturing and optimization. He published 100 more research articles in Internationally Peer-reviewed journals, one Patent grant, and three published patents.He has 16 more years of academic and research experiences in the various Engineering Colleges in Tamilnadu, India.

Geetha C. is currently working as a Professor in Department of Computer Science and Engineering at R.M. K. Engineering College. She has around 25 years of teaching experience. She has completed Ph.D. in Computer Science and Engineering from Manonmaniam Sundaranar University, Tirunelveli. She has published more than 25 papers in national and international journals. Her area of interest includes Sensor Networks, Computer Networks, IoT and Artificial Intelligence

Pankaj Dadheech received his Ph.D. degree in Computer Science & Engineering from Suresh Gyan Vihar University (Accredited by NAAC with 'A' Grade), Jaipur, Rajasthan, India. He received his M.Tech. degree in Computer Science & Engineering from Rajasthan Technical University, Kota and he has received his B.E. in Computer Science & Engineering from University of Rajasthan, Jaipur. He has more than 18 years of experience in teaching. He is currently working as an Associate Professor & Dy. HOD in the Department of Computer Science & Engineering (NBA Accredited), Swami Keshvanand Institute of Technology, Management & Gramothan (SKIT), Jaipur, Rajasthan, India. He has published 25 Patents at Intellectual Property India, Office of the Controller General of Patents, Design and Trade Marks, Department of Industrial Policy and Promotion, Ministry of Commerce and Industry, Government of India. He has published 8 International Patents (USA, South African, Australian, Germany) & 2 Copyrights. He has 72 publications in various International & National Journals, 62 papers in various National & International conferences. He has published 9 Books & 22 Book Chapters.

Mahmoud M. Elwaheidi is a professor at the Dep. of Geology & Geophysics, King Saud University (KSA). He was awarded his Ph.D. in Env. & Eng. Geophysics from the University of Genova (Italy). He received his M.Sc. in Applied Geophysics from the University of Jordan and his B.Sc. in Earth and Environmental Sciences from the Univ. of Yarmouk (Jordan). Elwaheidi has about 25 years of academic experience in applied geosciences. His research is focused on environmental and engineering issues including climate change. He published numerous research articles in the fields of environmental research and applied geosciences. He is a co-author of the book Concise Handbook of Waste Treatment Technologies (ISBN: 9780367631307). Dr. Elwaheidi is also acting as a reviewer for several prestigious scientific journals.

Gagandeep is currently working as Assistant Professor of Marketing in Chitkara University. He received his M.B.A. in Marketing and Ph.D. in Marketing & International Business from Panjab University's University Business School. He has more than 14 years of teaching and research experience. His area of interest in teaching are Marketing Research, Research Methodology, Management Science, Applied Operations Research, Marketing Management and Strategic Management. Dr Gagandeep has presented papers in various National and International conferences and has published more than 10 research papers. His area of interest in research are Empirical modeling of marketing problems, Service Quality, New product development and Emerging trends and issues in marketing.

Swapna H. R. is a Professor at the School of Commerce- PG studies JAIN (Deemed-to-be University) Bangalore, India.

Monika Hans is working as Assistant Professor at Govt PG College for Women Gandhi Nagar, Jammu-180003 Jammu & Kashmir, India

Sudheer Hanumanthakari is at the Department of Electronics and Communication Engineering, Faculty of Science and Technology, IFHE, Hyderabad, Andhra Pradesh India.

S. Hariram is with the Department of Mechanical Engineering, Rajalakshmi Engineering College, Chennai, India.

Uday Kanike is currently a research scholar at Georgia State University.

Jaskiran Kaur is an academician with more than five years' experience, UGC-NET qualified holding a doctorate degree. She has handled different profiles throughout her career, ranging from career counselling to Industrial Relation coordinator to Training and Placement Officer to MS-Excel Trainer. She has been a resource person in faculty development programmes and different expert sessions. Throughout her studies, she was awarded special merit for securing one of the top ten positions in Punjab Technical University (PTU), Jalandhar. As a researcher, she has presented 6 research papers in National & International Seminars and Conferences. She has published various research papers in Journal of Information and Optimization Sciences, Taylor & Francis (WoS indexed); Administrative Sciences, MDPI (Scopus indexed). She has attended a plethora of courses and received certifications in Business Analytics and Intelligence, HR Management and Analytics, Fundamentals of R, Embracing Change, Excel Skills for Business: Essentials, AI for Everyone. Currently, working as an Assistant Professor at Lovely Professional University, Jalandhar. She is also involved in developing E-content for LPU Online students.

Gurpreet Kaur has completed her PhD on the topic "Determinants of Liquidity Risk and Credit Risk-An Empirical Study on Indian Banks" and is working as Assistant Professor (Finance and accounting) for more than ten years. On her academic front, She has qualified UGC (net) and cleared executive level of Company Secretary and has also done M.Com, M.F.C from Punjab University, Chandigarh. She is awarded for her academic performance. Her areas of specialization includes accounting and finance related subjects (taxation, operation research, cost accounting, management accounting). She has published/presented research papers in journals/conferences (national & international).

Milind M. Kulkarni, presently working as Professor in Mechanical Department, School of Engineering, Ajeenkya D Y Patil University, Pune, Maharashtra. His research area mainly focused on refrigeration and air-conditioning. Dr. M.M. Kulkarni possesses 32 years of academic experience with a special focus on commercial research in mechanical engineering applications with special reference to application fields in refrigeration and air conditioning. He has strong connections with industry, allowing him to share valuable knowledge with the students and make them industry ready. He is a member of the ASHRAE and ISHRAE societies.

Sunil Kumar does research focusing on identifying lung diseases with the use of machine learning and imaging modalities. He is an Assistant Professor in the Department of Information Technology at the School of Engineering and Technology (UIET), CSJM University, Kanpur. He has over 12 years of teaching in computer science and engineering.

Geetika Madaan is working as Assistant Professor, at the University centre for research and development Chandigarh University, Mohali, India.

Lakshmana Phanendra Magulur was awarded a Ph.D. in Computer Science and Engineering from Annamalai University, Chidambaram. Post Graduation in Information Technology from Gandhi Institute of Technology and Management, Vishakapatnam Under Graduation in Computer Science and Engineering from Seshadri Rao Gudlavalleru Engineering College formerly known as Gudlavalleru Engineering College Automation Anywhere Certified Advanced RPA Professional 2022 ServiceNow Certified System Administrator Microsoft Certified: DevOps Engineer Expert Microsoft Certified: Azure Developer Associate Microsoft Certified: Azure Data Fundamentals Microsoft Certified: Azure AI Fundamentals Microsoft Certified: Azure Fundamentals Oracle Cloud Infrastructure Foundations 2021 Associate AWS Certified Cloud Practitioner TensorFlow Developer Certificate UI-PATH Certified RPA Developer Member of Professional Bodies of Computer Society of India (CSI), Indian Society for Technical Education (ISTE), Institution of Engineers (India) [IEI], Institution of Electronics and Telecommunication Engineers (IETE). Currently Working as an Associate Professor at Koneru Lakshmaiah Educational Foundation Formally Worked as Assistant Professor in the CSE Dept. of Gudlavalleru Engineering College, Gudlavalleru from July-2013-May-2016 (2.8 years) JNTUK Ratified. Authorized 39 International/National Research papers which are indexed in reputed databases like Scopus, Web of Science, IEEE Xplore, Springer Link, and ACM Digital Library. Sound Laboratory Experience in C, Oops through C++, Python, and R Programming. Passion to learn Cutting Edge technologies and transfer the knowledge as a part of the teaching-learning process. Have an aptitude and understanding of technology. Self-starter, Life long learner and take ownership of assignments. Drive to learn new technologies and turn around quickly to execute identified work.

Firdous Ahmad Malik graduated with a Ph.D. degree from Babasaheb Bhimrao Ambedkar University in Lucknow, India. He holds an MPhil degree in Economics from Babasaheb Bhimrao Ambedkar University in Lucknow and a Master's degree in Economics from the University of Kashmir. His research area of interest is microfinance, financial inclusion, and financial literacy. He has published more than 20 papers in Springer, Elsevier, Taylor & Francis, and other Scopus-indexed journals. He has published his latest book with a springer, "Financial Inclusion Schemes in India." Notion Press published his second book on "Financial Behavior of Urban Destitutes in India". His third book, "Asymmetry of Information and

Lending Risk: Livelihood Patterns of Street Vendors in India," was published by Booksclinic. Shineeks published his fourth book, "Health System in Jammu and Kashmir: Challenges and Opportunities." His fourth book, "Repatriation Management and Competency Transfer in a Culturally Dynamic World," is under contract with Springer. Prior to joining NIPFP he has worked with the Jindal Center for the Global South at O.P. Jindal Global University in Sonipat, Haryana.

Biswadip Basu Mallik is presently a Senior Assistant Professor of Mathematics in the Department of Basic Sciences & Humanities at Institute of Engineering & Management, Kolkata, India. He has been involved in teaching and research for almost 21 years and has published several research papers and book chapters in various International and National journals and publishers. He has authored five books at undergraduate levels in the areas of Engineering Mathematics, Quantitative Methods and Computational Intelligence. He has also published five Indian patents along with nine edited books. His fields of research work are Computational fluid Dynamics, Mathematical modelling and Optimization. Prof. Basu Mallik is a Managing editor of Journal of Mathematical Sciences & Computational Mathematics (JMSCM), USA. He is also the Editorial board member and reviewer of several international journals. He is a senior life member of Operational Research Society of India (ORSI), a life member of Calcutta Mathematical Society (CMS), Indian Statistical Institute (ISI), Indian Science Congress Association (ISCA), International Association of Engineers (IAENG) and an academic professional member of Society for Data Science (S4DS).

Manoj B. Mandake is with the Department of Chemical Engineering, Bharati Vidyapeeth College of Engineering.

Ataul Gani Osmani is an Assistant Professor of Economics at Varendra University Rajshahi. He is now on study leave and studying international master's degree in resource economics and Sustainable Development at the University of Bologna, Rimini, Italy. His research interests include economics, econometrics, agricultural sustainability, development, rural-urban issues, and climate change.

Binay Kumar Pandey currently working as an Assistant Professor in Department of Information Technology of Govind Ballabh Pant University of Agriculture and Technology Pantnagar Uttrakhand, India. He obtained his M. Tech with Specialization in Bioinformatics from Maulana Azad National Institute of Technology Bhopal M. P. India, in 2008 . He obtained his First Degree B. Tech at the IET Lucknow (Uttar Pradesh Technical University, Uttar Pradesh and Lucknow) India, in 2005. In 2010, he joined Department of Information Technology of College of Technology in Govind Ballabh Pant University of Agriculture and Technology Pantnagar as an Assistant Professor and worked for various UG and PG projects till date. He has more than ten years of experience in the field of teaching and research. He has more than 40 publications in reputed peer journal reputed journal Springer,Inderscience,(sci and socopus indexed journal and others) and 3 patent.He has many awards such PM Scholarship etc . He session chair in IEEE International Conference on Advent Trends in Multidisciplinary Research and Innovation (ICATMRI-2020) on December 30, 2020 organized by Pankaj Laddhad Institute of Technology and Management Studies; Buldhana, Maharashtra, India

Digvijay Pandey is a Lecturer at the Department of technical Education, Research Scholar IET Lucknow, India.

Lakshmana Phanendra Magulur is with the Department of Electronics and Communication Engineering, Aditya College of Engineering and Technology, Surampalem, India.

P.V. Pramila is with the Department of Computing Technologies, School of Computing, SRM Institute of Science and Technology, Chengalpattu District, India.

Maryum Sajid Raja is a PhD candidate in Technology Management at South West Jiaotong University. Her research interests lie in the intersection of sustainable development, information and communication technology (ICT), and logistics. Maryum's passion for sustainable development and its impact on technology and logistics drive her to pursue a PhD in Technology Management. Her research focuses on finding ways to integrate sustainable practices into technology and logistics industries, thereby contributing to the overall sustainable development of society. Maryum is dedicated to her work and is committed to making a positive impact on the world through her research."

P. S. Ranjit is with the Department of Mechanical Engineering, Aditya Engineering College, Surampalem, India.

Murugan S. is with the Department of Electronics and Communication Engineering, CK College of Engineering and Technology, Cuddalore, India.

Amara Gopala Gupta is at the Department of Computer Science and Engineering, Koneru Lakshmaiah Education Foundation, Vaddeswaram, India.

Satbir Singh Sehgal is Dean Research and Head TBI University Centre for Research and Development, Chandigarh University.

S. Selvakumar is with the Department of Computer Science Engineering Aarupadi Veedu Institute of Technology, Vinayaka Mission Research Foundation, Paiyanur, India.

Kumar Shalender is a Post-Doctoral Fellow of Global Institute of Flexible Systems Management and Doctor of Philosophy in Strategic Management. He has more than 14-year experience in domain of Business Policy, Strategic Management, and Business Model Development and a total of 70 Publications including presentations in international/national conferences and book chapters to his credit. His current research areas include the field of Metaverse, Blockchain Technology, and Sustainable Development with special focus on sustainable cities and mobility ecosystem in India.

Shankar R. has completed his Bachelor of Engineering in Electronics and Communication Engineering (ECE) from Bharathidasan University, Tiruchirapalli, Tamil Nadu, India. He acquired his Master of Technology in ECE from Pondicherry Engineering College. He accomplished his Doctor of Philosophy in ECE from Pondicherry University, Pondicherry. He has around 22 years of in the field of teaching and research as of the year 2023. He has published, various research papers in 12 International Conferences, 4 National Conferences, 23 indexed international journals, 8 patents and 6 books. He has been awarded Best Performing Faculty Award for producing 100% results, six times, in university examinations in the following subjects, 1. Embedded Systems, 2. Electromagnetic waves and waveguides, 2. Data Communications and Computer Networks, 4. Cellular Mobile Communication, 5. Microwave and Optical Engineering and 6. Professional Ethics. He is currently working as head of, the Internal Quality Assurance Cell, the Research and Development Centre and a Professor in the Department of ECE at Teegala Krishna Reddy Engineering College (Autonomous), Hyderabad 500097, India. His areas of interest are Electromagnetics, Microprocessors, Mobile Communication, Computer Networks, Wireless Communication (WC), 6G and beyond WC, Soft computing, IoT, Cyber Security, Machine and Deep Learning.

Sneha Sharma is graduate student at Chitkara Business School, Chitkara University.

Meenakshi Sharma was awarded a Ph.D. in 2015, Assistant Professor-SS at University of Petroleum & Energy Studies, Dehradun having 12 years of experience in Academics and 6 years in the Industry. Area of interest being Consumer Behavior & Market Research, Marketing Management and Customer Sustainability. Is Six Sigma Green Belt. Expertise in SPSS,AMOS,Tableau. Published more than 15 publications including Research papers in Scopus, International and National Journals. Many papers presented in International and National Conferences. Conducted full day sessions in organizations like HPCL,Tata Power, UGVNL etc. Completed various certification from Coursera.

Sandhir Sharma is Ph.D. in Strategic Management with nearly 22 years of experience in higher education and 5 years in Telecom industry. With more than 43 research papers to his credit published in various journals and conferences at national & international level, Dr. Sharma has developed his core expertise in the area of Strategy formulation. His research has widely published in various newspapers. Currently, he is serving as Dean, Chitkara Business School & visiting faculty to Binus University, Indonesia and University of Applied Sciences, Osnabruck, Germany. He has been instrumental in scaling Chitkara Business School operations from 60 students & one program to 1200 plus students & more than 12 programs with in span of 8 years. In his last assignment, Dr. Sharma set up a B-School in Ludhiana and was instrumental in taking that B-School to pinnacle of glory in the region.

Renuka Sharmin is a masters student in economics at the Varendra University, Rajshahi Bangladesh. She is doing her masters thesis. Her research interests include economics, waste management, and development.

Jashandeep Singh is a gregarious, enthusiastic, and promising young man who has a passion for teaching. He obtained his PhD from M.M. University, Haryana and MBA in Marketing Management from RIMT, Punjab. He possesses 12 years of work experience in the field of academics, academic administration and research in premier institutions of Punjab and Haryana. He is IKGPTU's certified resource person in Human Values and Ethics. He has been trained at IIM Kozhikode, SPJIMR, IIM Ahmedabad, and IIM Rohtak, and attended several workshops and development programs at various institutions of repute. He has published numerous research papers and a book as well. He has also been appointed as an editorial board member and reviewer for numerous refereed journals. He has also served as a resource person at MGSIPA, Punjab and SEBI for imparting financial literacy among various sections of society. He is a resource person for Value Ideas Investment and a freelance Personality development trainer too. He has filed 10 patents so far. He has completed many Pro-Bono projects for the upliftment of society. He has been a member of Nasha Mukt Abhiyan, Punjab (Punjab Govt run) team.

Babita Singla is an academician par excellence and has been involved in generating new ideas and devising feasible solutions to broadly relevant problems. She enjoys embracing the lessons learned from failure, standing up, and continuing to grow. Her areas of expertise are Marketing Management, E-commerce, and Omnichannel.

D. Suganthi is with the Department of Computer Science, Saveetha College of Liberal Arts and Sciences, SIMATS, Chennai, India.

Ashish Suri is at the School of Electronics and Communication Engineering, Shri Mata Vaishno Devi University, Katra, India.

Gopal Upadhyay is a multi-faceted professional with over 26 years of extensive experience in Teaching, Research and Administration in Top level Institutions as well as University in different regions of India. Served as 15 years in the position of Dean & Director in Engineering & Technology institutes & 8 years as Director General in Engineering, Management, Education & Allied medical Institutions. Currently working as Dean- Academic Affairs of University & Dean- Faculty of Engineering & Technology, Assam down town University (AdtU), Guwahati, Assam (NAAC Accredited University). Plays an active role in the promotion of Technical Education and Produces a vision to provide technically sound campus for excellence in education and research. Authored one book of Solid State, published by Laxmi publication, Ansari road Delhi and published more than 58 research papers & articles in National & International journals & Published 02 Patents & one book chapter by IGI Global Publishers. Serving as an active researcher, have attended more than 64 national & international conferences, FDP's, webinars, seminars, symposiums & orientation programmes. & Participated in 21 national level quizes. Acted as a session chair and convenor for various National and International Seminars and Conferences. Active in several different committees including serving as Member of Academic council, Member of Board of Studies, Dean Academics, Chairman of University Research committee, Chairman of convocation committee, Chairman Admission committee, Chairman of proctorial board & Chairman Examination committee of Institute & university and has also served as Ph.D. Supervisor, Expert Examiner, internal & external examiner for several universities in India. Serving as Member of Indian Physics Association, BARC Bombay, Member of Indian Physical society Kolkata, Member of International Association Computer Sc. & IT (IACSIT) Singapore, Member of PAS India, Life member of Indian Association of Teachers Educators & Life Member of International Association of Engineers (IAENG). Skills in handling & commissioning activities involving resource planning, in- process inspection, team building and co-ordination with internal / external departments. Talent for conducting out various audits & inspections to ensure that educational institutes are in compliance with pre-set specifications/criteria to be get NBA & NAAC accreditations, approvals and rankings in today's competitive era of technical education. Effective communicator & negotiator with strong analytical, problem solving & organizational abilities.

Jyoti Verma is a faculty at Chitkara Business School, Chitkara University, Punjab, India. With more than 20 research papers to her credit published in various journals and conferences at the national & international levels, Dr. Verma has developed her core expertise in the area of service quality, blockchain, and behavioral finance. She has authored 1 book titled 'Accelerating the Development of Quality of IT-Enabled Banking Services' by international publishers and published many book chapters in edited books at the international level. She has published five patents and out of which three patents are granted. She has prepared and delivered many E-Content programs on Research Methodology (MOOCs funded projects).

Index

F

G

H

I

L

M

N

O

P

Q

R

S

T

U

V

W

Y

www.ingramcontent.com/pod-product-compliance
Lightning Source LLC
LaVergne TN
LVHW062111300926
849161LV00010B/2431